Petre P. Teodorescu, Wilhelm W. Kecs,
and Antonela Toma

Distribution Theory

Related Titles

Forbes, C., Evans, M., Hastings, N., Peacock, B.

Statistical Distributions

2011
ISBN: 978-0-470-39063-4

Fujita, S., Godoy, S. V.

Mathematical Physics

2009
ISBN: 978-3-527-40808-5

Masujima, M.

Applied Mathematical Methods in Theoretical Physics

2009
ISBN: 978-3-527-40936-5

Velten, K.

Mathematical Modeling and Simulation
Introduction for Scientists and Engineers

2009
ISBN: 978-3-527-40758-3

Bayin, S. S.

Essentials of Mathematical Methods in Science and Engineering

2008
ISBN: 978-0-470-34379-1

Vaughn, M. T.

Introduction to Mathematical Physics

2007
ISBN: 978-3-527-40627-2

Kusse, B., Westwig, E. A.

Mathematical Physics
Applied Mathematics for Scientists and Engineers

2006
ISBN: 978-3-527-40672-2

Johnson, N. L., Kemp, A. W., Kotz, S.

Univariate Discrete Distributions

2005
ISBN: 978-0-471-27246-5

Trigg, G. L. (ed.)

Mathematical Tools for Physicists

2005
ISBN: 978-3-527-40548-0

Petre P. Teodorescu, Wilhelm W. Kecs, and Antonela Toma

Distribution Theory

With Applications in Engineering and Physics

WILEY-VCH

WILEY-VCH Verlag GmbH & Co. KGaA

The Authors

Prof. Petre P. Teodorescu
University of Bucharest
Faculty of Mathematics
Bucharest, Romania

Prof. Wilhelm W. Kecs
University of Petrosani
Faculty of Science
Petrosani, Hunedoara, Romania
wwkecs@yahoo.com

Prof. Antonela Toma
University 'Politehnica'
Bucharest, Romania

Cover Picture
SpieszDesign, Neu-Ulm

Library of Congress Card No.:
applied for

British Library Cataloguing-in-Publication Data:
A catalogue record for this book is available from the British Library.

Bibliographic information published by the Deutsche Nationalbibliothek
The Deutsche Nationalbibliothek lists this publication in the Deutsche Nationalbibliografie; detailed bibliographic data are available on the Internet at http://dnb.d-nb.de.

Print ISBN 978-3-527-41083-5
ePDF ISBN 978-3-527-65364-5
ePub ISBN 978-3-527-65363-8
mobi ISBN 978-3-527-65362-1
oBook ISBN 978-3-527-65361-4

Cover Design Grafik-Design Schulz, Fußgönheim
Typesetting le-tex publishing services GmbH, Leipzig
Printing and Binding Markono Print Media Pte Ltd, Singapore

Printed in Singapore

Printed on acid-free paper

Contents

Preface

The solution to many theoretical and practical problems is closely connected to the methods applied, and to the mathematical tools which are used. In the mathematical description of mechanical and physical phenomena, and in the solution of the corresponding boundary value and limit problems, difficulties may appear owing to additional conditions. Sometimes, these conditions result from the limited range of applicability of the mathematical tool which is involved; in general, such conditions may be neither necessary nor connected to the mechanical or physical phenomenon considered.

The methods of classical mathematical analysis are usually employed, but their applicability is often limited. Thus, the fact that not all continuous functions have derivatives is a severe restriction imposed on the mathematical tool; it affects the unity and the generality of the results. For example, it may lead to the conclusion of the nonexistence of the velocity of a particle at any moment during the motion, a conclusion which obviously is not true.

On the other hand, the development of mechanics, of theoretical physics and particularly of modern quantum mechanics, the study of various phenomena of electromagnetism, optics, wave propagation and the solution of certain boundary value problems have all brought about the introduction of new concepts and computations, which cannot be justified within the frame of classical mathematical analysis.

In this way, in 1926 Dirac introduced the delta function (denoted by δ), which from a physical point of view, represents the density of a load equal to unity located at one point. A formalism has been worked out for the function, and its use justifies and simplifies various results. Except for a small number of incipient investigations, it was only during the 1960s that the theory of distributions was included as a new chapter of functional analysis. This theory represents a mathematical tool applicable to a large class of problems, which cannot be solved with the aid of classical analysis. The theory of distributions thus eliminates the restrictions which are not imposed by the physical phenomenon and justifies procedure and results, e.g., those corresponding to the continuous and discontinuous phenomena, which can thus be stated in a unitary and general form.

This monograph presents elements of the theory of distributions, as well as theorems with possibility of application. While respecting the mathematical rigor, a large number of applications of the theory of distributions to problems of general Newtonian mechanics, as well as to problems pertaining to the mechanics of deformable solids, are presented in a systematic manner; special stress is laid upon the introduction of corresponding mathematical models.

Some notions and theorems of Newtonian mechanics are stated in a generalized form; the effect of discontinuities on the motion of a particle and its mechanical interpretation is thus emphasized.

Particular stress is laid upon the mathematical representation of concentrated and distributed loads; in this way, the solution of the problems encountered in the mechanics of deformable solids may be obtained in a unitary form.

Newton's fundamental equation, the equations of equilibrium and of motion of the theory of elasticity are presented in a modified form, which includes the boundary and the initial conditions. In this case, the Fourier and the Laplace transforms may be easily applied to obtain the fundamental solutions of the corresponding differential equations; the use of the convolution product allows the expression of the boundary-value problem solutions for an arbitrary load.

Concerning the mechanics of deformable solids, not only have classical elastic bodies been taken into consideration, but also viscoelastic ones, that is, stress is put into dynamical problems: vibrations and propagation of waves.

Applications in physics have been described (acoustics, optics and electrostatics), as well as in electrotechnics.

The aim of the book is to draw attention to the possibility of applying modern mathematical methods to the study of mechanical and physical phenomena and to be useful to mathematicians, physicists, engineers and researchers, which use mathematical methods in their field of interest.

P.P. Teodorescu, W.W. Kecs, A. Toma *Bucharest, 22 August 2012*

1
Introduction to the Distribution Theory

1.1
Short History

The theory of distributions, or of generalized functions, constitutes a chapter of functional analysis that arose from the need to substantiate, in terms of mathematical concepts, formulae and rules of calculation used in physics, quantum mechanics and operational calculus that could not be justified by classical analysis. Thus, for example, in 1926 the English physicist P.A.M. Dirac [1] introduced in quantum mechanics the symbol $\delta(x)$, called the Dirac delta function, by the formulae

$$\delta(x) = \begin{cases} 0\,, & x \neq 0 \\ \infty\,, & x = 0 \end{cases}\,, \quad \int_{-\infty}^{\infty} \delta(x)\mathrm{d}x = 1\,. \tag{1.1}$$

By this symbol, Dirac mathematically described a material point of mass density equal to the unit, placed at the origin of the coordinate axis.

We notice immediately that $\delta(x)$, called the impulse function, is a function not in the sense of mathematical analysis, as being zero everywhere except the origin, but that its integral is null and not equal to unity.

Also, the relations $x\delta(x) = 0, \mathrm{d}H(x)/\mathrm{d}x = \delta(x)$ do not make sense in classical mathematical analysis, where

$$H(x) = \begin{cases} 0\,, & x < 0 \\ 1\,, & x \geq 0 \end{cases}$$

is the Heaviside function, introduced in 1898 by the English engineer Oliver Heaviside.

The created formalism regarding the use of the function δ and others, although it was in contradiction with the concepts of mathematical analysis, allowed for the study of discontinuous phenomena and led to correct results from a physical point of view.

All these elements constituted the source of the theory of distributions or of the generalized functions, a theory designed to justify the formalism of calculation used in various fields of physics, mechanics and related techniques.

Distribution Theory, First Edition. Petre P. Teodorescu, Wilhelm W. Kecs, and Antonela Toma

In 1936, S.L. Sobolev introduced distributions (generalized functions) in an explicit form, in connection with the study of the Cauchy problem for partial differential equations of hyperbolic type.

The next major event took place in 1950–1951, when L. Schwartz published a treatise in two volumes entitled "Théory des distributions" [2]. This book provided a unified and systematic presentation of the theory of distributions, including all previous approaches, thus justifying mathematically the calculation formalisms used in physics, mechanics and other fields.

Schwartz's monograph, which was based on linear functionals and on the theory of locally convex topological vector spaces, motivated further development of many chapters of mathematics: the theory of differential equations, operational calculus (Fourier and Laplace transforms), the theory of Fourier series and others.

Properties in the sense of distributions, such as the existence of the derivative of any order of a distribution and in particular of the continuous functions, the convergence of Fourier series and the possibility of term by term derivation of the convergent series of distributions, led to important technical applications of the theory of distributions, thus removing some restrictions of classical analysis.

The distribution theory had a significant further development as a result of the works developed by J. Mikusiński and R. Sikorski [3], M.I. Guelfand and G.E. Chilov [4, 5], L. Hörmander [6, 7], A. H. Zemanian [8], and so on.

Unlike the linear and continuous functionals method used by Schwartz to define distributions, J. Mikusiński and R. Sikorski introduced the concept of distribution by means of fundamental sequences of continuous functions.

This method corresponds to the spirit of classical analysis and thus it appears clearly that the concept of distribution is a generalization of the notion of function, which justifies the term generalized function, mainly used by the Russian school.

Other mathematicians, such as H. König, J. Korevaar, Sebastiano e Silva, and I. Halperin have defined the notion of distribution by various means (axiomatic, derivatives method, and so on).

Today the notion of distribution is generalized to the concept of a hyperfunction, introduced by M. Sato, [9, 10], in 1958. The hyperdistributions theory contains as special cases the extensions of the notion of distribution approached by C. Roumieu, H. Komatsu, J.F. Colombeu and others.

1.2
Fundamental Concepts and Formulae

For the purpose of distribution theory and its applications in various fields, we consider some function spaces endowed with a convergence structure, called fundamental spaces or spaces of test functions.

1.2.1
Normed Vector Spaces: Metric Spaces

We denote by Γ either the body $\mathbb{R}$ of real numbers or the body $\mathbb{C}$ of complex numbers and by $\mathbb{R}_+$, $\mathbb{R}^+$, $\mathbb{N}_0$ the sets $\mathbb{R}_+ = [0,\infty)$, $\mathbb{R}^+ = (0,\infty)$, $\mathbb{N}_0 = \{0,1,2,\ldots,n,\ldots\}$.

Let E, F be sets of abstract objects. We denote by $E \times F$ the direct product (Cartesian) of those two sets; where the symbol "$\times$" represents the direct or Cartesian product.

Definition 1.1 The set E is called a vector space with respect to Γ, and is denoted by (E,Γ), if the following two operations are defined: the sum, a mapping $(x,y) \to x+y$ from $E \times F$ into E, and the product with scalars from Γ, the mapping $(\lambda,x) \to \lambda x$ from $\Gamma \times E$ into E, having the following properties:

1. $\forall x,y \in E\,, \quad x+y = y+x\,;$
2. $\forall x,y,z \in E\,, \quad (x+y)+z = x+(y+z)\,;$
3. $\exists 0 \in E\,, \quad \forall x \in E\,, \quad x+0 = x\,,$ (0 is the null element);
4. $\forall x \in E\,, \quad \exists x' = -x \in E\,, \quad x+(-x) = 0\,;$
5. $\forall x \in E\,, \quad 1 \cdot x = x\,;$
6. $\forall \lambda,\mu \in \Gamma\,, \quad \forall x \in E\,, \quad \lambda(\mu x) = (\lambda\mu)x\,;$
7. $\forall \lambda,\mu \in \Gamma\,, \quad \forall x \in E\,, \quad (\lambda+\mu)x = \lambda x + \mu x\,;$
8. $\forall \lambda \in \Gamma\,, \quad \forall x,y \in E\,, \quad \lambda(x+y) = \lambda x + \lambda y\,.$

The vector space (E,Γ) is real if $\Gamma = \mathbb{R}$ and it is complex if $\Gamma = \mathbb{C}$. The elements of (E,Γ) are called points or vectors.

Let X be an upper bounded set of real numbers, hence there is $M \in \mathbb{R}$ such that for all $x \in X$ we have $x \le M$. Then there exists a unique number $M^* = \sup X$, which is called the lowest upper bound of X, such that

1. $\forall x \in X\,, \quad x \le M^*\,;$
2. $\forall a \in \mathbb{R}\,, \quad a < M^*\,, \quad \exists x \in X$ such that $x \in (a, M^*]\,.$

Similarly, if Y is a lower bounded set of real numbers, that is, if there is $m \in \mathbb{R}$ such that for all $x \in Y$ we have $x \ge m$, then there exists a unique number $m^* = \inf X$, which is called the greatest lower bound of Y, such that

1. $\forall x \in Y\,, \quad x \ge m^*\,;$
2. $\forall b \in \mathbb{R}\,, \quad b > m^*\,, \quad \exists x \in Y$ such that $x \in [m^*, b)\,.$

Example 1.1 The vector spaces $\mathbb{R}^n$, $\mathbb{C}^n$, $n \ge 2$ Let us consider the n-dimensional space $\mathbb{R}^n = \mathbb{R} \times \cdots \times \mathbb{R}$ (n times). Two elements $x, y \in \mathbb{R}^n$, $x = (x_1,\ldots,x_n)$, $y = (y_1,\ldots,y_n)$, are said to be equal, $x = y$, if $x_i = y_i$, $i = \overline{1,n}$.

Denote $x+y = (x_1+y_1, x_2+y_2, \ldots, x_n+y_n)$, $\alpha x = (\alpha x_1, \alpha x_2, \ldots, \alpha x_n)$, $\alpha \in \mathbb{R}$, then $\mathbb{R}^n$ is a real vector space, also called n-dimensional real arithmetic space.

The n-dimensional complex space $\mathbb{C}^n$ may be defined in a similar manner. The elements of this space are ordered systems of n complex numbers. The sum and product operations performed on complex numbers are defined similarly with those in $\mathbb{R}^n$.

Definition 1.2 Let (X, Γ) be a real or complex vector space. A norm on (X, Γ) is a function $\|\cdot\| : X \to [0, \infty)$ satisfying the following three axioms:

1. $\forall x \in X\,, \quad \|x\| > 0$ for $x \neq 0$, $\|0\| = 0\,;$
2. $\forall \lambda \in \Gamma\,, \quad \forall x \in X\,, \quad \|\lambda x\| = |\lambda|\|x\|\,;$
3. $\forall x, y \in X\,, \quad \|x + y\| \le \|x\| + \|y\|\,.$

The vector space (X, Γ) endowed with the norm $\|\cdot\|$ will be called a normed vector space and will be denoted as $(X, \Gamma, \|\cdot\|)$.

The following properties result from the definition of the norm:

$$\|x\| \ge 0\,, \quad \forall x \in X\,,$$

$$|\|x_1\| - \|x_2\|| \le \|x_1 - x_2\|\,, \quad \forall x_1, x_2 \in X\,,$$

$$\forall \alpha_i \in \Gamma\,, \quad \forall x_i \in X\,, \quad \|\alpha_1 x_1 + \cdots + \alpha_n x_n\| \le |\alpha_1|\|x_1\| + \cdots + |\alpha_n|\|x_n\|\,.$$

Definition 1.3 Let (X, Γ) be a vector space. We call an inner product on (X, Γ) a mapping $\langle\cdot,\cdot\rangle : E \to \Gamma$ that satisfies the following properties:

1. Conjugate symmetry: $\forall x \in X$, $\langle x, y\rangle = \overline{\langle y, x\rangle}$;
2. Homogeneity: $\forall \alpha \in \Gamma, \forall x, y \in E, \langle \alpha x, y\rangle = \alpha\langle y, x\rangle$;
3. Additivity: $\forall x, y, z \in X, \langle x + y, z\rangle = \langle x, z\rangle + \langle y, z\rangle$;
4. Positive-definiteness: $\forall x \in X, \langle x, x\rangle \ge 0$ and $\langle x, x\rangle = 0 \Leftrightarrow x = 0$.

An inner product space $(X, \langle\cdot,\cdot\rangle)$ is a space containing a vector space (X, Γ) and an inner product $\langle\cdot,\cdot\rangle$.

Conjugate symmetry and linearity in the first variable gives

$$\langle x, a y\rangle = \overline{\langle a y, x\rangle} = \overline{a}\overline{\langle y, x\rangle} = \overline{a}\langle x, y\rangle\,,$$

$$\langle x, y + z\rangle = \overline{\langle y + z, x\rangle} = \overline{\langle y, x\rangle} + \overline{\langle z, x\rangle} = \langle x, y\rangle + \langle x, z\rangle\,,$$

so an inner product is a sesquilinear form. Conjugate symmetry is also called Hermitian symmetry.

In the case of $\Gamma = \mathbb{R}$, conjugate-symmetric reduces to symmetric, and sesquilinear reduces to bilinear. Thus, an inner product on a real vector space is a positive-definite symmetric bilinear form.

Proposition 1.1 In any inner product space $(X, \langle\cdot,\cdot\rangle)$ the Cauchy–Schwarz inequality holds:

$$|\langle x, y\rangle| \le \sqrt{\langle x, x\rangle} \cdot \sqrt{\langle y, y\rangle}\,, \quad \forall x, y \in X\,, \tag{1.2}$$

with equality if and only if x and y are linearly dependent.

This is also known in the Russian mathematical literature as the Cauchy–Bunyakowski–Schwarz inequality.

Lemma 1.1 The inner product is antilinear in the second variable, that is $\langle x, y + z\rangle = \langle x, y\rangle + \langle x, z\rangle$ for all $x, y, z \in \Gamma$ and $\langle x, ay\rangle = \overline{a}\langle x, y\rangle$.

Note that the convention in physics is often different. There, the second variable is linear, whereas the first variable is antilinear.

Definition 1.4 Let X be a nonempty set. We shall call metric (distance) on X any function $d : X \times X \to \mathbb{R}$, which satisfies the properties:

D_1 $d(x, x) = 0, \forall x \in X; d(x, y) > 0, \forall x, y \in X, x \neq y$,
D_2 $\forall x, y \in X, d(x, y) = d(y, x)$,
D_3 $\forall x, y, z \in X, d(x, z) \leq d(x, y) + d(y, z)$.

The real number $d(x, y) \geq 0$ represents the distance between x and y, and the ordered pair (X, d) a *metric space* (whose elements are called *points*).

Let (X, d) be a metric space. We shall call an *open ball* in X a ball of radius $r > 0$ centered at the point $x_0 \in X$, usually denoted $B_r(x_0)$ or $B(x_0; r)$, the set

$$B_r(x_0) = \{x \in X \mid d(x, x_0) < r\} . \tag{1.3}$$

The closed ball, which will be denoted by $\overline{B}_r(x_0)$ is defined by

$$\overline{B}_r(x_0) = \{x \in X \mid d(x, x_0) \leq r\} . \tag{1.4}$$

Note, in particular, that a ball (open or closed) always includes x_0 itself, since the definition requires $r > 0$. We shall call a sphere of radius $r > 0$ centered at the point $x_0 \in X$, usually denoted $S_r(x_0)$, the set

$$S_r(x_0) = \{x \in X \mid d(x, x_0) = r\} . \tag{1.5}$$

Proposition 1.2 Any normed vector space is a metric space by defining the distance by the formula

$$d(x, y) = \|x - y\| , \quad \forall x, y \in X . \tag{1.6}$$

Proposition 1.3 Any inner product space $(X, \langle\cdot, \cdot\rangle)$ is a normed vector space if we define the norm by

$$\|x\| = \sqrt{\langle x, x\rangle} , \quad \forall x \in X . \tag{1.7}$$

An inner product space is also called a pre-Hilbert space, since its completion with respect to the metric induced by its inner product, is a Hilbert space.

The real vector space $\mathbb{R}^n$ endowed with the inner product

$$\langle x, y\rangle = \sum_{i=1}^{n} x_i y_i, x = (x_1, \ldots, x_n) , \quad y = (y_1, \ldots, y_n) \in \mathbb{R}^n \tag{1.8}$$

is called the *n-dimensional Euclidean real space*.

The norm in $\mathbb{R}^n$ is called *the Euclidean norm* and is defined as

$$\|x\| = \langle x, x\rangle^{1/2} = \left(\sum_{i=1}^{n} x_i^2\right)^{1/2} , \tag{1.9}$$

whereas the metric associated to this norm is given by

$$d(x, y) = \|x - y\| = \left(\sum_{i=1}^{n} (x_i - y_i)^2\right)^{1/2} . \tag{1.10}$$

1.2.2
Spaces of Test Functions

Let $x = (x_1, \ldots, x_n) \in \mathbb{R}^n$ be a generic point in the n-dimensional Euclidean real space and let $\alpha = (\alpha_1, \ldots, \alpha_n) \in \mathbb{N}_0^n$ be a multiindex of order n; we denote by $|\alpha| = \alpha_1 + \cdots + \alpha_n$ the length of the multiindex. If $\alpha = (\alpha_1, \ldots, \alpha_n), \beta = (\beta_1, \ldots, \beta_n) \in \mathbb{N}_0^n$, then we use the following notations:

$$\alpha \le \beta \text{ if } \alpha_i \le \beta_i, i = \overline{1, n} ; \tag{1.11}$$

$$\binom{\beta}{\alpha} = \frac{\beta!}{\alpha!(\beta - \alpha)!} , \quad \text{where } \alpha! = \alpha_1!\alpha_2! \ldots \alpha_n! , \tag{1.12}$$

$$x^\alpha = x_1{}^{\alpha_1} x_2{}^{\alpha_2} \ldots x_n{}^{\alpha_n} . \tag{1.13}$$

We denote by $\mathrm{D}^\alpha f$ the partial derivative of order $|\alpha| = \alpha_1 + \cdots + \alpha_n$ of a function $f : \Omega \subset \mathbb{R}^n \to \Gamma$,

$$\mathrm{D}^\alpha f = \frac{\partial^{|\alpha|}}{\partial x_1^{\alpha_1} x_2^{\alpha_2} \ldots x_n^{\alpha_n}} f , \quad \mathrm{D}^\alpha = \mathrm{D}_1^{\alpha_1} \mathrm{D}_2^{\alpha_2} \ldots \mathrm{D}_n^{\alpha_n} , \quad \mathrm{D}_j = \frac{\partial}{\partial x_j} , \quad j = \overline{1, n} .$$

If $|\alpha| = 0$, then $\alpha_i = 0$, $i = \overline{1, n}$, that is, $\mathrm{D}^0 f = f$.

If the function f has continuous partial derivatives up to the order $|\alpha + \beta|$ inclusively, then

$$\mathrm{D}^{\alpha+\beta} f = \mathrm{D}^\alpha (D^\beta f) = \mathrm{D}^\beta (\mathrm{D}^\alpha f).$$

We shall denote by $C^m(\Omega)$ the set of functions $f : \Omega \subset \mathbb{R}^n \to \Gamma$ with continuous derivatives of order m, that is, $\mathrm{D}^\alpha f$ is continuous on Ω for every α with $|\alpha| \le m$. When $m = 0$ we have the set $C^0(\Omega)$ of continuous functions on Ω; $C^\infty(\Omega)$ is the set of functions on Ω with continuous derivatives of all orders. Clearly, we have $C^\infty(\Omega) \subset C^m(\Omega) \subset C^0(\Omega)$.

These sets are vector spaces over Ω with respect to the usual definition of addition of functions and multiplication by scalars from Ω. The null element of these spaces is the identically zero function on Ω and it will be denoted by 0.

Definition 1.5 We call the support of the function $f : \mathbb{R}^n \to \Gamma$ the set

$$\mathrm{supp}(f) = \overline{\{x \in \mathbb{R}^n, f(x) \neq 0\}}\,, \tag{1.14}$$

hence the closure of the set of points where the function is not zero.

If $x_0 \in \mathrm{supp}(f)$, then $\forall B_{x_0}(r)$, $\exists x \in \mathbb{R}^n$ thus that $f(x) \neq 0$. In particular, if $\mathrm{supp}(f)$ is bounded, then, since $\mathrm{supp}(f)$ is a closed set, it is also compact.

Proposition 1.4 If $f, g : \mathbb{R}^n \to \Gamma$, then:

$$\mathrm{supp}(f + g) \subset \mathrm{supp}(f) \cup \mathrm{supp}(g)\,, \tag{1.15}$$

$$\mathrm{supp}(f \cdot g) \subset \mathrm{supp}(f) \cap \mathrm{supp}(g)\,, \tag{1.16}$$

$$\mathrm{supp}(\lambda f) = \mathrm{supp}(f)\,, \quad \lambda \neq 0\,. \tag{1.17}$$

Proof: If $x_0 \in \mathrm{supp}(f + g)$, then $\forall B_r(x_0) \subset \mathbb{R}^n, \exists x \in B_r(x_0)$ such that $(f + g)(x) \neq 0$, from which results $f(x) \neq 0$ or $g(x) \neq 0$. Consequently, either $x_0 \in \mathrm{supp}(f)$ or $x_0 \in \mathrm{supp}(g)$, hence $x_0 \in \mathrm{supp}(f) \cup \mathrm{supp}(g)$. Regarding relation (1.16), we notice that $x_0 \in \mathrm{supp}(f \cdot g)$ implies $(fg)(x) \neq 0$, $x \in B_r(x_0)$; hence $f(x) \neq 0$ and $g(x) \neq 0$. Consequently, $x_0 \in \mathrm{supp}(f)$ and $x_0 \in \mathrm{supp}(g)$, hence $\mathrm{supp}(f) \cap \mathrm{supp}(g)$. Because relation (1.17) is obvious, the proof is complete. □

Proposition 1.5 If the functions $f, g \in C^p(\Omega)$, $\Omega \subset \mathbb{R}^n$, then $fg \in C^p(\Omega)$ and we have

$$\mathrm{D}^\alpha(f \cdot g) = \sum_{\beta+\gamma} \frac{\alpha!}{\beta!\gamma!} \mathrm{D}^\beta f \cdot \mathrm{D}^\gamma g\,, \quad \mathrm{D}^\alpha = \mathrm{D}_1^{\alpha_1} \mathrm{D}_2^{\alpha_2} \ldots \mathrm{D}_n^{\alpha_n}\,, \tag{1.18}$$

where $\alpha = (\alpha_1, \ldots, \alpha_n) \in \mathbb{N}_0^n, |\alpha| \leq p$.
The proof of this formula is accomplished through induction.

Definition 1.6 A function $f : A \subset \mathbb{R}^n \to \mathbb{R}$ is said to be uniformly continuous on A if for any $\varepsilon > 0$ there is $\delta > 0$ such that for any $x, y \in A$ satisfying the condition $\|x - y\| < \delta(\varepsilon)$ the inequality $|f(x) - f(y)| < \varepsilon$ holds.

We mention that a uniformly continuous function on $A \subset \mathbb{R}^n$ is continuous at each point of the set A. It follows that the continuity is a local (more precisely, pointwise) property of a function f, while the uniform continuity is a global property of f. In the study of the properties of spaces of test functions, the notion of uniformly convergent sequence plays an important role.

Definition 1.7 We consider the sequence of functions $(f_n)_{n\geq 1}$, $f_n : A \subset \mathbb{R}^n \to \mathbb{R}$ and the function $f : A \subset \mathbb{R}^n \to \mathbb{R}$. We say that the sequence of functions $(f_n)_{n\geq 1}$, $x \in A$ is uniformly convergent towards f, $x \in A$, and we write $f_n \overset{u}{\to} f$, $x \in A \subset \mathbb{R}^n$, if for every $\varepsilon > 0$ there exists a natural number $N(\varepsilon)$ such that for all $x \in A$ and all $n \geq N(\varepsilon)$ the inequality $|f_n(x) - f(x)| < \varepsilon$ holds.

In the case of uniform convergence, the natural number $N(\varepsilon)$ depends only on $\varepsilon > 0$, being the same for all $x \in A$, while in the case of pointwise convergence the natural number N depends on ε and $x \in A$. Therefore the uniform convergence implies pointwise convergence $f_n \xrightarrow{s} f$. The converse is not always true.

Definition 1.8 We say that the function $f : A \subset \mathbb{R}^n \to \mathbb{C}$ is absolutely integrable on A if the integral $\int_A |f(x)|\mathrm{d}x$ is finite, hence $\int_A |f(x)|\mathrm{d}x < \infty$. The integral can be considered either in the sense of Riemann, or in the sense of Lebesgue.

If the integral is considered in the sense of Lebesgue, then the existence of the integral $\int_A |f(x)|\mathrm{d}x$ implies the existence of the integral $\int_A f(x)\mathrm{d}x$.

The set of the Lebesgue integrable functions on A will be denoted $L^1(A)$.

If f is absolutely integrable on any bounded domain $A \subset \mathbb{R}^n$, then we say that f is a *locally integrable function*. We shall use $L^1_{\text{loc}}(A)$ to denote the space of locally integrable functions on A.

The set $A \subset \mathbb{R}^n$ is said to be *negligible* or of *null Lebesgue measure* if for any $\varepsilon > 0$ there is a sequence $(B_i)_{i\geq 1}$, $B_i \subset \mathbb{R}^n$, such that $\cup_{i=1}^{\infty} B_i \supset A$ and the summed volume of the open ball B_i is less than ε.

The function $f : A \subset \mathbb{R}^n \to \Gamma$ is said to be *null a.e.* (almost everywhere) on the set A if the set $\{x \in A,\ f(x) \neq 0\}$ is of null Lebesgue measure.

Thus, the functions $f, g : A \subset \mathbb{R}^n \to \Gamma$ are *a.e. equal* (almost everywhere equal), denoted by $f = g$ *a.e.*, $x \in A$, if the set $\{x \in A,\ f(x) \neq g(x)\}$ is of null Lebesgue measure.

The function $f : A \subset \mathbb{R}^n \to \Gamma$ is *p-integrable* on A, $1 \leq p < \infty$, if $|f|^p \in L^1(A)$. The set of *p-integrable* functions on A is denoted by $\mathcal{L}^p(A)$. In this set we can introduce the equivalence relation $f \sim g$ if $f(x) = g(x)$ a.e. The set of all the equivalence classes is denoted by $L^p(A)$.

The space $L^p(A)$ is a vector space over Γ. The spaces $L^p(A)$ and $L^q(A)$ for which we have $p^{-1} + q^{-1} = 1$ are called *conjugate*. For these spaces, we have Hölder's inequality

$$\int_A |f(x)g(x)|\mathrm{d}x \leq \left(\int_A |f(x)|^p \mathrm{d}x\right)^{1/p} \cdot \left(\int_A |g(x)|^q \mathrm{d}x\right)^{1/q} . \tag{1.19}$$

In particular, for $p = 2$, we have $q = 2$, that is, $L^2(A)$ is self-conjugated and Schwarz's inequality holds

$$\int_A |f(x)g(x)|\mathrm{d}x \leq \left(\int_A |f(x)|^2 \mathrm{d}x\right)^{1/2} \cdot \left(\int_A |g(x)|^2 \mathrm{d}x\right)^{1/2} . \tag{1.20}$$

The norm of the space $L^p(A)$ is defined as

$$\|f\|_p = \left(\int_A |f(x)|^p \mathrm{d}x\right)^{1/p} . \tag{1.21}$$

We notice that the space $L^p(A)$ is normed.

1.2.2.1 The Space $\mathcal{D}^m(\Omega)$

Definition 1.9 Let $\Omega \subset \mathbb{R}^n$ be a given compact set and consider the functions $\varphi : \mathbb{R}^n \to \Gamma$. The set of functions $\mathcal{D}^m(\Omega) = \{\varphi | \varphi \in C^m(\mathbb{R}^n), \text{supp}(\varphi) \subset \Omega\}$ is called the space of test functions $\mathcal{D}^m(\Omega)$.

We notice that $\varphi \in C^m(\mathbb{R}^n)$ with $\text{supp}(\varphi) \subset \Omega$ implies $\text{supp}(\mathrm{D}^\alpha \varphi(x)) \subset \text{supp}(\varphi) \subset \Omega, |\alpha| \leq m$. Consequently, all functions $\varphi \in C^m(\Omega)$ together with all their derivatives up to order m inclusive are null outside the compact Ω. We notice that $\mathcal{D}^m(\Omega)$ is a vector space with respect to Γ. The null element of this space is the identically null function, denoted by 0, $\forall x \in \mathbb{R}^n, \varphi(x) = 0$.

Definition 1.10 We say that the sequence of functions $(\varphi_i)_{i\geq 1} \subset \mathcal{D}^m(\Omega)$ converges towards $\varphi \in \mathcal{D}^m(\Omega)$, and we write $\varphi_i \xrightarrow{\mathcal{D}^m(\Omega)} \varphi$ if the sequence of functions $(\mathrm{D}^\alpha \varphi_i(x))_{i\geq 1}$ converges uniformly towards $\mathrm{D}^\alpha \varphi(x)$ in Ω, hence $\mathrm{D}^\alpha \varphi_i(x) \xrightarrow{u} \mathrm{D}^\alpha \varphi(x)$, $0 \leq |\alpha| \leq m, \forall x \in \Omega$.

We note that the space $\mathcal{D}^m(\Omega)$ becomes a normed vector space if we define the norm by

$$\|\varphi\|_{\mathcal{D}^m} = \sup_{|\alpha|\leq m,\, x\in\Omega} |\mathrm{D}^\alpha \varphi(x)| = \sup_{0\leq|\alpha|\leq m} \sup_{x\in\Omega} |\mathrm{D}^\alpha \varphi(x)|, \alpha \in \mathbb{N}_0^n . \quad (1.22)$$

In particular, for $m = 0$, the space $\mathcal{D}^0(\Omega)$ will be denoted by $C_C^0(\Omega)$. This is the space of complex (real) functions of class $C^0(\mathbb{R}^n)$, the supports of which are contained in the compact set $\Omega \subset \mathbb{R}^n$. The test functions space $C_C^0(\Omega)$ is a normed vector space with the norm

$$\|\varphi\|_{C_C^0} = \sup_{x\in\Omega} |\varphi(x)| . \quad (1.23)$$

The sequence $(\varphi_i)_{i\geq 1} \subset C_C^0(\Omega)$ converges towards $\varphi \in C_C^0(\Omega)$ if $\lim_i \sup_{x\in\Omega} |\varphi_i - \varphi| = 0$, that is, if $(\varphi_i)_{i\geq 1}$ converges uniformly towards φ in Ω.

An example of functions from the space $\mathcal{D}^m(\Omega)$ is the function

$$\varphi(x) = \begin{cases} \prod_{i=1}^{n} \sin^{m+1} \dfrac{x_i - a_i}{b_i - a_i}\pi , & x \in [a_1, b_1] \times \cdots \times [a_n, b_n] = \prod_{i=1}^{n} \times[a_i, b_i] \\ 0 , & x \notin [a_1, b_1] \times \cdots \times [a_n, b_n] \end{cases}$$

where

$$\Omega \supset \prod_{i=1}^{n} \times[a_i, b_i] .$$

It is immediately verified that $\varphi \in C^m(\mathbb{R}^n)$ and $\text{supp}(\varphi) = \prod_{i=1}^{n} \times[a_i, b_i]$.

Also the function $\varphi : \mathbb{R} \to \mathbb{R}$, where

$$\varphi = \begin{cases} (x-a)^\alpha (b-x)^\beta , & x \in [a, b] \\ 0 , & x \notin [a, b] \end{cases} , \quad \alpha, \beta > m ,$$

is a function from $\mathcal{D}^m([c,d])$, $[c,d] \supset [a,b]$, because $\varphi \in C^m([c,d])$ and $\operatorname{supp}(\varphi) = [a,b]$.

Let us consider the sequence of functions $(\varphi_n)_{n\geq 1} \subset \mathcal{D}^m(\Omega)$, defined by

$$\varphi_n(x) = \begin{cases} \dfrac{1}{n}\sin^{m+1}\dfrac{x+a}{2a}\pi\,, & x \in [-a,a]\,, \\ 0\,, & x \notin [-a,a]\,. \end{cases}$$

We have $\operatorname{supp}\varphi_n(x) = [-a,a] = \Omega$ for any n. This sequence, with its derivatives up to order m inclusive, converges uniformly towards zero in Ω. So we can write $\varphi_n(x) \xrightarrow{\mathcal{D}(\Omega)} 0$ in Ω.

Even if the sequence of functions

$$\varphi_n(x) = \begin{cases} \dfrac{1}{n}\sin^{m+1}\dfrac{a+x/n}{2a}\pi\,, & \dfrac{x}{n} \in [-a,a]\,, \\ 0\,, & \dfrac{x}{n} \notin [-a,a]\,, \end{cases}$$

converges uniformly towards zero, together with all their derivatives up to order m inclusive, it is not convergent towards zero in the space $\mathcal{D}^m(\Omega)$. This is because $\operatorname{supp}[\varphi_n(x)] = [-na, na]$, thus the supports of the functions $\varphi_n(x)$ are not bounded when $n \to \infty$, hence $\varphi_n(x)$, $x \in \mathbb{R}$, $n \in \mathbb{N}$, are not test functions from $\mathcal{D}^m(\Omega)$.

1.2.2.2 The Space $\mathcal{D}(\Omega)$

Definition 1.11 Let $\Omega \subset \mathbb{R}^n$ be a given compact set and consider the functions $\varphi : \mathbb{R}^n \to \Gamma$. The set of functions

$$\mathcal{D}(\Omega) = \{\varphi | \varphi \in C^\infty(\mathbb{R}^n), \operatorname{supp}(\varphi) \subset \Omega\}$$

is called the space of test functions $\mathcal{D}(\Omega)$.

The space $\mathcal{D}(\Omega)$ is a vector space over Γ like $\mathcal{D}^m(\Omega)$.

Definition 1.12 We say that the sequence $(\varphi_i)_{i\geq 1} \subset \mathcal{D}(\Omega)$ converges towards $\varphi \in \mathcal{D}(\Omega)$, and we write $\varphi_i \xrightarrow{\mathcal{D}(\Omega)} \varphi$, if the sequence of derivative $(\mathrm{D}^\alpha \varphi_i(x))_{i\geq 1}$ converges uniformly towards $\mathrm{D}^\alpha\varphi(x)$ in Ω, $\forall \alpha \in \mathbb{N}_0^n$, hence $\mathrm{D}^\alpha\varphi_i(x) \xrightarrow{u} \mathrm{D}^\alpha\varphi(x)$, $\forall x \in \Omega$, $\forall \alpha \in \mathbb{N}_0^n$.

We remark that the test space $\mathcal{D}(\Omega)$ is not a normed vector space.

Example 1.2 If $\Omega = \{x | x \in \mathbb{R}^n, \|x\| \leq 2a\}$, then the function $\varphi : \mathbb{R}^n \to \mathbb{R}$, having the expression

$$\varphi(x) = \begin{cases} \exp\left(-\dfrac{a^2}{a^2 - \|x\|^2}\right)\,, & \|x\| < a \\ 0\,, & \|x\| \geq a \end{cases}\,, \quad a > 0\,, \tag{1.24}$$

is an element of the space $\mathcal{D}(\Omega)$, since $\varphi \in C^{\infty}(\mathbb{R}^n)$ and $\operatorname{supp}(\varphi) = \{x | x \in \mathbb{R}^n, \|x\| \leq a\} \subset \Omega$.

The sets Ω and $\operatorname{supp}(\varphi)$ are compact sets of $\mathbb{R}^n$, representing closed balls with centers at the origin and radii $2a$ and a, respectively.

Unlike the function φ, the function $\psi : \mathbb{R} \to \mathbb{R}$,

$$\psi(x) = \begin{cases} 0\,, & x \leq 0\,, \\ \exp(-x^2)\,, & x > 0\,, \end{cases} \tag{1.25}$$

does not belong to the space $\mathcal{D}(\Omega)$.

This function is infinitely differentiable, so $\psi \in C^{\infty}(\mathbb{R}^n)$, but the support is not a compact set because $\operatorname{supp}(\psi) = (0, \infty)$.

1.2.2.3 The Space $\mathcal{E}$

Definition 1.13 The functions set

$$\mathcal{E} = \{\varphi | \varphi : \mathbb{R}^n \to \Gamma, \varphi \in C^{\infty}(\mathbb{R}^n)\}\,. \tag{1.26}$$

having arbitrary support is called the space of test functions $\mathcal{E} = \mathcal{E}(\mathbb{R}^n)$.

With respect to the usual sum and scalar product operation, the space $\mathcal{E}$ is a vector space over Γ.

Thus, the functions $\varphi(x) = 1, \varphi(x) = x^2, \varphi(x) = \exp(x^2), x \in \mathbb{R}$ are elements of $\mathcal{E}(\mathbb{R}^n)$.

As regards the convergence in the space $\mathcal{E}$ this is given:

Definition 1.14 The sequence $(\varphi_i)_{i \geq 1} \subset \mathcal{E}$ is said to converge towards $\varphi \in \mathcal{E}$, and we write $\varphi_i \xrightarrow{\mathcal{E}} \varphi$, if the sequence of functions $(\mathrm{D}^{\alpha}\varphi_i)_{i \geq 1} \subset \mathcal{E}$ converges uniformly towards $\mathrm{D}^{\alpha}\varphi(x) \in \mathcal{E}$ on any compact of $\mathbb{R}^n$, $\forall \alpha \in \mathbb{N}_0^n$, that is, $\mathrm{D}^{\alpha}\varphi_i \xrightarrow{u} \mathrm{D}^{\alpha}\varphi$.

The function (1.25) belongs to the space $\mathcal{E}$ since $\psi \in C^{\infty}(\mathbb{R}^n)$, its supports being the unbounded set $(0, \infty)$.

1.2.2.4 The Space $\mathcal{D}$ (the Schwartz Space)

Definition 1.15 The space $\mathcal{D} = \mathcal{D}(\mathbb{R}^n)$ consists of the set of functions

$$\mathcal{D} = \{\varphi | \varphi : \mathbb{R}^n \to \Gamma, \varphi \in C^{\infty}(\mathbb{R}^n), \operatorname{supp}(\varphi) = \Omega = \text{compact}\}\,. \tag{1.27}$$

Since $\forall \varphi \in \mathcal{D}$, it belongs to a certain $\mathcal{D}(\Omega)$, it follows that $\mathcal{D}$ is the reunion of spaces $\mathcal{D}(\Omega)$ over the compacts $\Omega \subset \mathbb{R}^n$. Consequently, we can write the following relations:

$$\mathcal{D} = \bigcup_{\Omega} \mathcal{D}(\Omega), \mathcal{D}(\Omega) \subset \mathcal{D} \subset \mathcal{E}\,.$$

With respect to the usual sum and scalar product operations, $\mathcal{D}$ is a vector space on Γ, its null element being the identically zero function. The support of this function is the empty set.

The convergence in the space $\mathcal{D}$ is defined as:

Definition 1.16 The sequence of functions $(\varphi_i)_{i\geq 1} \subset \mathcal{D}$ converges towards $\varphi \in \mathcal{D}$, and we write $\varphi_i \overset{\mathcal{D}}{\longrightarrow} \varphi$, if the following conditions are satisfied:

1. $\forall i \in \mathbb{N}$, there is a compact $\Omega \subset \mathbb{R}^n$ such that $\operatorname{supp}(\varphi_i), \operatorname{supp}(\varphi) \subset \Omega$;
2. $\forall \alpha \in \mathbb{N}_0^n$, $\mathrm{D}^\alpha \varphi_i$ converges uniformly towards $\mathrm{D}^\alpha \varphi$ on Ω, that is, $\mathrm{D}^\alpha \varphi_i \overset{u}{\to} \mathrm{D}^\alpha \varphi$ on Ω.

Thus, the convergence in the space $\mathcal{D}$ is reduced to the convergence in the space $\mathcal{D}(\Omega)$.

The vector space $\mathcal{D}(\mathbb{R}^n)$ endowed with the convergence structure defined above is called the space of test functions or the Schwartz space. Every element of the space $\mathcal{D}$ will be called a test function.

Example 1.3 The function $\varphi_a : \mathbb{R}^n \to \mathbb{R}$, $a > 0$, defined by

$$\varphi_a(x) = \begin{cases} \exp\left(-\dfrac{a^2}{a^2 - \|x\|^2}\right), & \|x\| < a\,, \\ 0\,, & \|x\| \geq a\,, \end{cases} \tag{1.28}$$

is an element of $\mathcal{D}(\mathbb{R}^n)$, since $\varphi_a \in C^\infty(\mathbb{R}^n)$ and $\operatorname{supp}(\varphi_a) = \{x | x \in \mathbb{R}^n, \|x\| \leq a\} = \text{compact}$.

Example 1.4 Let $\varphi : \mathbb{R} \to \mathbb{R}$ be a function defined by

$$\varphi(x) = \begin{cases} \exp\left(-\dfrac{|ab|}{(x-a)(b-x)}\right), & x \in (a, b)\,, \\ 0\,, & x \notin (a, b)\,. \end{cases} \tag{1.29}$$

It is noted that $\varphi \in C^\infty(\mathbb{R})$ has compact support $[a, b]$. At the points a and b, the function φ and with its derivatives of any order are zero. Consequently, $\varphi \in \mathcal{D}(\mathbb{R})$. The graph of the function is shown in Figure 1.1.

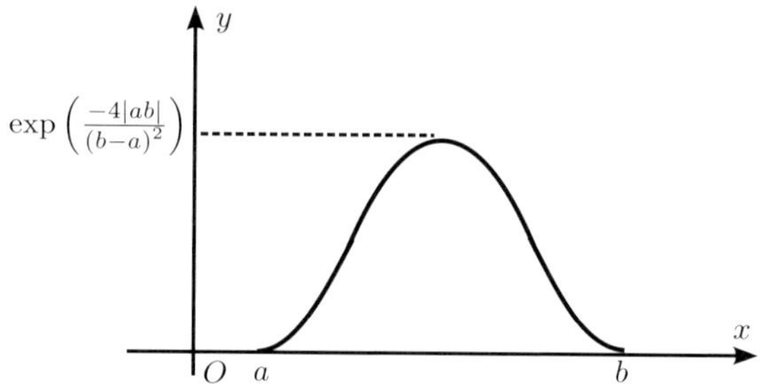

Figure 1.1

Also, the function $\varphi : \mathbb{R}^n \to \mathbb{R}$, where

$$\varphi(x_1, \ldots, x_n) = \begin{cases} \prod_{i=1}^{n} \exp\left(-\dfrac{|a_i b_i|}{(x_i - a_i)(b_i - x_i)}\right), & x_i \in (a_i, b_i)\,, \\ 0\,, & x_i \notin (a_i, b_i)\,, \end{cases} \tag{1.30}$$

is a function of the space $\mathcal{D}(\mathbb{R}^n)$, with the compact support $\Omega_n = [a_1, b_1] \times [a_2, b_2] \times \cdots \times [a_n, b_n]$.

Example 1.5 Let $(\varphi_n)_{n \geq 1} \subset \mathcal{D}(\mathbb{R})$ be a sequence of functions

$$\varphi_n(x) = \frac{1}{n}\varphi_a(x) = \begin{cases} \dfrac{1}{n}\exp\left(-\dfrac{a^2}{a^2 - x^2}\right), & |x| < a, a > 0\,, \\ 0\,, & |x| \geq a, a > 0\,. \end{cases} \tag{1.31}$$

We have $\varphi_n \xrightarrow{\mathcal{D}(\mathbb{R})} 0$, that is, the sequence $(\varphi_n)_{n \geq 1} \subset \mathcal{D}(\mathbb{R})$ converges towards $\varphi = 0 \in \mathcal{D}(\mathbb{R})$ in the space $\mathcal{D}(\mathbb{R})$, because $\forall n \in \mathbb{N}, \mathrm{supp}(\varphi_n) \subset \mathrm{supp}(\varphi_a) =$ compact and $(\mathrm{d}^\alpha/\mathrm{d}x^\alpha)\varphi_n(x) \xrightarrow{u} 0, \forall \alpha \in \mathbb{N}_0, |x| \leq a$.

Definition 1.17 We say that the function $\psi : \mathbb{R}^n \to \Gamma$ is a multiplier for the space $\mathcal{D}$ if $\forall \varphi \in \mathcal{D}$ the mapping $\varphi \to \psi\varphi$ is continuous from $\mathcal{D}$ in $\mathcal{D}$.

Hence, if ψ is a multiplier for space $\mathcal{D}$, then $\psi\varphi \in \mathcal{D}, \forall \varphi \in \mathcal{D}$ and $\varphi_i \xrightarrow{\mathcal{D}} \varphi$ implies $\psi\varphi_i \xrightarrow{\mathcal{D}} \psi\varphi$.

We can easily check that any function $\psi \in C^\infty(\mathbb{R}^n)$ is a multiplier for space $\mathcal{D}$.

Indeed, since $\psi \in C^\infty(\mathbb{R}^n)$ and $\varphi \in C^\infty(\mathbb{R}^n)$, $\varphi \in \mathcal{D}(\mathbb{R}^n)$, we apply formula (1.18) and have

$$\mathrm{D}^\alpha(\psi\varphi) \in \sum_{\beta+\gamma=\alpha} \frac{\alpha!}{\beta!\gamma!}\mathrm{D}^\beta\psi\mathrm{D}^\gamma\varphi, \mathrm{D}^\alpha = \mathrm{D}_1^{\alpha_1}\ldots\mathrm{D}_n^{\alpha_n}, \alpha = (\alpha_1, \ldots, \alpha_n) \in \mathbb{N}_0^n\,, \tag{1.32}$$

from which it results that $\psi\varphi \in C^\infty(\mathbb{R}^n)$.

On the other hand, we have $\mathrm{supp}(\psi\varphi) \subset \mathrm{supp}(\psi) \cap \mathrm{supp}(\varphi) \subset \mathrm{supp}(\varphi) = \Omega =$ compact.

Next, we show that $\varphi_i \xrightarrow{\mathcal{D}} \varphi$ implies $\psi\varphi_i \xrightarrow{\mathcal{D}} \psi\varphi$. From the expression of the derivative $\mathrm{D}^\alpha(\psi\varphi)$ it results

$$|\mathrm{D}^\alpha\psi(\varphi_i - \varphi)| \leq \sum_{\|\gamma\| \leq \|\alpha\|} A_\gamma |\mathrm{D}^\gamma(\varphi_i - \varphi)|\,, \quad A_\gamma > 0 \text{ constants}\,.$$

Since $\mathrm{D}^\alpha(\varphi_i - \varphi) \xrightarrow{\mathcal{D}} 0$, we obtain $|\mathrm{D}^\alpha\psi(\varphi_i - \varphi)| \xrightarrow{\mathcal{D}} 0$, hence $\psi\varphi_i \xrightarrow{\mathcal{D}} \psi\varphi$.

Theorem 1.1 The partition of unity If $\varphi \in \mathcal{D}$ and $U_i, i = 1, 2, \ldots, p$, are open and bounded sets, which form a finite covering of the support function φ, then there exist the functions $e_i \in \mathcal{D}, i = 1, 2, \ldots, p$, with the properties:

1. $e_i(x) \in [0,1]$, $\operatorname{supp}(e_i) \subset U_i$;
2. $\sum_{i=1}^{p} e_i(x) = 1$, $x \in \operatorname{supp}(\varphi)$;
3. $\varphi(x) = \sum_{i=1}^{p} e_i(x)\varphi(x)$.

We note that the partition theorem is frequently used to demonstrate the local properties of distributions, as well as the operations with them.

1.2.2.5 The Space $\mathcal{S}$ (the Space Functions which Decrease Rapidly)

Definition 1.18 We call the test function space $\mathcal{S} = \mathcal{S}(\mathbb{R}^n)$ the set of functions $\varphi : \mathbb{R}^n \to \Gamma$, infinitely differentiable, which for $\|x\| \to \infty$ approach zero together with all their derivatives of any order, faster than any power of $\|x\|^{-1}$.

If $\varphi \in \mathcal{S}$, then $\forall k \in \mathbb{N}$ and $\forall \beta \in \mathbb{N}_0^n$ we have

$$\lim_{\|x\|\to\infty} \|x\|^k \mathrm{D}^\beta \varphi = 0 .$$

This means that $\forall \varphi \in \mathcal{S}$, we have $\varphi \in C^\infty(\mathbb{R}^n)$ and $\forall \alpha, \beta \in \mathbb{N}_0^n$, $\lim_{\|x\|\to\infty} |x^\alpha \mathrm{D}^\beta \varphi| = 0$, that is, $|x^\alpha \mathrm{D}^\beta \varphi| < C_{\alpha,\beta}$, where $C_{\alpha,\beta}$ are constants.

Example 1.6 An example of a function in $\mathcal{S}$ is $\varphi(x) = \exp(-a\|x\|^2)$, $a > 0$, $x \in \mathbb{R}^n$. On the other hand, the function $\varphi(x) = \exp(-x)$, $x \in \mathbb{R}$, does not belong to the space $\mathcal{S}(\mathbb{R})$, since $\lim_{\|x\|\to-\infty} |x^\alpha \varphi^{(n)}(x)| = \lim_{x\to-\infty} |x|^\alpha \exp(-x) = \infty, \forall \alpha \in \mathbb{N}$, although $\lim_{\|x\|\to+\infty} |x^\alpha \varphi^{(n)}(x)| = \lim_{x\to+\infty} |x|^\alpha \exp(-x) = 0, \forall \alpha \in \mathbb{N}_0$.
Also, the functions $\varphi_1(x) = \exp(x), \varphi_2(x) = \exp(-|x|)$, $x \in \mathbb{R}$ do not belong to the space $\mathcal{S}(\mathbb{R})$ because the function $\varphi_1(x)$ does not tend to zero when $x \to \infty$, and the function $\varphi_2(x)$ is not differentiable at the origin.

Obviously, the space $\mathcal{S}$ is a vector space over Γ, having as null element $\varphi = 0, \forall x \in \mathbb{R}^n$. Between the spaces $\mathcal{D}, \mathcal{S}, \mathcal{E}$ there exist the relations $\mathcal{D} \subset \mathcal{S} \subset \mathcal{E}$.

Definition 1.19 Let $\varphi \in \mathcal{S}$ and consider the sequence $(\varphi_i)_{i\geq 1} \subset \mathcal{S}$. We say that the sequence of functions $(\varphi_i)_{i\geq 1}$ converges towards φ and write $\varphi_i \xrightarrow{\mathcal{S}} \varphi$ if

$$\forall \alpha, \beta \in \mathbb{N}_0^n, x^\beta \mathrm{D}^\alpha \varphi_i \xrightarrow{u} x^\beta \mathrm{D}^\alpha \varphi , \quad x \in \mathbb{R}^n . \tag{1.33}$$

Consequently, if $\varphi_i \xrightarrow{S} \varphi$, then $\forall \alpha, \beta \in \mathbb{N}_0^n$ on any compact from $\mathbb{R}^n$ we have $x^\beta \mathrm{D}^\alpha \varphi_i \xrightarrow{u} x^\beta \mathrm{D}^\alpha \varphi$.
Comparing the convergence of the spaces $\mathcal{D}$ and $\mathcal{S}$, $\mathcal{D} \subset \mathcal{S}$, we can state:

Proposition 1.6 The convergence in space $\mathcal{D}$ is stronger than the convergence in space $\mathcal{S}$.

Indeed, if $\varphi_i \xrightarrow{\mathcal{D}} \varphi$, then there is $\mathcal{D}(\Omega) \subset \mathcal{D}$ so that $\varphi_i \xrightarrow{\mathcal{D}(\Omega)} \varphi$, hence $x^\beta \mathrm{D}^\alpha \varphi_i$ converges uniformly towards $x^\beta \mathrm{D}^\alpha \varphi$ on any compact from $\mathbb{R}^n$, that is, $\varphi_i \xrightarrow{\mathcal{S}} \varphi$.

Proposition 1.7 The space $\mathcal{D}$ is dense in $\mathcal{S}$.

This means that $\forall \varphi \in \mathcal{S}$ there is $(\varphi_i)_{i \geq 1} \subset \mathcal{D}$ such that $\varphi_i \xrightarrow{\mathcal{S}} \varphi$.

Also, we can prove that the space $\mathcal{D}$ is dense in $\mathcal{E}$.

Regarding the multipliers of the space $\mathcal{S}$, we note that not every infinitely differentiable function is a multiplier.

Thus, the function $a(x) = \exp(\|x\|^2)$ belongs to the class $C^\infty(\mathbb{R}^n)$, but it is not a multiplier of the space $\mathcal{S}$, because considering $\varphi(x) = \exp(-\|x\|^2) \in \mathcal{S}$, we then have $a(x)\varphi(x) \equiv 1 \notin \mathcal{S}$.

We note O_M the functions of class $C^\infty(\mathbb{R}^n)$ such that the function and all its derivatives do not increase at infinity faster than a polynomial does, hence if $\psi \in O_M$, then we have

$$\forall \alpha \in \mathbb{N}_0^n, |D^\alpha \psi| \leq c_\alpha (1 + \|x\|)^{m_\alpha} , \tag{1.34}$$

where $c_\alpha > 0$, $m_\alpha \geq 0$ are constants.

It follows that O_M is the space of multipliers for $\mathcal{S}$, because if $\psi \in O_M$ and $\forall \varphi \in \mathcal{S}$, then $\psi\varphi \in \mathcal{S}$ and $\varphi_i \xrightarrow{\mathcal{S}} \varphi$ involve $\psi\varphi_i \xrightarrow{\mathcal{S}} \psi\varphi$.

Thus, the functions $f_1(x) = \cos x$, $f_2(x) = \sin x$, $P(x)$ (polynomial in x), $x \in \mathbb{R}$, are multipliers for the space $S(\mathbb{R})$.

Consequently, if $\varphi \in \mathcal{S}$ then $\forall \alpha, \beta \in \mathbb{N}_0^n$, $x^\beta D^\alpha \varphi \in \mathcal{S}$ is bounded and integrable on $\mathbb{R}^n$, hence $\mathcal{S} \subset L^p$, $p \geq 1$.

The spaces of functions with convergence $\mathcal{D}^m(\Omega), \mathcal{D}(\Omega), \mathcal{D}, \mathcal{E}$ and $\mathcal{S}$ will be called test function spaces, and the functions of these spaces, test functions.

Let Φ be a test function space, so $\Phi \in \{\mathcal{D}^m(\Omega), \mathcal{D}(\Omega), \mathcal{D}, \mathcal{E}, \mathcal{S}\}$.

We note that the function $h(x) = e^x$, $x \in \mathbb{R}$ is not a multiplier of the space $\mathcal{S}(\mathbb{R})$, because it increases to infinity faster than a polynomial.

1.2.3 Spaces of Distributions

The concept by which one introduces the notion of distribution is the linear functional one. This method, used by Schwartz, has been proved useful, with wide applications in various fields of mathematics, mechanics, physics and technology.

Let (E, Γ), (Y, Γ) be two vector spaces over the same scalar body Γ and let $X \subset E$ be a subspace of (E, Γ). We shall call the mapping $T : X \to Y$ operator defined on X with values in Y. The value of the operator T at the point $x \in X$ will be denoted by $(T, x) = T(x) = y \in Y$.

Definition 1.20 The operator $T : X \to Y$ is called linear if and only if

$$T(\alpha_1 x_1 + \alpha_2 x_2) = \alpha_1 T(x_1) + \alpha_2 T(x_2), \forall \alpha_1, \alpha_2 \in \Gamma, \forall x_1, x_2 \in X . \tag{1.35}$$

Thus, if we denote $E = C^n(\Omega)$ and $Y = C^0(\Omega)$, $\Omega \subset \mathbb{R}$, then the application $T : E \to Y$ defined by

$$(T, f) = a_0 D^n f + a_1 D^{n-1} f + \cdots + a_{n-1} D f + a_n f , \tag{1.36}$$

where $f(x) \in E, \mathrm{D}^k = \mathrm{d}^k/\mathrm{d}x^k, a_k(x) \in C^0(\Omega), k = 0, 1, 2, \dots, n$ is a linear operator on E.

The operator (1.36) expressed by means of derivatives D^j is called linear differential operator with variable coefficients or polynomial differential operator and we also note $P(\mathrm{D})$.

The operator $T : C^0[a, b] \to C^1[a, b]$ defined by

$$(T, f) = \int_a^x f(t)\mathrm{d}t\,, \quad x \in [a, b]\,, \tag{1.37}$$

is an integral operator. It is shown that it is an integral operator.

A particular class of operators is formed by functionals. Thus, if the domain Y in which the linear operator T takes values is Γ, $Y = \Gamma$, then the operator

$$T : X \subset E \to \Gamma \tag{1.38}$$

will be called functional.

The functional T will be called real or complex as its value (T, x) at the point $x \in X$ is a real or complex number.

We say that the functional (1.38) is linear if it satisfies the condition of linearity of an operator (1.35).

Definition 1.21 A continuous linear functional defined on a space of test functions $\Phi \in \{\mathcal{D}^m(\Omega), \mathcal{D}(\Omega), \mathcal{D}, \mathcal{E}, \mathcal{S}\}$ is called distribution.

This definition involves the fulfillment of the following conditions:

1. To any function $\varphi \in \Phi$ we associate according to some rule f, a complex number $(f, \varphi) \in \Gamma$;
2. $\forall \lambda_1, \lambda_2 \in \Gamma, \forall \varphi_1, \varphi_2 \in \Phi, (f, \lambda_1\varphi_1 + \lambda_2\varphi_2) = \lambda_1(f, \varphi_1) + \lambda_2(f, \varphi_2)$;
3. If $(\varphi_i)_{i \geq 1} \in \Phi, \varphi \in \Phi$ and $\varphi_i \xrightarrow{\Phi} \varphi$, then $\lim_i(f, \varphi_i) = (f, \varphi)$.

The first condition expresses the fact that it is a functional, the second condition corresponds to the linearity of the functional, whereas the third condition expresses its continuity.

The set of distributions defined on Φ is denoted by Φ' and can be organized as a vector space over the field of scalars Γ.

For this purpose, we define the sum of two distributions and the product of a distribution with a scalar as follows:

$$\forall f\,, g \in \Phi'\,, \quad \forall \varphi \in \Phi\,, \; (f + g, \varphi) = (f, \varphi) + (g, \varphi)\,, \tag{1.39}$$

$$\forall \alpha \in \Gamma\,, \; \forall \varphi \in \Phi\,, \quad \forall f \in \Phi'\,, \; (\alpha f, \varphi) = \alpha(f, \varphi)\,. \tag{1.40}$$

It can be verified immediately that the functional $\alpha f + \beta g$ is linear and continuous, hence it is a distribution from Φ'.

Definition 1.22 Let $f \in \Phi'$ and consider the sequence $(f_i)_{i \geq 1} \subset \Phi'$. We say that the sequence $(f_i)_{i \geq 1}$ converges towards the distribution f and we shall write $\lim_i f_i = f$ if and only if $\forall \varphi \in \Phi$ we have $\lim_i (f_i, \varphi) = (f, \varphi)$.

This convergence is called weak convergence.

The vector space of distributions Φ' endowed with the structure of weak convergence is called distributions space and will be noted by Φ'.

It can be shown that the space Φ' is a complete space with respect to the weak convergence introduced.

If the sequence of distributions $(f_i)_{i \geq 1} \subset \Phi'$ is such that, for any $\varphi \in \Phi$ the numerical sequence (f_i, φ) has a limit, then there is a single distribution $f \in \Phi'$ for which we have $\lim_i (f_i, \varphi) = (f, \varphi)$.

The linearity and continuity properties of a distribution allow us to state:

Proposition 1.8 Let $f \in \Phi'(\mathbb{R}^n)$ be the distribution and $\varphi_a(x) \in \Phi(\mathbb{R}^n)$ the test function, depending on the parameter $a \in I \subset \mathbb{R}$. If $\partial \varphi_a(x)/\partial a$ exists and $\partial \varphi_a(x)/\partial a \in \Phi$, and $(\varphi_{a+h}(x) - \varphi_a(x))/h \overset{\Phi}{\longrightarrow}_{h \to 0} (\partial)/(\partial a) \varphi_a(x), \forall a \in I \subset \mathbb{R}$, then the following relation occurs

$$\frac{\mathrm{d}}{\mathrm{d}a}(f(x), \varphi_a(x)) = \left(f(x), \frac{\partial \varphi_a(x)}{\partial a} \right) , \quad a \in I \subset \mathbb{R} . \tag{1.41}$$

As an application of this proposition, we have:

Proposition 1.9 Let $\varphi \in \Phi = \mathcal{D}(\mathbb{R}^n)$ and the distribution $f \in \Phi' = \mathcal{D}'(\mathbb{R}^n)$. Then, we have

$$\frac{\mathrm{d}}{\mathrm{d}a} \left(f(x_1, \ldots, x_n), \varphi\left(\frac{x_1}{a}, \ldots, \frac{x_n}{a}\right) \right) = \left(f(x), -\sum_{i=1}^{n} \frac{x_i}{a^2} \frac{\partial}{\partial x_i} \varphi\left(\frac{x}{a}\right) \right) , \quad a > 0 . \tag{1.42}$$

Let $f \in \Phi'$ be the distribution and the function $\psi \in C^\infty(\mathbb{R}^n)$, multiplier of the test function space Φ. Then the product ψf is defined by the formula

$$(\psi f, \varphi) = (f, \psi\varphi) , \quad \forall \varphi \in \Phi . \tag{1.43}$$

Obviously, $(\psi f) \in \Phi'$ is a distribution, because ψ being the multiplier for Φ we have $\psi \phi \in \Phi$.

Various spaces of distributions are obtained by customizing the test function space Φ. Thus, the distributions defined on $\mathcal{D}$ are called Schwartz distributions and we note $\mathcal{D}' = \mathcal{D}'(\mathbb{R}^n)$.

If $\Phi = \mathcal{D}^m$ then the distribution $\mathcal{D}'^m$ are called distributions of finite order $\leq m$, and the distributions defined on $\mathcal{D}^0 = C^0_C(\mathbb{R}^n)$ are called measures.

Also, the distributions defined on the test functions space $\mathcal{S} = \mathcal{S}(\mathbb{R}^n)$ are called tempered distributions.

Because $\mathcal{D} \subset \mathcal{S}$ and the convergence in the space $\mathcal{D}$ is stronger than the convergence in the space $\mathcal{S}$, then between the spaces of distributions $\mathcal{S}'$ and $\mathcal{D}'$ the relation $\mathcal{S}' \subset \mathcal{D}'$ occurs.

Because $\mathcal{S} \subset \mathcal{E}$ and the convergence of a sequence from $\mathcal{S}$ implies the convergence in the space $\mathcal{E}$, it follows that between the spaces $\mathcal{E}'$ and $\mathcal{S}'$ there exists the relation $\mathcal{E}' \subset \mathcal{S}'$.

Consequently, any distribution from $\mathcal{E}'$ is a distribution with compact support and at the same time it is a tempered distribution.

Thus, the dependence of the spaces of test functions $\mathcal{D}, \mathcal{E}, \mathcal{S}$ is $\mathcal{D} \subset \mathcal{S} \subset \mathcal{E}$ and between the corresponding spaces of distributions occur the inclusions $\mathcal{E}' \subset \mathcal{S}' \subset \mathcal{D}'$.

Let $\varphi \in \Phi$ be a complex-valued function test, hence $\varphi(x) \in \mathbb{C}$, $f \in \Phi'$ representing a complex-valued distribution, that is, $(f, \varphi) \in \mathbb{C}$. Then, the product of, complex distribution f and, complex function $a : \mathbb{R}^n \to \mathbb{C}$ is defined by the relation

$$(a f, \varphi) = \overline{a}(f, \varphi) = (f, \overline{a}\varphi) , \tag{1.44}$$

with the assumption that $\overline{a}\varphi \in \Phi$, where $\overline{a}$ represents the complex conjugate of the function a.

We note that to the complex-valued distribution f can be associated a complex conjugate distribution $\overline{f}$ by the relation

$$(\overline{f}, \varphi) = \overline{(f, \overline{\varphi})} . \tag{1.45}$$

As well, to each locally integrable complex-valued function $f \in L^1_{\text{loc}}(\mathbb{R}^n)$ corresponds a distribution from $\mathcal{D}'$, $T_f = f \in \mathcal{D}'$, defined by the formula

$$(f, \varphi) = \int_{\mathbb{R}^n} \overline{f(x)}\varphi(x)\mathrm{d}x , \tag{1.46}$$

where $\varphi \in \mathcal{D}$ represents a complex-valued function test.

An important distribution in mathematical physics is the Dirac delta distribution $\delta_a = \delta(x - a)$, $x, a \in \mathbb{R}^n$, which can be defined on any test function space by the relation

$$(\delta_a, \varphi(x)) = \varphi(a), \forall \varphi \in \Phi . \tag{1.47}$$

One can easily verify, taking into account the uniform convergence properties, that the functional δ_a defined by (1.47) is a distribution.

We say that the Dirac delta distribution δ_a is concentrated at the point $a \in \mathbb{R}^n$.

If the distribution δ_a is defined on the space $\mathcal{D}^0(\mathbb{R}^n) = C^0_C(\mathbb{R}^n)$ of continuous functions with compact support and if $\psi \in C^0(\mathbb{R}^n)$, then the product $\psi(x)\delta(x - a)$ makes sense and we can write

$$\forall \varphi \in \mathcal{D}^0 , \quad (\psi(x)\delta(x - a), \varphi(x)) = (\delta(x - a), \varphi(x)\psi(x)) = \psi(a)\varphi(a) . \tag{1.48}$$

Instead, the functional T defined on the space of test functions Φ by the formula

$$(T, \varphi(x)) = |\varphi(a)|, \varphi \in \Phi , \tag{1.49}$$

is not a distribution from Φ', because although the functional T is continuous, it is not linear.

An important class distribution are the distributions of function type or regular distributions, which are generated by locally integrable functions.

We shall show now that to every locally integrable function $f \in L^1_{\text{loc}}(\mathbb{R}^n)$ corresponds a distribution from $\mathcal{D}'(\mathbb{R}^n)$ denoted by T_f or f, if it does not lead to confusion.

We consider the functional $T_f : \mathcal{D} \to \Gamma$ defined by the formula

$$(T_f, \varphi) = \int_{\mathbb{R}^n} f(x)\varphi(x)\mathrm{d}x, \varphi \in \mathcal{D} . \tag{1.50}$$

The linearity of the functional being obvious, we show its continuity. If the sequence $\varphi_i \xrightarrow{\mathcal{D}} \varphi$, then there is the compact set Ω so that the sequence $\varphi_i \xrightarrow{\mathcal{D}(\Omega)} \varphi$; it results that $\operatorname{supp}(\varphi_i) \subset \Omega, \operatorname{supp}(\varphi) \subset \Omega$.

Taking into account (1.50), we get

$$|(T_f, \varphi_i) - (T_f, \varphi)| = |(T_f, \varphi_i - \varphi)| \leq \sup_{\Omega} |\varphi_i - \varphi| \int_{\Omega} |f(x)|\mathrm{d}x . \tag{1.51}$$

Because $\varphi_i \xrightarrow{\mathcal{D}(\Omega)} \varphi$ we have $\lim_i \sup_\Omega |\varphi_i - \varphi| = 0$, hence $\lim_i (T_f, \varphi_i) = (T_f, \varphi)$, which reflects the continuity of T_f.

Therefore, the functional T_f associated by (1.50) with the locally integrable function $f \in L^1_{\text{loc}}(\mathbb{R}^n)$ is a distribution on $\mathcal{D}'$, called a function type distribution or regular distribution.

In general, if $f \in L^1_{\text{loc}}$ such that $|f(x)| \leq A\|x\|^k, k > 0$, for $\|x\| \to \infty$, and if $\varphi \in \mathcal{S}$, that is, $|\varphi(x)| \leq B\|x\|^{-(k+2)}$, then the functional T_f given by (1.50) defines a regular distribution on the space of test functions $\mathcal{S}$, because $|f(x)\varphi(x)| \leq AB\|x\|^{-2}$ and thus the integral (1.50) exists.

Consequently, T_f is a regular tempered distribution, from which $T_f \in \mathcal{S}'$.

The distributions which cannot be represented in the integral form (1.50) are called singular distributions.

Such distributions cannot be identified with locally integrable functions. For example, the Dirac delta distribution δ_a defined by (1.47) is a singular distribution.

The function $H : \mathbb{R}^n \to \mathbb{R}$ where

$$H(x) = \begin{cases} 1 & \text{for } x_1 \geq 0, x_2 \geq 0, \ldots, x_n \geq 0 , \\ 0 & \text{otherwise} , \end{cases} \tag{1.52}$$

is called the Heaviside function and obviously generates a distribution of function type that we denote by H and which acts according to the rule

$$(H(x), \varphi(x)) = \int_0^\infty \cdots \int_0^\infty \varphi(x_1, x_2, \ldots, x_n)\mathrm{d}x_1 \ldots \mathrm{d}x_n , \quad \varphi \in \mathcal{D} . \tag{1.53}$$

We remark that this regular distribution can be represented as

$$H(x_1, \ldots, x_n) = H(x_1) \ldots H(x_n) , \tag{1.54}$$

where $H(x_i)$ represents the Heaviside distribution of one variable, namely

$$H(x_i) = \begin{cases} 0 , & x_i < 0 \\ 1 , & x_i \geq 0 \end{cases} , \quad x_i \in \mathbb{R} . \tag{1.55}$$

Proposition 1.10 Let $(f_i)_{i \geq 1} \subset L^1_{\text{loc}}(\mathbb{R}^n)$ be a sequence of locally integrable functions, uniformly convergent towards the function $f : \mathbb{R}^n \to \Gamma$ on any compact $\Omega \subset \mathbb{R}^n$; then $f \in L^1_{\text{loc}}(\mathbb{R}^n)$ and $T_{f_i} \xrightarrow{\mathcal{D}'} T_f$.

Proof: Since f_i converges uniformly to f on any compact Ω, then $f \in L^1_{\text{loc}}$ and $\int_\Omega f_i \mathrm{d}x \to \int_\Omega f \mathrm{d}x$. For $\forall \varphi \in \mathcal{D}$ with $\text{supp}(\varphi) \subset \Omega$ we have

$$\begin{aligned} |(T_{f_i}, \varphi) - (T_f, \varphi)| &\leq \int_\Omega |\varphi| \cdot |f_i - f| \mathrm{d}x \\ &\leq \text{mes}(\Omega) \sup_\Omega |\varphi(x)| \cdot \sup_\Omega |f_i(x) - f(x)| , \end{aligned} \tag{1.56}$$

where $\text{mes}(\Omega)$ denotes the measure of Ω. Since $\text{mes}(\Omega)$, $\sup_\Omega |\varphi(x)|$ are bounded and $\lim_i \sup_\Omega |f_i(x) - f(x)| = 0$, it follows that $\lim_i (T_{f_i}, \varphi) = (T_f, \varphi)$, that is, T_{f_i} converges towards T_f on $\mathcal{D}'$. □

1.2.3.1 Equality of Two Distributions: Support of a Distribution

Definition 1.23 The distribution $f \in \mathcal{D}'$ is said to be null on the open set $A \subset \mathbb{R}^n$ if $\forall \varphi \in \mathcal{D}$ with $\text{supp}(\varphi) \subset A$ we have $(f, \varphi) = 0$; we write $f = 0, x \in A$.

Also, we say that the distributions $f, g \in \mathcal{D}'$ are equal on the open set A, and we write $f = g, x \in A$, if $\forall \varphi \in \mathcal{D}$ with $\text{supp}(\varphi) \subset A$ we have $(f - g, \varphi) = 0$.

Hence, in particular, $f = g$ on $\mathbb{R}^n$ if the condition

$$(f, \varphi) = (g, \varphi), \forall \varphi \in \mathcal{D} , \tag{1.57}$$

is satisfied.

Definition 1.24 We call support of the distribution $f \in \mathcal{D}'(\mathbb{R}^n)$ and we note $\text{supp}(f)$ the complement of the reunion of open sets which nullify the distribution f.

If the support of a distribution is bounded, and since it is closed, then we say that the distribution is with compact support.

Hence, if $x_0 \in \text{supp}(f)$, then the distribution f is not nullified on any open neighborhood of x_0.

If $x_0 \notin \text{supp}(f)$, then there exists a neighborhood of point x_0 where $f = 0$.

For example, for the Dirac delta distribution δ_a given by formula (1.47) it follows: $\text{supp}(\delta_a) = \{a\}$, so δ_a is a distribution with compact support, the support being formed from a single point $a \in \mathbb{R}^n$ in which we say that the distribution is concentrated.

From the definition of equality of two distributions on an open set it results that $\delta_a = 0$ for $x \neq a$.

Indeed, $\forall \varphi \in \mathcal{D}(\mathbb{R}^n)$ with the property $a \notin \text{supp}(\varphi)$ we have $(\delta_a, \varphi) = \varphi(a) = 0$, hence the distribution δ_a is zero on the set $A = \mathbb{R}^n - \{a\}$.

The complement of it is $\complement_{\mathbb{R}^n} A = \{a\}$, that is, $\text{supp}(\delta_a) = \{a\}$. In other words, the distribution δ_a does not vanish on any neighborhood of the point $a \in \mathbb{R}^n$, hence $\delta_a = 0$, $x \neq a$.

From the physical point of view, the distribution δ_a expresses the density of a material point of mass equal to the unit and placed at the point $a \in \mathbb{R}^n$.

An important property of the distribution $\delta_a = \delta(x-a)$, $x \in \mathbb{R}^n$, called the filter property of the Dirac delta distribution, is given by the relation

$$\psi(x)\delta(x-a) = \psi(a)\delta(x-a)\,, \tag{1.58}$$

where ψ is a continuous function in the vicinity of the origin.

Indeed, we have

$$\begin{aligned}(\psi(x)\delta(x-a), \varphi(x)) &= (\delta(x-a), \varphi(x)\psi(x)) = \varphi(a)\psi(a)\\ &= \psi(a)(\delta(x-a), \varphi(x)) = (\psi(a)\delta(x-a), \varphi)\,, \quad \forall \varphi \in \mathcal{D}\,,\end{aligned} \tag{1.59}$$

from which follows (1.58).

We will show now that the Dirac delta distribution δ_a is a singular distribution that cannot be identified with a locally integrable function.

Indeed, otherwise there is $f \in L^1_{\text{loc}}$ such that

$$\int_{\mathbb{R}^n} f(x)\varphi(x)\mathrm{d}x = (\delta(x-a), \varphi(x)) = \varphi(a)\,, \quad \forall \varphi \in \mathcal{D}\,. \tag{1.60}$$

Because $\varphi \in \mathcal{D}$ is arbitrary, in its place we consider the function $\|x-a\|^2\psi(x)$, where $\psi \in \mathcal{D}$ is arbitrary; from (1.60) we obtain

$$\int_{\mathbb{R}^n} f(x)\psi(x)\|x-a\|^2\mathrm{d}x = 0, \forall \psi \in \mathcal{D}\,. \tag{1.61}$$

It follows that $\|x-a\|^2 f(x) = 0$ almost everywhere on $\mathbb{R}^n$, which implies $f(x) = 0\,a.e.$, from which we have $f(x)\varphi(x) = 0\,a.e.$ But $f(x)\varphi(x) = 0\,a.e.$ involves $\int_{\mathbb{R}^n} f(x)\varphi(x)\mathrm{d}x = 0$, which contradicts relation (1.60).

If T_f is a distribution of function type generated by a continuous function f, then their supports coincide, that is, we have

$$\text{supp}(T_f) = \text{supp}(f) = \overline{\{x \in \mathbb{R}^n, f(x) \neq 0\}}\,. \tag{1.62}$$

Regarding the Heaviside function $H(x), x \in \mathbb{R}^n$, defined by (1.52), we have $\text{supp}(f) = \mathbb{R}^n_+ = [0, \infty) \times [0, \infty] \times \cdots \times [0, \infty) = \text{supp}(T_H)$, where $T_H \in \mathcal{D}'(\mathbb{R}^n)$ represents the distribution generated by the Heaviside function, which is a locally integrable function.

Let the function $f : \mathbb{R} \to \mathbb{R}$, $f(x) = 0, x \in \mathbb{R}\backslash\{x_1\}$.

This function is piecewise continuous and its support is

$$\text{supp}(f) = \overline{\{x \in \mathbb{R}, f(x) \neq 0\}} = \overline{\{x_1\}} = \{x_1\} \, . \tag{1.63}$$

The support of the distribution function type $T_f \in \mathcal{D}'(\mathbb{R})$, generated by the function f, is the empty set $\emptyset$, that is, $\text{supp}(T_f) = \emptyset$. Indeed, $\forall \varphi \in \mathcal{D}(\mathbb{R})$ we have $(T_f, \varphi) = \int_{\mathbb{R}} f(x)\varphi(x)\text{d}x = 0$, hence $T_f = 0$ on $\mathbb{R}$.

An analogue of the Dirac delta distribution δ_a is the distribution $\delta_S = \delta(S)$, where $S \subset \mathbb{R}^n$ is a piecewise smooth hypersurface.

The functional $\delta_S : \mathcal{D} \to \mathbb{C}$, acting according to the formula

$$(\delta_S, \varphi) = \int_S \varphi(x)\text{d}S, \forall \varphi \in \mathcal{D} \, , \tag{1.64}$$

represents the Dirac delta distribution concentrated on the hypersurface S, where $\text{d}S$ is the differential area element on $S \subset \mathbb{R}^n$.

For any $\varphi \in \mathcal{D}$ whose support does not contain points from S, the distribution δ_S is null, that is, $\delta_S = 0, x \notin S$. The support of this distribution is the set of all points of S.

From the physical point of view, the distribution δ_S expresses a mass density equal to unity, distributed on the hypersurface S.

For this reason, the distribution δ_S is called Dirac delta distribution concentrated on $S \subset \mathbb{R}^n$.

If $S = S_1 \cup S_2$, then from (1.63) we obtain

$$\delta_{S_1 \cup S_2} = \delta_{S_1} + \delta_{S_2} \, . \tag{1.65}$$

Indeed, we have

$$\begin{aligned}(\delta_{S_1 \cup S_2}, \varphi) &= \int_{S_1 \cup S_2} \varphi(x)\text{d}S = \int_{S_1} \varphi(x)\text{d}S_1 + \int_{S_2} \varphi(x)\text{d}S_2 \\ &= (\delta_{S_1} + \delta_{S_2}, \varphi) \, , \quad \forall \varphi \in \mathcal{D} \, ,\end{aligned} \tag{1.66}$$

from which, on the basis of the equality of two distributions, we get (1.65).

Obviously, $\text{supp}(\delta_S) = S$ because if $x \notin S$ then $\delta_S = 0$. In general, if f is a piecewise continuous function, given on the surface S, we have

$$(f\delta_S, \varphi) = \int_S f(x)\varphi(x)\text{d}S, \forall \varphi \in \mathcal{D}(\mathbb{R}^n) \, . \tag{1.67}$$

In addition to the distribution $\delta_S \in \mathcal{D}'(\mathbb{R}^n)$ concentrated on piecewise smooth surface $S \subset \mathbb{R}^n$, the distribution $\delta_{S\times\mathbb{R}} \in \mathcal{D}(\mathbb{R}^n \times \mathbb{R})$ associated to the surface $S \subset \mathbb{R}^n$ and to the temporal variable $t \in \mathbb{R}$ is important in mechanics.

This distribution is defined by the formula

$$(\delta_{S\times\mathbb{R}}, \varphi(x,t)) = \int_{\mathbb{R}} \mathrm{d}t \int_{S} \varphi(x,t)\mathrm{d}S, \forall\varphi \in \mathcal{D}(\mathbb{R}^n \times \mathbb{R}), \tag{1.68}$$

where $\int_S$ is the surface integral, and $\mathrm{d}S$ the differential element of area on $S \subset \mathbb{R}^n$.

For the continuous real function $f(x,t) \in C^0(\mathbb{R}^n \times \mathbb{R})$ the distribution $f(x,t)\delta_{S\times\mathbb{R}} \in \mathcal{D}'(\mathbb{R}^n \times \mathbb{R})$ acts according to the formula

$$(f(x,t)\delta_{S\times\mathbb{R}}, \varphi(x,t)) = (\delta_{S\times\mathbb{R}}, f(x,t)\varphi(x,t)) = \int_{\mathbb{R}} \mathrm{d}t \int_{S} f(x,t)\varphi(x,t)\mathrm{d}S . \tag{1.69}$$

We note that $\delta_{S\times\mathbb{R}} = 0$, for $x \notin S, \forall t \in \mathbb{R}$, hence $\mathrm{supp}(\delta_{S\times\mathbb{R}}) = S \times \mathbb{R}$.

In general, the local nonintegrable functions cannot be associated with distributions. However, in some cases, by the regularization process, we can correspond local nonintegrable functions with distributions, on which we can apply linear differential operators.

To illustrate this point we will consider the following.

Example 1.7 Let $\lambda \in \mathbb{R}$ and the function $f_\lambda : \mathbb{R}\backslash\{0\} \to \mathbb{R}$, where

$$f_\lambda(x) = \frac{\cos\lambda x}{x} . \tag{1.70}$$

We show that the functional $T_{f_\lambda} : \mathcal{D} \to \mathbb{C}$ defined by the formula

$$(T_{f_\lambda}, \varphi) = \text{p.v.} \int_{-\infty}^{\infty} \frac{\cos\lambda x}{x}\varphi(x)\mathrm{d}x, \varphi \in \mathcal{D}(\mathbb{R}) \tag{1.71}$$

is a distribution of first order which satisfies the relation

$$\lim_{\lambda\to+\infty} f_\lambda(x) = 0 , \tag{1.72}$$

where the notation p.v. represents the Cauchy principal value. The distribution T_{f_λ} will be denoted as p.v. $(\cos\lambda x)/x \in \mathcal{D}'(\mathbb{R})$.

Proof: We note that the function f_λ is not integrable in the neighborhood of the origin, hence $f_\lambda \notin L^1_{\mathrm{loc}}(\mathbb{R})$ and the integral (1.71) is considered in the sense of Cauchy principal value; we thus have

$$\text{p.v.} \int_{-\infty}^{\infty} \frac{\cos\lambda x}{x}\varphi(x)\mathrm{d}x = \lim_{\varepsilon\to+0} \left[\int_{-\infty}^{-\varepsilon} \frac{\cos\lambda x}{x}\varphi(x)\mathrm{d}x + \int_{\varepsilon}^{\infty} \frac{\cos\lambda x}{x}\varphi(x)\mathrm{d}x \right] . \tag{1.73}$$

Since $x \to (\cos \lambda x)/x$ is an odd function we obtain

$$\text{p.v.} \int_{-\infty}^{\infty} \frac{\cos \lambda x}{x} \mathrm{d}x = 0 . \tag{1.74}$$

Therefore, relation (1.71) can be written as

$$(T_{f_\lambda}, \varphi) = \text{p.v.} \int_{-\infty}^{\infty} \cos \lambda x \frac{[\varphi(x) - \varphi(0)]}{x} \mathrm{d}x , \quad \varphi \in \mathcal{D}(\mathbb{R}) . \tag{1.75}$$

T_{f_λ} is obviously a linear functional.

We shall prove its continuity. Applying *the mean value formula*, we can write

$$\varphi(x) - \varphi(0) = x\varphi'(\xi_x) , \quad \xi_x \in (0, x) , \quad \text{or} \quad \xi_x \in (x, 0) . \tag{1.76}$$

Therefore, considering $\operatorname{supp}(\varphi) \subset [-a, a]$, $a > 0$, from (1.75) we obtain

$$\begin{aligned} \left|(T_{f_\lambda}, \varphi)\right| &= \left|\text{p.v.} \int_{-a}^{a} \cos \lambda x \frac{[\varphi(x) - \varphi(0)]}{x} \mathrm{d}x\right| \\ &\leq \text{p.v.} \int_{-a}^{a} |\cos \lambda x| \left|\frac{\varphi(x) - \varphi(0)}{x}\right| \mathrm{d}x \leq 2a \sup_{x \in [-a,a]} |\varphi'(x)| . \end{aligned} \tag{1.77}$$

Hence, $\forall \varphi \in \mathcal{D}(\mathbb{R})$ with $\operatorname{supp}(\varphi) \subset [-a, a]$ and we have

$$\left|(T_{f_\lambda}, \varphi)\right| \leq c \sup_{x \in [-a,a]} |\varphi'(x)|, c = 2a ; \tag{1.78}$$

the relation shows that the linear functional T_{f_λ} defined by (1.71) is a first-order distribution, hence $T_{f_\lambda} = f_\lambda \in \mathcal{D}'^1(\mathbb{R})$.

The distribution p.v. $(\cos \lambda x)/x$ is a regularization of the function $f_\lambda = (\cos \lambda x)/x$.

We note that for $\varphi \in \mathcal{D}(\mathbb{R})$ one can write

$$\begin{aligned} &\left(x \text{ p.v.} \frac{\cos \lambda x}{x}, \varphi\right) = \left(\text{p.v.} \frac{\cos \lambda x}{x}, x\varphi\right) \\ &= \text{p.v.} \int_{\mathbb{R}} \varphi(x) \cos(\lambda x) \mathrm{d}x = \int_{\mathbb{R}} \cos(\lambda x)\varphi(x) \mathrm{d}x = (\cos \lambda x, \varphi(x)) , \end{aligned} \tag{1.79}$$

from which we obtain

$$x \text{ p.v.} \frac{\cos \lambda x}{x} = \cos \lambda x . \tag{1.80}$$

Hence, for $x \neq 0$ the distribution p.v. $(\cos \lambda x)/x$ coincides with the function $(\cos \lambda x)/x$.

To show that f_λ converges to zero on $\mathcal{D}'(\mathbb{R})$ when $\lambda \to +\infty$ we note that we can write

$$\varphi(x) - \varphi(0) = \int_0^x \varphi'(t)\mathrm{d}t \,. \tag{1.81}$$

Making the change of variable $t = xu$, relation (1.81) becomes

$$\varphi(x) - \varphi(0) = x \int_0^1 \varphi'(xu)\mathrm{d}u \,. \tag{1.82}$$

We denote

$$\psi(x) = \int_0^1 \varphi'(xu)\mathrm{d}u \,, \tag{1.83}$$

which is a function from $\mathcal{D}(\mathbb{R})$, because $\varphi \in \mathcal{D}(\mathbb{R})$ and $\mathrm{supp}(\psi) \subset [-a, a]$.

Taking into account (1.75) and (1.83), we obtain

$$\begin{aligned}(T_{f_\lambda}, \varphi) &= \text{p.v.} \int_{-a}^{a} \psi(x)\cos(\lambda x)\mathrm{d}x \\ &= \int_{-a}^{a} \psi(x)\cos(\lambda x)\mathrm{d}x \,, \quad \forall \varphi \in \mathcal{D} \,, \quad \mathrm{supp}(\varphi) \subset [-a, a] \,.\end{aligned} \tag{1.84}$$

Integrating by parts, we have

$$(T_{f_\lambda}, \varphi) = \frac{1}{\lambda}\left\{(\psi(x)\sin(\lambda x))|_{-a}^{a} - \int_{-a}^{a} \sin(\lambda x) \cdot \psi'(x)\mathrm{d}x\right\} \,, \tag{1.85}$$

from which the inequality $|(T_{f_\lambda}, \varphi)| \leq A/|\lambda|$, where A is a positive constant which depends on $a > 0$; therefore, $\lim_{\lambda\to\infty}(T_{f_\lambda}, \varphi) = 0, \forall \varphi \in \mathcal{D}, \mathrm{supp}(\varphi) \subset [-a, a]$, hence $\lim_{\lambda\to\infty} T_{f_\lambda} = 0$.

The last relation shows that the family of distributions $T_{f_\lambda} = f_\lambda$ converges to zero on $\mathcal{D}'(\mathbb{R})$ when $\lambda \to \pm\infty$. □

Example 1.8 We consider the function $f(x) = 1/x^2$, $x \in \mathbb{R}\setminus\{0\}$, to which we assign the functional p.v. $1/x^2 : \mathcal{D} \to \mathbb{C}$ defined by the relation

$$\begin{aligned}\left(\text{p.v.}\frac{1}{x^2}, \varphi\right) &= \text{p.v.} \int_{\mathbb{R}} \frac{\varphi(x) - \varphi(0)}{x^2}\mathrm{d}x \\ &= \lim_{\varepsilon\to+0}\left[\int_{-\infty}^{-\varepsilon} \frac{\varphi(x) - \varphi(0)}{x^2}\mathrm{d}x + \int_{\varepsilon}^{\infty} \frac{\varphi(x) - \varphi(0)}{x^2}\mathrm{d}x\right] \,.\end{aligned} \tag{1.86}$$

Let us show that the functional p.v. $1/x^2$ is a second-order distribution from $\mathcal{D}'(\mathbb{R})$.

Because the linearity of the functional is evident we shall test only its continuity. Thus, taking into account that $\varphi(x) - \varphi(0) = x\varphi'(0) + x^2\varphi''(\xi_x)/2$, $\xi_x \in (0, x)$ or $\xi_x \in (x, 0)$, we have

$$\left(\text{p.v.}\,\frac{1}{x^2}, \varphi\right) = \text{p.v.}\int\limits_{\mathbb{R}} \left(\frac{\varphi'(0)}{x} + \frac{\varphi''(\xi_x)}{2}\right) \mathrm{d}x = \frac{1}{2}\int\limits_{\mathbb{R}} \varphi''(\xi_x)\mathrm{d}x\,, \tag{1.87}$$

because p.v. $\int_{\mathbb{R}}(\mathrm{d}x/x) = 0$.

Consequently, considering $\operatorname{supp}(\varphi) \subset [-a, a]$, $a > 0$, the previous relation becomes

$$\left|\left(\text{p.v.}\,\frac{1}{x^2}, \varphi\right)\right| = \frac{1}{2}\left[\int\limits_{-a}^{a} \varphi''(\xi_x)\mathrm{d}x\right] \le a \sup_{x\in[-a,a]} |\varphi''(x)|\,, \tag{1.88}$$

from which results the continuity of the functional p.v. $1/x^2$ and that p.v. $1/x^2$ is a second-order distribution from $\mathcal{D}'(\mathbb{R})$.

For $\forall\varphi \in \mathcal{D}(\mathbb{R})$ we have

$$\left(x^2\text{p.v.}\,\frac{1}{x^2}, \varphi\right) = \left(\text{p.v.}\,\frac{1}{x^2}, x^2\varphi\right) = \text{p.v.}\int\limits_{\mathbb{R}} \varphi(x)\mathrm{d}x = \int\limits_{\mathbb{R}} \varphi(x)\mathrm{d}x = (1, \varphi)\,, \tag{1.89}$$

hence x^2 p.v. $1/x^2 = 1$, which shows that, except at the origin, the distribution p.v. $1/x^2$ coincides with the function $1/x^2$.

We associate to the function $f = 1/x^2$, $x \in \mathbb{R}\backslash\{0\}$, the functional $Pf1/x^2 : \mathcal{D}(\mathbb{R}) \to \mathbb{C}$, called a pseudofunction, defined by the relation

$$\left(Pf\frac{1}{x^2}, \varphi\right) = \lim_{\varepsilon\to+0}\left[\int\limits_{-\infty}^{-\varepsilon} \frac{\varphi(x)}{x^2}\mathrm{d}x + \int\limits_{\varepsilon}^{\infty} \frac{\varphi(x)}{x^2}\mathrm{d}x - 2\frac{\varphi(0)}{\varepsilon}\right]. \tag{1.90}$$

One can show as in the previous case that $Pf1/x^2$ is a second-order distribution from $\mathcal{D}'(\mathbb{R})$ and that $x^2 Pf1/x^2 = 1$. Also, the distributions $PfH(x)/x$ and $PfH(-x)/x$ are defined by the relations

$$\left[Pf\frac{H(x)}{x}, \varphi\right] = \lim_{\varepsilon\to+0}\left[\int\limits_{\varepsilon}^{\infty} \frac{\varphi(x)}{x}\mathrm{d}x + \varphi(0)\ln\varepsilon\right], \tag{1.91}$$

$$\left[Pf\frac{H(-x)}{x}, \varphi\right] = \lim_{\varepsilon\to+0}\left[\int\limits_{-\infty}^{-\varepsilon} \frac{\varphi(x)}{x}\mathrm{d}x - \varphi(0)\ln\varepsilon\right]. \tag{1.92}$$

One easily verifies the relation p.v. $1/x = PfH(x)/x + PfH(-x)/x$.

We remark that the concept of finite part of an integral and the concept of pseudofunction were introduced by J. Hadamard.

1.2.4
Characterization Theorems of Distributions

To test whether a functional defined on the test space Φ is a distribution, we check the linearity and continuity of the functional. In general, the linearity of the functional is easy to verify, but we have difficulties verifying the continuity, because it involves the use of the convergence introduced on the test space Φ.

In the following we give a condition equivalent to the continuity of the linear functional defined on Φ, which is particularly useful in applications, which can be considered also as definition of the distribution of the space Φ' [6, 11, 12].

Theorem 1.2 The linear functional $T : \mathcal{D} \to \Gamma$ is a distribution of $\mathcal{D}'$ if and only if for any compact $\Omega \subset \mathbb{R}^n$ there exist the constants $C(\Omega)$ and $m(\Omega) \in \mathbb{N}_0$ such that

$$|(T, \varphi)| \leq C(\Omega) \sup_{\substack{|\alpha| \leq m(\Omega) \\ x \in \Omega}} |\mathrm{D}^\alpha \varphi| , \quad \forall \varphi \in \mathcal{D}(\Omega) . \tag{1.93}$$

Theorem 1.3 The linear functional $T : \mathcal{D} \to \Gamma$ is a distribution of $\mathcal{D}'^m$ if and only if, for any compact $\Omega \subset \mathbb{R}^n$, there exists the constant $c(\Omega) > 0$ such that

$$|(T, \varphi)| \leq C(\Omega) \sup_{\substack{|\alpha| \leq m \\ x \in \Omega}} |\mathrm{D}^\alpha| \varphi , \quad |\forall \varphi \in \mathcal{D}(\Omega) . \tag{1.94}$$

Theorem 1.4 The linear functional $T : \mathcal{D} \to \Gamma$ is a measure if and only if any are the compact $\Omega \subset \mathbb{R}^n$ there exists the constant $c(\Omega) > 0$ such that

$$|(T, \varphi)| \leq c(\Omega) \sup_{x \in \Omega} |\varphi(x)| , \quad \forall \varphi \in \mathcal{D}(\Omega) . \tag{1.95}$$

Below, we give some applications of the characterization theorems of distributions.

Example 1.9 We consider the function $f \in L^1_{\text{loc}}(\mathbb{R}^n)$ and the functional $T_f : \mathcal{D} \to \mathbb{C}$ defined by the formula

$$(T_f, \varphi) = \int_{\mathbb{R}^n} f(x)\varphi(x)\mathrm{d}x , \quad \varphi \in \mathcal{D} . \tag{1.96}$$

Using Theorem 1.2 of the characterization of distributions of $\mathcal{D}'$, we shall show that the functional T_f associated to the locally integrable function f is a distribution of $\mathcal{D}'$, referred to as distribution of function type.

The functional T_f is, obviously, linear and we have

$$|(T_f, \varphi)| \leq \int_{\mathbb{R}^n} |f(x)||\varphi(x)|\mathrm{d}x \leq \sup_{x \in \mathbb{R}^n} |\varphi(x)| \int_{\operatorname{supp}(\varphi)} |f(x)|\mathrm{d}x . \tag{1.97}$$

For $\varphi \in \mathcal{D}(\Omega) \subset \mathcal{D}$, the previous relation becomes

$$|(T_f, \varphi)| \leq \sup_{x \in \Omega} |\varphi(x)| \int_{\Omega} |f(x)| \mathrm{d}x \, . \tag{1.98}$$

Putting $c(\Omega) = \int_{\Omega} |f(x)|\mathrm{d}x$, we have

$$|(T_f, \varphi)| \leq c(\Omega) \sup_{x \in \Omega} |\varphi(x)| \, , \quad \forall \varphi \in \mathcal{D}(\Omega) \, . \tag{1.99}$$

If $f \neq 0$, and taking into account Theorem 1.4, it follows that T_f is a zero-order distribution and that the constant $c(\Omega) > 0$ depends on the compact $\Omega \in \mathbb{R}^n$. Hence, $T_f = f \in \mathcal{D}'^0$.

For $f = 0$ we have $c(\Omega) = 0$, and relation (1.99) becomes $|(T_f, \varphi)| \leq 0 \leq \sup_{x \in \Omega} |\varphi(x)|$, which shows that, in this case, any positive number can be considered as constant $c(\Omega)$.

In conclusion, any locally integrable function $f \in L^1_{\text{loc}}(\mathbb{R}^n)$ can be identified with a zero-order distribution, hence $T_f = f \in \mathcal{D}'^0$.

In general, let $f \in L^1_{\text{loc}}(\Omega)$, $\Omega \in \mathbb{R}^n$ be a compact set, and the linear functional

$$(T_f, \varphi) = \int_{\Omega} f(x) \mathrm{D}^{\alpha} \varphi(x) \mathrm{d}x \, , \quad \forall \varphi \in \mathcal{D}(\Omega) \, , \quad \alpha \in \mathbb{N}_0^n \, , \tag{1.100}$$

where D^{α} is the operator of derivation.

For $\forall \varphi \in \mathcal{D}(\Omega_1) \subset \mathcal{D}(\Omega)$, $\forall \Omega_1$ compact set and for $\Omega_1 \subset \Omega$ from (1.100), we obtain

$$|(T_f, \varphi)| \leq \sup_{x \in \Omega_1} |\mathrm{D}^{\alpha} \varphi(x)| \cdot \int_{\Omega_1} |f(x)| \mathrm{d}x \, , \tag{1.101}$$

hence,

$$|(T_f, \varphi)| \leq c \sup_{x \in \Omega_1} |\mathrm{D}^{\alpha} \varphi(x)| \, , \quad \forall \varphi \in \mathcal{D}(\Omega_1) \, , \tag{1.102}$$

where $c = \int_{\Omega} |f(x)|\mathrm{d}x$.

From relation (1.102) it follows that the linear functional T_f associated with the function $f \in L^1_{\text{loc}}(\Omega)$, by formula (1.96), is a distribution of function type of order $k = |\alpha|$.

Example 1.10 Let $\delta_a : \mathrm{C}^0_{\mathrm{C}}(\mathbb{R}^n) \to \mathbb{C}$, $a \in \mathbb{R}^n$, defined by the formula

$$(\delta_a, \varphi) = \varphi(a) \, , \quad \varphi \in \mathrm{C}^0_{\mathrm{C}}(\mathbb{R}^n) \, , \tag{1.103}$$

where $\mathrm{C}^0_{\mathrm{C}} = \mathcal{D}^0(\mathbb{R}^n)$ is the continuous function space with compact support. Obviously, $\delta_a = \delta(x - a)$ represents the Dirac delta distribution concentrated at the point $a \in \mathbb{R}^n$. Using Theorem 1.4 we shall show that functional δ_a is a zero-order distribution of $\mathcal{D}'$.

The linearity of the functional δ_a is evident. To prove the continuity of the functional, from (1.103) we have

$$|(\delta_a, \varphi)| = |\varphi(a)| \leq \sup_{x\in\Omega} |\varphi(x)| \,, \quad \forall \varphi \in \mathrm{C}^0_{\mathrm{C}}(\Omega) \,. \tag{1.104}$$

This relation shows that the linear functional δ_a is a zero-order distribution, because $k = 0$ and $c(\Omega) = 1$.

Example 1.11 Let the functional δ_S be associated to the piecewise smooth hypersurface $S \subset \mathbb{R}^n$, by the formula

$$(\delta_S, \varphi) = \int_S \varphi(x) \mathrm{d}S \,, \quad \varphi \in \mathcal{D} \,. \tag{1.105}$$

The functional $\delta_S : \mathcal{D} \to \Gamma$ is called the Dirac delta distribution concentrated on S.

We will show that δ_S is a zero-order distribution. Since the linearity of the functional δ_S results from the surface integral linearity, we shall show the continuity of it.

From (1.105), we obtain

$$|(\delta_S, \varphi)| \leq \sup_{x\in\Omega} |\varphi(x)| \cdot \int_{S\cap \mathrm{supp}(\varphi)} \mathrm{d}S \,, \quad \forall \varphi \in \mathcal{D}(\Omega) \,. \tag{1.106}$$

Noting $c(\Omega) = \int_{S\cap \mathrm{supp}(\varphi)} \mathrm{d}S$, the previous relation becomes

$$|(\delta_S, \varphi)| \leq c \sup_{x\in\Omega} |\varphi(x)| \,, \quad \forall \varphi \in \mathcal{D}(\Omega) \,, \tag{1.107}$$

which shows that δ_S is a zero-order distribution.

Theorem 1.5 The linear functional $T : \mathcal{S} \to \Gamma$ is a distribution of $\mathcal{S}'$ if and only if there exist the constant $c > 0$ and the integers $m, \ell \in \mathbb{N}_0$ such that

$$|(T, \varphi)| \leq c \sup_{\substack{|\alpha|\leq m \\ x\in\mathbb{R}^n}} \left|\left(1 + \|x\|^2\right)^{\ell} \mathrm{D}^{\alpha} \varphi(x)\right| \,, \quad \forall \varphi \in \mathcal{S} \,. \tag{1.108}$$

Theorem 1.6 The linear functional $T : \mathcal{E} \to \Gamma$ is a distribution of $\mathcal{E}'$ if and only if there exist a compact $\Omega \subset \mathbb{R}^n$ and the constants $c > 0$, $m \in \mathbb{N}_0$ such that

$$|(T, \varphi)| \leq c \sup_{|\alpha|\leq m} \sup_{x\in\Omega} |\mathrm{D}^{\alpha} \varphi(x)| \,, \quad \forall \varphi \in \mathcal{E}(\mathbb{R}^n) \,. \tag{1.109}$$

Example 1.12 Let $\delta_a^{(p)} : \mathcal{S}(\mathbb{R}^n) \to \Gamma$ be a functional, $p \in \mathbb{N}$, $a \in \mathbb{R}^n$, defined by the formula

$$\left(\delta_a^{(p)}, \varphi\right) = (-1)^p \mathrm{D}^{\alpha} \varphi(a) \,, \quad \forall \varphi \in \mathcal{S}(\mathbb{R}^n) \,, \quad |\alpha| = p \,, \tag{1.110}$$

where D^{α} is the derivation operator.

The linearity of the functional results from the relation

$$\left(\delta_a^{(p)}, \alpha_1\varphi_1 + \alpha_2\varphi_2\right) = (-1)^p \mathrm{D}^\alpha(\alpha_1\varphi_1 + \alpha_2\varphi_2)$$
$$= \alpha_1(-1)^p \mathrm{D}^\alpha\varphi_1 + \alpha_2(-1)^p \mathrm{D}^\alpha\varphi_2 = \alpha_1\left(\delta_a^{(p)}, \varphi_1\right) + \alpha_2\left(\delta_a^{(p)}, \varphi_2\right), \quad (1.111)$$

$\forall \varphi_1, \varphi_2 \in \mathcal{S}(\mathbb{R}^n)$, and $\forall \alpha_1, \alpha_2 \in \Gamma$.

Regarding the continuity of the functional $\delta_a^{(p)}$ on $\mathcal{S}(\mathbb{R}^n)$ we have, from (1.110),

$$\left|\left(\delta_a^{(p)}, \varphi\right)\right| = |\mathrm{D}^\alpha\varphi(a)| \leq \sup_{\substack{x\in\Omega \\ |\alpha|=p}} |\mathrm{D}^\alpha\varphi(x)|, \quad \forall\varphi \in S(\mathbb{R}^n), \quad (1.112)$$

which, on the basis of Theorem 1.5, show that the linear functional $\delta_a^{(p)}$ is continuous, hence $\delta_a^{(p)} \in \mathcal{E}'$.

A particular class of tempered distributions consists of locally integrable functions with slow growth to infinity, that is, the functions $f \in L^1_{\text{loc}}(\mathbb{R}^n)$ that satisfy to infinity the relation $|f(x)| \leq c(1 + \|x\|)^k$, $c \geq 0$, $k \geq 0$.

In this case we associate the functional $T_f = f$ defined by the formula

$$(f, \varphi) = \int_{\mathbb{R}^n} f(x)\varphi(x)\mathrm{d}x, \varphi \in \mathcal{S}, \quad (1.113)$$

to the function f, from which we obtain

$$|(f, \varphi)| \leq \int_{\mathbb{R}^n} |f(x)||\varphi(x)|\mathrm{d}x \leq \int_{\mathbb{R}^n} c(1 + \|x\|)^k|\varphi(x)|\mathrm{d}x \leq A \sup_{x\in\mathbb{R}^n} |\varphi|, \quad A > 0, \quad (1.114)$$

because $(1 + \|x\|)^k|\varphi(x)| \in L^1(\mathbb{R}^n)$.

Since the linear functional f defined by (1.113) is bounded, according to Theorem 1.5 of characterization of distributions of $\mathcal{S}'$, it means that $T_f = f$ is a distribution of function type of $\mathcal{S}'$.

Also, the absolutely integrable functions, $f \in L^1(\mathbb{R}^n)$, and the functions with polynomial growth $f(x) = x^m = x_1^{m_1}x_2^{m_2}\ldots x_n^{m_n}$, $x \in \mathbb{R}^n$, $m_i \geq 0$, generate temperate distribution of function type on $\mathcal{S}'(\mathbb{R}^n)$.

Conversely, the locally integrable function $f(x) = \mathrm{e}^x\cos\mathrm{e}^x$, $x \in \mathbb{R}$, $f \in L^1_{\text{loc}}(\mathbb{R}^n)$, although it is not with slow growth (polynomial) generates a temperate distribution of function type, which is defined by the formula

$$(f(x), \varphi(x)) = \int_{\mathbb{R}} \mathrm{e}^x\cos\mathrm{e}^x\varphi(x)\mathrm{d}x, \quad \forall\varphi \in \mathcal{S}(\mathbb{R}). \quad (1.115)$$

Indeed, the integral on the right-hand side does exist and we have

$$\left|\int_{\mathbb{R}} e^x \cos e^x \varphi(x)dx\right| = \left|\int_{\mathbb{R}} \varphi(x)(\sin e^x)' dx\right|$$

$$= \left|\varphi(x)\sin e^x|_{-\infty}^{\infty} - \int_{\mathbb{R}} \sin e^x \varphi'(x)dx\right| = \left|\int_{\mathbb{R}} \varphi'(x)\sin e^x dx\right|$$

$$\leq \int_{\mathbb{R}} |\varphi'(x)|dx = \int_{\mathbb{R}} (1+x^2)|\varphi'|(1+x^2)^{-1} dx \leq A \sup_{x\in\mathbb{R}} (1+x^2)|\varphi'| , \quad (1.116)$$

where we take into consideration that $\varphi(x)\sin e^x \in \mathcal{S}(\mathbb{R})$, hence $\lim_{|x|\to\infty} \sin e^x \varphi(x) = 0$ and $\exists \int_{\mathbb{R}} |\varphi'(x)|dx < \infty$. According to Theorem 1.5, $f \in \mathcal{S}'(\mathbb{R})$.

1.3 Operations with Distributions

1.3.1 The Change of Variables in Distributions

In geometry, mechanics and mathematical analysis the transformations of independent variables are frequently used [13], so as to simplify the calculations and interpretation of the results.

These changes, applied to functions, lead to new functions.

The methodology of these changes of variables can be extended from functions to distributions.

Let $T : \mathbb{R}^n \to \mathbb{R}^n$ be an application defined by the relation $x = h(u)$, hence

$$x_i = h_i(u_1, \ldots, u_n) , \quad i = \overline{1, n} , \quad (1.117)$$

which represents a transformation from Cartesian coordinates $(x_1, x_2, \ldots, x_n) \in \mathbb{R}^n$ to the coordinates $(u_1, u_2, \ldots, u_n) \in \mathbb{R}^n$.

We see that the functions $h_i, i = \overline{1, n}$, are of class $C^\infty(\mathbb{R}^n)$ and that the punctual transformation is bijective. Therefore, the transformation (1.117) allows for the inverse punctual transform $T^{-1} : \mathbb{R}^n \to \mathbb{R}^n$, defined by the formula

$$u = h^{-1}(x) \Leftrightarrow u_i = h_i^{-1}(x_1, \ldots, x_n) . \quad (1.118)$$

The Jacobians of the transformations T and T^{-1} are $\partial(x)/\partial(u)$ and $\partial(u)/\partial(x)$ for which we have $\partial(x)/\partial(u) = (\partial(u)/\partial(x))^{-1}$, $(\partial(u)/\partial(x)) \neq 0$.

To see how to approach the definition of the change of variables, we shall consider the case of a locally integrable function which can be identified with a distribution of function type.

Let $f \in L^1_{\text{loc}}(\mathbb{R}^n)$ and $\varphi \in \mathcal{D}(\mathbb{R}^n)$. We have

$$(f(h(u)), \varphi(u)) = \int_{\mathbb{R}^n} f(h(u))\varphi(u)du = \int_{\mathbb{R}^n} f(x)\varphi(h^{-1}(x)) \left|\frac{\partial(u)}{\partial(x)}\right| dx . \quad (1.119)$$

This equality can be transcribed as

$$(f(x), \psi(x)) = (f(h(u)), \varphi(u)) , \tag{1.120}$$

where $\psi(x) \in \mathcal{D}(\mathbb{R}^n)$, and has the expression

$$\psi(x) = \varphi(h^{-1}(x)) \left| \frac{\partial(u)}{\partial(x)} \right| . \tag{1.121}$$

Noting $(Tf)(u) = f(h(u))$, relation (1.120) becomes

$$(f(x), \psi(x) = ((Tf)(u), \varphi(u)) \tag{1.122}$$

and sets the dependence between the function given in the variable u, namely $f(h(u))$, and its correspondent $(Tf)(x)$, obtained by means of the punctual transformation (1.117).

Since ψ and φ are functions on D, relation (1.122) is adopted for defining the change of variables in the case of distributions.

Definition 1.25 Let $f(x) \in \mathcal{D}'(\mathbb{R}^n)$ be a distribution in the variable $x \in \mathbb{R}^n$. Then, the corresponding distribution in the variable $u \in \mathbb{R}^n$, defined by the transformation (1.117), will be denoted $(Tf)(u) \in \mathcal{D}'(\mathbb{R}^n)$ and is given by the formula

$$(f(x), \psi(x)) = ((Tf)(u), \varphi(u)) , \quad \varphi \in \mathcal{D}(\mathbb{R}^n) , \tag{1.123}$$

where $\psi(x) \in \mathcal{D}(\mathbb{R}^n)$ and has the expression

$$\psi(x) = \varphi(u(x)) \frac{1}{|\partial(x)/\partial(u)|} . \tag{1.124}$$

We note that if the punctual transformation is not bijective, hence $\partial(x)/\partial(u) = 0$ at some points, then the change of variable formula (1.123) is inapplicable. Such cases will be analyzed for the transition to spherical coordinates on $\mathbb{R}^n$ and for the transition to cylindrical coordinates on $\mathbb{R}^3$.

To illustrate the change of variables for the Dirac delta distribution $\delta_0 = \delta(x_1, \ldots, x_n)$ concentrated at the origin. According to formula (1.123), we have

$$(\delta(x), \psi(x)) = (\delta(h(u)), \varphi(u)) = \psi(0) , \quad \varphi \in \mathcal{D} , \tag{1.125}$$

where

$$\psi(x) = \frac{\varphi(u(x))}{|\partial(h)/\partial(u)|} , \quad \frac{\partial(h)}{\partial(u)} = \frac{\partial(h_1, \ldots, h_n)}{\partial(u_1, \ldots, u_n)} . \tag{1.126}$$

From (1.126), we obtain

$$\psi(0) = \varphi(u_0) \left[\left| \frac{\partial(h)}{\partial(u)} \right|_{u=u_0} \right]^{-1} , \tag{1.127}$$

where $0 = h(u_0)$ and $x = h(u)$, $h \in C^\infty(\mathbb{R}^n)$, is the punctual bijective transformation.

Taking into account (1.125) and (1.127), we can write

$$(\delta(x), \psi(x)) = (\delta(h(u)), \varphi(u)) = \left(\frac{\delta(u-u_0)}{|\partial(h)/\partial(u)|_{u=u_0}}, \varphi(u) \right) , \tag{1.128}$$

from which we obtain the formula

$$(T\delta)(u) = \delta(h(u)) = \delta(h_1(u), \ldots, h_n(u)) = \frac{\delta(u-u_0)}{|\partial(h)/\partial(u)|_{u=u_0}} . \tag{1.129}$$

In particular, for the Dirac delta distribution of a variable, we obtain, from (1.129),

$$\delta(h(u)) = \frac{\delta(u-u_0)}{|h'(u_0)|} , \quad u \in \mathbb{R} . \tag{1.130}$$

Thus, we have $\delta(\mathrm{e}^{au} - 1) = \delta(u)/|a|$, $a \neq 0$, because $x = h(u) = \mathrm{e}^{au} - 1$ and $h(u) = 0 \Rightarrow u = 0$, $\partial h/\partial u = a\mathrm{e}^{u}$.

We note that formula (1.130) can be generalized to the case in which the equation $h(u) = 0$ allows a finite or infinite number of simple roots, that is, $\alpha_1, \alpha_2, \ldots, \alpha_p, \ldots$ and $h'(\alpha_i) \neq 0$.

By definition, we write

$$\delta(h(u)) = \sum_p \frac{\delta(u-u_p)}{|h'(u_p)|} . \tag{1.131}$$

For example

$$\delta(u^2 - a^2) = \frac{1}{2a}[\delta(u-a) + \delta(u+a)] , \quad a > 0 , \tag{1.132}$$

$$\delta(\cos u) = \sum_n \delta\left(u - (2n+1)\frac{\pi}{2}\right) . \tag{1.133}$$

Let us look at the case of the punctual transformation $x = \rho\cos\theta$, $y = \rho\sin\theta$, $\rho \geq 0$, $\theta \in [0, 2\pi)$, which expresses the transition from Cartesian coordinates $(x, y) \in \mathbb{R}^2$ to polar coordinates (ρ, θ), and the Jacobian of the transformation is $J(\rho, \theta) = \partial(x, y)/\partial(\rho, \theta) = \rho$.

In all points $(x, y) \in \mathbb{R}^2$ where $J(\rho, \theta) = \rho \neq 0$, the considered transformation is locally bijective, that is, except at the origin $(0, 0)$ for which $\rho = 0$, and θ is arbitrary. Consequently, the considered punctual transformation is inapplicable to the Dirac delta distribution $\delta(x, y) \in \mathcal{D}'(\mathbb{R}^2)$, which is concentrated (has the support) at the point $(0, 0)$, but it may be applied to the Dirac delta distribution $\delta(x-a, y-b)$ which has the support at the point $(a, b) \neq (0, 0)$.

Passing to polar coordinates in the formulae (1.123) and (1.124), we can write

$$(\delta(x-a, y-b), \psi(x, y)) = \psi(a, b) = (\delta(\rho\cos\theta - a, \rho\sin\theta - b), \varphi(\rho, \theta)) , \tag{1.134}$$

where $\psi(x, y) = \varphi(\rho, \theta)/\rho$, $\varphi \in \mathcal{D}(\mathbb{R}^2)$.

From here, we obtain $\psi(a, b) = \varphi(\rho_0, \theta_0)/\rho_0\rho_0$, because (ρ_0, θ_0) represents the polar coordinates of the point (a, b), hence $a = \rho_0 \cos\theta_0$, $b = \rho_0 \sin\theta_0$, $\rho_0 = \sqrt{a^2 + b^2}$, $\theta_0 \in [0, 2\pi)$, $\tan\theta_0 = b/a$.

Ultimately, we obtain the relation

$$\begin{aligned}\psi(a, b) &= \frac{\varphi(\rho_0, \theta_0)}{\rho_0} = \left(\frac{\delta(\rho - \rho_0, \theta - \theta_0)}{\rho_0}, \quad \varphi(\rho, \theta)\right) \\ &= (\delta(\rho\cos\theta - a, \rho\sin\theta - b), \varphi(\rho, \theta)) , \end{aligned} \tag{1.135}$$

resulting in the formula

$$\delta(\rho\cos\theta - a, \rho\sin\theta - b) = \frac{\delta(\rho - \rho_0, \theta - \theta_0)}{\rho_0} . \tag{1.136}$$

Forwards, we shall treat the transition from the Cartesian coordinates $(x, y, z) \in \mathbb{R}^3$ to the spherical coordinates (r, φ, θ), given by the formulae

$$x = r\sin\varphi\cos\theta , \quad y = r\sin\varphi\sin\theta , \quad z = r\cos\varphi , \tag{1.137}$$

where $r \geq 0$, $\varphi \in [0, \pi]$, $\theta \in [0, 2\pi)$.

The Jacobian of the transformation is $J(r, \varphi, \theta) = \partial(x, y, z)/\partial(r, \varphi, \theta) = r^2 \sin\varphi$ and shows that the punctual transformation is locally bijective everywhere on $\mathbb{R}^3$ with the exception of the points on the Oz-axis, for which $\varphi = 0$ or $\varphi = \pi$. At the origin $(0, 0, 0)$ we can consider $r = 0$ and φ arbitrary.

Consequently, the transition to spherical coordinates cannot be achieved for the Dirac delta distribution $\delta(x, y, z) \in \mathcal{D}'(\mathbb{R}^3)$ concentrated at the point $(0, 0, 0)$.

For the Dirac delta distribution $\delta(x - a, y - b, z - c)$ concentrated at the point $(a, b, c) \notin Oz$, where $a^2 + b^2 > 0$, we can apply the considered transformation since $J(r, \varphi, \theta) \neq 0$. On the basis of the formulae (1.123) and (1.124), we have

$$\begin{aligned}&(\delta(x - a, y - b, z - c), \psi(x, y, z)) \\ &= (\delta(r\sin\varphi\cos\theta - a, r\sin\varphi\sin\theta - b, r\cos\varphi - c), \varphi(r, \varphi, \theta)) , \end{aligned} \tag{1.138}$$

where $\psi(x, y, z) = \varphi(r, \varphi, \theta)/r^2 \sin\varphi$, $\varphi \in \mathcal{D}(\mathbb{R}^3)$.

If $(r_0, \varphi_0, \theta_0)$ represents the spherical coordinates of the point $(a, b, c) \notin Oz$, $a^2 + b^2 > 0$, then we have the relations

$$a = r_0 \sin\varphi_0 \cos\theta_0 , \quad b = r_0 \sin\varphi_0 \sin\theta_0 , \quad c = r_0 \cos\varphi_0 , \tag{1.139}$$

$$r_0 = \sqrt{a^2 + b^2 + c^2} , \ \tan\theta_0 = \frac{b}{a} , \ \theta_0 \in [0, 2\pi) , \ \varphi_0 = \arccos\frac{c}{r_0} , \ \varphi_0 \in [0, \pi] , \tag{1.140}$$

and thus we obtain

$$\begin{aligned}\psi(a, b, c) &= \frac{\varphi(r_0, \varphi_0, \theta_0)}{r^2 \sin\varphi_0} = \left(\frac{\delta(r - r_0, \varphi - \varphi_0, \theta - \theta_0)}{r_0^2 \sin\varphi_0}, \varphi(r, \varphi, \theta)\right) \\ &= (\delta(r\sin\varphi\cos\theta - a, r\sin\varphi\sin\theta - b, r\cos\varphi - c), \varphi(r, \varphi, \theta)) . \end{aligned} \tag{1.141}$$

There results the formula

$$\delta(r\sin\varphi\cos\theta - a, r\sin\varphi\sin\theta - b, r\cos\varphi - c) = \frac{\delta(r - r_0, \varphi - \varphi_0, \theta - \theta_0)}{r_0^2\sin\varphi_0} . \tag{1.142}$$

The punctual transformation $(\rho, \theta, z) \to (x, y, z)$, given by formulae

$$x = \rho\cos\theta , \quad y = \rho\sin\theta , \quad z = z , \quad \rho \geq 0 , \quad \theta \in [0, 2\pi) , \quad z \in \mathbb{R} , \tag{1.143}$$

expresses the transition from Cartesian coordinates $(x, y, z) \in \mathbb{R}^3$ to cylindrical coordinates $(\rho, \theta, z) \in \mathbb{R}^3$.

The Jacobian of this transformation is

$$J(\rho, \theta, z) = \frac{\partial(x, y, z)}{\partial(\rho, \theta, z)} = \begin{vmatrix} \frac{\partial x}{\partial \rho} & \frac{\partial x}{\partial \theta} & \frac{\partial x}{\partial z} \\ \frac{\partial y}{\partial \rho} & \frac{\partial y}{\partial \theta} & \frac{\partial y}{\partial z} \\ \frac{\partial z}{\partial \rho} & \frac{\partial z}{\partial \theta} & \frac{\partial z}{\partial z} \end{vmatrix} = \begin{vmatrix} \cos\theta & -\rho\sin\theta & 0 \\ \sin\theta & \rho\cos\theta & 0 \\ 0 & 0 & 1 \end{vmatrix} = \rho . \tag{1.144}$$

Consequently, the punctual transformation is locally bijective everywhere on $\mathbb{R}^3$ with the exception of the points on the Oz-axis, where $\rho = 0$.

Thus, for the Dirac delta distribution $\delta(x, y, z)$ concentrated at the point $(0, 0, 0)$, we cannot apply the considered punctual transformation, but we may apply it to the Dirac delta distribution $\delta(x - a, y - b, z - c)$, concentrated at the point $(a, b, c) \notin Oz$, hence $a^2 + b^2 > 0$.

We can write

$$\begin{aligned} &(\delta(x - a, y - b, z - c), \psi(x, y, z)) \\ &= (\delta(\rho\cos\theta - a, \rho\sin\theta - b, z - c), \varphi(\rho, \theta, z)) , \end{aligned} \tag{1.145}$$

where $\psi(x, y, z) = (\varphi(\rho, \theta, z)/J(\rho, \theta, z)) = (\varphi(\rho, \theta, z)/\rho)$, $\varphi \in \mathcal{D}(\mathbb{R}^3)$.

Noting with (ρ_0, θ_0, z_0) the cylindrical coordinates of the point $(a, b, c) \notin Oz$, we have $a = \rho_0\cos\theta_0$, $b = \rho_0\sin\theta_0$, $c = z_0$, $\rho_0 = \sqrt{a^2 + b^2}$, $\tan\theta_0 = b/a$, $\theta_0 \in [0, 2\pi]$, $z_0 = c \in \mathbb{R}$ and thus we obtain

$$\begin{aligned} \psi(a, b, c) &= \frac{\varphi(\rho_0, \theta_0, z_0)}{\rho_0} = \left(\frac{\delta(\rho - \rho_0, \theta - \theta_0, z - z_0)}{\rho_0}, \varphi(\rho, \theta, z)\right) \\ &= (\delta(\rho\cos\theta - a, \rho\sin\theta - b, z - c), \varphi(\rho, \theta, z)) . \end{aligned} \tag{1.146}$$

It results in the formula

$$\frac{\delta(\rho - \rho_0, \theta - \theta_0, z - c)}{\rho_0} = \delta(\rho\cos\theta - a, \rho\sin\theta - b, z - c) . \tag{1.147}$$

1.3.2
Translation, Symmetry and Homothety of Distributions

In applications, it is important to consider the punctual linear transformation defined by the equation

$$x = au + b\,, \quad x, u, b \in \mathbb{R}^n\,, \tag{1.148}$$

that is,

$$x_i = \sum_{k=1}^{n} a_{ik} u_k + b_i\,, \tag{1.149}$$

where the transformation matrix is

$$a = (a_{ij}) = \begin{pmatrix} a_{11} & a_{12} & \dots & a_{1n} \\ a_{21} & a_{22} & \dots & a_{2n} \\ \dots & \dots & \dots & \dots \\ a_{n1} & a_{n2} & \dots & a_{nn} \end{pmatrix}, \quad \det a \neq 0\,, \tag{1.150}$$

while a^{-1} is the inverse matrix.

Because $\det a \neq 0$, the linear transformation (1.148) is bijective and therefore the conditions of application of the formula (1.123) are satisfied.

Consequently, if $f(x) \in \mathcal{D}'(\mathbb{R}^n)$, in accordance with (1.123), then we have

$$(f(x), \psi(x)) = (f(au + b), \varphi(u))\,, \tag{1.151}$$

$$\psi(x) = \frac{\varphi(u(x))}{|\det a|} = \frac{\varphi(a^{-1} \cdot (x - b))}{|\det a|}\,, \tag{1.152}$$

because $\partial(x)/\partial(u) = \det a$.

By customizing the formula (1.151), we obtain the transformations: translation, symmetry, homothety.

Translation Let $a = (\delta_{ij})$ be the diagonal matrix, where

$$\delta_{ij} = \begin{cases} 1\,, & i = j\,, \\ 0\,, & i \neq j\,, \end{cases} \quad i, j = 1, 2, \dots, n \tag{1.153}$$

represents Kronecker's symbol.

In this case, the linear transformation (1.148) takes the particular form $x = u + b$ which represents the translation of the variable u by the vector $b \in \mathbb{R}^n$ and for which $\det a = 1$.

As a consequence, the formula (1.151) becomes

$$(f(x), \psi(x)) = (f(u + b), \varphi(u))\,, \quad \varphi \in \mathcal{D}\,, \tag{1.154}$$

where $\psi(x) = \varphi(x - b)$.

We shall note with τ_b the symbol of the operator of translation by the vector $b \in \mathbb{R}^n$. Then $\tau_b f = f(x+b)$, $\tau_b \varphi = \varphi(x+b)$ represent the translation, by the vector b, of the distribution f and of the test function $\varphi \in \mathcal{D}$, respectively. With the formula (1.154) it can be written

$$(\tau_b f, \varphi) = (f, \tau_{-b}\varphi)\,, \tag{1.155}$$

that is,

$$(f(x+b), \varphi(x)) = (f(x), \varphi(x-b))\,. \tag{1.156}$$

The equivalent formulae (1.155) and (1.156) express the translation formula of the distribution $f(x) \in \mathcal{D}'(\mathbb{R}^n)$. From (1.156), there results the manner of application of the translation operator τ_b, namely

$$\tau_{\alpha+\beta} f = \tau_\alpha(\tau_\beta f) = \tau_\beta(\tau_\alpha f)\,, \tag{1.157}$$

hence

$$\tau_{\alpha+\beta} = \tau_\alpha(\tau_\beta)\,. \tag{1.158}$$

For the Dirac delta distribution we have

$$(\delta(x-a), \varphi(x)) = (\delta(x), \varphi(x+a)) = \varphi(a)\,, \quad \varphi \in \mathcal{D}\,, \tag{1.159}$$

hence

$$(\tau_{-a}\delta, \varphi) = (\delta, \tau_a\varphi)\,, \quad \tau_{-a}\delta = \delta(x-a) = \delta_a\,. \tag{1.160}$$

If $f \in \mathcal{D}'(\mathbb{R}^n)$ and $\psi \in C^\infty(\mathbb{R}^n)$, then we have the relation

$$\tau_b(\psi f) = (\tau_b\psi)(\tau_b f)\,, \quad b \in \mathbb{R}^n\,. \tag{1.161}$$

Indeed, we have

$$(\tau_b(\psi f), \varphi) = (f, \psi\tau_{-b}\varphi) = (f, \tau_{-b}(\varphi\tau_b\psi)) = (\tau_b f, \varphi\tau_b\psi)\,, \tag{1.162}$$

hence

$$(\tau_b(\psi f), \varphi) = (\tau_b f \tau_b\psi, \varphi)\,, \tag{1.163}$$

from which results the relation.

By means of translation, we can define the periodic distributions.

Let $f \in \mathcal{D}'(\mathbb{R}^n)$. We say that the distribution f is a periodic distribution if there is $T \in \mathbb{R}^n$, $T \neq 0$, with the property $\tau_T f = f$. The vector $T \in \mathbb{R}^n$ is called the distribution period.

Based on this definition, any periodic distribution $f \in \mathcal{D}'$ satisfies the relation

$$(f(x), \varphi(x)) = (\tau_T f, \varphi) = (f(x), \varphi(x-T))\,. \tag{1.164}$$

It is immediately verified that the periodic distributions satisfy the relation

$$\tau_T f = \tau_{-T} f\,. \tag{1.165}$$

Indeed, we have $(\tau_T f, \varphi) = (f, \tau_{-T}(\tau_T\varphi)) = (\tau_{-T} f, \tau_{-T}\varphi) = (f, \tau_{-T}\varphi)$, which gives the required relation.

Symmetry towards the origin of the coordinates The symmetry towards the origin of the function $f : \mathbb{R}^n \to \Gamma$ will be noted f^ν and defined by the relation

$$(f^\nu)(x) = f(-x) , \quad x \in \mathbb{R}^n . \tag{1.166}$$

From (1.166) we obtain the properties

$$(f^\nu)^\nu = f , \quad \operatorname{supp}(f^\nu) = \operatorname{supp}(f) . \tag{1.167}$$

If $f \in \mathcal{D}'(\mathbb{R}^n)$, then the symmetry of this distribution is given by the relation

$$(f^\nu, \varphi) = (f, \varphi^\nu) , \quad \varphi \in \mathcal{D} , \tag{1.168}$$

hence

$$(f(-x), \varphi(x)) = (f(x), \varphi(-x)) . \tag{1.169}$$

This formula is obtained from (1.151) by considering $a = (-\delta_{ij})$ and $b = 0$, for which $\det a = (-1)^n$.

Therefore, the formula (1.151) takes the form $(f(x), \psi(x)) = (f(-u), \varphi(u))$, $\psi(x) = \varphi(-x)$, that is,

$$(f(x), \varphi(-x)) = (f(-x), \varphi(x)) , \quad \varphi \in \mathcal{D}(\mathbb{R}^n) . \tag{1.170}$$

We say that the distribution f is even if $(f, \varphi^\nu) = (f, \varphi)$ and odd if $(f, \varphi^\nu) = -(f, \varphi)$.

For example, in the case of the Dirac delta distribution $\delta(x) \in \mathcal{D}'(\mathbb{R}^n)$ we have

$$(\delta(-x), \varphi(x)) = (\delta(x), \varphi(-x)) = \varphi(0) = (\delta(x), \varphi(x)) , \quad \forall \varphi \in \mathcal{D} , \tag{1.171}$$

which leads to the relation $\delta(-x) = \delta(x)$, which shows that the Dirac delta distribution $\delta(x)$ is an even distribution.

If $f \in \mathcal{D}'(\mathbb{R}^n)$ and $\psi \in C^\infty(\mathbb{R}^n)$, then we have

$$(\psi f)^\nu = \psi^\nu \cdot f^\nu . \tag{1.172}$$

Indeed, $((\psi f)^\nu, \varphi) = (\psi f, \varphi^\nu) = (f, \psi \varphi^\nu) = (f, (\varphi \psi^\nu)^\nu)$, hence

$$((\psi f)^\nu, \varphi) = (f^\nu, \varphi \psi^\nu) = (f^\nu \psi^\nu, \varphi) , \quad \forall \varphi \in \mathcal{D} . \tag{1.173}$$

Homothety The transformation through homothety is obtained from the linear transformation (1.148) considering $b = 0$ and the matrix transformation of the form $a = (a_{ij} \delta_{ij})$.

By specifications, the homothety transformation takes the form

$$x_i = a_{ii} u_i , \quad i = \overline{1, n} , \tag{1.174}$$

and the determinant of the transformation has the value $\det a = a_{11} a_{22} \ldots a_{nn}$.

Taking into account (1.174) and the formula (1.151), the homothety transformation of the distribution $f(x) \in \mathcal{D}'(\mathbb{R}^n)$ is given by

$$(f(x), \psi(x)) = (f(au), \varphi(u)) , \tag{1.175}$$

where

$$\psi(x) = \frac{\varphi(a^{-1} \cdot x)}{|\det a|} , \quad \det a = \prod_{i=1}^{n} a_{ii} \neq 0 , \tag{1.176}$$

that is, we have

$$\frac{1}{|\det a|}(f(x), \varphi(a^{-1} \cdot x)) = (f(au), \varphi(u)) . \tag{1.177}$$

In particular, if $a_{ii} = \beta \neq 0$, $i = \overline{1, n}$, then $\det a = \beta^n$, and

$$x = au \Leftrightarrow x = \beta u , \quad a^{-1}x = \frac{1}{\beta}x ; \tag{1.178}$$

thus, the formula (1.177) becomes

$$(f(\beta x), \varphi(x)) = \frac{1}{|\beta|^n}\left(f(x), \varphi\left(\frac{x}{\beta}\right)\right) . \tag{1.179}$$

For the Dirac delta distribution $\delta = \delta(x) \in \mathcal{D}(\mathbb{R}^n)$ we obtain

$$(\delta(\beta x), \varphi(x)) = \frac{1}{|\beta|^n}\left(\delta(x), \varphi\left(\frac{x}{\beta}\right)\right) = \frac{\varphi(0)}{|\beta|^n} = \left(\frac{\delta(x)}{|\beta|^n}, \varphi(x)\right) , \tag{1.180}$$

leading to the relation

$$\delta(\beta x) = \frac{1}{|\beta|^n}\delta(x) . \tag{1.181}$$

The homothety transformation allows for the introduction of the notion of homogeneous distribution.

Let $f(x) \in \mathcal{D}'(\mathbb{R}^n)$ be a distribution and $\alpha > 0$. We say that the distribution f is homogeneous and of degree $\lambda \in \mathbb{R}$ if it satisfies the relation

$$f(\alpha x) = \alpha^{\lambda} f(x) . \tag{1.182}$$

Substituting this in (1.179), we obtain the formula for characterizing the homogeneous distributions of degree λ, namely

$$(f(x), \varphi(x)) = \alpha^{-n-\lambda}\left(f(x), \varphi\left(\frac{x}{\alpha}\right)\right) , \quad \varphi \in \mathcal{D} . \tag{1.183}$$

Taking into account (1.181), it results that the Dirac delta distribution $\delta(x) \in \mathcal{D}'(\mathbb{R}^n)$ is a homogeneous distribution of degree $-n$.

Obviously, the homogeneous and locally integrable functions in the ordinary sense will be particular cases of homogeneous distributions of function type having the degree of homogeneity equal to the locally integrable function.

1.3.3
Differentiation of Distributions

Among the distribution operations, the operation of derivation has a special importance because of its effectiveness. Unlike the functions that do not always allow for derivatives, the distributions have derivatives of any order. Therefore, any locally integrable function considered as a regular distribution and, in particular, the continuous functions will have derivatives of any order in the sense of distributions.

This essentially changes the issues of the series of functions and of the Fourier series; this is because on the space of distributions any convergent series of locally integrable function can be differentiated term by term and the Fourier series are always convergent.

To find the natural way of introducing the concept of derivative, we consider the distribution function $f \in C^1(\mathbb{R})$, which obviously generates a regular distribution $T_f = f \in \mathcal{D}'(\mathbb{R})$.

To the function $f' \in C^0(\mathbb{R})$ also corresponds a regular distribution $T_{f'} = f' \in \mathcal{D}'(\mathbb{R})$, which is defined by

$$(f', \varphi) = \int_{\mathbb{R}} f'(x)\varphi(x)\mathrm{d}x \,, \quad \forall \varphi \in \mathcal{D}(\mathbb{R}) \,. \tag{1.184}$$

Integrating by parts, we can write

$$(f', \varphi) = f(x)\varphi(x)|_{-\infty}^{\infty} - \int_{\mathbb{R}} f(x)\varphi'(x)\mathrm{d}x \,, \tag{1.185}$$

but as the function φ has compact support, the first term on the right-hand side is zero and thus we obtain

$$(f', \varphi) = -(f, \varphi') \,, \quad \forall \varphi \in \mathcal{D}(\mathbb{R}) \,. \tag{1.186}$$

Relation (1.186) is adopted for the definition of the first-order derivative of a distribution from $\mathcal{D}'(\mathbb{R})$.

Hence, if $f \in \mathcal{D}'(\mathbb{R})$, then the functional f' given by (1.186) is called a derivative of the distribution f.

It is immediately verified that the new functional f', defined on $\mathcal{D}(\mathbb{R})$, is linear and continuous, hence $f' \in \mathcal{D}'(\mathbb{R})$ is a distribution.

Thus, if H is the Heaviside function of one variable, then we have, in the sense of distributions,

$$\frac{\mathrm{d}H(x)}{\mathrm{d}x} = \delta(x) \,, \tag{1.187}$$

where $\delta(x) \in \mathcal{D}'(\mathbb{R})$ is the Dirac delta distribution concentrated at the origin.

Indeed, since

$$H(x) = \begin{cases} 0 \,, & x < 0 \,, \\ 1 \,, & x \geq 0 \,, \end{cases} \tag{1.188}$$

we have

$$\left(\frac{\mathrm{d}H}{\mathrm{d}x}, \varphi\right) = -(H(x), \varphi'(x)) = -\int_{\mathbb{R}} H(x)\varphi'(x)\mathrm{d}x$$

$$= -\int_0^\infty \varphi'(x)\mathrm{d}x = \varphi(0) = (\delta(x), \varphi(x)), \quad \forall \varphi \in \mathcal{D}(\mathbb{R}), \quad (1.189)$$

resulting in the formula (1.187).

We note that, in the ordinary sense, the function H is differentiable everywhere, except the point $x = 0$, where there is a first type discontinuity.

Noting with $\tilde{\mathrm{d}}/\mathrm{d}x$ the derivative in the classical sense, we have $\tilde{\mathrm{d}}H/\mathrm{d}x = 0$, $x \neq 0$.

We note that, due to the translation operator τ_h, the derivative f' of the distribution $f \in \mathcal{D}'(\mathbb{R})$ is given by the following limit on the space $\mathcal{D}'$

$$f'(x) = \lim_{h \to 0} \frac{\tau_h f - f}{h} = \lim_{h \to 0} \frac{f(x+h) - f(x)}{h} . \quad (1.190)$$

This definition of the derivative of a distribution coincides with the classical one. Indeed, because of the continuity of the functional f, we have

$$\left(\frac{f(x+h) - f(x)}{h}, \varphi(x)\right) = \left(f(x), \frac{\varphi(x-h) - \varphi(x)}{h}\right) \underset{h \to 0}{\to} (f(x), -\varphi'(x)) \quad (1.191)$$

giving

$$\lim_{h \to 0} \frac{\tau_h f - f}{h} = f'(x) . \quad (1.192)$$

As a generalization of the differentiation formula (1.186) of one-variable distributions, we have the following definition.

Definition 1.26 Let $f \in \mathcal{D}'(\mathbb{R}^n)$ and $\mathrm{D}^\alpha = \mathrm{D}^{|\alpha|}/\partial x_1^{\alpha_1} \ldots \partial x_n^{\alpha_n}$ the derivative operator of order $|\alpha| = \sum_{i=1}^n \alpha_i$. We call the derivative of order $|\alpha|$ of the distribution f, the distribution denoted $\mathrm{D}^\alpha f$ and given by the relation

$$(\mathrm{D}^\alpha f, \varphi) = (-1)^{|\alpha|}(f, \mathrm{D}^\alpha \varphi), \quad \forall \varphi \in \mathcal{D}(\mathbb{R}^n) . \quad (1.193)$$

We note that this definition is correct, because the functional given on $\mathrm{D}^\alpha f$ by the formula (1.193) is linear and continuous, which is easily checked. On the other hand, formula (1.193) contains, as a particular case, formula (1.186).

We emphasize that the distributions' derivative does not depend on the order of derivation, so that there is the relation

$$\mathrm{D}^{\alpha+\beta} f = \mathrm{D}^\alpha(\mathrm{D}^\beta f) = \mathrm{D}^\beta(\mathrm{D}^\alpha f), \quad f \in \mathcal{D}' . \quad (1.194)$$

Let $H(x, y), (x, y) \in \mathbb{R}^2$, be the Heaviside function on $\mathbb{R}^2$, namely:

$$H(x) = \begin{cases} 1\,, & x \geq 0, y \geq 0\,, \\ 0\,, & \text{otherwise}\,. \end{cases} \tag{1.195}$$

We have

$$\frac{\partial^2 H(x, y)}{\partial x \partial y} = \delta(x, y)\,, \tag{1.196}$$

where $\delta(x, y) \in \mathcal{D}'(\mathbb{R}^2)$ is the Dirac delta distribution concentrated at the origin.

Indeed, $\forall \varphi \in \mathcal{D}$ we can write

$$\begin{aligned} \left[\frac{\partial^2 H}{\partial x \partial y}, \varphi(x, y)\right] &= \left(H, \frac{\partial^2 \varphi}{\partial x \partial y}\right) = \iint\limits_{\mathbb{R}^2} H(x, y) \frac{\partial^2 \varphi(x, y)}{\partial x \partial y} \mathrm{d}x \mathrm{d}y \\ &= \int_0^\infty \int_0^\infty \frac{\partial}{\partial x}\left(\frac{\partial \varphi}{\partial y}\right) \mathrm{d}x \mathrm{d}y = \int_0^\infty \mathrm{d}y \left(\frac{\partial \varphi}{\partial y}\right)\Bigg|_0^\infty = -\int_0^\infty \frac{\partial \varphi(0, y)}{\partial y} \mathrm{d}y \\ &= -\varphi(0, y)|_0^\infty = \varphi(0, 0) = (\delta(x, y), \varphi(x, y)) \end{aligned} \tag{1.197}$$

giving the formula (1.196).

The general formula can be proved in the same way, that is,

$$\frac{\partial^n H(x_1, x_2, \ldots, x_n)}{\partial x_1 \partial x_2 \ldots \partial x_n} = \delta(x_1, \ldots, x_n)\,, \tag{1.198}$$

where $\delta(x_1, \ldots, x_n) = \delta \in \mathcal{D}'(\mathbb{R}^n)$ is the Dirac distribution and $H(x)$ is the Heaviside function

$$H(x) = \begin{cases} 1\,, & x_1 \geq 0, x_2 \geq 0, \ldots, x_n \geq 0\,, \\ 0, & \text{otherwise}\,. \end{cases} \tag{1.199}$$

Below we will denote D^α the derivation operator in the sense of distributions and $\tilde{\mathrm{D}}^\alpha$ the derivation operator in the usual sense, where it exists for the regular distributions.

Thus, in the case of the Heaviside function, we have

$$\frac{\tilde{\partial}^n H(x_1, \ldots, x_n)}{\partial x_1 \ldots \partial x_n} = 0\,, \quad \frac{\partial^n H(x_1, \ldots, x_n)}{\partial x_1 \ldots \partial x_n} = \delta(x_1, x_2, \ldots, x_n)\,, \tag{1.200}$$

also,

$$(\mathrm{D}^\alpha \delta(x - a), \varphi(x)) = (-1)^{|\alpha|}(\mathrm{D}^\alpha \varphi)(a)\,, \quad \forall \varphi \in \mathcal{D}(\mathbb{R}^n)\,, \tag{1.201}$$

where $\delta_a = \delta(x - a)$ is the Dirac delta distribution concentrated at the point $a \in \mathbb{R}^n$.

Now, we have the function $a \in C^\infty(\mathbb{R}^n)$ and the distribution $f \in \mathcal{D}'(\mathbb{R}^n)$; the derivation formula of a product, known in the classical case, remains valid, that is, we have

$$\frac{\partial}{\partial x_i}(af) = \frac{\partial a}{\partial x_i} f + a\frac{\partial f}{\partial x_i}\,, \quad i = 1, 2, \ldots, n\,. \tag{1.202}$$

For distributions defined on $\mathcal{D}^m$ and $\mathcal{S}$ the formula is the same so long as the product between a function and a distribution makes sense.

Example 1.13 Let the function $f(x) = \ln|x|$, $x \in \mathbb{R}\backslash\{0\}$. This function is locally integrable, so $f \in L^1_{\text{loc}}(\mathbb{R})$. To the considered function we assign the linear functional defined by the formula

$$(\ln|x|, \varphi(x)) = \int_{\mathbb{R}} \varphi(x) \ln|x| \mathrm{d}x\,, \quad \varphi \in \mathcal{D}(\mathbb{R})\,. \tag{1.203}$$

We shall show that the functional $T_f = \ln|x|$ is a distribution on $\mathcal{D}'(\mathbb{R})$.
We have

$$\begin{aligned}(\ln|x|, \varphi) &= \int_{\mathbb{R}} (x) \ln|x| \mathrm{d}x = \lim_{\varepsilon \to +0} \left[\int_{-\infty}^{-\varepsilon} \varphi(x) \ln|x| \mathrm{d}x + \int_{\varepsilon}^{\infty} \varphi(x) \ln|x| \mathrm{d}x \right] \\ &= \lim_{\varepsilon \to +0} \int_{\varepsilon}^{\infty} [\varphi(x) - \varphi(-x)] \ln x \mathrm{d}x\,. \end{aligned} \tag{1.204}$$

Considering $\text{supp}(\varphi) \subset [-a, a]$, $a > 0$, then we can write

$$|(\ln|x|, \varphi)| \le \lim_{\varepsilon \to +0} \int_{\varepsilon}^{a} |\varphi(x) - \varphi(-x)|\, |\ln x| \mathrm{d}x\,. \tag{1.205}$$

Because

$$\lim_{x \to +0} [\varphi(x) - \varphi(-x)] \ln x = 2 \lim_{x \to +0} \left[\varphi'(\xi_x) x \ln x\right] = 2\varphi'(0) \lim_{x \to +0} (x \ln x) = 0\,, \tag{1.206}$$

we have

$$|(\ln|x|, \varphi)| \le 2 \sup_{x \in [0,a]} |x \ln x| \sup_{x \in [0,a]} |\varphi'(x)|\,, \tag{1.207}$$

hence

$$|(\ln|x|, \varphi)| \le c \sup_{x \in [0,a]} |\varphi'(x)|\,, \text{ where } c = 2 \sup_{x \in [0,a]} |x \ln x|\,. \tag{1.208}$$

Consequently, the functional T_f is continuous and as it is obviously a linear functional, it results that $T_f = \ln|x|$ is a first-order distribution on $\mathcal{D}'(\mathbb{R})$.

The derivative of this function is given by

$$\frac{d}{dx}\ln|x| = \text{p.v.}\ \frac{1}{x}\ . \tag{1.209}$$

Indeed, we have

$$\begin{aligned}((\ln|x|)',\varphi) &= -(\ln|x|,\varphi') = -\int_{\mathbb{R}}\varphi'(x)\ln|x|dx \\ &= -\lim_{\varepsilon\to+0}\int_{\varepsilon}^{\infty}[\varphi'(x)-\varphi'(-x)]\ln x dx \\ &= \lim_{\varepsilon\to 0}[\varphi(\varepsilon)-\varphi(-\varepsilon)]\ln\varepsilon + \lim_{\varepsilon\to+0}\int_{\varepsilon}^{\infty}\frac{\varphi(x)-\varphi(-x)}{x}dx\ .\end{aligned} \tag{1.210}$$

Because $\lim_{\varepsilon\to 0}[\varphi(\varepsilon)-\varphi(-\varepsilon)]\ln\varepsilon = 0$ and

$$\lim_{\varepsilon\to 0}\int_{\varepsilon}^{\infty}\frac{\varphi(x)-\varphi(-x)}{x}dx = \left(\text{p.v.}\ \frac{1}{x},\varphi\right) = \text{p.v.}\int_{\mathbb{R}}\frac{\varphi(x)}{x}dx$$

we can write

$$((\ln|x|)',\varphi) = \left(\text{p.v.}\ \frac{1}{x},\varphi\right)$$

from which we get the formula $(\ln|x|)' = \text{p.v.}\ 1/x$.

Example 1.14 Let the distribution p.v. $1/x^2 \in \mathcal{D}'(\mathbb{R})$, defined by the formula

$$\left(\text{p.v.}\ \frac{1}{x^2},\varphi\right) = \text{p.v.}\int_{\mathbb{R}}\frac{\varphi(x)-\varphi(0)}{x^2}dx = \lim_{\varepsilon\to+0}\int_{|x|\geq\varepsilon}\frac{\varphi(x)-\varphi(0)}{x^2}dx\ , \quad \varphi\in\mathcal{D}(\mathbb{R})\ . \tag{1.211}$$

The relation d/dx p.v. $1/x = -$p.v. $1/x^2$, where the distribution p.v. $1/x$ is defined by

$$\left(\text{p.v.}\ \frac{1}{x},\varphi\right) = \text{p.v.}\int_{\mathbb{R}}\frac{\varphi(x)}{x}dx = \lim_{\varepsilon\to+0}\int_{|x|\geq\varepsilon}\frac{\varphi(x)}{x}dx\ . \tag{1.212}$$

Indeed, since $\forall\varphi\in\mathcal{D}(\mathbb{R})$ we have

$$\left(\frac{d}{dx}\text{p.v.}\ \frac{1}{x},\varphi\right) = -\left(\text{p.v.}\ \frac{1}{x},\varphi'\right) = -\text{p.v.}\int_{\mathbb{R}}\frac{\varphi'(x)}{x}dx = -\lim_{\varepsilon\to+0}\int_{|x|\geq\varepsilon}\frac{\varphi'(x)}{x}dx\ . \tag{1.213}$$

Integrating by parts, we obtain

$$\left(\frac{\mathrm{d}}{\mathrm{d}x}\,\mathrm{p.v.}\,\frac{1}{x},\varphi\right) = -\lim_{\varepsilon\to+0}\left[\left.\frac{\varphi(x)}{x}\right|_{-\infty}^{-\varepsilon} + \left.\frac{\varphi(x)}{x}\right|_{\varepsilon}^{\infty} + \int\limits_{|x|\geq\varepsilon}\frac{\varphi(x)}{x^2}\mathrm{d}x\right]$$

$$= -\lim_{\varepsilon\to+0}\left[-\frac{\varphi(-\varepsilon)}{\varepsilon} - \frac{\varphi(\varepsilon)}{\varepsilon} - \left.\frac{\varphi(0)}{x}\right|_{-\infty}^{-\varepsilon} - \left.\frac{\varphi(0)}{x}\right|_{\varepsilon}^{\infty} + \int\limits_{|x|\geq\varepsilon}\frac{\varphi(x)-\varphi(0)}{x^2}\mathrm{d}x\right]. \tag{1.214}$$

Therefore, we can write

$$\left(\frac{\mathrm{d}}{\mathrm{d}x}\,\mathrm{p.v.}\,\frac{1}{x},\varphi\right) = \lim_{\varepsilon\to+0}\left[\frac{\varphi(\varepsilon)-\varphi(0)}{\varepsilon} - \frac{\varphi(-\varepsilon)-\varphi(0)}{-\varepsilon}\right] - \lim_{\varepsilon\to+0}\int\limits_{|x|\geq\varepsilon}\frac{\varphi(x)-\varphi(0)}{x^2}\mathrm{d}x\,. \tag{1.215}$$

Because $\lim_{\varepsilon\to+0}((\varphi(\varepsilon)-\varphi(0))/\varepsilon - (\varphi(-\varepsilon)-\varphi(0))/(-\varepsilon)) = \varphi'(0)-\varphi'(0) = 0$ the previous relation becomes

$$\left(\frac{\mathrm{d}}{\mathrm{d}x}\,\mathrm{p.v.}\,\frac{1}{x},\varphi\right) = -\lim_{\varepsilon\to+0}\int\limits_{|x|\geq\varepsilon}\frac{\varphi(x)-\varphi(0)}{x^2}\mathrm{d}x = \left(-\mathrm{p.v.}\,\frac{1}{x^2},\varphi\right), \tag{1.216}$$

resulting in $\mathrm{d}/\mathrm{d}x\,(\mathrm{p.v.}\,1/x) = -\mathrm{p.v.}\,1/x^2$.

1.3.3.1 Properties of the Derivative Operator

Let D^α be the derivative operator in the sense of distributions, and τ_a, ν the operator of translation and the symmetry operator, respectively.

The following properties occur:

$$\mathrm{D}^\alpha(\lambda f + \mu g) = \lambda\,\mathrm{D}^\alpha f + \mu\,\mathrm{D}^\alpha g\,,\quad f,g\in\mathcal{D}'\,,\quad \lambda,\mu\in\Gamma\,, \tag{1.217}$$

$$f_i \xrightarrow{\mathcal{D}'} f \Rightarrow \mathrm{D}^\alpha f_i \xrightarrow{\mathcal{D}'} \mathrm{D}^\alpha f\,,\quad (f_i)_{i\in N}\subset D'\,,\quad f\in\mathcal{D}'\,, \tag{1.218}$$

$$\mathrm{supp}(\mathrm{D}^\alpha f)\subset\mathrm{supp}(f)\,,\quad f\in\mathcal{D}'\,, \tag{1.219}$$

$$\mathrm{D}^\alpha(\tau_a f) = \tau_a(\mathrm{D}^\alpha f)\,,\quad \forall a\in\mathbb{R}^n\,,\quad \mathrm{D}^\alpha(f^\nu) = (-1)^{|\alpha|}(\mathrm{D}^\alpha f)^\nu\,,\quad f\in\mathcal{D}'. \tag{1.220}$$

The first relation expresses the linearity of the operator of derivation and its demonstration is easy. Relation (1.218) shows the continuity of the operator and, for its justification, we can write $(\mathrm{D}^\alpha f_i,\varphi) = (-1)^{|\alpha|}(f_i,\mathrm{D}^\alpha\varphi)$.

On the basis of the completeness theorem of the space $\mathcal{D}'$, we obtain

$$\lim_i(\mathrm{D}^\alpha f_i,\varphi) = (-1)^{|\alpha|}\lim_i(f_i,\mathrm{D}^\alpha\varphi) = (-1)^{|\alpha|}(f,\mathrm{D}^\alpha\varphi) = (\mathrm{D}^\alpha f,\varphi)\,, \tag{1.221}$$

hence $\lim_i \mathrm{D}^\alpha f_i = \mathrm{D}^\alpha f$. For the proof of formula (1.219), we allow for $f = 0$, $x \in A \subset \mathbb{R}^n$. Then, for any $\varphi \in \mathcal{D}$ with $\mathrm{supp}(\varphi) \subset A$, we have $(\mathrm{D}^\alpha f, \varphi) = (-1)^{|\alpha|}(f, \mathrm{D}^\alpha \varphi) = 0$, because $\mathrm{supp}(\mathrm{D}^\alpha \varphi) \subset A$. From the last equality, it follows $\mathrm{D}^\alpha f = 0$, $x \in A$, and considering the complementaries of the sets on which the distributions f and $\mathrm{D}^\alpha f$ vanished, we get the required relation.

Example 1.15 We have

$$\begin{array}{ll} f(x) = H(x)\,, & \mathrm{supp}(f) = [0,\infty] \\ f'(x) = \delta(x)\,, & \mathrm{supp}(f') = \{0\} \end{array} \Rightarrow \begin{array}{l} \mathrm{supp}(f') = \mathrm{supp}(\delta) = \{0\} \\ \subset \mathrm{supp}(f) = [0,\infty]\,. \end{array} \tag{1.222}$$

As to formulae (1.220), the first equation shows the commutativity of the derivation operator with respect to the operator of translation τ_a, $a \in \mathbb{R}^n$.

For this, we can write

$$(\mathrm{D}^\alpha(\tau_a f), \varphi) = (-1)^{|\alpha|}(f, \tau_{-a}\mathrm{D}^\alpha \varphi) = (-1)^{|\alpha|}(f, \mathrm{D}^\alpha(\tau_{-a}\varphi))\,, \tag{1.223}$$

because $\mathrm{D}^\alpha(\tau_a\varphi) = \tau_a(\mathrm{D}^\alpha \varphi)$.

The second formula of (1.220) expresses the relation between the derivation operator and the symmetry operator ν with respect to the origin of the coordinates.

Because $\forall\varphi \in \mathcal{D}$, we have $\mathrm{D}^\alpha(\varphi^\nu) = (-1)^{|\alpha|}(\mathrm{D}^\alpha \varphi)^\nu$, hence we can write

$$(\mathrm{D}^\alpha(f^\nu), \varphi) = (f, \mathrm{D}^\alpha(\varphi^\nu)) = ((-1)^{|\alpha|}(\mathrm{D}^\alpha f)^\nu, \varphi)\,, \tag{1.224}$$

which proves the required formula.

Proposition 1.11 Let the function $\psi \in C^\infty(\mathbb{R}^n)$, the distribution $f \in \mathcal{D}'(\mathbb{R}^n)$, and $\alpha \in \mathbb{N}_0^n$. Thus, the formula occurs

$$\mathrm{D}^\alpha(\psi f) = \sum_{\beta+\gamma=\alpha} \frac{\alpha!}{\beta!\gamma!}\mathrm{D}^\beta \psi \cdot \mathrm{D}^\gamma f = \sum_{\beta\le\alpha} \frac{\alpha!}{\beta!(\alpha-\beta)!}\mathrm{D}^\beta \psi \cdot \mathrm{D}^{\alpha-\beta} f\,. \tag{1.225}$$

The formula (1.225) represents the *Leibniz's formula for the derivation of a product in the space of distributions.*

In applications, the point of interest is the calculation of the derivatives of the function type distributions which have discontinuities of the first type, distributed at some points or on certain types of manifolds of $\mathbb{R}^n$.

We can state the following.

Proposition 1.12 Let f be a function of the class $C^1(\mathbb{R})$ except at the point x_0, where it has a discontinuity of the first order with the jump

$$s_0(f) = f(x_0+0) - f(x_0-0)\,, \tag{1.226}$$

where $f(x_0+0) = \lim_{x\to x_0+0} f(x)$, $f(x_0-0) = \lim_{x\to x_0-0} f(x)$. The formula occurs

$$f' = \tilde{f}' + s_0(f)\delta_{x_0}\,. \tag{1.227}$$

Proof: For any $\varphi \in \mathcal{D}(\mathbb{R})$, we can write

$$(f',\varphi) = -(f,\varphi') = -\lim_{\varepsilon\to+0}\left[\int_{-\infty}^{x_0-\varepsilon} f\varphi' \mathrm{d}x + \int_{x_0+\varepsilon}^{\infty} f\varphi' \mathrm{d}x\right]$$
$$= s_0(f)\varphi(x_0) + (\tilde{f}',\varphi) = (\tilde{f}' + (s_0 f)\delta_{x_0},\varphi)\,. \tag{1.228}$$

□

According to the adopted convention, $f' = \mathrm{d}f/\mathrm{d}x$ is the derivative in the sense of distributions and $\tilde{f}' = \tilde{\mathrm{d}}\,f/\mathrm{d}x$ represents the function type distribution corresponding to the derivative f in the ordinary sense.

Generalizing the formula (1.227), we obtain the following.

Corollary 1.1 Let $f \in C^1(\mathbb{R})$ except for the points x_i, $i = 1, 2, \ldots, p$, where it has discontinuities of the first type with the jumps $s_{x_i}(f) = f(x_i+0) - f(x_i-0)$, then

$$f' = \tilde{f}' + \sum_{i=1}^{p} s_{x_i}(f)\delta_{x_i}\,. \tag{1.229}$$

Corollary 1.2 Let $f \in C^\ell(\mathbb{R})$ except the point x_0 where both the function and its derivatives up to order $\ell - 1$ have discontinuities of the first type with the jumps $s_{x_0}(\tilde{f}^{(i)})$, corresponding to the function $\tilde{f}^{(i)}$, $i = 0, 1, 2, \ldots, \ell - 1$. Then, we have the formula

$$f^{(p)} = \tilde{f}^{(p)} + \sum_{i=0}^{p-1} s_{x_0}(\tilde{f}^{(i)})\delta_{x_0}^{(p-i-1)}\,, \quad p = 1, 2, \ldots, \ell\,. \tag{1.230}$$

This last formula is obtained from (1.227) by successive derivation.

Example 1.16 Let the operator be $P(\mathrm{D}) = (\mathrm{d}^2/\mathrm{d}x^2) + (\mathrm{d}/\mathrm{d}x) - 2$ and the function type distribution

$$f(x) = \begin{cases} -\mathrm{e}^{x}/3\,, & x < 0\,, \\ -\mathrm{e}^{-2x}/3\,, & x \ge 0\,. \end{cases} \tag{1.231}$$

The relation then follows

$$P(\mathrm{D})f = f'' + f' - 2f = \delta(x)\,. \tag{1.232}$$

We note that f is a continuous function on $\mathbb{R}$, hence $f \in C^0(\mathbb{R})$ and, in accordance with the formula (1.227), we have

$$f'(x) = \tilde{f}'(x) = \begin{cases} -\mathrm{e}^{x}/3\,, & x < 0\,, \\ 2\mathrm{e}^{-2x}/3\,, & x > 0\,. \end{cases} \tag{1.233}$$

The distribution $f' = \tilde{f}'$ is of function type which has at the origin a discontinuity of the first order with the jump $s_0(\tilde{f}') = (2/3) - (-1/3) = 1$.

Taking into account that

$$\tilde{f}'' = \frac{\tilde{\mathrm{d}}}{\mathrm{d}x} f' = \frac{\tilde{\mathrm{d}}^2}{\mathrm{d}x^2} f = \begin{cases} -\mathrm{e}^x/3 \,, & x < 0 \,, \\ -4\mathrm{e}^{-2x}/3 \,, & x > 0 \,, \end{cases} \tag{1.234}$$

and applying (1.227), we obtain $f'' = \tilde{f}'' + \delta(x)$, from which we obtain that $\mathrm{P(D)} f = \delta(x)$.

Let $\Omega \subset \mathbb{R}^n$ be a bounded domain and $u : \Omega \times [0, \infty) \to \mathbb{R}$ a real function of the class $C^2(\Omega \times [0, \infty))$. We define the function type distribution $u^* \in \mathcal{D}'(\mathbb{R}^n \times \mathbb{R})$, that is,

$$u^*(x, t) = \begin{cases} u(x, t) \,, & (x, t) \in \Omega \times [0, \infty) \,, \\ 0 \,, & \text{otherwise} \,. \end{cases} \tag{1.235}$$

This function type distribution can be written in the form

$$u^*(x, t) = \chi(x) H(t) u(x, t) \,, \quad (x, t) \in \mathbb{R}^n \times \mathbb{R} \,, \tag{1.236}$$

where

$$\chi(x) = \begin{cases} 1 \,, & x \in \Omega \subset \mathbb{R}^n \,, \\ 0 \,, & x \notin \Omega \,, \end{cases} \tag{1.237}$$

is the characteristic function corresponding to the domain $\Omega \subset \mathbb{R}^n$, while H is the Heaviside function.

We can state the following.

Proposition 1.13 The formulae

$$\frac{\partial u^*}{\partial t} = \frac{\tilde{\partial} u^*}{\partial t} + u_0^*(x) \times \delta(t) \,, \tag{1.238}$$

$$\frac{\partial^2 u^*}{\partial t^2} = \frac{\tilde{\partial}^2 u^*}{\partial t^2} + \dot{u}_0^*(x) \times \delta(t) + u_0^*(x) \times \dot{\delta}(t) \,, \tag{1.239}$$

exist for the function type distribution $u^* \in \mathcal{D}'(\mathbb{R}^n \times \mathbb{R})$, where

$$u_0^*(x) = u^*(x, t)|_{t=+0} = \chi(x) u_0(x) \,, \quad \dot{u}_0^*(x) = \left.\frac{\tilde{\partial} u^*(x, t)}{\partial t}\right|_{t=+0} = \chi(x) \dot{u}_0(x) \,, \tag{1.240}$$

and where the conditions $u_0(x) = u(x, t)|_{t=+0} \in C^0(\Omega)$, $\dot{u}_0(x) = [\tilde{\partial} u(x, t)/\partial t]|_{t=+0} \in C^0(\Omega)$ are satisfied.

Proof: For any $\varphi(x,t) \in \mathcal{D}(\mathbb{R}^n \times \mathbb{R})$ we have

$$\left(\frac{\partial u^*}{\partial t}, \varphi\right) = -\left(u^*, \frac{\tilde{\partial}\varphi}{\partial t}\right) = -\int_{\mathbb{R}^n\times\mathbb{R}} u^*(x,t)\frac{\tilde{\partial}\varphi}{\partial t}\mathrm{d}x\mathrm{d}t$$

$$= -\int_{\mathbb{R}^n} \mathrm{d}x \int_{\mathbb{R}} u^*(x,t)\frac{\tilde{\partial}\varphi}{\partial t}\mathrm{d}t = -\int_{\mathbb{R}^n} \chi(x)\mathrm{d}x \int_{\mathbb{R}} H(t)u(x,t)\frac{\tilde{\partial}\varphi}{\partial t}\mathrm{d}t\,. \tag{1.241}$$

On the other hand, we can write

$$\int_{\mathbb{R}} H(t)u(x,t)\frac{\tilde{\partial}\varphi}{\partial t}\mathrm{d}t = \int_0^\infty u(x,t)\frac{\tilde{\partial}\varphi}{\partial t}\mathrm{d}t = u\varphi|_0^\infty - \int_0^\infty \frac{\tilde{\partial}u}{\partial t}\varphi(x,t)\mathrm{d}t$$

$$= -u(x,0+0)\varphi(x,0) - \int_0^\infty \frac{\tilde{\partial}u}{\partial t}\varphi(x,t)\mathrm{d}t = -u_0(x)\varphi(x,0) - \int_0^\infty \frac{\tilde{\partial}u}{\partial t}\varphi(x,t)\mathrm{d}t\,. \tag{1.242}$$

Thus, the previous relation becomes

$$\left(\frac{\partial u^*}{\partial t}, \varphi\right) = \int_{\mathbb{R}^n} \chi(x)u_0(x)\varphi(x,0)\mathrm{d}x + \int_{\mathbb{R}^n\times\mathbb{R}} \chi(x)H(t)\frac{\tilde{\partial}u}{\partial t}\mathrm{d}x\mathrm{d}t\,. \tag{1.243}$$

But we have

$$\int_{\mathbb{R}^n} \chi(x)u_0(x)\varphi(x,0)\mathrm{d}x = ((\chi(x)u_0(x)) \times \delta(t), \varphi(x,t))\,, \tag{1.244}$$

because

$$((\chi(x)u_0(x)) \times \delta(t), \varphi(x,t)) = (u_0(x)\chi(x), (\delta(t), \varphi(x,t)))$$
$$= (u_0(x)\chi(x), \varphi(x,0)) = \int_{\mathbb{R}^n} u_0(x)\chi(x)\varphi(x,0)\mathrm{d}x\,. \tag{1.245}$$

Consequently, we obtain

$$\left(\frac{\partial u^*}{\partial t}, \varphi\right) = ((\chi(x)u_0(x)) \times \delta(t), \varphi(x,t)) + \left(\chi(x)H(t)\frac{\tilde{\partial}u}{\partial t}, \varphi\right)$$
$$= \left(u_0^*(x) \times \delta(t) + \frac{\tilde{\partial}u^*(x,t)}{\partial t}, \varphi\right)\,, \tag{1.246}$$

from which results the first formula of the sentence, because $\chi(x)H(t)(\tilde{\partial}u/\partial t) = (\tilde{\partial}u^*/\partial t)$ and $u_0(x)\chi(x) = u_0^*(x)$.

To get the second relation, we can write

$$\frac{\partial^2 u^*}{\partial t^2} = \frac{\partial}{\partial t}\left[\frac{\tilde{\partial} u^*}{\partial t} + u_0^*(x) \times \delta(t)\right] = \frac{\partial}{\partial t}\left[\frac{\tilde{\partial} u^*}{\partial t}\right] + u_0^*(x) \times \delta'(t)$$

$$= \frac{\tilde{\partial}}{\partial t}\left(\frac{\tilde{\partial} u^*}{\partial t}\right) + \frac{\tilde{\partial}}{\partial t u^*}\bigg|_{t=+0} \times \delta(t) + u_0^*(x) \times \delta'(t)\,, \tag{1.247}$$

hence

$$\frac{\partial^2}{u^*}\partial t^2 = \frac{\tilde{\partial}^2}{u^*}\partial t^2 + \dot{u}_0^*(x) \times \delta(t) + u_0^*(x) \times \delta'(t)\,, \tag{1.248}$$

and thus the proposition is demonstrated. □

We note that this proposition generalizes the formula (1.227) and has important applications in mechanics [14–18], because the variable may be interpreted as a time variable.

Remark 1.1 For the functions $u^*(x,t) = \chi(x)H(t)u(x,t)$ and

$$\frac{\tilde{\partial} u^*}{\partial t} = \chi(x)H(t)\frac{\tilde{\partial} u(x,t)}{\partial t}\,, \quad (x,t) \in \mathbb{R}^n \times \mathbb{R}\,,$$

the hyperplane $t = 0, (x,t) \in \mathbb{R}^n \times \mathbb{R}$ of $\mathbb{R}^n \times \mathbb{R}$ represents a discontinuity hyperplane. When crossing the hyperplane in the direction of the increasing of the variable $t \in \mathbb{R}$, the jumps of the two functions are:

$$u_0^*(x) = u^*(x,0+0) - u^*(x,0-0) = u_0^*(x,0+0) = \chi(x)u_0(x)\,, \tag{1.249}$$

$$\dot{u}_0^*(x) = \frac{\tilde{\partial} u^*}{\partial t}(x,0+0) - \frac{\tilde{\partial} u^*}{\partial t}(x,0-0) = \frac{\tilde{\partial} u^*}{\partial t}(x,0+0) = \chi(x)\dot{u}_0(x)\,, \tag{1.250}$$

where $x \in \mathbb{R}^n$.

Particularly, if $\Omega = \mathbb{R}^n$, the obtained formulae are simplified, because $\chi(x) = 1$; we have $u_0^*(x) = u_0(x)$ and $\dot{u}_0^*(x) = \dot{u}_0(x)$, where $u_0, \dot{u}_0 \in C^0(\mathbb{R}^n)$.

In connection with the distributions supports, we have the following properties:

1. $$\operatorname{supp}(\alpha f + \beta g) \subset \operatorname{supp}(f) \cup \operatorname{supp}(g)\,, \quad f, g \in \mathcal{D}'(\mathbb{R}^n)\,, \quad \alpha, \beta \in \mathbb{R}\,; \tag{1.251}$$
2. $$\operatorname{supp}\left(\sum_{i=1}^m \alpha_i f_i\right) \subset \bigcup_{i=1}^m \operatorname{supp}(f_i)\,, \quad f_i \in \mathcal{D}'(\mathbb{R}^n)\,, \quad \alpha_i \in \mathbb{R}\,; \tag{1.252}$$
3. $$\operatorname{supp}(D^\alpha f) \subset \operatorname{supp}(f)\,; \tag{1.253}$$
4. $$\operatorname{supp}(a(x)\, D^\alpha f) \subset \operatorname{supp}(f)\,, \quad a(x) \in C^\infty(\mathbb{R}^n)\,; \tag{1.254}$$

5. Let

$$P(x,\mathrm{D})=\sum_{|\alpha|=0}^{m}a_\alpha(x)\,\mathrm{D}^\alpha\,,\qquad a_\alpha(x)\in C^\infty(\mathbb{R}^n)$$

a linear differential operator with variable coefficients. Then, we have

$$\operatorname{supp}(P(x,\mathrm{D}))\subset\operatorname{supp}(f)\,. \tag{1.255}$$

Proposition 1.14 Let f be a function of the class C^1 on $\mathbb{R}^n$, except for a piecewise smooth hypersurface S, where it has a first discontinuity; we have

$$\frac{\partial f}{\partial x_i}=\frac{\tilde{\partial} f}{\partial x_i}+\sigma_i\cos\alpha_i\delta_S\,, \tag{1.256}$$

where δ_S is the Dirac delta distribution concentrated on the hypersurface S, σ_i is the jump function across the hypersurface in the positive direction of the Ox_i-axis, and α_i is the angle between the Ox_i-axis and the normal to the hypersurface oriented in the direction of its crossing.
To establish the formula, we see that f is locally integrable and thus we have

$$\begin{aligned}\left(\frac{\partial f}{\partial x_i},\varphi\right)&=-\left(f,\frac{\partial\varphi}{\partial x_i}\right)=-\int_{\mathbb{R}^n}f(x)\frac{\partial\varphi}{\partial x_i}\mathrm{d}x\\&=(-1)^i\int_{\mathbb{R}^{n-1}}\mathrm{d}x_1\ldots\mathrm{d}x_{i-1}\mathrm{d}x_{x+1}\ldots\mathrm{d}x_n\int_{-\infty}^{\infty}f(x)\frac{\partial\varphi}{\partial x_i}\mathrm{d}x_i\,.\end{aligned} \tag{1.257}$$

But, we can write

$$\begin{aligned}\int_{-\infty}^{\infty}f(x)\frac{\partial\varphi}{\partial x_i}\mathrm{d}x_i&=\lim_{\varepsilon\to+0}\left[\int_{-\infty}^{x_i^*-\varepsilon}f(x)\frac{\partial\varphi}{\partial x_i}\mathrm{d}x_i+\int_{x_i^*+\varepsilon}^{\infty}f(x)\frac{\partial\varphi}{\partial x_i}\mathrm{d}x_i\right]\\&=\lim_{\varepsilon\to+0}\left[(f(x)\varphi(x))|_{-\infty}^{x_i^*-\varepsilon}+(f(x)\varphi(x))|_{x_i^*+\varepsilon}^{\infty}-\int_{-\infty}^{x_i^*-\varepsilon}\varphi(x)\frac{\tilde{\partial}}{\partial x_i}f(x)\mathrm{d}x_i\right.\\&\left.-\int_{x_i^*+\varepsilon}^{\infty}\varphi(x)\frac{\tilde{\partial}}{\partial x_i}f(x)\mathrm{d}x_i\right]=-\sigma_i\varphi(x^*)-\int_{-\infty}^{\infty}\varphi(x)\frac{\tilde{\partial}f(x)}{\partial x_i}\mathrm{d}x_i\,,\end{aligned} \tag{1.258}$$

where $\sigma_i=f(x_1^*,\ldots,x_{i-1}^*,x_i^*+0,x_{i+1}^*,\ldots,x_n^*)-f(x_1^*,\ldots,x_{i-1}^*,x_i^*-0,x_{i+1}^*,\ldots,x_n^*)$ is the jump of the function f at the point $x^*\in S$, when crossing the hypersurface S in the positive direction of the Ox_i-axis.

Thus, we get

$$\left(\frac{\partial f}{\partial x_i}, \varphi\right) = \int_{\mathbb{R}^n} \frac{\tilde{\partial} f(x)}{\partial x_i} \varphi(x) \mathrm{d}x + (-1)^{i-1} \int_S \sigma_i \varphi(x) \mathrm{d}x_1 \ldots \mathrm{d}x_{i-1} \mathrm{d}x_{i+1} \ldots \mathrm{d}x_n . \tag{1.259}$$

If we note with α_i the angle formed by the Ox_i-axis with the normal to the hypersurface directed for increasing x_i, then

$$(-1)^{i-1} \int_S \sigma_i \varphi(x) \mathrm{d}x_1 \ldots \mathrm{d}x_{i-1} \mathrm{d}x_{i+1} \ldots \mathrm{d}x_n = \int_S \sigma_i \varphi(x) \cos \alpha_i \mathrm{d}S , \tag{1.260}$$

where $\mathrm{d}S$ is the area element.

Taking into account the definition of the Dirac delta distribution concentrated on the hypersurface S, the previous formula becomes

$$\int_S \sigma_i \varphi(x) \cos \alpha_i \mathrm{d}S = (\sigma_i \cos \alpha_i \delta_S, \varphi) . \tag{1.261}$$

Substituting this in (1.259), we obtain

$$\left(\frac{\partial f}{\partial x_i}, \varphi\right) = \left(\frac{\tilde{\partial} f}{\partial x_i}, \varphi\right) + (\sigma_i \cos \alpha_i \delta_S, \varphi) , \tag{1.262}$$

from which we obtain formula (1.256).

We notice that this formula is a generalization of the formula (1.227) established for the case $n = 1$.

In particular, for the real function f of class $C^1(\mathbb{R}^3)$, with the exception of the piecewise smooth surface $S \subset \mathbb{R}^3$, where it has discontinuities of the first order (Figure 1.2), we have the formula

$$\frac{\partial f(x, y, z)}{\partial z} = \frac{\tilde{\partial} f(x, y, z)}{\partial z} + \sigma_3 \cos \alpha_3 \delta_S , \tag{1.263}$$

where α_3 is the angle between $\boldsymbol{n}$ and $\boldsymbol{k}$, $\sigma_3 = f(x, y, z+0) - f(x, y, z-0)$ and

$$(\sigma_3 \cos \alpha_3 \delta_S, \varphi) = \int_S \sigma_3 \cos \alpha_3 \varphi(x, y, z) \mathrm{d}S , \quad \varphi \in \mathcal{D}(\mathbb{R}^3) . \tag{1.264}$$

Regarding formula (1.256), an important case in mechanics is when the hypersurface $S \subset \mathbb{R}^n$ is a cylindrical hypersurface.

Let $\Gamma \subset \mathbb{R}^2$ be a piecewise smooth curve in the Oxy-plane and let us denote by $S = \Gamma \times \mathbb{R} \subset \mathbb{R}^3$ the cylindrical surface with generators parallel to Oz with respect to the orthogonal reference system $Oxyz$ (Figure 1.3).

The curve Γ of the Oxy-plane is the director curve of the cylindrical surface $S = \Gamma \times \mathbb{R} \subset \mathbb{R}^3$ and the normal unit vector $\boldsymbol{n}$ at the point $M(x, y, z) \in S$ is equal to the normal unit vector $\boldsymbol{n}^*$ at the point $P(x, y) \in \Gamma$. Hence, between the differential element of area $\mathrm{d}S$ of the cylindrical surface $S = \Gamma \times \mathbb{R}$ and the differential element $\mathrm{d}s$ of the curve arc Γ leads to the relation $\mathrm{d}S = \mathrm{d}z\mathrm{d}s$.

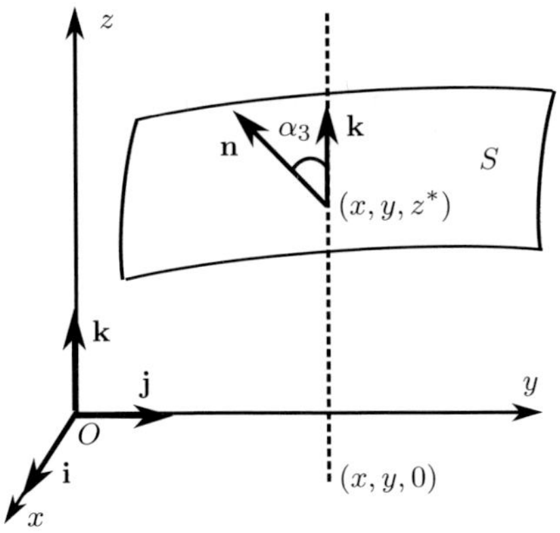

Figure 1.2

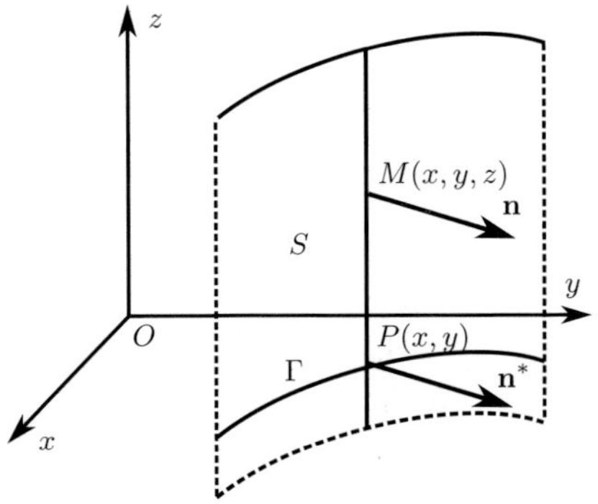

Figure 1.3

Proposition 1.15 Let the function $f : \mathbb{R}^3 \to \mathbb{R}$ of the class $C^1(\mathbb{R}^3)$, except for cylindrical surface $S = \Gamma \times \mathbb{R} \subset \mathbb{R}^3$, where it has a first-order discontinuity. Then, the following formula results

$$\frac{\partial f(x,y,z)}{\partial y} = \frac{\tilde{\partial} f(x,y,z)}{\partial y} + \sigma_y \cos(\boldsymbol{n}, y)\delta_S , \tag{1.265}$$

where $\partial/\partial y, \tilde{\partial}/\partial y$ are the derivatives in the sense of distributions and in the ordinary sense, respectively, δ_S is the Dirac delta distribution concentrated on the cylindrical surface $S = \Gamma \times \mathbb{R}$, $\sigma_y = f(x, y+0, z) - f(x, y-0, z) = \lim_{\varepsilon \to +0} f(x, y+\varepsilon, z) - \lim_{\varepsilon \to +0} f(x, y-\varepsilon, z)$, $(x, y, z) \in S$, is the jump of the function f at the crossing of the cylindrical surface in the positive direction of the Oy-axis, and $\cos(\boldsymbol{n}, y)$ is the cosine of the angle between the Oy-axis with the normal to the surface S, oriented in the direction of its crossing.
The Dirac delta distribution $\delta_S = \delta_{\Gamma \times \mathbb{R}} \in \mathcal{D}'(\mathbb{R}^3)$ acts according to the formula

$$(\delta_S, \varphi) = \int_S \varphi(x,y,z)\mathrm{d}S = \int_{\mathbb{R}} \mathrm{d}z \int_{\Gamma} \varphi(x,y,z)\mathrm{d}s\,, \quad \forall \varphi \in \mathcal{D}(\mathbb{R}^3)\,. \tag{1.266}$$

Hence, $\forall \varphi \in \mathcal{D}(\mathbb{R}^3)$ we have

$$(\sigma_y \cos(\boldsymbol{n}, y)\delta_S, \varphi) = (\delta_S, \cos(\boldsymbol{n}, y)\varphi) = \int_{\mathbb{R}} \mathrm{d}z \int_{\Gamma} \cos(\boldsymbol{n}, y)\varphi(x, y, z)\mathrm{d}s\,. \quad (1.267)$$

Proof: For any $\varphi(x, y, z) \in \mathcal{D}(\mathbb{R}^3)$ we get

$$\left(\frac{\partial f}{\partial y}, \varphi\right) = -\left(f, \frac{\tilde{\partial}\varphi}{\partial y}\right) = -\int_{\mathbb{R}^3} f\varphi'_y \mathrm{d}x\mathrm{d}y\mathrm{d}z = -\int_{\mathbb{R}} \mathrm{d}z \int_{\mathbb{R}^2} f(x, y, z)\varphi'_y \mathrm{d}x\mathrm{d}y\,. \quad (1.268)$$

On the other hand, we can write

$$\int_{\mathbb{R}^2} f\varphi'_y \mathrm{d}x\mathrm{d}y = \int_{\mathbb{R}} \mathrm{d}x \int_{\mathbb{R}} f\varphi'_y \mathrm{d}y = \int_{\mathbb{R}} \mathrm{d}x \left\{ \lim_{\varepsilon \to +0} \left[\int_{-\infty}^{y'-\varepsilon} f\varphi'_y \mathrm{d}y + \int_{y'+\varepsilon}^{\infty} f\varphi'_y \mathrm{d}y \right] \right\}. \quad (1.269)$$

There results

$$\begin{aligned}
\lim_{\varepsilon \to +0} \left[\int_{-\infty}^{y'-\varepsilon} f\varphi'_y \mathrm{d}y + \int_{y'+\varepsilon}^{\infty} f\varphi'_y \mathrm{d}y \right] &= \lim_{\varepsilon \to +0} \left[f(x, y, z)\varphi(x, y, z)\Big|_{-\infty}^{y'-\varepsilon} \right.\\
&\quad \left. + f(x, y, z)\varphi(x, y, z)\Big|_{y'+\varepsilon}^{\infty} - \int_{-\infty}^{y'-\varepsilon} \varphi \frac{\tilde{\partial} f}{\partial y} \mathrm{d}y - \int_{y'+\varepsilon}^{\infty} \varphi \frac{\tilde{\partial} f}{\partial y} \mathrm{d}y \right]\\
&= -\varphi(x, y', z)[f(x, y' + 0, z) - f(x, y' - 0, z)] - \int_{\mathbb{R}} \varphi(x, y, z) \frac{\tilde{\partial} f}{\partial y} \mathrm{d}y\\
&= -\sigma_y \varphi(x, y', z) - \int_{\mathbb{R}} \varphi(x, y, z) \frac{\tilde{\partial} f}{\partial y} \mathrm{d}y\,,
\end{aligned} \quad (1.270)$$

where

$$\begin{aligned}
\sigma_y &= \sigma_y(x, y', z) = f(x, y' + 0, z) - f(x, y' - 0, z)\\
&= \lim_{\varepsilon \to +0} f(x, y' + \varepsilon, z) - \lim_{\varepsilon \to +0} f(x, y' - \varepsilon, z)\,,
\end{aligned} \quad (1.271)$$

represents the jump of the function f at the crossing of the cylindrical surface $S = \Gamma \times \mathbb{R} \subset \mathbb{R}^3$ in the positive direction of the Oy-axis. Thus, we get

$$\int_{\mathbb{R}^2} f\varphi'_y \mathrm{d}x\mathrm{d}y = -\int_{\mathbb{R}^2} \varphi \frac{\tilde{\partial} f}{\partial y} \mathrm{d}x\mathrm{d}y - \int_{\mathbb{R}} \sigma_y \varphi(x, y', z)\mathrm{d}x\,, \quad (1.272)$$

and the following relation occurs

$$\left(\frac{\partial f}{\partial y},\varphi\right)=\int_{\mathbb{R}}\mathrm{d}z\int_{\mathbb{R}}\sigma_y\varphi(x,y',z)\mathrm{d}x+\int_{\mathbb{R}^3}\varphi(x,y,z)\frac{\tilde{\partial} f}{\partial y}\mathrm{d}x\mathrm{d}y\mathrm{d}z\,. \tag{1.273}$$

If we denote with $\mathrm{d}s$ the element of arc on the curve Γ, then $\mathrm{d}x=\mathrm{d}s\cos(\boldsymbol{n},y)$, where $(\boldsymbol{n},y)$ is the angle between the Oy-axis and the normal to the curve Γ at the point (x,y').

Consequently, we have

$$\int_{\mathbb{R}}\mathrm{d}z\int_{\mathbb{R}}\sigma_y\varphi(x,y',z)\mathrm{d}x=\int_{\mathbb{R}}\mathrm{d}z\int_{\Gamma}\sigma_y\varphi(x,y',z)\cos(\boldsymbol{n},y)\mathrm{d}s\,, \tag{1.274}$$

which allows us to define the Dirac delta distribution $\delta_{\Gamma\times\mathbb{R}}\in\mathcal{D}'(\mathbb{R}^3)$, concentrated on the cylindrical surface $S=\Gamma\times\mathbb{R}\subset\mathbb{R}^3$, by the formula

$$(\delta_{\Gamma\times\mathbb{R}},\varphi(x,y,z))=\int_{\mathbb{R}}\mathrm{d}z\int_{\Gamma}\varphi(x,y,z)\mathrm{d}s\,,\quad\forall\varphi\in\mathcal{D}(\mathbb{R}^3)\,. \tag{1.275}$$

Consequently, we can write

$$\begin{aligned}(\sigma_y\cos(\boldsymbol{n},y)\delta_{\Gamma\times\mathbb{R}},\varphi)&=\int_S\sigma_y\cos(\boldsymbol{n},y)\varphi(x,y,z)\mathrm{d}S\\&=\int_{\mathbb{R}}\mathrm{d}z\int_{\Gamma}\sigma_y\cos(\boldsymbol{n},y)\varphi(x,y,z)\mathrm{d}s\,.\end{aligned} \tag{1.276}$$

With these results, one gets

$$\left(\frac{\partial f}{\partial y},\varphi\right)=\left(\frac{\tilde{\partial} f}{\partial y}+\sigma_y\cos(\boldsymbol{n},y)\delta_S,\varphi\right)\,, \tag{1.277}$$

giving the requested formula. □

The obtained derivation formula can be generalized as

$$\frac{\partial f(x,z)}{\partial x_i}=\frac{\tilde{\partial} f(x,z)}{\partial x_i}+\sigma_{x_i}\cos(\boldsymbol{n},x_i)\delta_\Sigma\,, \tag{1.278}$$

where $\Sigma=S\times\mathbb{R}\subset\mathbb{R}^{n+1}$ is the cylindrical surface with generators parallel to the Oz-axis with respect to the orthogonal reference system $Ox_1x_2\ldots x_nz$ and $S\subset\mathbb{R}^n$ is a piecewise smooth surface.

The function $f:\mathbb{R}^{n+1}\to\mathbb{R}$ is considered to be of class $C^1(\mathbb{R}^{n+1})$ except for the cylindrical surface $\Sigma=S\times\mathbb{R}$, where it has a first-order discontinuity.

Obviously, σ_i is the jump of the function f at the crossing of the cylindrical surface Σ in the positive direction of the Ox_i-axis and $(\boldsymbol{n},x_i)$ the angle between the Ox_i-axis and the normal to the surface Σ, oriented in the direction of its crossing.

The Dirac delta distribution $\delta_\Sigma = \delta_{S\times\mathbb{R}}$, concentrated on the cylindrical surface Σ, acts according to the formula

$$(\delta_\Sigma, \varphi) = \int_\Sigma \varphi \, \mathrm{d}S_\Sigma = \int_{\mathbb{R}} \mathrm{d}z \int_S \varphi \, \mathrm{d}S \,, \quad \forall \varphi \in \mathcal{D}(\mathbb{R}^{n+1}) \,. \tag{1.279}$$

Example 1.17 Let the function $f : \mathbb{R}^2 \to \mathbb{R}$

$$f(x, y) = -\frac{a}{2} H(y) H(ay - |x|) = \begin{cases} -a/2 & \text{for } |x| \le ay, y > 0 \,, \quad a > 0 \,, \\ 0 & \text{otherwise} \,. \end{cases} \tag{1.280}$$

We shall demonstrate that the function type distribution $T_f = f \in D'(\mathbb{R}^2)$ is the fundamental solution of the operator

$$P(\mathrm{D}) = \frac{\partial^2}{\partial x^2} - \frac{1}{a^2}\frac{\partial^2}{\partial y^2} \,, \quad a > 0 \,, \tag{1.281}$$

that is, $P(\mathrm{D}) f(x, y) = \delta(x, y)$, by using formula (1.256).

We observe that the function f has the value $-a/2$ inside the cone Γ^+ (Figure 1.4) and is zero outside it. The frontier of the cone Γ^+ is the curve Γ which consists of the branches Γ_1 and Γ_2, $\Gamma = \Gamma_1 \cup \Gamma_2$, defined by the parametric equations

$$\begin{aligned} &\Gamma_1 : x = at \,, \quad y = -t \,, \quad t \in (-\infty, 0] \,, \\ &\Gamma_2 : x = at \,, \quad y = t \,, \quad t \in [0, \infty) \,. \end{aligned} \tag{1.282}$$

The curve $\Gamma = \Gamma_1 \cup \Gamma_2$ represents the discontinuity curve at the crossing of which the function f has a first-order discontinuity. The derivatives in the ordinary sense will be

$$\frac{\tilde{\partial} f}{\partial x} = 0 \,, \quad \frac{\tilde{\partial} f}{\partial y} = 0 \,. \tag{1.283}$$

Applying the formula (1.256), we obtain

$$\begin{aligned} \frac{\partial f}{\partial x} &= \sigma_x|_{\Gamma_1} \cos(\boldsymbol{n}_1, x)|_{\Gamma_1} \delta_{\Gamma_1} + \sigma_x|_{\Gamma_2} \cos(\boldsymbol{n}_2, x)|_{\Gamma_2} \delta_{\Gamma_2} \,, \\ \frac{\partial f}{\partial y} &= \sigma_y|_{\Gamma_1} \cos(\boldsymbol{n}_1, x)|_{\Gamma_1} \delta_{\Gamma_1} + \sigma_y|_{\Gamma_2} \cos(\boldsymbol{n}_2^*, x)|_{\Gamma_2} \delta_{\Gamma_2} \,. \end{aligned} \tag{1.284}$$

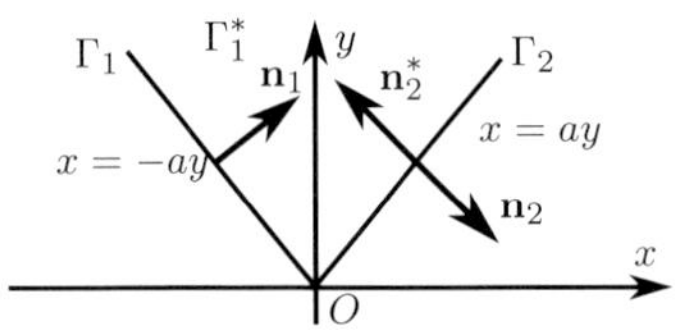

Figure 1.4

In Figure 1.4, $\boldsymbol{n}_1$ and $\boldsymbol{n}_2$ are the normals to Γ_1 and Γ_2, respectively, oriented in the rising sense of the variable x, and $\boldsymbol{n}_1$, $\boldsymbol{n}_2^*$ are the normal to Γ_1 and Γ_2, respectively, oriented in the rising sense of the variable y. Taking this into account we obtain for the function jumps and the directors cosines the values

$$\sigma_x|_{\Gamma_1} = -\frac{a}{2}\,,\ \sigma_x|_{\Gamma_2} = \frac{a}{2}\,,\ \cos(\boldsymbol{n}_1, x)|_{\Gamma_1} = \frac{1}{\sqrt{1+a^2}}\,,\ \cos(\boldsymbol{n}_2, x)|_{\Gamma_2} = \frac{1}{\sqrt{1+a^2}}\,,$$
$$\sigma_y|_{\Gamma_1} = \sigma_y|_{\Gamma_2} = -\frac{a}{2}\,,\ \cos(\boldsymbol{n}_1, y)|_{\Gamma_1} = \frac{a}{\sqrt{1+a^2}}\,,\ \cos(\boldsymbol{n}_2^*, y)|_{\Gamma_2} = \frac{a}{\sqrt{1+a^2}}\,, \tag{1.285}$$

hence

$$\frac{\partial f}{\partial x} = \frac{-a}{2\sqrt{1+a^2}}\delta_{\Gamma_1} + \frac{a}{2\sqrt{1+a^2}}\delta_{\Gamma_2}\,,\quad \frac{\partial f}{\partial y} = \frac{-a^2}{2\sqrt{1+a^2}}\delta_{\Gamma_1} - \frac{a^2}{2\sqrt{1+a^2}}\delta_{\Gamma_2}\,. \tag{1.286}$$

For the second derivative we have

$$\frac{\partial^2 f}{\partial x^2} = \frac{-a}{2\sqrt{1+a^2}}\frac{\partial}{\partial x}\delta_{\Gamma_1} + \frac{a}{2\sqrt{1+a^2}}\frac{\partial}{\partial x}\delta_{\Gamma_2}\,, \tag{1.287}$$

wherefrom, $\forall \varphi \in \mathcal{D}(\mathbb{R}^2)$ it results

$$\begin{aligned}\left(\frac{\partial^2 f}{\partial x^2}, \varphi\right) &= \frac{a}{2\sqrt{1+a^2}}\left(\delta_{\Gamma_1}, \frac{\partial \varphi}{\partial x}\right) - \frac{a}{2\sqrt{1+a^2}}\left(\delta_{\Gamma_2}, \frac{\partial \varphi}{\partial x}\right) \\ &= \frac{a}{2\sqrt{1+a^2}}\int_{\Gamma_1}\frac{\partial \varphi}{\partial x}\mathrm{d}s_1 - \frac{a}{2\sqrt{1+a^2}}\int_{\Gamma_2}\frac{\partial \varphi}{\partial x}\mathrm{d}s_2\,.\end{aligned} \tag{1.288}$$

Taking into account the parametric representations (1.282) of the curves Γ_1 and Γ_2, we obtain

$$\mathrm{d}s_1 = \sqrt{1+a^2}\mathrm{d}t\,,\quad \mathrm{d}s_2 = \sqrt{1+a^2}\mathrm{d}t\,. \tag{1.289}$$

Therefore, the expression (1.288) becomes

$$\begin{aligned}\left(\frac{\partial^2 f}{\partial x^2}, \varphi\right) &= \frac{a}{2}\int_{-\infty}^{0}\frac{\partial \varphi(at, -t)}{\partial x}\mathrm{d}t - \frac{a}{2}\int_{0}^{\infty}\frac{\partial \varphi(at, t)}{\partial x}\mathrm{d}t \\ &= \frac{a}{2}\int_{0}^{\infty}\left[\frac{\partial \varphi(-at, t)}{\partial x} - \frac{\partial \varphi(at, t)}{\partial x}\right]\mathrm{d}t\,.\end{aligned} \tag{1.290}$$

Proceeding similarly, we have

$$\frac{\partial^2 f}{\partial y^2} = \frac{-a^2}{2\sqrt{1+a^2}}\frac{\partial}{\partial y}\delta_{\Gamma_1} - \frac{a^2}{2\sqrt{1+a^2}}\frac{\partial}{\partial y}\delta_{\Gamma_2}\,, \tag{1.291}$$

hence

$$\left(\frac{\partial^2 f}{\partial y^2}, \varphi\right) = \frac{a^2}{2}\int\limits_{-\infty}^{0} \frac{\partial}{\partial y}\varphi(at, -t)\mathrm{d}t + \frac{a^2}{2}\int\limits_{0}^{\infty} \frac{\partial}{\partial y}\varphi(at, t)\mathrm{d}t$$
$$= \frac{a^2}{2}\int\limits_{0}^{\infty}\left[\frac{\partial}{\partial y}\varphi(-at, t) + \frac{\partial}{\partial y}\varphi(at, t)\right]\mathrm{d}t\,. \tag{1.292}$$

Because

$$\frac{\mathrm{d}}{\mathrm{d}t}\varphi(at, t) = a\frac{\partial\varphi(at, t)}{\partial x} + \frac{\partial\varphi(at, t)}{\partial y}\,,$$
$$\frac{\mathrm{d}\varphi(-at, t)}{\mathrm{d}t} = -a\frac{\partial\varphi(-at, t)}{\partial x} + \frac{\partial\varphi(-at, t)}{\partial y}\,, \tag{1.293}$$

from (1.288) and (1.292) we get

$$\left(\frac{\partial^2 f}{\partial x^2} - \frac{1}{a^2}\frac{\partial^2 f}{\partial y^2}, \varphi\right) = -\frac{1}{2}\int\limits_{0}^{\infty}\frac{\mathrm{d}}{\mathrm{d}t}[\varphi(-at, t) + \varphi(at, t)]\mathrm{d}t$$
$$= \varphi(0, 0) = (\delta(x, y), \varphi(x, y))\,, \tag{1.294}$$

that is,

$$P(\mathrm{D})f = \frac{\partial^2 f}{\partial x^2} - \frac{1}{a^2}\frac{\partial^2 f}{\partial y^2} = \delta(x, y)\,. \tag{1.295}$$

1.3.4
The Fundamental Solution of a Linear Differential Operator

Let the linear differential operator with constant coefficients be $P(\mathrm{D}) : \mathcal{D}'(\mathbb{R}^n) \to \mathcal{D}'(\mathbb{R}^n)$, having the expression

$$P(\mathrm{D}) = \sum_{|\alpha|\le \ell} a_\alpha \mathrm{D}^\alpha\,, \quad \alpha \in \mathbb{N}_0^n\,, \quad x \in \mathbb{R}^n\,, \tag{1.296}$$

where the scalars $a_\alpha \in \Gamma$ represent the operator coefficients.

Definition 1.27 We say that the distribution $E(x) \in \mathcal{D}'(\mathbb{R}^n)$ is the fundamental solution for the operator $P(\mathrm{D})$ if it satisfies the following relation:

$$P(\mathrm{D})E(x) = \delta(x)\,. \tag{1.297}$$

Based on this definition, we can say that the distribution of function type given by (1.231) is the fundamental solution for the operator $P(\mathrm{D}) = (\mathrm{d}^2/\mathrm{d}x^2) + (\mathrm{d}/\mathrm{d}x) -$

2. It is verified that the distribution of function type $f_1 = H(x)(\mathrm{e}^x - \mathrm{e}^{-2x})/3$, $x \in \mathbb{R}$ is the fundamental solution for the same operator.

It follows that the fundamental solution of an operator is generally not unique. Thus, if $f \in \mathcal{D}'(\mathbb{R}^n)$ satisfies the equation $P(\mathrm{D}) f = 0$ and E is a fundamental solution for $P(\mathrm{D})$, then $E_1 = f + E$ is the fundamental solution, because on the basis of linearity of $P(\mathrm{D})$ we can write

$$P(\mathrm{D}) E_1 = P(\mathrm{D})(f + E) = P(\mathrm{D}) f + P(\mathrm{D}) E = \delta \,. \tag{1.298}$$

Proposition 1.16 Let there be a linear differential operator with constant coefficients $P(\mathrm{D})$ having the expression

$$P(\mathrm{D}) = a_0 \frac{\mathrm{d}^n}{\mathrm{d}x^n} + a_1 \frac{\mathrm{d}^{n-1}}{\mathrm{d}x^{n-1}} + \cdots + a_{n-1} \frac{\mathrm{d}}{\mathrm{d}x} + a_n \,, \quad a_0 \neq 0 \,. \tag{1.299}$$

Then, the distribution of function type $E \in \mathcal{D}'(\mathbb{R}^n)$, that is,

$$E(x) = H(x) Y(x) \,, \tag{1.300}$$

is the fundamental solution for $P(\mathrm{D})$, where H is the Heaviside function and Y the solution of the homogeneous equation $P(\mathrm{D}) Y = 0$, verifying the initial conditions

$$Y(0) = 0,\ Y'(0) = 0, \ldots,\ Y^{(n-2)}(0) = 0,\ Y^{(n-1)}(0) = \frac{1}{a_0} \,, \quad a_0 \neq 0 \,. \tag{1.301}$$

Proof: We note that the function Y is infinitely differentiable and HY is also infinitely differentiable, except at the origin where it has a first-order discontinuity with the jumps

$$s_0(HY^{(p)}) = Y^{(p)}(0) \,, \quad p = 0, 1, 2, \ldots, n-2 \,, \quad s_0(HY^{(n-1)}) = Y^{(n-1)}(0) = \frac{1}{a_0} \,. \tag{1.302}$$

We can write

$$(HY)^{(p)} = HY^{(p)} + \sum_{i=0}^{p-1} s_0(HY^{(i)}) \delta^{(p-i-1)} \,, \quad p = 1, 2, \ldots, n \,. \tag{1.303}$$

Consequently, we have

$$(HY)^{(p)} = HY^{(p)} \,, \quad p = 1, 2, \ldots, n-1 \,, \quad (HY)^{(n)} = HY^{(n)} + \frac{1}{a_0}\delta \,. \tag{1.304}$$

Because $P(\mathrm{D}) Y = 0$, we have $P(\mathrm{D})(HY) = HP(\mathrm{D}) Y + \delta = \delta$.

The fundamental solution $E = HY$ is a function of class $C^\infty[0, \infty)$ and is unique because of the uniqueness of the solution Y of the Cauchy problem for the equation $P(\mathrm{D}) Y = 0$. Thus, the proposition is proved. □

For example, for the operator

$$P(\mathrm{D}) = \frac{\mathrm{d}^2}{\mathrm{d}x^2} + \omega^2 , \quad \omega \in \mathbb{R}\backslash\{0\} \tag{1.305}$$

we have $Y = (\sin \omega x)/x$, because $P(\mathrm{D})\, Y = 0$ and $Y(0) = 0$, $Y'(0) = 1$ and thus the fundamental solution is

$$E = H(x)\, Y(x) = \begin{cases} 0, & x < 0 , \\ \dfrac{\sin(\omega x)}{\omega} , & x \geq 0 . \end{cases} \tag{1.306}$$

Example 1.18 Let there be the linear differential operator

$$P(\mathrm{D}) = -\frac{\partial^2}{\partial x^2} + \frac{\partial^2}{\partial y^2} + 2\frac{\partial}{\partial y} + 1 , \quad (x, y) \in \mathbb{R}^2 , \tag{1.307}$$

and the distribution of function type $E(x, y) \in \mathcal{D}'(\mathbb{R}^2)$,

$$E(x, y) = \begin{cases} 0 , & y < 0 , \quad x \in \mathbb{R} , \\ \frac{1}{2}\mathrm{e}^{-y}[H(x + y) - H(x - y)] , & y \geq 0 , \quad x \in \mathbb{R} , \end{cases} \tag{1.308}$$

where $H \in \mathcal{D}'(\mathbb{R})$ is the Heaviside distribution.

We will show that the distribution $E \in \mathcal{D}'(\mathbb{R}^2)$ is the fundamental solution of the operator $P(\mathrm{D})$, namely $P(\mathrm{D})\, E(x, y) = \delta(x, y)$.

We notice that $E(x, y)$ has the value $\mathrm{e}^{-y}/2$ inside the cone Γ^+ and is zero outside it (Figure 1.5).

For any $\varphi \in \mathcal{D}(\mathbb{R}^2)$ we have

$$(P(\mathrm{D})\, E, \varphi) = -\left(E, \frac{\partial^2 \varphi}{\partial x^2}\right) + \left(E, \frac{\partial^2 \varphi}{\partial y^2}\right) - 2\left(E, \frac{\partial \varphi}{\partial y}\right) + (E, \varphi)$$
$$= -\int_{\Gamma^+} E\frac{\partial^2 \varphi}{\partial x^2}\mathrm{d}x\mathrm{d}y + \int_{\Gamma^+} E\frac{\partial^2 \varphi}{\partial y^2}\mathrm{d}x\mathrm{d}y - 2\int_{\Gamma^+} E\frac{\partial \varphi}{\partial y}\mathrm{d}x\mathrm{d}y + \int_{\Gamma^+} E\varphi\mathrm{d}x\mathrm{d}y .$$

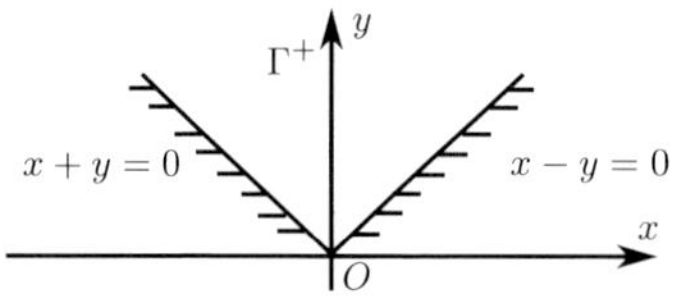

Figure 1.5

By calculation of the four integrals, we obtain:

$$\left(-E, \frac{\partial^2 \varphi}{\partial x^2}\right) = -\int_{\Gamma+} E \frac{\partial^2 \varphi}{\partial x^2} \mathrm{d}x \mathrm{d}y = -\frac{1}{2} \int_0^\infty \mathrm{e}^{-y} \mathrm{d}y \int_{-y}^{y} \frac{\partial^2 \varphi}{\partial x^2} \mathrm{d}x$$

$$-\frac{1}{2} \int_0^\infty \mathrm{e}^{-y} \left[\frac{\partial \varphi(y, y)}{\partial x} - \frac{\partial \varphi(-y, y)}{\partial x}\right] \mathrm{d}y$$

$$= -\frac{1}{2} \int_0^\infty \mathrm{e}^{-t} \left[\frac{\partial \varphi(t, t)}{\partial x} - \frac{\partial \varphi(-t, t)}{\partial x}\right] \mathrm{d}t ,$$

$$\left(E, \frac{\partial^2 \varphi}{\partial y^2}\right) = \int_{\Gamma+} E \frac{\partial^2 \varphi}{\partial y^2} \mathrm{d}x \mathrm{d}y = \int_{-\infty}^{0} \mathrm{d}x \int_{-x}^{\infty} \frac{1}{2} \mathrm{e}^{-y} \frac{\partial^2 \varphi}{\partial y^2} \mathrm{d}y + \int_0^\infty \mathrm{d}x \int_x^\infty \frac{1}{2} \mathrm{e}^{-y} \frac{\partial^2 \varphi}{\partial y^2} \mathrm{d}y$$

$$= \int_{-\infty}^{0} \left[-\frac{1}{2} \mathrm{e}^{x} \frac{\partial \varphi(x, -x)}{\partial y} - \frac{1}{2} \mathrm{e}^{x} \varphi(x, -x) + \int_{-x}^{\infty} \frac{1}{2} \mathrm{e}^{-y} \varphi \mathrm{d}y\right] \mathrm{d}x$$

$$+ \int_0^\infty \left[-\frac{1}{2} \mathrm{e}^{-x} \frac{\partial \varphi(x, x)}{\partial y} - \frac{1}{2} \mathrm{e}^{-x} \varphi(x, x) + \int_x^\infty \frac{1}{2} \mathrm{e}^{-y} \varphi \mathrm{d}y\right] \mathrm{d}x$$

$$= \int_0^\infty -\frac{1}{2} \mathrm{e}^{-t} \left[\frac{\partial \varphi(-t, t)}{\partial y} + \frac{\partial \varphi(t, t)}{\partial y} + \varphi(-t, t) + \varphi(t, t)\right] \mathrm{d}t$$

$$+ \int_{\Gamma+} \frac{1}{2} \mathrm{e}^{-y} \varphi(x, y) \mathrm{d}x \mathrm{d}y ,$$

$$\left(-2E, \frac{\partial \varphi}{\partial y}\right) = -2 \int_{\Gamma+} \frac{1}{2} \mathrm{e}^{-y} \frac{\partial \varphi}{\partial y} \mathrm{d}x \mathrm{d}y = -\int_{-\infty}^{0} \mathrm{d}x \int_{-x}^{\infty} \mathrm{e}^{-y} \frac{\partial \varphi}{\partial y} \mathrm{d}y$$

$$-\int_0^\infty \mathrm{d}x \int_x^\infty \mathrm{e}^{-y} \frac{\partial \varphi}{\partial} y \mathrm{d}y = \int_0^\infty \mathrm{e}^{-t} [\varphi(-t, t) + \varphi(t, t)] \mathrm{d}t$$

$$-\int_{\Gamma+} \mathrm{e}^{-y} \varphi(x, y) \mathrm{d}x \mathrm{d}y ,$$

$$(E, \varphi) = \int_{\Gamma+} \frac{1}{2} \mathrm{e}^{-y} \varphi(x, y) \mathrm{d}x \mathrm{d}y .$$

Consequently, we obtain

$$(P(\mathrm{D}) E, \varphi) = \int_0^\infty \frac{1}{2} \mathrm{e}^{-t} \left[\varphi(t, t) + \varphi(-t, t) - \frac{\partial \varphi(t, t)}{\partial x} \right.$$

$$\left. - \frac{\partial \varphi(t, t)}{\partial y} + \frac{\partial \varphi(-t, t)}{\partial x} - \frac{\partial \varphi(-t, t)}{\partial y}\right] \mathrm{d}t . \qquad (1.309)$$

Taking into account the relations

$$\frac{\mathrm{d}\varphi(-t,t)}{\mathrm{d}t} = -\frac{\partial\varphi(-t,t)}{\partial x} + \frac{\partial\varphi(-t,t)}{\partial y}, \frac{\mathrm{d}\varphi(t,t)}{\mathrm{d}t} = \frac{\partial\varphi(t,t)}{\partial x} + \frac{\partial\varphi(t,t)}{\partial y}, \quad (1.310)$$

the previous one becomes

$$(P(\mathrm{D}),\varphi) = -\frac{1}{2}\int_0^\infty \frac{\mathrm{d}}{\mathrm{d}t}[\mathrm{e}^{-t}(\phi(t,t) + \phi(-t,t))]\mathrm{d}t = \varphi(0,0) = (\delta(x,y),\varphi(x,y)), \quad (1.311)$$

namely $P(\mathrm{D})E = \delta(x,y)$.

1.3.5
The Derivation of the Homogeneous Distributions

The distribution $f \in \mathcal{D}'(\mathbb{R}^n)$ is homogeneous and of degree λ if for $\alpha > 0$ it satisfies the relation

$$f(\alpha x) = \alpha^\lambda f(x), \quad x \in \mathbb{R}^n, \quad (1.312)$$

or, equivalently

$$\alpha^{\lambda+n}(f(x),\varphi(x)) = \left(f(x),\varphi\left(\frac{x}{\alpha}\right)\right), \quad \varphi \in \mathcal{D}(\mathbb{R}^n). \quad (1.313)$$

It is known that the homogeneous functions of degrees λ and of class $C^1(\mathbb{R}^n)$ satisfy the Euler equation

$$\sum_{i=1}^n x_i \frac{\partial f}{\partial x_i} = \lambda f. \quad (1.314)$$

In fact, (1.314) fully characterizes the homogeneous functions of degree λ, since this equation represents the necessary and sufficient condition for a function to be homogeneous of degree λ.

This result is valid for functions which translate to the homogeneous distributions. Suppose that $f \in \mathcal{D}'(\mathbb{R}^n)$ is a homogeneous distribution of degree λ.

Then, we can derive the equality (1.313) with respect to $\alpha > 0$, on the basis of Proposition 1.9, and we obtain

$$(n+\lambda)\alpha^{n+\lambda-1}(f(x),\varphi(x)) = -\frac{1}{\alpha^2}\left(f(x_1,\dots,x_n), \sum_{i=1}^n x_i \frac{\partial\varphi}{\partial x_i}\left(\frac{x_1}{\alpha},\dots,\frac{x_n}{\alpha}\right)\right). \quad (1.315)$$

Considering $\alpha = 1$, the relation (1.315) becomes

$$(n+\lambda)(f,\varphi) = \left(\sum_{i=1}^n \frac{\partial}{\partial x_i}(x_i f),\varphi\right), \quad \forall\varphi \in \mathcal{D}(\mathbb{R}^n), \quad (1.316)$$

wherefrom

$$\sum_{i=1}^{n} \frac{\partial}{\partial x_i}(x_i f) = (n + \lambda) f , \tag{1.317}$$

hence

$$\sum_{i=1}^{n} x_i \frac{\partial f}{\partial x_i} = \lambda f , \quad f \in \mathcal{D}'(\mathbb{R}^n) . \tag{1.318}$$

Let us now show the converse. We acknowledge that the distribution $f \in \mathcal{D}'(\mathbb{R}^n)$ satisfies (1.316). Differentiating with respect to $\alpha > 0$ the fraction

$$\frac{1}{\alpha^{\lambda+n}} \left(f(x), \varphi\left(\frac{x_1}{\alpha}, \ldots, \frac{x_n}{\alpha}\right)\right) \tag{1.319}$$

and taking into account (1.316), it results that the derivative is zero. This means that the fraction is reduced to a constant.

Hence, we can write

$$\left(f(x), \varphi\left(\frac{x_1}{\alpha}, \ldots, \frac{x_n}{\alpha}\right)\right) = c\alpha^{\lambda+n} . \tag{1.320}$$

To determine the value of the constant, we take $\alpha = 1$ and we obtain $c = (f(x), \varphi(x))$, hence

$$\left(f(x), \varphi\left(\frac{x}{\alpha}\right)\right) = \alpha^{\lambda+n}(f(x), \varphi(x)) , \quad \forall \varphi \in \mathcal{D} , \tag{1.321}$$

which shows that the distribution $f \in \mathcal{D}'(\mathbb{R}^n)$ is homogeneous and of degree λ.

The homogeneous distributions with the singularities generated by homogeneous locally integrable functions are of interest in applications.

Let there be the homogeneous function $f : \mathbb{R}^n \backslash \{0\} \to \Gamma$ of degree λ, with a singularity (discontinuity) at the point $x = 0$, to which we assign the functional $T_f : \mathcal{D}(\mathbb{R}^n) \to \Gamma$ by the formula

$$(T_f, \varphi(x)) = \lim_{\varepsilon \to 0} \int_{\|x\| \geq \varepsilon} f(x)\varphi(x) \mathrm{d}x , \quad \forall \varphi \in \mathcal{D}(\mathbb{R}^n) . \tag{1.322}$$

Obviously, in the case of the convergence of the integral (1.322), the functional T_f is linear and continuous, hence $T_f \in \mathcal{D}'(\mathbb{R}^n)$.

We note by $F_\varepsilon(\varphi)$ the integral

$$F_\varepsilon(\varphi) = \int_{\varepsilon \leq \|x\| \leq a} f(x)\varphi(x) \mathrm{d}x , \quad \varphi \in \mathcal{D}(\mathbb{R}^n) , \quad \mathrm{supp}(\varphi) \subset B_a , \tag{1.323}$$

where B_a is the open sphere of radius a, centered at the origin of the coordinates.

We pass to spherical coordinates, expressed by the relations

$$\begin{aligned}
x_1 &= r \sin\theta_1 \sin\theta_2 \ldots \sin\theta_{n-2} \sin\theta_{n-1} ,\\
x_2 &= r \sin\theta_1 \sin\theta_2 \ldots \sin\theta_{n-2} \cos\theta_{n-1} ,\\
x_3 &= r \sin\theta_1 \sin\theta_2 \ldots \sin\theta_{n-3} \cos\theta_{n-2} ,\\
&\vdots\\
x_{n-2} &= r \sin\theta_1 \sin\theta_2 \cos\theta_3 ,\\
x_{n-1} &= r \sin\theta_1 \cos\theta_2 ,\\
x_n &= r \cos\theta_1 ,
\end{aligned} \tag{1.324}$$

where

$$r \geq 0, \theta_i \in [0, \pi] , \quad i = \overline{1, n-2} , \quad \theta_{n-1} \in [0, 2\pi) , \tag{1.325}$$

and where the Jacobian of the transformation is

$$\begin{aligned}
J(r, \theta_1, \theta_2, \ldots, \theta_{n-1}) &= \frac{\partial(x_1, x_2, \ldots, x_n)}{\partial(r, \theta_1, \theta_2, \ldots, \theta_{n-1})}\\
&= r^{n-1} \sin^{n-2}\theta_1 \sin^{n-3}\theta_2 \ldots \sin\theta_{n-2} ;
\end{aligned} \tag{1.326}$$

the expression (1.323) becomes

$$F_\varepsilon(\varphi) = \int_\varepsilon^a \underbrace{\int_0^\pi \cdots \int_0^\pi}_{n-2} \int_0^{2\pi} f^*(r, \theta_1, \ldots, \theta_{n-1}) \varphi^*(r, \theta_1, \ldots, \theta_{n-1}) r^{n-1} \mathrm{d}r \mathrm{d}s_1 , \tag{1.327}$$

where $\mathrm{d}s_1$ is the element of area of the unit sphere, and f^*, φ^* are the expressions in polar coordinates of the functions f and φ.

Because f is a homogeneous function of degree λ, the formula (1.327) becomes

$$F_\varepsilon(\varphi) = \int_\varepsilon^a \underbrace{\int_0^\pi \cdots \int_0^\pi}_{n-2} \int_0^{2\pi} r^{\lambda+n-1} f^*(1, \theta_1, \ldots, \theta_{n-1}) \varphi^*(r, \theta_1, \ldots, \theta_{n-1}) r^{n-1} \mathrm{d}r \mathrm{d}s_1 . \tag{1.328}$$

Taking into account that the functions $f^*(1, \theta_1, \ldots, \theta_{n-1})$ and $\varphi^*(r, \theta_1, \ldots, \theta_{n-1})$ are bounded, it results that the integral (1.322) is convergent together with the integral $\lim_{\varepsilon\to 0} \int_\varepsilon^a r^{\lambda+n-1} \mathrm{d}r$.

Thus, the functional T_f, given by (1.322), does exist if $\lambda + n - 1 > -1$, namely $\lambda > -n$.

In particular, if $\lambda \geq -n+1$, then the homogeneous distribution of function type T_f exists.

We shall establish the formula for the derivative of the homogeneous distribution T_f of degree $\lambda \geq -n+1$, taking into account (1.322), we have

$$\left(\frac{\partial T_f}{\partial x_i}, \varphi\right) = -\left(T_f, \frac{\partial \varphi}{\partial x_i}\right) = -\lim_{\varepsilon \to +0} \int\limits_{\|x\| \geq \varepsilon} f(x) \frac{\partial \varphi}{\partial x_i} \mathrm{d}x$$

$$= -\lim_{\varepsilon \to +0} \left[\int\limits_{\|x\| \geq \varepsilon} \frac{\tilde{\partial}}{\partial x_i} (f\varphi) \mathrm{d}x - \int\limits_{\|x\| \geq \varepsilon} \frac{\tilde{\partial} f}{\partial x_i} \varphi \mathrm{d}x \right], \tag{1.329}$$

where $\tilde{\partial}/\partial x_i$ is the derivative in the usual sense.

We consider $\varphi \in D(\mathbb{R}^n)$ with $\mathrm{supp}(\varphi) \subset B_a$, where B_a is the open sphere of radius a centered at the origin. Because $\varepsilon > 0$ is arbitrary, we will take $\varepsilon < a$. Thus, the integrals of (1.329) are performed on a spherical crown $\varepsilon < r < a$, so that we can apply the Gauss–Ostrogradski formula.

Therefore, we can write

$$\int\limits_{\|x\| \geq \varepsilon} \frac{\tilde{\partial}}{\partial x_i} (f\varphi) \mathrm{d}x = -\int\limits_{S_\varepsilon} f\varphi \cos \alpha_i \mathrm{d}S_\varepsilon, \tag{1.330}$$

where S_ε is the sphere centered at the origin, of radius $\varepsilon > 0$, $\mathrm{d}S_\varepsilon$ is the corresponding area element and α_i is the angle formed by the outer normal to S_ε and the Ox_i-axis.

We mention that the functions φ and $\partial\varphi/\partial x_i$ are zero on the sphere B_a and beyond it.

Substituting (1.330) in (1.329), we obtain

$$\left(\frac{\partial T_f}{\partial x_i}, \varphi\right) = \lim_{\varepsilon \to +0} \int\limits_{\|x\| \geq \varepsilon} \frac{\tilde{\partial} f}{\partial x_i} \varphi \mathrm{d}x + \lim_{\varepsilon \to +0} \int\limits_{S_\varepsilon} f\varphi \cos \alpha_i \mathrm{d}S_\varepsilon$$

$$= \left(\frac{\tilde{\partial} f}{\partial x_i}, \varphi\right) + \lim_{\varepsilon \to +0} \int\limits_{S_\varepsilon} f\varphi \cos \alpha_i \mathrm{d}S_\varepsilon. \tag{1.331}$$

In connection with the evaluation of the second term from the right-hand side of the formula (1.331) we introduce:

Definition 1.28 We call residue of the homogeneous function f of degree $\lambda \geq -n+1$ at the singular point $x = 0$, corresponding to the Ox_i-axis, the number given by the expression:

$$(\mathrm{res}\, f)_i(0) = \int\limits_{S_1} f(x) \cos \alpha_i \mathrm{d}S_1. \tag{1.332}$$

Proposition 1.17 Let the homogeneous function $f : \mathbb{R}^n \backslash \{0\} \to \Gamma$ of degree λ have the origin, $x = 0$, as a singular point. Also, if $\lambda \geq -n+1$, then the derivative of the homogeneous distribution of function type T_f given by (1.322) is calculated according to the formula

$$\frac{\partial T_f}{\partial x_i} = \begin{cases} \frac{\tilde{\partial} f}{\partial x_i}\,, & \lambda > -n+1\,, \\ \frac{\tilde{\partial} f}{\partial x_i} + \delta(x)(\operatorname{res} f)_i(0)\,, & \text{if } \lambda = -n+1\,. \end{cases} \tag{1.333}$$

Indeed, passing to polar coordinates and taking into account that $\mathrm{d}S_\varepsilon = \varepsilon^{n-1}\mathrm{d}S_1$ and that f is homogeneous, we can write

$$\begin{aligned} &\lim_{\varepsilon\to+0} \int\limits_{S_\varepsilon} f\varphi \cos\alpha_i \mathrm{d}S_\varepsilon \\ &= \lim_{\varepsilon\to+0} \int\limits_0^\pi \cdots \int\limits_0^\pi \int\limits_0^{2\pi} f^*(\varepsilon,\theta_1,\theta_2,\ldots,\theta_{n-1})\varphi^*(\varepsilon,\theta_1,\theta_2,\ldots,\theta_{n-1}) \cos\alpha_i \varepsilon^{n-1}\mathrm{d}S_1 \\ &= \lim_{\varepsilon\to+0} \int\limits_{S_1} \varepsilon^{\lambda+n-1} f(x)\varphi(\varepsilon x)\cos\alpha_i \mathrm{d}S_1\,, \end{aligned} \tag{1.334}$$

where $x \in S_1$, S_1 is the unit radius sphere.

Since the integral from the right of (1.334) is taken on the unit sphere and the functions $\cos\alpha_i$, $f(x)$ do not depend on the radius ε of the sphere S_ε, from (1.334), we obtain

$$\lim_{\varepsilon\to+0} \int\limits_{S_\varepsilon} f(x)\cos\alpha_i \mathrm{d}S_\varepsilon = \begin{cases} 0\,, & \text{if } \lambda > -n+1\,, \\ \varphi(0)\int\limits_{S_1} f(x)\cos\alpha_i \mathrm{d}S_1\,, & \text{if } \lambda = -n+1\,, \end{cases} \tag{1.335}$$

namely

$$\lim_{\varepsilon\to+0} \int\limits_{S_\varepsilon} f(x)\varphi(x)\cos\alpha_i \mathrm{d}S_\varepsilon = \begin{cases} 0, & \text{if } \lambda > -n+1\,, \\ (\delta(x)(\operatorname{res} f)_i(0), \varphi)\,, & \text{if } \lambda = -n+1\,. \end{cases} \tag{1.336}$$

Substituting (1.336) in (1.331), we obtain the formula (1.333) and the proposition is proved.

We note that, using the exterior product $\wedge$, the Gauss–Ostrogradski formula can be written

$$\begin{aligned} &\int\limits_\Omega \sum_{i=1}^n \frac{\partial a_i(x)}{\partial x_i} \mathrm{d}x_1 \wedge \cdots \wedge \mathrm{d}x_n \\ &= \int\limits_{\partial\Omega} (-1)^{i-1} a_i(x) \mathrm{d}x_1 \wedge \cdots \wedge \mathrm{d}x_{i-1} \wedge \mathrm{d}x_{i+1} \wedge \cdots \wedge \mathrm{d}x_n\,, \end{aligned} \tag{1.337}$$

where $\Omega \subset \mathbb{R}^n$ is a bounded domain, and $\partial\Omega$ is its border.

Using the Gauss–Ostrogradski formula, the residue (1.332) can be written in the form

$$(\operatorname{res} f)_i(0) = (-1)^{i-1} \int_{\overline{U}_1} f(x) \mathrm{d}x_1 \wedge \cdots \wedge \mathrm{d}x_{i-1} \wedge \mathrm{d}x_{i+1} \wedge \cdots \wedge \mathrm{d}x_n \,, \quad (1.338)$$

where $\overline{U}_1$ is the closed sphere of the unit radius centered at the coordinates origin, because $\cos \alpha_i \mathrm{d}S_1 = (-1)^{i-1} \mathrm{d}x_1 \wedge \cdots \wedge \mathrm{d}x_{i-1} \wedge \mathrm{d}x_{i+1} \wedge \cdots \wedge \mathrm{d}x_n$.

Example 1.19 To illustrate the application of the formula (1.333), we establish the following relations

$$\Delta \frac{1}{r^{n-2}} = -(n-2)\delta(x) S_1 = -(n-2)\delta(x) \frac{2\pi^{n/2}}{\Gamma(n/2)} \,, \qquad x \in \mathbb{R}^n \,, \qquad n \geq 3 \,, \quad (1.339)$$

$$\Delta \ln r = 2\pi\delta(x) \,, \qquad n = 2 \,, \quad (1.340)$$

where $r = \|x\|$, $\Gamma(z) = \int_0^\infty \mathrm{e}^{-t} t^{z-1} \mathrm{d}t$, $\operatorname{Re} z > 0$, is the Euler gamma function, $z = x + \mathrm{i}y$, S_1 is the area of the unit radius sphere in $\mathbb{R}^n$ and Δ is the Laplace operator.

We consider the function $f(x) = 1/r^{n-2}$, $x \in \mathbb{R}^n \backslash \{0\}$, $n \geq 3$, which is homogeneous and of the degree $\lambda = -n + 2$.

Taking into account (1.333), the function f is locally integrable and we have

$$\frac{\partial f}{\partial x_i} = \frac{\tilde{\partial} f}{\partial x_i} = (-n+2) \frac{x_i}{r^n} \,. \quad (1.341)$$

We observe that the function $g(x) = x_i / r^n$ is also homogeneous and of degree $-n + 1$.

Consequently, we can apply the formula (1.333), thus we may write

$$\begin{aligned} \frac{\partial^2 f}{\partial x^2} &= \frac{\tilde{\partial}^2 f}{\partial x_i^2} + (-n+2)\delta(x) \int_{S_1} \frac{x_i}{r^n} \cos \alpha_i \mathrm{d}S_1 \\ &= (-n+2) \frac{r^2 - n x_i^2}{r^{n+2}} + (-n+2)\delta(x) \int_{S_1} x_i^2 \mathrm{d}S_1 \,. \end{aligned} \quad (1.342)$$

From (1.342) we obtain, by summing,

$$\Delta \frac{1}{r^{n-2}} = \tilde{\Delta} \frac{1}{r^{n-2}} + (-n+2)\delta(x) \int_{S_1} \mathrm{d}S_1 = (-n+2) S_1 \delta(x) \,, \quad (1.343)$$

because $\tilde{\Delta} = 1/r^{n-2} = 0$, $\tilde{\Delta} = (\tilde{\partial}^2/\partial x_1^2) + \cdots + (\tilde{\partial}/\partial x_n^2)$.

Observing that $S_1 = 2\pi^{n/2}/\Gamma(n/2)$, we obtain the formula (1.339).

Particularly, for $n = 3$ we have $\Delta 1/r = -4\pi\delta(x)$.

As regards the formula (1.340), we consider the locally integrable function $h(x) = \ln r$, $r = \sqrt{x^2 + y^2}$, $(x, y) \in \mathbb{R}^2 \backslash \{0\}$, for which we can write

$$\frac{\partial}{\partial x} \ln(x^2 + y^2) = \frac{\tilde{\partial}}{\partial x} \ln(x^2 + y^2) = \frac{2x}{x^2 + y^2} \,. \quad (1.344)$$

Because the function $F(x) = 2x/(x^2 + y^2)$ is a homogeneous function of degree $\lambda = -1$, we can apply the formula (1.333) and read

$$\frac{\partial^2}{\partial x^2}\ln(x^2+y^2) = \frac{\tilde{\partial}^2}{\partial x^2}\ln(x^2+y^2)+2\delta(x,y)\int\limits_{S_1}\frac{x}{x^2+y^2}\cos\alpha\,\mathrm{d}S_1\,, \quad (1.345)$$

where S_1 is the unit circle and $\cos\alpha = x/\sqrt{x^2+y^2}$.

Consequently, we get

$$\frac{\partial^2\ln(x^2+y^2)}{\partial x^2} = \frac{\tilde{\partial}^2}{\partial x^2}\ln(x^2+y^2) + 2\delta(x,y)\int\limits_{S_1}x^2\mathrm{d}S_1\,. \quad (1.346)$$

Similarly, we have

$$\frac{\partial^2\ln(x^2+y^2)}{\partial y^2} = \frac{\tilde{\partial}^2\ln(x^2+y^2)}{\partial y^2} + 2\delta(x,y)\int\limits_{S_1}y^2\mathrm{d}S_1\,. \quad (1.347)$$

By summing up, and because $\tilde{\Delta}\ln(x^2+y^2) = 0$, we obtain the formula (1.340).

The function $f(x,y) = y^3/(x^2+y^2)^2$, $(x,y) \in \mathbb{R}^2\backslash\{0\}$ has an integrable singularity at the origin of coordinates and is a homogeneous function of degree $\lambda = -1$. Consequently, we can apply the formula (1.333) and get

$$\frac{\partial f}{\partial x} = -\frac{4xy^3}{(x^2+y^2)^3}\,, \quad (1.348)$$

$$\frac{\partial f}{\partial y} = \frac{y^2(3x^2-y^2)}{(x^2+y^2)^3} + \frac{3\pi}{4}\delta(x,y)\,, \quad (1.349)$$

because $(\mathrm{res}\,f)_x(0) = 0$ and $(\mathrm{res}\,f)_y(0) = 3\pi/4$.

The formula (1.339) can be derived using the Green second formula. In this case, because the function $f(x) = 1/r^{n-2}$, $x \in \mathbb{R}^n\backslash\{0\}$, $n \geq 3$, $r = \|x\|$ is locally integrable, for $\varphi \in \mathcal{D}(\mathbb{R}^n)$ we have

$$\left(\Delta\frac{1}{r^{n-2}},\varphi\right) = \left(\frac{1}{r^{n-2}},\Delta\varphi\right) = \int\limits_{\mathbb{R}^n}\frac{1}{r^{n-2}}\Delta\varphi\mathrm{d}x = \lim_{\varepsilon\to+0}\int\limits_{r\geq\varepsilon}\frac{1}{r^{n-2}}\Delta\varphi\mathrm{d}x\,. \quad (1.350)$$

Further, we shall apply the Green second formula

$$\int\limits_{\Omega}(f\Delta\varphi - \varphi\Delta f)\mathrm{d}x = \int\limits_{S}\left(f\frac{\partial\varphi}{\partial\boldsymbol{n}} - \varphi\frac{\partial f}{\partial\boldsymbol{n}}\right)\mathrm{d}S\,, \quad (1.351)$$

where $\boldsymbol{n}$ is the outer normal to the surface S, bordering the bounded domain $\Omega \subset \mathbb{R}^n$, S being the spherical crown $\varepsilon < r < a$, with $\mathrm{supp}(\varphi) \subset B_a$; the support of φ included in the sphere centered at O and of radius a.

We have

$$\int_{r\geq\varepsilon}\frac{1}{r^{n-2}}\Delta\varphi\mathrm{d}x=\int_{\varepsilon<r<a}\frac{1}{r^{n-2}}\Delta\varphi\mathrm{d}x=-\int_{S_\varepsilon}\left[\frac{1}{r^{n-2}}\frac{\partial\varphi}{\partial r}-\varphi\frac{\partial}{\partial r}\left(\frac{1}{r^{n-2}}\right)\right]\mathrm{d}S\,,\tag{1.352}$$

where S_ε is the sphere centered at the origin and of radius ε.

This occurs because $\Delta(1/r^{n-2}) = 0$ for $r > \varepsilon$, and $\varphi, \partial\varphi/\partial r$ are zero on the sphere U_a and beyond. We notice that $(\partial/\partial \boldsymbol{n})\varphi = \operatorname{grad}\varphi \cdot \boldsymbol{n} = (\partial/\partial r)\varphi$ and $(\partial/\partial\boldsymbol{n})(1/r^{n-2}) = (\partial/\partial r)(1/r^{n-2})$.

Consequently, we get

$$\int_{r\geq\varepsilon}\frac{1}{r^{n-2}}\Delta\varphi\mathrm{d}x=-\left[\frac{1}{\varepsilon^{n-2}}\frac{\partial\varphi(\xi)}{\partial r}+\frac{n-2}{\varepsilon^{n-1}}\varphi(\xi)\right]S_\varepsilon\,,\quad \xi\in U_\varepsilon\,,\tag{1.353}$$

namely

$$\int_{r\geq\varepsilon}\frac{1}{r^{n-2}}\Delta\varphi\mathrm{d}x=-\frac{\varepsilon\pi^{n/2}}{\Gamma(n/2)}\frac{\partial\varphi(\xi)}{\partial r}-\frac{(n-2)\pi^{n/2}}{\Gamma(n/2)}\varphi(\xi)\,,\tag{1.354}$$

because $S_\varepsilon = \varepsilon^{n-1}\pi^{n/2}/\Gamma(n/2)$ is the area of the sphere of radius ε from $\mathbb{R}^n$.

For $\varepsilon \to 0$ we have $\xi \to 0$, hence $(\Delta(1/r^{n-2}), \varphi) = -((n-2)\pi^{n/2}/\Gamma(n/2))\varphi(0) = (-(n-2)(\pi^{n/2}/\Gamma(n/2))\delta(x), \varphi)$ giving the formula (1.339).

We note that this result is correct because the distribution $\Delta(1/r^{n-2})$ has as support the origin.

1.3.6
Dirac Representative Sequences: Criteria for the Representative Dirac Sequences

The Dirac delta distribution $\delta(x) \in \mathcal{D}'(\mathbb{R}^n)$ plays an important role in operational calculus, in the theory of electrical systems and the construction of fundamental solutions of linear differential operators with constant coefficients.

In many theoretical or practical problems (Fourier series, Fourier integral, elasticity problems, and so on) occur sequences of locally integrable functions that are convergent in the sense of convergence in the distribution space, having as a limit the Dirac delta distribution δ. Such sequences of the locally integrable functions are called the representative sequences. The functions which form the representative sequences δ are also called the impulsive functions.

We can say that any term of such a sequence represents a certain approximation of the Dirac delta distribution δ; this is very important from the practical point of view. Indeed, assuming that we want to obtain numerical values in a problem in which the results are expressed as distributions, we can substitute – for the calculations – a singular distribution by a term of a corresponding representative sequence. Thus, the obtained formulae can be used in calculations by the computer, obtaining a desired approximation, depending on the chosen term of the representative sequence.

Definition 1.29 Let $f_i : \mathbb{R}^n \to \mathbb{C}, i \in \mathbb{N}$, be a sequence of locally integrable functions. We say that $(f_i)_{i\geq 1}$ is a Dirac representative sequence if on the space $\mathcal{D}'(\mathbb{R}^n)$ we have $\lim_{i\to\infty} f_i(x) = \delta(x)$, that is,

$$\forall\varphi \in \mathcal{D}(\mathbb{R}^n) \Rightarrow \lim_{i\to\infty} (f_i(x), \varphi(x)) = (\delta(x), \varphi(x)) = \varphi(0) . \tag{1.355}$$

We will show that continuous functions and locally integrable functions with certain properties allow for the construction of the representative Dirac sequences. Thus, we mention the following proposition.

Proposition 1.18 Let there be $f \in C^0(\mathbb{R}^n)$, $f : \mathbb{R}^n \to \mathbb{C}$ with the property $\int_{\mathbb{R}^n} f(x)\mathrm{d}x = 1$; then the family of functions $f_\varepsilon, \varepsilon > 0$, having the expression

$$f_\varepsilon(x) = \frac{1}{\varepsilon^n} f\left(\frac{x}{\varepsilon}\right) , \quad x \in \mathbb{R}^n , \quad \varepsilon > 0 , \tag{1.356}$$

forms a representative Dirac family; hence

$$\lim_{\varepsilon\to+0} f_\varepsilon(x) = \delta(x) . \tag{1.357}$$

Proof: For any $\varphi \in \mathcal{D}(\mathbb{R}^n)$ we have

$$(f_\varepsilon(x), \varphi(x)) = \frac{1}{\varepsilon^n} \int_{\mathbb{R}^n} f\left(\frac{x}{\varepsilon}\right) \varphi(x)\mathrm{d}x , \quad \varepsilon > 0 . \tag{1.358}$$

Performing the change of variable $x = \varepsilon u$, $x_k = \varepsilon u_k$, $k = \overline{1, n}$, the Jacobian of the transformations is

$$J(u) = \frac{\partial(x_1, x_2, \ldots, x_n)}{\partial(u_1, u_2, \ldots, u_n)} = \begin{vmatrix} \frac{\partial x_1}{\partial u_1} & \frac{\partial x_1}{\partial u_2} & \cdots & \frac{\partial x_1}{\partial u_n} \\ \vdots & \vdots & \vdots & \vdots \\ \frac{\partial x_n}{\partial u_1} & \frac{\partial x_n}{\partial u_2} & \cdots & \frac{\partial x_n}{\partial u_n} \end{vmatrix}$$

$$= \begin{vmatrix} \varepsilon & 0 & \ldots & 0 & 0 \\ 0 & \varepsilon & \ldots & 0 & 0 \\ \vdots & \vdots & \vdots & \vdots & \vdots \\ 0 & 0 & \ldots & \varepsilon & 0 \\ 0 & 0 & \ldots & 0 & \varepsilon \end{vmatrix} = \varepsilon^n ; \tag{1.359}$$

thus we can write

$$(f_\varepsilon(x), \varphi(x)) = \int_{\mathbb{R}^n} f(u)\varphi(\varepsilon u)\mathrm{d}u . \tag{1.360}$$

Hence, it follows that

$$\lim_{\varepsilon\to 0}(f_\varepsilon(x), \varphi(x)) = \lim_{\varepsilon\to 0}\int\limits_{\mathbb{R}^n} f(u)\varphi(\varepsilon u)\mathrm{d}u$$

$$= \int\limits_{\mathbb{R}^n} f(u)\varphi(0)\mathrm{d}u = \varphi(0) = (\delta(x), \varphi(x)) . \tag{1.361}$$

□

Example 1.20 Let there be the function

$$f(x) = \frac{n}{S_1}\frac{1}{(\|x\|^2+1)^{(n+2)/2}} , \quad x \in \mathbb{R}^n , \quad n \geq 2 , \tag{1.362}$$

where S_1 represents the area of the unit radius sphere from $\mathbb{R}^n$. Obviously, f is a continuous function, hence $f \in C^0(\mathbb{R}^n)$.

Using the polar coordinates $(r, \theta_1, \theta_2, \ldots, \theta_{n-1}) \in \mathbb{R}^n$, whose connection with the Cartesian coordinates $(x_1, x_2, \ldots, x_n) \in \mathbb{R}^n$ is expressed by the relations

$$\begin{aligned}
x_1 &= r\sin\theta_1\sin\theta_2\ldots\sin\theta_{n-2}\sin\theta_{n-1}\\
x_2 &= r\sin\theta_1\sin\theta_2\ldots\sin\theta_{n-2}\cos\theta_{n-1}\\
x_3 &= r\sin\theta_1\sin\theta_2\ldots\sin\theta_{n-3}\cos\theta_{n-2}\\
&\vdots\\
x_{n-2} &= r\sin\theta_1\sin\theta_2\cos\theta_3\\
x_{n-1} &= r\sin\theta_1\cos\theta_2\\
x_n &= r\cos\theta_1 ,
\end{aligned} \tag{1.363}$$

where

$$r \geq 0 , \quad \theta_i \in [0, \pi] , \quad i = \overline{1, n-2} , \quad \theta_{n-1} \in [0, 2\pi) , \tag{1.364}$$

we have

$$\int\limits_{\mathbb{R}^n} f(x)\mathrm{d}x = \frac{n}{S_1}\int\limits_0^\infty \frac{r^{n-1}}{(r^2+1)^{(n+2)/2}}\left\{\int\limits_\Omega \mathrm{d}\Omega\right\}\mathrm{d}r = n\int\limits_0^\infty \frac{r^{n-1}}{(r^2+1)^{(n+2)/2}}\mathrm{d}r , \tag{1.365}$$

where $\mathrm{d}\Omega$ represents the area element of the unit radius sphere from $\mathbb{R}^n$, centered at the origin.

We observe that the Jacobian of the transformation (1.363) is

$$\begin{aligned}
J(r, \theta_1, \theta_2, \ldots, \theta_{n-1}) &= \frac{\partial(x_1, x_2, \ldots, x_n)}{\partial(r, \theta_1, \theta_2, \ldots, \theta_{n-1})}\\
&= r^{n-1}\sin^{n-2}\theta_1\sin^{n-3}\theta_2\ldots\sin\theta_{n-2} ,
\end{aligned} \tag{1.366}$$

so that $\mathrm{d}\Omega$ has the expression

$$\mathrm{d}\Omega = \frac{J}{r^{n-1}}\mathrm{d}\theta_1\mathrm{d}\theta_2\ldots\mathrm{d}\theta_{n-1}\,. \tag{1.367}$$

Regarding the volume element $\mathrm{d}x = \mathrm{d}x_1\ldots\mathrm{d}x_n$ in polar coordinates (1.363) it has the expression

$$\mathrm{d}x = \mathrm{d}v = J(r,\theta_1,\ldots,\theta_{n-1})\mathrm{d}r\mathrm{d}\theta_1\ldots\mathrm{d}\theta_{n-1} = r^{n-1}\mathrm{d}r\mathrm{d}\Omega\,. \tag{1.368}$$

Making the change of variable $t = 1/r^2$, $t \in \mathbb{R}$, we have

$$\begin{aligned} I &= \int_0^\infty \frac{r^{n-1}}{(r^2+1)^{(n+2)/2}}\mathrm{d}r = \int_0^\infty \frac{r^{n-1}}{r^{n+2}\left(1+(1/r^2)\right)^{(n+2)/2}}\mathrm{d}r \\ &= \int_0^\infty \frac{\mathrm{d}r}{r^3\left(1+(1/r^2)\right)^{(n+2)/2}} = -\int_\infty^0 \frac{\mathrm{d}t}{2(1+t)^{(n+2)/2}} \\ &= \frac{1}{2}\int_0^\infty (1+t)^{-(n+2)/2}\mathrm{d}t = \frac{1}{2}\cdot\frac{1}{(-n)/2}(1+t)^{-n/2}\Big|_0^\infty = \frac{1}{n}\,. \end{aligned} \tag{1.369}$$

Taking into account (1.365) we can write $\int_{\mathbb{R}^n} f(x)\mathrm{d}x = 1$ and as $f \in C^0(\mathbb{R}^n)$ follows that the two conditions of Proposition 1.18 are fulfilled. With this on the basis of (1.356) we have

$$\begin{aligned} f_\varepsilon(x) &= \frac{1}{\varepsilon^n}\cdot\frac{n}{S_1}f\left(\frac{\varepsilon}{n}\right) = \frac{1}{\varepsilon^n}\cdot\frac{n}{S_1}\cdot\varepsilon^{n+2}\frac{1}{\left(\|x\|^2+\varepsilon^2\right)^{(n+2)/2}} \\ &= \frac{n}{S_1}\frac{\varepsilon^2}{\left(\|x\|^2+\varepsilon^2\right)^{(n+2)/2}}\,, \qquad \varepsilon > 0\,, \end{aligned} \tag{1.370}$$

and thus $\lim_{\varepsilon\to+0} f_\varepsilon(x) = \delta(x)$, namely $f_\varepsilon \xrightarrow[\varepsilon\to+0]{\mathcal{D}'(\mathbb{R}^n)} \delta(x)$.

Thus, the family of functions $f_\varepsilon(x)$, $\varepsilon > 0$, $x \in \mathbb{R}^n$ is a representative Dirac family.

Particularly, for $n = 2$ we obtain

$$f_\varepsilon(x,y) = \frac{2}{2\pi}\frac{\varepsilon^2}{\left(x^2+y^2+\varepsilon^2\right)^2}\,, \qquad \varepsilon > 0\,, \tag{1.371}$$

hence

$$\lim_{\varepsilon\to+0}\frac{1}{\pi}\frac{\varepsilon^2}{\left(x^2+y^2+\varepsilon^2\right)^2} = \delta(x,y)\,. \tag{1.372}$$

Example 1.21 Let there be the function

$$f(x) = \frac{1}{\left(\sqrt{\pi}\right)^n}\mathrm{e}^{-\|x\|^2}\,, \qquad x \in \mathbb{R}^n\,. \tag{1.373}$$

Obviously, $f \in C^0(\mathbb{R}^n)$ and

$$\int_{\mathbb{R}^n} f(x)\mathrm{d}x = \frac{1}{(\sqrt{\pi})^n} \prod_{k=1}^{n} \int_{\mathbb{R}} \exp\left(-x_k^2\right) \mathrm{d}x_k = 1.$$

According to Proposition 1.18, we have

$$f_\varepsilon(x) = \frac{1}{\varepsilon^n} \frac{1}{(\sqrt{\pi})^n} \exp\left(-\frac{\|x\|^2}{\varepsilon^2}\right), \quad x \in \mathbb{R}^n, \tag{1.374}$$

and $\lim_{\varepsilon \to +0} f_\varepsilon(x) = \delta(x)$.

In the Fourier integral theory, the Dirichlet function is used in the form

$$\lim_{n \to \infty} \frac{\sin nx}{\pi x} = \delta(x).$$

Particularly, for $n = 1$ we obtain

$$\lim_{\varepsilon \to +0} \frac{1}{\varepsilon} \frac{1}{\sqrt{\pi}} \exp\left(\frac{-x^2}{\varepsilon^2}\right) = \delta(x). \tag{1.375}$$

Particular forms of Dirac sequences of one variable have been used in connection with Fourier integrals, heat propagation, wave theory of light, representation of concentrated loads, and so on.

Thus, G.R. Kirchhoff, formulating Huygens principle in the wave theory of light, mentions the function

$$f_n(x) = \sqrt{\frac{n}{2\pi}} \exp\left(-n\frac{x^2}{2}\right), \tag{1.376}$$

which is obtained from (1.375), thus $\lim_{n \to \infty} f_n(x) = \delta(x)$.

Lord Kelvin used this function to represent the point heat sources in the form

$$q_t(x) = \frac{1}{2\sqrt{\pi k t}} \exp\left(-\frac{x^2}{4kt}\right), \quad k > 0, \quad t > 0, \quad \lim_{t \to 0} q_t(x) = \delta(x). \tag{1.377}$$

We also mention the impulsive function of Stieltjes

$$f_n(x) = \frac{2}{\pi} \frac{n}{\cosh nx}, \quad n \in \mathbb{N} \tag{1.378}$$

and the Cauchy impulsive function

$$q_\varepsilon(x) = \frac{1}{\pi} \frac{\varepsilon}{x^2 + \varepsilon^2}, \quad \varepsilon > 0. \tag{1.379}$$

For these functions we have

$$\lim_{n \to \infty} f_n(x) = \delta(x), \quad \lim_{\varepsilon \to 0} q_\varepsilon(x) = \delta(x). \tag{1.380}$$

Example 1.22 Let there be the function

$$f : \mathbb{R}^2 \to \mathbb{R}\,, \quad f(x, y) = \frac{p-2}{2\pi} \cdot \frac{1}{(x^2 + y^2 + 1)^{p/2}}\,, \quad p > 2\,. \tag{1.381}$$

This function is obviously continuous, thus $f \in C^0(\mathbb{R}^2)$ and we obtain

$$\int_{\mathbb{R}^2} f(x, y)\mathrm{d}x\mathrm{d}y = 1\,. \tag{1.382}$$

Indeed, passing to polar coordinates $x = \rho\cos\theta$, $y = \rho\sin\theta$, $\theta \in [0, 2\pi]$, $\rho \geq 0$ we can write

$$\begin{aligned}\int_{\mathbb{R}^2} f(x, y)\mathrm{d}x\mathrm{d}y &= \frac{p-2}{2\pi}\int_{\mathbb{R}^2} \frac{\mathrm{d}x\mathrm{d}y}{(x^2 + y^2 + 1)^{p/2}} \\ &= \frac{p-2}{2\pi}\int_0^{2\pi} \mathrm{d}\theta \int_0^{\infty} \frac{\rho\mathrm{d}\rho}{(\rho^2+1)^{p/2}} = \frac{(p-2)}{2}\int_0^{\infty} (\rho^2+1)^{-p/2}(\rho^2+1)'\mathrm{d}\rho \\ &= \frac{p-2}{2}\left.\frac{(\rho^2+1)^{-(p/2)+1}}{1-p/2}\right|_0^{\infty} = \frac{p-2}{2}\cdot\frac{1}{-1+p/2} = 1\,. \end{aligned} \tag{1.383}$$

Thus, the conditions of Proposition 1.18 are fulfilled and we can build the function

$$\begin{aligned} f_\varepsilon(x, y) &= \frac{1}{\varepsilon^2} f\left(\frac{x}{\varepsilon}, \frac{y}{\varepsilon}\right) = \frac{p-2}{2\pi\cdot\varepsilon^2}\cdot\frac{\varepsilon^p}{(x^2+y^2+\varepsilon^2)^{p/2}} \\ &= \frac{p-2}{2\pi}\frac{\varepsilon^{p-2}}{(x^2+y^2+\varepsilon^2)^{p/2}}\,, \quad \varepsilon > 0\,, \quad p > 2\,. \end{aligned} \tag{1.384}$$

Consequently, the relation follows

$$\lim_{\varepsilon\to+0}\frac{p-2}{2\pi}\frac{\varepsilon^{p-2}}{(x^2+y^2+\varepsilon^2)^{p/2}} = \delta(x, y)\,, \quad p > 2\,. \tag{1.385}$$

We note that the family of Dirac representative functions (1.384) plays an important role in the construction of the fundamental solution of the elastic half-space problem [19].

Particularly, for $n = 3$ and $n = 5$ we obtain

$$\lim_{\varepsilon\to+0}\frac{\varepsilon}{2\pi}\cdot\frac{1}{(x^2+y^2+\varepsilon^2)^{3/2}} = \lim_{\varepsilon\to+0}\frac{3\varepsilon^3}{2\pi}\cdot\frac{1}{(x^2+y^2+\varepsilon^2)^{5/2}} = \delta(x, y)\,. \tag{1.386}$$

Also, taking into account improper integrals values

$$\int_{-\infty}^{\infty}\frac{\sin x}{x}\mathrm{d}x = 2\int_0^{\infty}\frac{\sin x}{x}\mathrm{d}x = \pi\,, \quad \int_{-\infty}^{\infty}\frac{\sin^2 x}{x^2}\mathrm{d}x = 2\int_0^{\infty}\frac{\sin^2 x}{x^2}\mathrm{d}x = \pi\,, \tag{1.387}$$

based on Proposition 1.18, we get

$$1. \quad \lim_{\varepsilon\to+0}\frac{1}{\pi x}\sin\frac{x}{\varepsilon}=\delta(x)\,, \qquad 2. \quad \lim_{\varepsilon\to+0}\frac{\varepsilon}{\pi x^2}\sin^2\frac{x}{\varepsilon}=\delta(x)\,. \quad (1.388)$$

Thus, for 1., the continuous function $f(x) = (\sin x)/(\pi x)$, $x \in \mathbb{R}$ is considered, resulting in $f(x/\varepsilon)/\varepsilon = (\sin /\varepsilon x)/(\pi x)$.

For 2., the continuous function $g(x) = \sin^2 x/(\pi x^2)$, $x \in \mathbb{R}$ is considered. It follows

$$g_\varepsilon(x)=\frac{1}{\varepsilon}g\left(\frac{x}{\varepsilon}\right)=\frac{\varepsilon}{\pi x^2}\sin^2\frac{x}{\varepsilon}\,, \qquad \varepsilon>0\,.$$

Another criterion for the Dirac representative sequences is given by the following.

Proposition 1.19 Let there be $f_\varepsilon \in L^1_{\text{loc}}(\mathbb{R}^n)$, $\varepsilon > 0$, a family of locally integrable functions with the properties:

1. $f_\varepsilon(x) \ge 0\,, \quad \forall x \in \mathbb{R}^n, \forall \varepsilon > 0\,,$
2. $\int_{\mathbb{R}^n} f_\varepsilon(x)\mathrm{d}x = 1\,,$
3. $\forall R > 0$ we have $\lim_{\varepsilon\to+0}\int_{\|x\|\ge R} f_\varepsilon(x)\mathrm{d}x = 0\,;$

then

$$\lim_{\varepsilon\to+0} f_\varepsilon(x)=\delta(x) \Leftrightarrow f_\varepsilon(x)\xrightarrow[\varepsilon\to+0]{\mathcal{D}'(\mathbb{R}^n)}\delta(x)\,. \tag{1.389}$$

Proof: Taking into account 1 and 2 for $\forall\varphi \in \mathcal{D}(\mathbb{R}^n)$ we have

$$\begin{aligned}|(f_\varepsilon,\varphi)-\varphi(0)| &= \left|\int_{\mathbb{R}^n} f_\varepsilon(x)\varphi(x)\mathrm{d}x-\varphi(0)\int_{\mathbb{R}^n} f_\varepsilon \mathrm{d}x\right| \\ &= \left|\int_{\mathbb{R}^n} f_\varepsilon(x)[\varphi(x)-\varphi(0)]\mathrm{d}x\right| \le \int_{\mathbb{R}^n}|\varphi(x)-\varphi(0)|\,f_\varepsilon(x)\mathrm{d}x\,.\end{aligned} \tag{1.390}$$

Based on the continuity of the function $\varphi(x) \in \mathcal{D}(\mathbb{R}^n)$ in the origin, we can write $|\varphi(x) - \varphi(0)| < \varepsilon_0/2$, $\|x\| < \eta_{\varepsilon_0}$.

Consequently, for the integral on the right-hand side of the relation (1.390) we obtain

$$\begin{aligned}\int_{\mathbb{R}^n}|\varphi(x)-\varphi(0)|\,f_\varepsilon(x)\mathrm{d}x &= \int_{\|x\|<\eta_{\varepsilon_0}}|\varphi(x)-\varphi(0)|\,f_\varepsilon(x)\mathrm{d}x \\ &+ \int_{\|x\|\ge\eta_{\varepsilon_0}}|\varphi(x)-\varphi(0)|\,f_\varepsilon(x)\mathrm{d}x \le \frac{\varepsilon_0}{2}+\int_{\|x\|\ge\eta_{\varepsilon_0}}|\varphi(x)-\varphi(0)|\,f_\varepsilon(x)\mathrm{d}x\,.\end{aligned} \tag{1.391}$$

But

$$\int_{\|x\|\geq\eta_{\varepsilon_0}} |\varphi(x)-\varphi(0)|\, f_\varepsilon(x)\mathrm{d}x$$

$$\leq \sup_{x\in\mathbb{R}^n} |\varphi(x)-\varphi(0)| \int_{\|x\|\geq\eta_{\varepsilon_0}} f_\varepsilon(x)\mathrm{d}x = M^* \int_{\|x\|\geq\eta_{\varepsilon_0}} f_\varepsilon(x)\mathrm{d}x\,, \tag{1.392}$$

where $M^* = \sup_{x\in\mathbb{R}^n} |\varphi(x)-\varphi(0)| > 0$ (which exists because it is continuous and has compact support).

Substituting (1.392) in (1.391), we obtain

$$\int_{\mathbb{R}^n} |\varphi(x)-\varphi(0)|\, f_\varepsilon(x)\mathrm{d}x < \frac{\varepsilon_0}{2} + M^* \int_{\|x\|\geq\eta_{\varepsilon_0}} f_\varepsilon(x)\mathrm{d}x\,. \tag{1.393}$$

On the other hand, condition 3 of the proposition means the following: $\forall\,\varepsilon_0 > 0$, $\exists M_{\varepsilon_0} > 0$ so that for $\varepsilon \leq M_{\varepsilon_0}$ we have

$$\int_{\|x\|\geq\mathbb{R}} f_\varepsilon(x)\mathrm{d}x < \frac{\varepsilon_0}{2M^*}\,. \tag{1.394}$$

The relation (1.393) becomes

$$\int_{\mathbb{R}^n} |\varphi(x)-\varphi(0)|\, f_\varepsilon(x)\mathrm{d}x < \frac{\varepsilon_0}{2} + \frac{\varepsilon_0}{2} = \varepsilon_0\,, \quad \|x\| < \eta_{\varepsilon_0}\,, \quad \text{and } 0 < \varepsilon \leq M_{\varepsilon_0}\,. \tag{1.395}$$

Taking into account (1.395) and (1.390), it follows that $\forall\,\varepsilon_0 > 0, \exists\eta_{\varepsilon_0} > 0$ and $M_{\varepsilon_0} > 0$ so that

$$|(f_\varepsilon,\varphi)-\varphi(0)| < \varepsilon_0\,, \quad \text{for } \|x\| < \eta_{\varepsilon_0}\,, \quad \text{and } 0 < \varepsilon \leq M_{\varepsilon_0}\,. \tag{1.396}$$

The last relation is equivalent to $\lim_{\varepsilon\to+0} |(f_\varepsilon,\varphi)-\varphi(0)| = 0$, thus $\lim_{\varepsilon\to+0} (f_\varepsilon,\varphi) = \varphi(0) = (\delta(x),\varphi(x))$, namely $\lim_{\varepsilon\to+0} f_\varepsilon(x) = \delta(x)$, and thus the proposition is proved. □

We consider the family of functions (1.384), namely

$$f_\varepsilon(x,y) = \frac{p-2}{2\pi}\cdot\frac{\varepsilon^{p-2}}{(x^2+y^2+\varepsilon^2)^{p/2}}\,, \quad \varepsilon > 0, p > 2\,. \tag{1.397}$$

We note that $f_\varepsilon(x,y) \geq 0$ and, according to (1.383), we have $\int_{\mathbb{R}^2} f_\varepsilon(x,y)\mathrm{d}x\mathrm{d}y = 1$, so that the conditions 1. and 2. of Proposition 1.19 are satisfied. We will show that condition 3. is satisfied, namely $\forall\,R > 0$ we have

$$\lim_{\varepsilon\to+0} \iint_{r=\sqrt{x^2+y^2}\geq R} f_\varepsilon(x,y)\mathrm{d}x\mathrm{d}y = 0\,. \tag{1.398}$$

Indeed, we have

$$\iint\limits_{r=\sqrt{x^2+y^2}\geq R} f_\varepsilon(x,y)\mathrm{d}x\mathrm{d}y = \lim_{r_1\to\infty} \iint\limits_{r_1\geq r\geq R} f_\varepsilon(x,y)\mathrm{d}x\mathrm{d}y$$

$$= \frac{p-2}{2\pi}\cdot\varepsilon^{p-2}\lim_{r_1\to\infty}\iint\limits_{R\leq r\leq r_1}\frac{\mathrm{d}x\mathrm{d}y}{(x^2+y^2+\varepsilon^2)^{p/2}}\,. \tag{1.399}$$

Using the polar coordinates $x = \rho\cos\theta$, $y = \rho\sin\theta$, we obtain

$$\iint\limits_{R<r\leq r_1}\frac{\mathrm{d}x\mathrm{d}y}{(x^2+y^2+\varepsilon^2)^{p/2}} = \int_R^{r_1}\int_0^{2\pi}\frac{\rho\mathrm{d}\rho\mathrm{d}\theta}{(\rho^2+\varepsilon^2)^{p/2}}$$

$$= \pi\int_R^{r_1}(\rho^2+\varepsilon^2)^{-p/2}(\rho^2+\varepsilon^2)'\mathrm{d}\rho = \frac{2\pi}{2-p}\left[(r_1^2+\varepsilon^2)^{(2-p)/2}-(R^2+\varepsilon^2)^{(2-p)/2}\right]. \tag{1.400}$$

Substituting in (1.399) we have

$$\iint\limits_{r=\sqrt{x^2+y^2}\geq R} f_\varepsilon(x,y)\mathrm{d}x\mathrm{d}y = \frac{p-2}{2\pi}\varepsilon^{p-2}\cdot\frac{2\pi}{2-p}\left[-(R^2+\varepsilon^2)^{(2-p)/2}\right]$$

$$= \varepsilon^{p-2}(R^2+\varepsilon^2)^{(2-p)/2} < \varepsilon^{p-2}(R^2)^{(2-p)/2} = \left(\frac{\varepsilon}{R}\right)^{p-2},\quad p>2\,, \tag{1.401}$$

giving $\lim_{\varepsilon\to+0}\iint_{r\geq R} f_\varepsilon(x,y)\mathrm{d}x\mathrm{d}y \leq 0$, hence $\lim_{\varepsilon\to+0}\iint_{r\geq R} f_\varepsilon(x,y)\mathrm{d}x\mathrm{d}y = 0$ because $f_\varepsilon \geq 0$.

We showed that the conditions of Proposition 1.19 are fulfilled, hence $\lim_{\varepsilon\to+0} f_\varepsilon(x,y) = \delta(x,y)$, a result that was obtained using Proposition 1.18.

Example 1.23 Let there be the sequence $(f_n(x))_{n\geq1}$ (Figure 1.6) where

$$f_n(x) = \begin{cases} n - n^2x/2\,, & x\in[0,2/n]\,,\\ 0\,, & x\notin[0,2/n]\,.\end{cases} \tag{1.402}$$

We note that the three conditions of Proposition 1.19 are satisfied.

Indeed, $f_n(x)\geq 0$, $\int_{\mathbb{R}} f_n(x)\mathrm{d}x = 1$ and $\lim_{n\to\infty}\int_{|x|\geq r>0} f_n(x)\mathrm{d}x = 0$, $\forall r > 0$, because

$$\int\limits_{|x|\geq r>0} f_n(x)\mathrm{d}x = \int_{-\infty}^{r} f_n(x)\mathrm{d}x + \int_r^{\infty} f_n(x)\mathrm{d}x = \int_r^{\infty} f_n(x)\mathrm{d}x\,, \tag{1.403}$$

and for $2/n < r$ it results $\int_r^\infty f_n(x)\mathrm{d}x = 0$, hence $\lim_{n\to\infty}\int_{|x|\geq r>0} f_n(x)\mathrm{d}x = 0$.

Consequently, we have $\lim_{n\to\infty} f_n(x) = \delta(x)$.

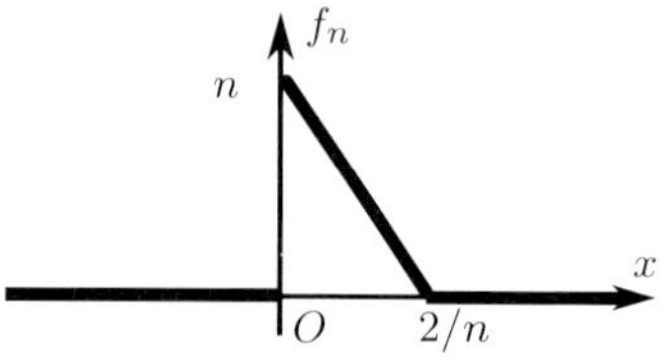

Figure 1.6

Example 1.24 Let there be $f_\varepsilon \in L^1_{\text{loc}}(\mathbb{R})$, $\varepsilon > 0$ a family of locally integrable functions, where $f_\varepsilon(x) = (2H(x)/\pi) \cdot (\varepsilon/(x^2 + \varepsilon^2))$, $\varepsilon > 0$ (Figure 1.7), and H is the Heaviside function.

We shall show that $f_\varepsilon \xrightarrow{\mathcal{D}'(\mathbb{R})}_{\varepsilon \to +0} \delta(x)$.

Indeed, we note that the first two conditions of Proposition 1.19 are fulfilled because:

1. $$f_\varepsilon \geq 0\,, \quad \forall \varepsilon > 0\,, \quad \forall x \in \mathbb{R}\,, \quad f_\varepsilon \in L^1_{\text{loc}}(\mathbb{R})\,;$$

2. $$\int_{\mathbb{R}} f_\varepsilon(x)\mathrm{d}x = \int_0^\infty \frac{2\varepsilon}{\pi(x^2 + \varepsilon^2)}\mathrm{d}x = \frac{2\varepsilon}{\pi} \cdot \frac{1}{\varepsilon} \arctan \left.\frac{x}{\varepsilon}\right|_0^\infty = \frac{2}{\pi} \cdot \frac{\pi}{2} = 1\,.$$

Regarding the third condition, we have

$$\lim_{\substack{\varepsilon \to +0 \\ \forall R > 0}} \int_{|x| \geq R} f_\varepsilon(x)\mathrm{d}x = \lim_{\varepsilon \to +0} \int_R^\infty \frac{2\varepsilon}{\pi(x^2 + \varepsilon^2)}\mathrm{d}x$$
$$= \lim_{\varepsilon \to +0} \frac{2\varepsilon}{\pi} \cdot \frac{1}{\varepsilon} \arctan \left.\frac{x}{\varepsilon}\right|_R^\infty = \lim_{\varepsilon \to +0} \frac{2}{\pi}\left(\frac{\pi}{2} - \arctan \frac{R}{\varepsilon}\right) = 0\,. \qquad (1.404)$$

Thus, the conditions of Proposition 1.19 being fulfilled, we have $\lim_{\varepsilon \to +0} f_\varepsilon(x) = \delta(x)$.

Example 1.25 Let there be the sequence of locally integrable functions $f_n(x) = nH(x)\mathrm{e}^{-nx}$, $x \in \mathbb{R}$ (Figure 1.8). We shall show that $\lim_{n \to +\infty} f_n(x) = \delta(x)$.

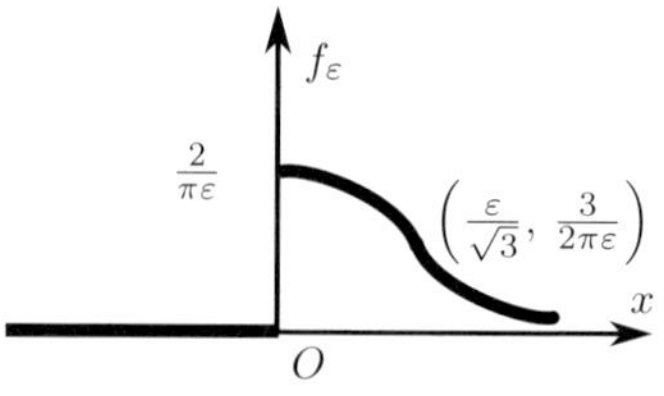

Figure 1.7

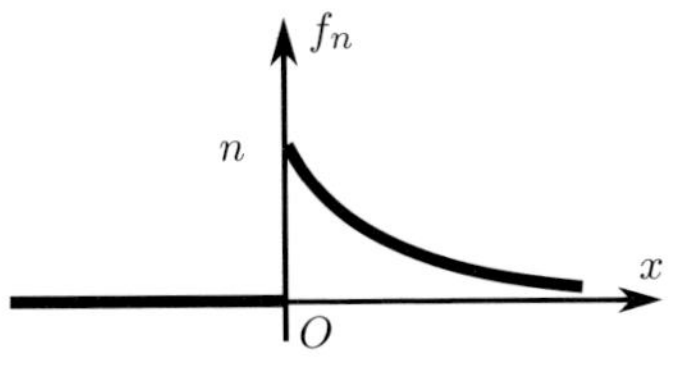

Figure 1.8

We will use Proposition 1.19, because the function f_n is discontinuous at the origin. We note that $f_n(x) \geq 0$, $\forall n \in \mathbb{N}, \forall x \in \mathbb{R}$. Also

$$\int_{\mathbb{R}} f_n(x)\mathrm{d}x = \int_0^\infty n\mathrm{e}^{-nx}\mathrm{d}x = \left.\frac{n\mathrm{e}^{-nx}}{-n}\right|_0^\infty = -\mathrm{e}^{-nx}|_0^\infty = 1\,. \tag{1.405}$$

Thus, the first two conditions of the proposition are met. We will check the third condition, namely $\lim_{\substack{n\to\infty \\ \forall R>0}} \int_{|x|\geq R} f_n(x)\mathrm{d}x = 0$. We can write

$$\lim_{n\to\infty} \int_{|x|\geq R} f_n(x)\mathrm{d}x = \lim_{n\to\infty} \int_R^\infty n\mathrm{e}^{-nx}\mathrm{d}x = \lim_{n\to\infty} \left.\frac{n\mathrm{e}^{-nx}}{-n}\right|_R^\infty$$
$$= \lim_{n\to\infty} [-\mathrm{e}^{-nx}]|_R^\infty = \lim_{n\to\infty} \mathrm{e}^{-nR} = 0\,. \tag{1.406}$$

Consequently, the conditions of the Proposition 1.19 are fulfilled, so we have $\lim_{n\to\infty} f_n(x) = \delta(x)$.

Example 1.26 We shall show that the sequence $(g_n(x))_{n\geq 1}$, $g_n(x) = xn^2 H(x)\mathrm{e}^{-nx}$, $x \in \mathbb{R}$ is a Dirac representative sequence, hence $\lim_{n\to\infty} g_n(x) = \delta(x)$.

Indeed, $f_n(x) \geq 0$, $\forall x \in \mathbb{R}$, and we have

$$\int_{\mathbb{R}} f_n(x)\mathrm{d}x = n^2 \int_0^\infty x\mathrm{e}^{-nx}\mathrm{d}x = \frac{n^2}{-n} \int_0^\infty x(\mathrm{e}^{-nx})'\mathrm{d}x$$
$$= -n\left[x\mathrm{e}^{-nx}|_0^\infty - \int_0^\infty \mathrm{e}^{-nx}\mathrm{d}x\right] = n\int_0^\infty \mathrm{e}^{-nx}\mathrm{d}x = \left.\frac{n\mathrm{e}^{-nx}}{-n}\right|_0^\infty = 1\,. \tag{1.407}$$

As regards the third condition of the Proposition 1.19 we have

$$\lim_{\substack{n\to\infty \\ \forall R>0}} \int_{|x|\geq R} g_n(x)\mathrm{d}x = \lim_{n\to\infty} n^2 \int_R^\infty x\mathrm{e}^{-nx}\mathrm{d}x$$
$$= \lim_{n\to\infty} (-n)\left[x\mathrm{e}^{-nx}|_R^\infty - \int_R^\infty \mathrm{e}^{-nx}\mathrm{d}x\right]$$
$$= \lim_{n\to\infty} (-n)\left[-R\exp(-Rn) - \left.\frac{\exp(-nx)}{-n}\right|_R^\infty\right]$$
$$= \lim_{n\to\infty} \left[\mathrm{e}^{-Rn} + Rn\mathrm{e}^{-Rn}\right] = 0\,. \tag{1.408}$$

With this, the conditions of the Proposition 1.19 are fulfilled, so we have $\lim_{n\to\infty} g_n(x) = \delta(x)$.

Below, we will give another criterion for Dirac representative sequences that complement the Proposition 1.19.

Proposition 1.20 Let there be the sequence of functions $(f_i(x))_{i\in\mathbb{N}} \subset \mathcal{D}(\mathbb{R}^n)$, which satisfies the conditions:

1. $$f_i(x) \geq 0\,, \quad \forall i \in \mathbb{N}\,, \quad \forall x \in \mathbb{R}^n\,;$$

2. $$\int_{\mathbb{R}^n} f_i(x)\mathrm{d}x = 1\,, \quad \forall i \in \mathbb{N}\,;$$

3. $$\forall i \in \mathbb{N}, \operatorname{supp}(f_i) \subset \overline{U}_{\varepsilon_i} = \{x | x \in \mathbb{R}^n, \|x\| \leq \varepsilon_i\} \text{ and } \lim_{i\to\infty} \varepsilon_i = 0\,.$$

Then $\lim_{i\to\infty} f_i(x) = \delta(x)$.

Proof: For any $\varphi \in \mathcal{D}(\mathbb{R}^n)$ we have

$$|(f_i, \varphi) - \varphi(0)| \leq \int_{\overline{U}_{\varepsilon_i}} f_i(x)|\varphi(x) - \varphi(0)|\mathrm{d}x \leq \sup_{\|x\|\leq\varepsilon_i} |\varphi(x) - \varphi(0)|\,. \quad (1.409)$$

On the basis of the uniform continuity of the function $\varphi \in \mathcal{D}(\mathbb{R}^n)$ and because if $i \to \infty$, then $\varepsilon_i \to 0$, we have $\sup_{\|x\|\leq\varepsilon_i} |\varphi(x) - \varphi(0)| \to 0$, hence $\lim_i (f_i, \varphi) = \varphi(0) = (\delta(x), \varphi(x))$, namely $\lim_i f_i(x) = \delta(x)$. □

Example 1.27 Let there be the family of sequences $\rho_\varepsilon(x)$, $\varepsilon > 0$, $x \in \mathbb{R}^n$, namely

$$\rho_\varepsilon(x) = \begin{cases} c_\varepsilon \exp\left(-\dfrac{\varepsilon^2}{\varepsilon^2 - \|x\|^2}\right), & \|x\| < \varepsilon\,, \\ 0, & \|x\| \geq \varepsilon\,, \end{cases} \quad (1.410)$$

where the constant c_ε has the expression

$$c_\varepsilon = \varepsilon^{-n}\left[\int_{\|x\|\leq 1} \exp\left(-\frac{1}{1 - \|x\|^2}\right)\mathrm{d}x\right]^{-1}.$$

We observe that the function $\rho_\varepsilon \geq 0$, $\varepsilon > 0$, is a test function of Schwartz's space, hence $\rho_\varepsilon \in \mathcal{D}(\mathbb{R}^n)$. The support of the function ρ_ε is the closed ball $\overline{B}_\varepsilon = \{x | x \in \mathbb{R}^n, \|x\| \leq \varepsilon\}$, hence $\operatorname{supp}(\rho_\varepsilon) = \overline{B}_\varepsilon$.

Due to the value chosen for the constant c_ε, the function $\rho_\varepsilon(x)$ has the property $\int_{\mathbb{R}^n} \rho_\varepsilon(x)\mathrm{d}x = 1$. Consequently, the family of functions ρ_ε, $\varepsilon > 0$, has the properties $\rho_\varepsilon \geq 0$, $\varepsilon > 0$, $\int_{\mathbb{R}^n} \rho_\varepsilon(x)\mathrm{d}x = 1$ and $\operatorname{supp}(\rho_\varepsilon) \subset \overline{B}_\varepsilon$, hence $\varepsilon \to +0$ involves $\operatorname{supp}(\rho_\varepsilon) \to 0$.

Thus, the conditions of the Proposition 1.20 are fulfilled and we can write $\lim_{\varepsilon\to+0}\rho_\varepsilon(x) = \delta(x)$; namely $\rho_\varepsilon, \varepsilon > 0$, is a family of Dirac representative sequences, but from the space of test functions $\mathcal{D}(\mathbb{R}^n)$.

Example 1.28 Let there be the family of sequences of polynomials $L_\varepsilon(x)$, $\varepsilon > 0$, $x \in \mathbb{R}^n$, where

$$L_\varepsilon(x) = \begin{cases} \dfrac{1}{c_\varepsilon}\left(\varepsilon^2 - \|x\|^2\right)^p , & \|x\| \le \varepsilon , \quad p \in \mathbb{N} \text{ fixed} , \\ 0 , & \|x\| > \varepsilon , \end{cases} \tag{1.411}$$

and $c_\varepsilon = \int_{\|x\|\le\varepsilon} (\varepsilon^2 - \|x\|^2)^p \mathrm{d}x$.

We shall show that $\lim_{\varepsilon\to+0} L_\varepsilon(x) = \delta(x)$.

Indeed, $L_\varepsilon \ge 0, \forall\varepsilon > 0, \int_{\mathbb{R}^n} L_\varepsilon(x)\mathrm{d}x = 1$, $\operatorname{supp}(L_\varepsilon(x)) = \overline{U}_\varepsilon = \{x | x \in\mathbb{R}^n, \|x\| \le \varepsilon\}$ and $\lim_{\varepsilon\to+0}\operatorname{supp}(L_\varepsilon(x)) = 0$.

The conditions of the proposition are fulfilled, hence $\lim_{\varepsilon\to+0} L_\varepsilon(x) = \delta(x)$.

We note that $L_\varepsilon(x)$ are polynomials of degree $2p$, but with compact support, which tends to zero when $\varepsilon \to 0$ [20, p. 51].

1.3.7
Distributions Depending on a Parameter

1.3.7.1 Differentiation of Distributions Depending on a Parameter

The different quantities encountered in the mathematical-physical problems are generally functions of space variable $x \in \mathbb{R}^n$, but they may also depend on certain parameters of real or complex variable, such as the temporal variable $t \in I \subset \mathbb{R}$.

This requires considerations on distributions depending on a real or complex parameter $t \in \Omega \subset \mathbb{C}^m$.

In the following, we consider the real parameter t, hence $t = (t_1, t_2, \ldots, t_m) \in \Omega \subset \mathbb{R}^m$. For example, the Dirac delta distribution $\delta(x - t) = \delta(x_1 - t_1, x_2 - t_2, x_3 - t_3) \in \mathcal{D}'(\mathbb{R}^3)$ depends on the real parameter $t = (t_1, t_2, t_3) \in \mathbb{R}^3$.

If for any $t \in \Omega \subset \mathbb{R}^m$ we can associate, after a certain rule, a single distribution $f_t(x) \in \mathcal{D}'(\mathbb{R}^n)$, we say that this distribution depends on the real parameter $t \in \Omega \subset \mathbb{R}^m$.

Definition 1.30 We say that the distribution $f \in \mathcal{D}'(\mathbb{R}^n)$ is the limit of the distribution $f_t \in \mathcal{D}'(\mathbb{R}^n), t \in \Omega \subset \mathbb{R}^m$, when $t \to t_0$, t_0 being the accumulation point for $\Omega \subset \mathbb{R}^m$, and we write $\lim_{t\to t_0} f_t(x) = f(x)$, if $\forall\varphi \in \mathcal{D}(\mathbb{R}^n)$ the function $(f_t(x), \varphi(x)), t \in \Omega \subset \mathbb{R}^m$, converges to $(f(x), \varphi(x))$, that is, we have

$$\lim_{t\to t_0} (f_t(x), \varphi(x)) = (f(x), \varphi(x)) . \tag{1.412}$$

Proposition 1.21 Let there be the distributions $f_t, g_t \in \mathcal{D}'(\mathbb{R}^n)$ depending on the parameter $t \in \Omega \subset \mathbb{R}^m$. If $\lim_{t\to t_0} f_t(x), \lim_{t\to t_0} g_t(x) \in \mathcal{D}'(\mathbb{R}^n)$ exist, then we have

$$\lim_{t\to t_0} (\alpha f_t + \beta g_t) = \alpha \lim_{t\to t_0} f_t + \beta \lim_{t\to t_0} g_t , \quad \forall\alpha, \beta \in \mathbb{R} . \tag{1.413}$$

Indeed, noting $\lim_{t\to t_0} f_t(x) = a(x), \lim_{t\to t_0} g_t(x) = b(x), \forall\varphi \in \mathcal{D}(\mathbb{R}^n)$, we can write

$$\begin{aligned}
&\lim_{t\to t_0} (\alpha f_t(x) + \beta g_t(x), \varphi(x)) = \lim_{t\to t_0} [\alpha(f_t, \varphi) + \beta(g_t, \varphi)] \\
&= \alpha \lim_{t\to t_0} (f_t, \varphi) + \beta \lim_{t\to t_0} (g_t, \varphi) \\
&= \alpha(a(x), \varphi) + \beta(b(x), \varphi) = (\alpha a(x) + \beta b(x), \varphi) , \qquad (1.414)
\end{aligned}$$

which leads to (1.413).

Definition 1.31 The distribution $f_t \in \mathcal{D}'(\mathbb{R}^n), t \in \Omega \subset \mathbb{R}^m$, is continuous with respect to the parameter t on the set $\Omega \subset \mathbb{R}^m$ if $\forall t_0 \in \Omega$, we have $\lim_{t\to t_0} f_t(x) = f_{t_0}(x)$, hence

$$\lim_{t\to t_0} (f_t(x), \varphi(x)) = (f_{t_0}(x), \varphi(x)), \forall\varphi \in \mathcal{D}(\mathbb{R}^n) . \qquad (1.415)$$

From the definition of continuity and the limit property with respect to the parameter $t \in \Omega \subset \mathbb{R}^m$, it follows that $\forall\alpha, \beta \in \mathbb{R}$ the distribution $\alpha f_t(x) + \beta g_t(x)$ is continuous on Ω if the distributions $f_t(x), g_t(x) \in \mathcal{D}'(\mathbb{R}^n)$ are continuous on Ω.

Definition 1.32 Let $f_t \in \mathcal{D}'(\mathbb{R}^n)$ be a distribution depending on the parameter $t \in \Omega \subset \mathbb{R}^m$. We call derivative of the distribution f_t with respect to $t_j \in \mathbb{R}$, $j = \overline{1, m}, t = (t_1, t_2, \ldots, t_m)$, the distribution $\partial f_t(x)/\partial t_j \in \mathcal{D}'(\mathbb{R}^n)$, defined by

$$\frac{\partial}{\partial t_j} f_t(x) = \lim_{\Delta t_j \to 0} \frac{f_{(t_1,t_2,\ldots,t_j+\Delta t_j,\ldots,t_m)}(x) - f_{(t_1,t_2,\ldots,t_j,\ldots,t_m)}(x)}{\Delta t_j} , \qquad (1.416)$$

if the limit exists and is unique.

This means that $\forall\varphi \in \mathcal{D}(\mathbb{R}^n)$ we have

$$\begin{aligned}
&\left(\frac{\partial}{\partial t_j} f_t(x), \varphi(x)\right) \\
&= \lim_{\Delta t_j \to 0} \frac{(f_{(t_1,\ldots,t_j+\Delta t_j,\ldots,t_m)}(x), \varphi(x)) - (f_{(t_1,\ldots,t_j,\ldots,t_m)}(x), \varphi(x))}{\Delta t_j} \\
&= \frac{\partial}{\partial t_j} \psi(t_1, \ldots, t_j, \ldots, t_m) = \frac{\partial}{\partial t_j} (f_t(x), \varphi(x)) , \qquad (1.417)
\end{aligned}$$

where

$$\psi(t_1, \ldots, t_j, \ldots, t_m) = (f_t(x), \varphi(x)) , \quad t \in \Omega \subset \mathbb{R}^m . \qquad (1.418)$$

Proposition 1.22 The necessary and sufficient condition that the derivative $\partial f_t(x)/\partial t_j \in \mathcal{D}'(\mathbb{R}^n)$ does exist is that the function $\psi(t) = (f_t(x), \varphi(x))$ be differentiable with respect to the variable t_j, $j = \overline{1, m}$.

We note that the existence of the limit (1.416) implies the existence of the limit

$$\lim_{\Delta t_j \to 0} \left(\frac{f_{(t_1,\ldots,t_j+\Delta t_j,\ldots,t_m)}(x) - f_{(t_1,\ldots t_j,\ldots,t_m)}(x)}{\Delta t_j}, \varphi(x) \right) \tag{1.419}$$

and, on the basis of the completeness theorem of the distribution space $\mathcal{D}'(\mathbb{R}^n)$, it defines a distribution depending on the parameter $t \in \Omega \subset \mathbb{R}^m$.

Consequently, $\partial f_t(x)/\partial t_j$ is a distribution from $\mathcal{D}'(\mathbb{R}^n)$ depending on the parameter $t \in \Omega$.

Proposition 1.23 If the derivative $\partial f_t(x)/\partial t_j \in \mathcal{D}'(\mathbb{R}^n)$, $t \in \Omega \subset \mathbb{R}^m$, then $(\partial/\partial t_j)(\partial f_t(x)/\partial x_i)$ exists and the following formula occurs

$$\frac{\partial}{\partial t_j}\left(\frac{\partial f_t(x)}{\partial x_i}\right) = \frac{\partial}{\partial x_i}\left(\frac{\partial}{\partial t_j} f_t(x)\right), \quad i = \overline{1,n} . \tag{1.420}$$

Proof: For any $\varphi \in \mathcal{D}(\mathbb{R}^n)$, we have

$$\left(\frac{\partial f_t(x)}{\partial x_i}, \varphi(x)\right) = \left(f_t(x), -\frac{\partial \varphi(x)}{\partial x_i}\right), \quad t \in \Omega \subset \mathbb{R}^m . \tag{1.421}$$

From the existence of the derivative $\partial f_t(x)/\partial t_j$ and taking into account (1.417) it results that the function defined by (1.418) is differentiable with respect to t_j; hence we get

$$\begin{aligned}\frac{\partial}{\partial t_j}\left(f_t(x), -\frac{\partial \varphi}{\partial x_i}\right) &= \left(\frac{\partial f_t(x)}{\partial t_j}, -\frac{\partial \varphi}{\partial x_i}\right) = \left(\frac{\partial}{\partial x_i}\frac{\partial f_t(x)}{\partial t_j}, \varphi(x)\right) \\ &= \frac{\partial}{\partial t_j}\left(\frac{\partial f_t}{\partial x_i}, \varphi\right) = \left(\frac{\partial}{\partial t_j}\frac{\partial f_t}{\partial x_i}, \varphi\right),\end{aligned} \tag{1.422}$$

wherefrom we obtain the relation (1.420). □

Proposition 1.24 Let there be the distribution $f_t \in \mathcal{D}'(\mathbb{R}^n), t \in \Omega \subset \mathbb{R}^m$. If $\lim_{t\to t_0} f_t(x)$ exists, then

$$\lim_{t\to t_0}\left(\frac{\partial}{\partial x_i} f_t(x)\right) = \frac{\partial}{\partial x_i}\left[\lim_{t\to t_0} f_t(x)\right]. \tag{1.423}$$

Indeed, if we note $\lim_{t\to t_0} f_t(x) = a(x) \in \mathcal{D}'(\mathbb{R}^n)$, then $\forall \varphi \in \mathcal{D}(\mathbb{R}^n)$ we have $\lim_{t\to t_0}(f_t(x), \varphi(x)) = (a(x), \varphi(x))$, and, consequently, we obtain

$$\begin{aligned}\lim_{t\to t_0}\left(\frac{\partial}{\partial x_i} f_t(x), \varphi(x)\right) &= \lim_{t\to t_0}\left(f_t(x), -\frac{\partial \varphi(x)}{\partial x_i}\right) \\ &= \left(a(x), -\frac{\partial \varphi(x)}{\partial x_i}\right) = \left(\frac{\partial a(x)}{\partial x_i}, \varphi(x)\right),\end{aligned} \tag{1.424}$$

namely

$$\lim_{t\to t_0}\frac{\partial}{\partial x_i} f_t(x) = \frac{\partial a(x)}{\partial x_i} = \frac{\partial}{\partial x_i}\left[\lim_{t\to t_0} f_t(x)\right]. \tag{1.425}$$

Proposition 1.25 Let there be the distribution $f_t(x) = F(u) \in \mathcal{D}'(\mathbb{R})$, $u = ax + \alpha(t)$, $x \in \mathbb{R}$, where $a \in \mathbb{R}\backslash\{0\}$ and $\alpha \in C^1(I)$, $I \subset \mathbb{R}$. We have

$$\frac{\partial}{\partial t} F(ax + \alpha(t)) = \alpha'(t) F'(ax + \alpha(t)) , \tag{1.426}$$

$$\frac{\partial^2}{\partial t \partial x} F(ax + \alpha(t)) = a\alpha'(t) F''(ax + \alpha(t)) . \tag{1.427}$$

Example 1.29 Let there be the distributions $f_t(x) = \delta(x - at) \in \mathcal{D}'(\mathbb{R})$, $g_t(x) = H(x - bt) \in \mathcal{D}'(\mathbb{R})$ depending on the parameter $t \in \mathbb{R}$, where $a, b \in \mathbb{R}$, and $H(u)$ is the Heaviside distribution. We have

$$\frac{\partial}{\partial t}\delta(x - at) = -a\delta'(x - at) , \quad \frac{\partial}{\partial t} H(x - bt) = -b\delta(x - bt) . \tag{1.428}$$

Indeed, the relations (1.428) are obtained directly by applying the formula (1.426), because $H'(u) = \delta(u)$.

1.3.7.2 Integration of Distributions Depending on a Parameter

For the distributions depending on a real parameter $t \in I \subset \mathbb{R}$ we can define the integral with respect to the corresponding parameter.

Let there be $f_t \in \mathcal{D}'(\mathbb{R}), t \in I \subset \mathbb{R}$, a distribution depending on the real parameter t. If the distribution f_t is continuous on $I \subset \mathbb{R}$ with respect to the parameter t, then, according to the continuity definition, $\forall \varphi \in \mathcal{D}(\mathbb{R})$, the function $\psi : I \subset \mathbb{R} \to \mathbb{R}$, having the expression,

$$\psi(t) = (f_t(x), \varphi(x)) , \tag{1.429}$$

is continuous on I.

Consequently, the functional $F : \mathcal{D}(\mathbb{R}) \to \mathbb{R}$, defined by

$$(F(x), \varphi(x)) = \int_a^b \psi(t) \mathrm{d}t = \int_a^b (f_t(x), \varphi(x)) \mathrm{d}t , \quad t \in [a, b] \subset I , \tag{1.430}$$

exists $\forall \varphi \in \mathcal{D}(\mathbb{R})$ and it represents a distribution from $\mathcal{D}'(\mathbb{R})$.

Indeed, according to the definition integral (1.430) we can write

$$(F, \varphi) = \int_a^b \psi(t) \mathrm{d}t = \lim_{\nu(\pi)\to 0} \sum_{i=1}^n \psi(\tau_i)\Delta t_i = \lim_{\nu(\pi)\to 0} \sum_{i=1}^n (f_{\tau_i}(x), \varphi(x))\Delta t_i , \tag{1.431}$$

where $\pi = \{t_0 = a, t_1, \ldots, t_i, \ldots, t_n = b\}$ is a partition of the interval $[a, b] \subset I$, with the norm $\nu(\pi) = \max_{1\le i\le n} \Delta t_i$, $\Delta t_i = t_i - t_{i-1}$, and $\tau_i \in [t_{i-1}, t_i]$, $i = \overline{1, n}$, are the intermediary points of the partition.

Denoting $\sigma_\pi(f_t) \in \mathcal{D}'(\mathbb{R})$ the distribution depending on the parameter $t \in [a, b]$, namely $\sigma_\pi(f_t) = \sum_{i=1}^{n} f_{\tau_i}(x)\Delta t_i$, then we have

$$(\sigma_\pi(f_t), \varphi(x)) = \sum_{i=1}^{n} (f_{\tau_i}(x)\Delta t_i, \varphi(x)) = \sum_{i=1}^{n} \psi(\tau_i)\Delta t_i \,, \quad \forall \varphi \in \mathcal{D}(\mathbb{R}) \,. \tag{1.432}$$

Thus, (1.431) becomes

$$(F, \varphi) = \lim_{\nu(\pi)\to 0} (\sigma_\pi(f_t), \varphi(x)) = \int_a^b \psi(t)\mathrm{d}t = \int_a^b (f_t(x), \varphi(x))\mathrm{d}t \,. \tag{1.433}$$

Because, the limit of (1.433) exists, according to the theorem of completeness of the distribution space $\mathcal{D}'(\mathbb{R})$, we obtain

$$\lim_{\nu(\pi)\to 0} \sigma_\pi(f_t(x)) = F(x) \,, \tag{1.434}$$

hence the functional $F(x)$ is a distribution from $\mathcal{D}'(\mathbb{R})$.

The distribution $F \in \mathcal{D}'(\mathbb{R})$ is denoted by

$$F(x) = \int_a^b f_t(x)\mathrm{d}t \,, \tag{1.435}$$

and will be called the integral of the distribution $f_t \in \mathcal{D}'(\mathbb{R})$ depending on the parameter $t \in [a, b] \subset I \subset \mathbb{R}$.

Obviously, the distribution (1.435) exists if $f_t \in \mathcal{D}'(\mathbb{R})$ is continuous for $t \in [a, b]$, and its mode of action on the test functions space $\mathcal{D}(\mathbb{R})$ is given by the formula (1.430), that is,

$$\left(\int_a^b f_t(x)\mathrm{d}t, \varphi(x) \right) = \int_a^b (f_t(x), \varphi(x))\mathrm{d}t \,, \quad \forall \varphi \in \mathcal{D}(\mathbb{R}) \,. \tag{1.436}$$

We note that the distribution (1.435) exists even if the distribution $f_t \in \mathcal{D}'(\mathbb{R})$ is not continuous, but the function ψ defined by (1.429) is integrable on $[a, b] \subset I$.

Proposition 1.26 Let there be the distribution $f_t \in \mathcal{D}'_+$ continuous for $t \in [a, b] \subset \mathbb{R}$. Then, we have $F(x) = \int_a^b f_t(x)\mathrm{d}t \in \mathcal{D}'_+$.

Proposition 1.27 If the distribution $f_t \in \mathcal{D}'(\mathbb{R})$ is continuous on $[a, b] \subset \mathbb{R}$, then we have

$$\frac{\mathrm{d}}{\mathrm{d}x} \int_a^b f_t(x)\mathrm{d}t = \int_a^b \frac{\partial}{\partial x} f_t(x)\mathrm{d}t \,. \tag{1.437}$$

Proposition 1.28 Let there be the distribution $f \in \mathcal{D}'(\mathbb{R})$ and the integrable function $g : [a, b] \subset \mathbb{R} \to \mathbb{R}$. Then, for the distribution $f_t(x) = f(x)g(t) \in \mathcal{D}'(\mathbb{R})$ depending on the parameter $t \in [a, b]$, the following results:

$$\int_a^b f_t(x)\mathrm{d}t = \int_a^b f(x)g(t)\mathrm{d}t = f(x)\int_a^b g(t)\mathrm{d}t\,. \tag{1.438}$$

Example 1.30 Let there be the distributions $\delta(x-t)$, $H(x-t) \in \mathcal{D}'(\mathbb{R})$, depending on the parameter $t \in \mathbb{R}$, where H is the Heaviside distribution. The following relations take place

$$F_1(x) = \int_a^b H(x-t)\mathrm{d}t = \begin{cases} 0\,, & x \le a < b\,, \\ b-a\,, & a < b < x\,, \\ x-a\,, & a \le x \le b\,, \end{cases} \tag{1.439}$$

$$F_2(x) = \int_a^b \delta(x-t)\mathrm{d}t = H(x-a) - H(x-b) = \begin{cases} 1\,, & x \in [a, b]\,, \\ 0\,, & x \notin [a, b]\,. \end{cases} \tag{1.440}$$

Obviously, $H(x-t) \in L^1_{\mathrm{loc}}(\mathbb{R})$, and, using the definition of the Heaviside function, we obtain

$$F_1(x) = \int_a^b H(x-t)\mathrm{d}t = \begin{cases} 0\,, & x \le a < b\,, \\ b-a\,, & a < b < x\,, \\ x-a\,, & a \le x \le b\,. \end{cases}$$

As regards the formula (1.440), this is obtained by applying the definition of the integral of a distribution depending on a parameter.

Thus, $\forall \varphi \in \mathcal{D}(\mathbb{R})$, we can write

$$\left(\int_a^b \delta(x-t)\mathrm{d}t, \varphi(x)\right) = \int_a^b (\delta(x-t), \varphi(x))\mathrm{d}t = \int_a^b \varphi(t)\mathrm{d}t\,. \tag{1.441}$$

Because φ has compact support, the expression (1.441) may be written as

$$\int_a^b \varphi(t)\mathrm{d}t = \int_a^\infty \varphi(t)\mathrm{d}t - \int_b^\infty \varphi(t)\mathrm{d}t = \int_0^\infty \varphi(x+a)\mathrm{d}x - \int_0^\infty \varphi(x+b)\mathrm{d}x$$

$$= \int_0^\infty [\varphi(x+a) - \varphi(x+b)]\mathrm{d}x = (H(x), \varphi(x+a) - \varphi(x+b))\,. \tag{1.442}$$

Consequently, (1.441) becomes

$$\left(\int_a^b \delta(x-t)\mathrm{d}t, \varphi(x)\right) = (H(x), \varphi(x+a) - \varphi(x+b))$$
$$= (H(x-a) - H(x-b), \varphi(x)) , \tag{1.443}$$

wherefrom results (1.440).

We note that the main properties of the defined integral are maintained for distributions depending on a parameter.

Proposition 1.29 Let there be the distributions $f_t, g_t \in \mathcal{D}'(\mathbb{R})$ depending on the real parameter $t \in [a, b] \subset \mathbb{R}$.

1. If $\partial f_t/\partial t \in \mathcal{D}'(\mathbb{R})$ exists, then we have

$$\int_a^b \frac{\partial}{\partial t} f_t(x)\mathrm{d}t = f_t(x)|_a^b = f_b(x) - f_a(x) , \tag{1.444}$$

which is an analogue of the Leibniz–Newton formula.

2. If $f_t \in \mathcal{D}'(\mathbb{R})$ is continuous on $[a, b] \subset \mathbb{R}$, $c \in [a, b]$, we have

$$\int_a^b f_t(x)\mathrm{d}t = \int_a^c f_t(x)\mathrm{d}t + \int_c^b f_t(x)\mathrm{d}t , \tag{1.445}$$

$$\int_a^b f_t(x)\mathrm{d}t = -\int_b^a f_t(x)\mathrm{d}t . \tag{1.446}$$

Particularly, $\int_a^a f_t(x)\mathrm{d}t = 0$ and

$$\frac{\partial}{\partial t}\int_a^t f_u(x)\mathrm{d}u = f_t(x) , \quad t \in [a, b] . \tag{1.447}$$

Example 1.31 The relation (1.440) is obtained by applying the formula (1.444). Indeed, because $\partial H(x-t)/\partial t = -H'(x-t) = -\delta(x-t)$, we have

$$F_2(x) = \int_a^b \delta(x-t)\mathrm{d}t = -\int_a^b \frac{\partial}{\partial t} H(x-t)\mathrm{d}t$$
$$= -H(x-t)|_a^b = H(x-a) - H(x-b). \tag{1.448}$$

1.3.8
Direct Product and Convolution Product of Functions and Distributions

The direct (or tensor) product of two distributions is a new operation with distributions which extends the usual product of two functions.

Let there be $\mathbb{R}^n$ and $\mathbb{R}^m$ two Euclidean spaces with the dimensions n, m, respectively, and let $x = (x_1, x_2, \ldots, x_n) \in \mathbb{R}^n$, $y = (y_1, y_2, \ldots, y_m) \in \mathbb{R}^m$ be points of these spaces. Then the Cartesian product of these spaces is $\mathbb{R}^n \times \mathbb{R}^m = \mathbb{R}^{n+m}$ and represents a new $n + m$-dimensional Euclidean space with generic point $(x, y) \in \mathbb{R}^{n+m}$.

Let f and g be two complex functions defined on $\mathbb{R}^n$ and $\mathbb{R}^m$, respectively, with the generic points $x \in \mathbb{R}^n$, $y \in \mathbb{R}^m$.

Definition 1.33 The function $f \times g : \mathbb{R}^{n+m} \to \Gamma$, defined by the relation $(f \times g)(x, y) = f(x)g(y)$ is called direct or tensor product of the function f by the function g.

So, the direct product of two numerical functions coincides with their usual product.

Proposition 1.30 Let there be the functions $f \in C^k(\mathbb{R}^n)$, $g \in C^k(\mathbb{R}^m)$. Then, the following properties occur

1. $f \times g \in C^k(\mathbb{R}^{n+m})$;
2. $D_x^p D_y^q(f(x) \times g(x)) = D_x^p f \times D_y^q g\,, \quad |p| \le k\,, \quad |q| \le k$,
   where D_x^p, D_y^q are derivation operators of orders $|p|, |q|$, respectively, in relation to the variables $x \in \mathbb{R}^n$ and $y \in \mathbb{R}^m$;
3. $\mathrm{supp}(f \times g) = \mathrm{supp}(f) \times \mathrm{supp}(g)$.

Proof: Indeed, we have $D_x^p D_y^q(f \times g)(x, y) = D_x^p f(x) D_y^q g(y) = (D_x^p f \times D_y^q g)$ (x, y), wherefrom results the properties 1. and 2.

Let there be $(x_0, y_0) \in \mathrm{supp}(f \times g)$. Then $\forall\, U_r(x_0, y_0), \exists (x, y) \in U_r(x_0, y_0)$, so that $(f \times g)(x, y) \neq 0$ which implies $f(x) \neq 0$ and $g(y) \neq 0$, hence $x_0 \in \mathrm{supp}(f)$ and $y_0 \in \mathrm{supp}\,(g)$, wherefrom results 3. □

We note with $\mathcal{D}(\mathbb{R}^n), \mathcal{D}(\mathbb{R}^m), \mathcal{D}(\mathbb{R}^n \times \mathbb{R}^m)$ the indefinitely derivable test functions spaces with compact support on $\mathbb{R}^n, \mathbb{R}^m, \mathbb{R}^{n+m}$, and with $\mathcal{D}'(\mathbb{R}^n), \mathcal{D}'(\mathbb{R}^m)$, $\mathcal{D}'(\mathbb{R}^{n+m})$ (the space $\mathcal{D}'$ of corresponding distributions).

We note that $\mathcal{D}(\mathbb{R}^n) \times \mathcal{D}(\mathbb{R}^m)$ is a vector subspace of $\mathcal{D}(\mathbb{R}^{n+m})$, generated by functions of the form $u \times v$, $u \in \mathcal{D}(\mathbb{R}^n)$, $v \in \mathcal{D}(\mathbb{R}^m)$.

Proposition 1.31 The space $\mathcal{D}(\mathbb{R}^n) \times \mathcal{D}(\mathbb{R}^m)$ is dense in $\mathcal{D}(\mathbb{R}^{n+m})$.
This means that $\forall \varphi(x, y) \in \mathcal{D}(\mathbb{R}^{n+m})$, there exists the sequence of functions $(\varphi_i(x, y))_{i \in \mathbb{N}}$ of the form $\varphi_i(x, y) = \sum_{k=1}^{p} u_{ik}(x) v_{ik}(y)$, with $u_{ik} \in \mathcal{D}(\mathbb{R}^n)$, $v_{ik} \in \mathcal{D}(\mathbb{R}^m)$ so that $\varphi_i \xrightarrow{\mathcal{D}(\mathbb{R}^{n+m})} \varphi$.

This result can be generalized, so that $\mathcal{D}(\mathbb{R}^n) \times \mathcal{D}(\mathbb{R}^m) \times \mathcal{D}(\mathbb{R}^\ell)$ is a vector subspace of $\mathcal{D}(\mathbb{R}^{n+m+\ell})$ and is dense in it.

Let f and g be locally integrable functions on $\mathbb{R}^n$ and $\mathbb{R}^m$, respectively. Then, their direct product $h(x, y) = f(x)g(y), (x, y) \in \mathbb{R}^{n+m}$ is a locally integrable function, which generates a regular distribution on the test functions space $\mathcal{D}(\mathbb{R}^{n+m})$. Then, $\forall\varphi \in \mathcal{D}(\mathbb{R}^{n+m})$, on the basis of Fubini's theorem of interchange of the order of integration, we can write

$$(f(x) \times g(y), \varphi(x, y)) = \int\limits_{\mathbb{R}^{n+m}} f(x)g(y)\varphi(x, y)\mathrm{d}x\mathrm{d}y$$

$$= \int\limits_{\mathbb{R}^n} f(x)\mathrm{d}x \int\limits_{\mathbb{R}^m} g(y)\varphi(x, y)\mathrm{d}y = (f(x), (g(y), \varphi(x, y))) , \tag{1.449}$$

namely

$$(f(x) \times g(y), \varphi(x, y)) = (f(x), (g(y), \varphi(x, y))) , \quad \forall\varphi \in \mathcal{D}(\mathbb{R}^{n+m}) . \tag{1.450}$$

This relation will be adopted as the definition of the direct product of two distributions.

Definition 1.34 Let there be the distributions $f \in \mathcal{D}'(\mathbb{R}^n)$ and $g \in \mathcal{D}'(\mathbb{R}^m)$. We call direct or tensor product of the distribution f with g, the functional $f \times g : \mathcal{D}(\mathbb{R}^{n+m}) \to \Gamma$ defined by the relation

$$(f(x) \times g(y), \varphi(x, y)) = (f(x), (g(y), \varphi(x, y))) , \quad \varphi \in \mathcal{D}(\mathbb{R}^{n+m}) . \tag{1.451}$$

Proposition 1.32 The direct product $f \times g$ of the distributions $f \in \mathcal{D}'(\mathbb{R}^n), g \in \mathcal{D}'(\mathbb{R}^m)$, defined through the relation (1.451), exists and is a distribution from $\mathcal{D}'(\mathbb{R}^{n+m})$, namely $f \times g \in \mathcal{D}'(\mathbb{R}^{n+m})$.

In particular, if $\varphi(x, y) \in \mathcal{D}(\mathbb{R}^{n+m})$ is of the form $\varphi(x, y) = \varphi_1(x)\varphi_2(y)$, where $\varphi_1 \in \mathcal{D}(\mathbb{R}^n), \varphi_2 \in \mathcal{D}(\mathbb{R}^m)$, then the formula (1.451) becomes

$$(f(x) \times g(y), \varphi_1(x)\varphi_2(y)) = (f(x), \varphi_1(x)) \cdot (g(y), \varphi_2(y)) . \tag{1.452}$$

From the formula (1.452) we obtain the following.

Proposition 1.33 The necessary and sufficient condition for $f \times g = 0, f \in \mathcal{D}'(\mathbb{R}^n), g \in \mathcal{D}'(\mathbb{R}^m)$ is that one of the factors be null.

Indeed, if $f(x) = 0$, then $\forall\varphi_1 \in \mathcal{D}(\mathbb{R}^n)$ we have $(f(x), \varphi_1(x)) = 0$ and, from (1.452), it results $f \times g = 0$.

Conversely, if $f \times g = 0$, then $\forall\varphi_1(x) \in \mathcal{D}(\mathbb{R}^n)$ and $\forall\varphi_2(y) \in \mathcal{D}(\mathbb{R}^m)$ we have

$$(f(x) \times g(y), \varphi_1(x)\varphi_2(y)) = (f(x), \varphi_1(x)) \cdot (g(y), \varphi_2(y)) = 0 , \tag{1.453}$$

wherefrom it results that one of the factors is zero. Thus, if $(f(x), \varphi_1(x)) = 0$ ($\varphi_1(x)$ is arbitrary), then we get $f(x) = 0$.

1.3.8.1 Properties of the Direct Product

Proposition 1.34 The direct product is commutative, associative and distributive with respect to the addition of distributions from the same space. Namely, we have

1. $f \times g = g \times f, \forall f \in \mathcal{D}'(\mathbb{R}^n), g \in \mathcal{D}'(\mathbb{R}^m)$;
2. $f \times (g \times h) = (f \times g) \times h, \forall f \in \mathcal{D}'(\mathbb{R}^n), g \in \mathcal{D}'(\mathbb{R}^m), h \in \mathcal{D}'(\mathbb{R}^\ell)$;
3. $f \times (\alpha g \times \beta h) = \alpha(f \times g) + \beta(f \times h), \forall f \in \mathcal{D}'(\mathbb{R}^n), g, h \in \mathcal{D}'(\mathbb{R}^m), \alpha, \beta \in \Gamma$.

Proposition 1.35 Let there be the distributions $f \in \mathcal{D}'(\mathbb{R}^n), g \in \mathcal{D}'(\mathbb{R}^m)$, ν the symmetry operator and $\mathrm{D}_x^\alpha, \mathrm{D}_y^\beta$ the derivation operators. The following relations take place:

$$
\begin{aligned}
&\operatorname{supp}(f \times g) = \operatorname{supp}(f) \times \operatorname{supp}(g)\,,\\
&\mathrm{D}_x^\alpha \mathrm{D}_y^\beta (f(x) \times g(y)) = \mathrm{D}_x^\alpha f(x) \times \mathrm{D}_y^\beta f(y)\\
&(f \times g)^\nu = f^\nu \times g^\nu\,,\\
&a(x)b(y)(f(x) \times g(y)) = a(x)f(x) \times b(y)g(y)\,,\\
&a(x) \in C^\infty(\mathbb{R}^n), b(y) \in C^\infty(\mathbb{R}^m)\,.
\end{aligned}
$$

Example 1.32 Let $H(x), x \in \mathbb{R}^n$ be the Heaviside distribution of n variables. By means of the direct product, it can be written as

$$H(x_1, \ldots, x_n) = H(x_1) \times H(x_2) \times \cdots \times H(x_n)\,. \tag{1.454}$$

Because $\mathrm{d}H(x_i)/\mathrm{d}x_i = \delta(x_i)$, we obtain

$$
\begin{aligned}
\frac{\partial^n H(x_1, \ldots, x_n)}{\partial x_1 \partial x_2 \ldots \partial x_n} &= \delta(x_1, x_2, \ldots, x_n)\\
&= \frac{\mathrm{d}H(x_1)}{\mathrm{d}x_1} \times \frac{\mathrm{d}H(x_2)}{\mathrm{d}x_2} \times \cdots \times \frac{\mathrm{d}H(x_n)}{\mathrm{d}x_n} = \delta(x_1) \times \cdots \times \delta(x_n)\,,
\end{aligned} \tag{1.455}
$$

hence

$$\delta(x_1, \ldots, x_n) = \delta(x_1) \times \delta(x_2) \times \cdots \times \delta(x_n)\,. \tag{1.456}$$

Definition 1.35 We say that the distribution $g(x, y) \in \mathcal{D}'(\mathbb{R}^{n+m})$ does not depend on the variable $y \in \mathbb{R}^m$ if it is of the form

$$g(x, y) = f(x) \times 1(y), f \in \mathcal{D}'(\mathbb{R}^n)\,. \tag{1.457}$$

This distribution will be denoted by $f(x) \in \mathcal{D}'(\mathbb{R}^{n+m})$ and should not be confused with $f(x) \in \mathcal{D}'(\mathbb{R}^n)$, which is defined on the test functions space $\mathcal{D}(\mathbb{R}^n)$.

For $\forall f \in \mathcal{D}'(\mathbb{R}^n)$, the relation follows

$$\int\limits_{\mathbb{R}^m} (f(x), \varphi(x, y))\mathrm{d}y = \left(f(x), \int\limits_{\mathbb{R}^m} \varphi(x, y)\mathrm{d}y \right) , \quad \forall \varphi \in \mathcal{D}(\mathbb{R}^{n+m}) . \quad (1.458)$$

Indeed, on the basis of the direct product, $\forall \varphi(x, y) \in \mathcal{D}(\mathbb{R}^{n+m})$ we can write

$$(f(x) \times 1(y), \varphi(x, y)) = (f(x), (1(y), \varphi(x, y))) = \left(f(x), \int\limits_{\mathbb{R}^m} \varphi(x, y)\mathrm{d}y \right)$$
$$= (1(y), (f(x), \varphi(x, y))) = \int\limits_{\mathbb{R}^m} (f(x), \varphi(x, y))\mathrm{d}y , \quad (1.459)$$

giving the formula (1.458).

If the distribution $g(x, y) \in \mathcal{D}'(\mathbb{R}^{n+m})$ does not depend on the variable $y \in \mathbb{R}^m$, then we have

$$\mathrm{D}_i g(x, y) = 0, \mathrm{D}_i = \frac{\partial}{\partial y_i} . \quad (1.460)$$

Indeed, $\mathrm{D}_i g(x, y) = \mathrm{D}_i(f(x) \times 1(y)) = f(x) \times \mathrm{D}_i 1(y) = f(x) \times 0 = 0$.

In general we have the following: The necessary and sufficient condition for the distribution $f \in \mathcal{D}'(\mathbb{R}^n)$ should not depend on the variable x_i is $\partial f/\partial x_i = 0$.

Proposition 1.36 The necessary and sufficient condition for the distribution $f \in \mathcal{D}'(\mathbb{R}^n)$ to be a constant is

$$\frac{\partial f}{\partial x_i} = 0 , \quad i = \overline{1, n} . \quad (1.461)$$

Proposition 1.37 Let there be the distribution $f(x) \in \mathcal{D}'(\mathbb{R})$ and the Dirac representative sequence $g_\ell(t) \xrightarrow{\mathcal{D}'} \delta(t)$, where $g_\ell \in L^1_{\mathrm{loc}}(\mathbb{R})$. Then, we have

$$\lim_{\ell \to \infty} (f(x) \times g_\ell(t)) = f(x) \times \delta(t) . \quad (1.462)$$

Indeed, $\forall \varphi(x, t) \in \mathcal{D}(\mathbb{R}^2)$ we can write

$$(f(x) \times g_\ell(t), \varphi(x, t)) = (g_\ell(t), (f(x), \varphi(x, t))) , \quad (1.463)$$

where $(f(x), \varphi(x, t)) = \psi(t) \in \mathcal{D}(\mathbb{R})$.

Consequently, we obtain

$$\lim_{\ell\to\infty}(f(x)\times g_\ell(t),\varphi(x,t)) = \lim_{\ell\to\infty}(g_\ell(t),\psi(t)) = (\delta(t),\psi(t))$$
$$= (\delta(x),(f(x),\varphi(x,t))) = (f(x)\times\delta(t),\varphi(x,t)) , \tag{1.464}$$

giving the formula (1.462).

In general, we can write the formulae

$$\lim_{\ell\to\infty}(f(x)\times g_\ell(t-t_k)) = f(x)\times\delta(t-t_k) ,$$
$$\lim_{\ell\to\infty}(f(x)\times g'_\ell(t-t_k)) = f(x)\times\delta'(t-t_k) , \tag{1.465}$$

because

$$\lim_{\ell\to\infty} g_\ell(t-t_k) = \delta(t-t_k) , \text{ and } \lim_{\ell\to\infty} g'_\ell(t-t_k) = \delta'(t-t_k) . \tag{1.466}$$

1.3.8.2 **The Convolution Product of Distributions**

In order to extend the convolution product to distribution, we will consider the functions $f, g \in L^1(\mathbb{R}^n)$. Then, $f * g \in L^1(\mathbb{R}^n)$, hence it is a regular distribution and, $\forall\varphi \in \mathcal{D}(\mathbb{R}^n)$, we have

$$((f*g)(x),\varphi(x)) = \int_{\mathbb{R}^n}(f*g)(x)\varphi(x)\mathrm{d}x = \int_{\mathbb{R}^n}\varphi(x)\left[\int_{\mathbb{R}^n} f(t)g(x-t)\mathrm{d}t\right]\mathrm{d}x . \tag{1.467}$$

Making the change of variables $u = x - t, v = t$, the previous relation becomes

$$((f*g)(x),\varphi(x)) = \int_{\mathbb{R}^{2n}} f(t)g(x-t)\varphi(x)\mathrm{d}x\mathrm{d}t$$
$$= \int_{\mathbb{R}^{2n}} f(v)g(u)\varphi(u+v)\mathrm{d}u\mathrm{d}v . \tag{1.468}$$

Taking into account that $f(v)g(u) = f(v)\times g(u)$, formula (1.468) can be written in the form

$$((f*g)(x),\varphi(x)) = (f(x)\times g(y),\varphi(x+y)) . \tag{1.469}$$

This relation is considered to be a definition formula of the convolution product of two distributions.

Definition 1.36 If $f, g \in \mathcal{D}'(\mathbb{R}^n)$, then their convolution product $f * g$ represents a new distribution from $\mathcal{D}'(\mathbb{R}^n)$, defined by the formula

$$(f*g,\varphi) = (f(x)\times g(y),\varphi(x+y)) , \quad \forall\varphi\in\mathcal{D}(\mathbb{R}^n) . \tag{1.470}$$

We note that the distribution $f * g \in \mathcal{D}'(\mathbb{R}^n)$ does not exist for any distributions $f, g \in \mathcal{D}'(\mathbb{R}^n)$.

Indeed, when $\varphi(x) \in \mathcal{D}(\mathbb{R}^n)$ the function $\varphi(x+y)$ is indefinitely differentiable on $\mathbb{R}^{2n}$, but does not have compact support, hence $\varphi(x+y) \notin \mathcal{D}(\mathbb{R}^{2n})$.

The convolution product $f * g$ exists if the sets $\text{supp}(f(x) \times g(y))$ and $\text{supp}(\varphi(x+y))$ have a compact intersection.

Proposition 1.38 Let there be $f, g \in \mathcal{D}'(\mathbb{R}^n)$. The convolution product $f * g \in \mathcal{D}'(\mathbb{R}^n)$ exists if one of the distributions f, g has compact support.

Proof: Let us assume that the distribution $f \in \mathcal{D}'(\mathbb{R}^n)$ has compact support, hence $\text{supp}(f) = \Omega =$ compact. We notice that $\forall \alpha \in \mathcal{D}'(\mathbb{R}^n)$ the function $\psi(y) = (\alpha(x), \varphi(x+y))$ is indefinitely differentiable. In particular, the function $(f(x), \varphi(x+y)) = \psi_1(y)$ has compact support, because $x \in \Omega = \text{supp}(f)$ is bounded; it means that, for $|y|$ large enough, $\varphi(x+y) = 0$. Hence, $(f(x), \varphi(x+y)) \in \mathcal{D}(\mathbb{R}^n)$.

Consequently, formula (1.469) makes sense and we can write

$$\begin{aligned}(f * g, \varphi) &= (f(x) \times g(y), \varphi(x+y)) = (g(y), (f(x), \varphi(x+y))) \\ &= (f(x), (g(y), \varphi(x+y)))\,, \quad \forall \varphi \in \mathcal{D}(\mathbb{R}^n)\,. \end{aligned} \tag{1.471}$$

□

Proposition 1.39 Let there be the distributions $\delta(x), f(x) \in \mathcal{D}'(\mathbb{R}^n)$ and D^α the derivation operator. Then, we have

$$D^\alpha \delta * f = D^\alpha f\,, \tag{1.472}$$

$$\delta(x-a) * f(x) = f(x-a)\,. \tag{1.473}$$

Proof: Because $D^\alpha \delta(x)$ has as support the origin, it means that the product $D^\alpha \delta(x) * f(x) \in \mathcal{D}'(\mathbb{R}^n)$ exists and for any $\varphi \in \mathcal{D}(\mathbb{R}^n)$ we have

$$\begin{aligned}(D^\alpha \delta * f, \varphi) &= (D^\alpha \delta(x) \times f(y), \varphi(x+y)) \\ &= (-1)^{|\alpha|}(f(y), (\delta(x), D^\alpha \varphi(x+y))) = (-1)^{|\alpha|}(f(y), D^\alpha \varphi(y)) \\ &= (D^\alpha f(y), \varphi(y)) = (D^\alpha f(x), \varphi(x))\,, \end{aligned} \tag{1.474}$$

giving the formula (1.472).

Also, we have

$$\begin{aligned}(\delta(x-a) * f(x), \varphi(x)) &= (\delta(x-a) \times f(y), \varphi(x+y)) \\ &= (f(y), (\delta(x-a), \varphi(x+y))) = (f(y), (\delta(x), \varphi(x+a+y))) \\ &= (f(y), \varphi(a+y)) = (f(y-a), \varphi(y))\,, \end{aligned} \tag{1.475}$$

hence

$$(\delta(x-a) * f(x), \varphi(x)) = (f(x-a), \varphi(x))\,; \tag{1.476}$$

thus, the relation (1.473) is proved. □

Particularly, for $a = 0$ we have

$$\delta(x) * f(x) = f(x) , \quad \forall f \in \mathcal{D}'(\mathbb{R}^n) . \tag{1.477}$$

This property shows that Dirac's delta distribution $\delta(x) \in \mathcal{D}'(\mathbb{R}^n)$ is the unit element with respect to the convolution product.

Corollary 1.3 Let there be $P(\mathrm{D}) = \sum_{|\alpha| \le l} a_\alpha \mathrm{D}^\alpha , \alpha \in \mathbb{N}_0^n, a_\alpha \in \mathbb{C}$ a linear differential operator with constant coefficients. Then, $\forall f \in \mathcal{D}'(\mathbb{R}^n)$ we have

$$P(\mathrm{D})\delta(x) * f(x) = P(\mathrm{D}) f(x) . \tag{1.478}$$

This result is obtained from property (1.472) and the linearity of the operator $P(\mathrm{D})$.

Thus, for $P(\mathrm{D}) = \Delta = \sum_{k=1}^{n} (\partial^2/\partial x_k^2)$ representing the Laplace operator in $\mathbb{R}^n$ and $\forall f \in \mathcal{D}'(\mathbb{R}^n)$ we have

$$\Delta\delta(x) * f = \Delta f . \tag{1.479}$$

For the operational calculus, of special importance are the distributions from $\mathcal{D}'(\mathbb{R})$, having bounded supports to the left, hence the supports on $[a, \infty)$.

Proposition 1.40 If $f, g \in \mathcal{D}'(\mathbb{R})$ and have supports bounded to the left, then the product $f * g$ exists.

Proof: Let there be $\mathrm{supp}(f) = \Omega_1, \mathrm{supp}(g) = \Omega_2$ and $\Omega_1, \Omega_2 \subset [a, \infty)$. Let there be $\mathrm{supp}(\varphi) = \Omega , \varphi \in \mathcal{D}(\mathbb{R})$. Then, for $x \in \Omega_1, y \in \Omega_2$, because $x + y \in \Omega$ and Ω is compact, it results $x \ge a, y \ge a, b_1 \le x + y \le b_2$. Hence, $x \le b_2 - a, y \le b_2 - a$ and therefore the set $(x, y) \in \mathbb{R}^2$ with $x \in \Omega_1, y \in \Omega_2, x + y \in \Omega$ is bounded and thus the product $f * g$ exists (Figure 1.9).

We note that the hatched trapezoid $ABCD$ is the intersection of the sets $\mathrm{supp}(f(x) \times g(y))$ and $\mathrm{supp}(\varphi(x + y))$, which is a bounded set for which there exists the convolution product $f * g \in \mathcal{D}'(\mathbb{R})$. Similarly, it is shows the existence of the product $f * g$ if the distributions $f, g \in \mathcal{D}'(\mathbb{R})$ have bounded supports to the right, hence $\mathrm{supp}(f), \mathrm{supp}(g) \subset (-\infty, b]$.

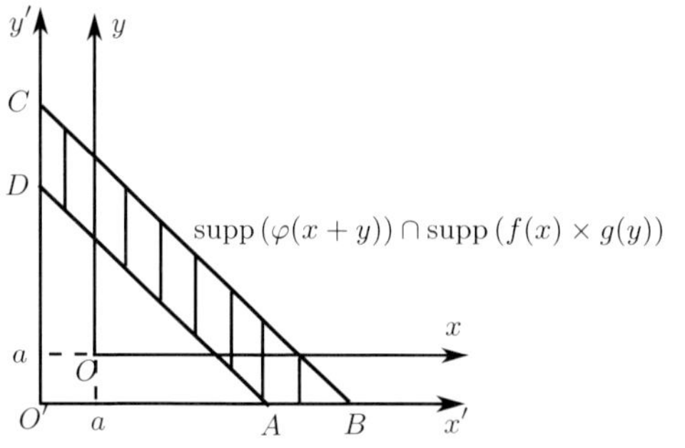

Figure 1.9

We note by $\mathcal{D}'_+$ all the distributions of $\mathcal{D}'(\mathbb{R})$ with the support on the half-axis $[0, \infty)$. □

Proposition 1.41 If $f, g \in \mathcal{D}'(\mathbb{R}^n)$ and $f * g$ exists, then we have

$$\mathrm{supp}(f * g) \subset \overline{\mathrm{supp}(f) + \mathrm{supp}(g)} \,. \tag{1.480}$$

Proof: Let there be $A = \mathrm{supp}(f)$, $B = \mathrm{supp}(g)$. To justify the sentence, it is enough to show that $\forall \varphi \in \mathcal{D}(\mathbb{R}^n)$, such that $\mathrm{supp}(\varphi) \subset \mathrm{C}_{\mathbb{R}^n}\overline{(A + B)}$, we have $(f * g, \varphi) = 0$.

Because $\mathrm{supp}(f(x) \times g(y)) = A \times B$ and $x + y \in \mathrm{supp}(\varphi) \subset \mathrm{C}_{\mathbb{R}^n}\overline{(A + B)}$, from $x \in A$ and $y \in B$ it results $x + y \in A + B$; but as $(A + B) \cap \mathrm{supp}(\varphi(x + y)) = \emptyset$, we have $(f * g, \varphi) = (f(x) \times g(y), \varphi(x + y)) = 0$ for $\mathrm{supp}(\varphi) \subset \mathrm{C}_{\mathbb{R}^n}\overline{(A + B)}$.

Taking into account that the closure of a set is a closed set, we deduce that $\mathrm{C}_{\mathbb{R}^n}\overline{(A + B)}$ is an open set, hence $f * g = 0$ in $\mathrm{C}_{\mathbb{R}^n}\overline{(A + B)}$, that is, $\mathrm{supp}(f * g) \subset \overline{A + B}$. □

Corollary 1.4 Let there be $f, g \in \mathcal{D}'(\mathbb{R})$, having $\mathrm{supp}(f) \subset [a, \infty)$ and $\mathrm{supp}(g) \subset [b, \infty)$. Then $f * g$ exists and we have

$$\mathrm{supp}(f * g) \subset [a + b, \infty) \,. \tag{1.481}$$

Indeed, on the basis of the Proposition 1.40, the product $f * g$ exists and, according to the Proposition 1.41, we have

$$\mathrm{supp}(f * g) \subset \overline{\mathrm{supp}(f) + \mathrm{supp}(g)} \subset [a, \infty) + [b, \infty) = [a + b, \infty) \,. \tag{1.482}$$

Particularly, if $f, g \in \mathcal{D}'_+$, that is, $\mathrm{supp}(f), \mathrm{supp}(g) \subset [0, \infty)$, then $\mathrm{supp}(f \times g) \subset [0, \infty)$, wherefrom it results $f * g \in \mathcal{D}'_+$.

Thus, for example, we have

$$H(x) * H(x) = x H(x) = \begin{cases} 0\,, & x < 0\,, \\ x\,, & x \geq 0\,, \end{cases} \tag{1.483}$$

where $H \in \mathcal{D}'_+$ is the Heaviside distribution.

Corollary 1.5 If one of the distributions $f, g \in \mathcal{D}'(\mathbb{R}^n)$ has compact support, then $\mathrm{supp}(f * g) \subset \mathrm{supp}(f) + \mathrm{supp}(g)$.

Proposition 1.42 Let there be the distributions $f, g \in \mathcal{D}'(\mathbb{R}^n)$. If $f * g$ exists, then we have $f * g = g * f$, that is, the convolution product is commutative.

Indeed, because $f(x) \times g(y) = g(y) \times f(x), \forall \varphi \in \mathcal{D}(\mathbb{R}^n)$, we have

$$\begin{aligned} (f * g, \varphi) &= (f(x) \times g(y), \varphi(x + y)) \\ &= (g(y) \times f(x), \varphi(x + y)) = (g * f, \varphi)\,, \end{aligned} \tag{1.484}$$

namely $f * g = g * f$.

Proposition 1.43 If two of the distributions $f, g, h \in \mathcal{D}'(\mathbb{R}^n)$ have compact support, then the convolution product is associative, that is,

$$f * (g * h) = (f * g) * h \,. \tag{1.485}$$

Proof: Suppose that the distributions f and g have compact supports. Then, the distributions $f * (g * h)$ and $(f * g) * h$ exist and $\forall \varphi \in \mathcal{D}(\mathbb{R}^n)$ we have

$$\begin{aligned}(f * (g * h), \varphi) &= (f(x) \times (g \times h)(y), \varphi(x + y)) \\ &= ((g * h)(y), (f_x, \varphi(x + y))) \\ &= (g(y) \times h(z), (f(x), \varphi(x + y + z))) \\ &= (f(x) \times g(y) \times h(z), \varphi(x + y + z)) \,. \end{aligned} \tag{1.486}$$

Proceeding analogously, we obtain

$$\begin{aligned}((f * g) * h, \varphi) &= ((f * g)(y) \times h(z), \varphi(y + z)) \\ &= (f(x) \times g(y), (h(z), \varphi(x + y + z))) \\ &= (f(x) \times g(y) \times h(z), \varphi(x + y + z)) \,. \end{aligned} \tag{1.487}$$

Comparing the two expressions, we obtain the associativity of the convolution product. □

In connection with the property of associativity of distributions bounded to the left, we can state the following.

Proposition 1.44 Let there be $f, g, h \in \mathcal{D}'(\mathbb{R})$ and $\text{supp}(f, g, h) \subset [a, \infty)$. Then, the convolution product of these distributions is associative.

Remark 1.2 Apart from the cases presented by associativity, we note that the convolution product $f * g * h$, $f, g, h \in \mathcal{D}'(\mathbb{R}^n)$ is associative if the following conditions are fulfilled: $f \in \mathcal{E}'(\mathbb{R}^n)$, hence it is a distribution with compact support and there exists the product $g * h \in \mathcal{D}'(\mathbb{R}^n)$.

Thus, for example, we have

$$(\delta(x) * H(x)) * H(x) = \delta(x) * (H * H) = x H(x) \,. \tag{1.488}$$

Instead, the product $1 * \delta'(x) * H(x)$, $x \in \mathbb{R}$ is not associative because the associativity conditions are not respected and we have

$$(1 * \delta'(x)) * H(x) = 0 * H = 0 \,, \tag{1.489}$$

$$1 * (\delta' * H) = 1 * \delta = 1 \,. \tag{1.490}$$

In connection with the distributivity of the convolution product with respect to the addition operation, we state the following.

Proposition 1.45 Let there be the distributions $f, g, h \in \mathcal{D}'(\mathbb{R}^n)$ and $\alpha, \beta \in \mathbb{C}$. If two of the products $f * (\alpha g + \beta h)$, $f * g$, $f * h$ exist, then the third one exists as well, and we have

$$f * (\alpha g + \beta h) = \alpha(f * g) + \beta(f * h) . \tag{1.491}$$

Corollary 1.6 If $f, g, h \in \mathcal{D}'(\mathbb{R}^n)$ and f has compact support, hence $f \in \mathcal{E}'(\mathbb{R}^n)$, then the convolution product is distributive, namely

$$f * (\alpha g + \beta h) = \alpha(f * g) + \beta(f * h) , \quad \alpha, \beta \in \mathbb{C} . \tag{1.492}$$

Indeed, because f has compact support, the three products exist and, on the basis of the Proposition 1.45, the distributivity property occurs.

Proposition 1.46 Let there be $f, g \in \mathcal{D}'(\mathbb{R}^n)$ and $P(\mathrm{D}) = \sum_{|\alpha| \le l} a_\alpha \mathrm{D}^\alpha$ a linear differential operator with constant coefficients. If $f * g$ exists, then the distributions $P(\mathrm{D}) f * g$, $f * P(\mathrm{D}) g$ exist as well, and we have

$$P(\mathrm{D})(f * g) = P(\mathrm{D}) f * g = f * P(\mathrm{D}) g . \tag{1.493}$$

Proof: Let there be $\delta(x) \in \mathcal{D}'(\mathbb{R}^n)$ the Dirac delta distribution. Then, the distribution $P(\mathrm{D})\delta \in \mathcal{E}'(\mathbb{R}^n)$ hence, it has compact support and therefore the product $P(\mathrm{D})\delta * (f * g)$ is commutative and associative. Based on the associative property and on the Corollary 1.3, we have

$$P(\mathrm{D})\delta * (f * g) = P(\mathrm{D})(f * g) = (P(\mathrm{D})\delta * f) * g = P(\mathrm{D}) f * g . \tag{1.494}$$

Similarly, using the commutativity of the convolution product and the previous formula, we have

$$P(\mathrm{D})\delta * (f * g) = P(\mathrm{D})\delta * (g * f) = P(\mathrm{D}) g * f , \tag{1.495}$$

wherefrom the required result is obtained. □

Example 1.33 Let $H(x) \in \mathcal{D}'_+$ be the Heaviside distribution. Then, we can write

$$(H * H)' = H' * H = \delta * H = H , \tag{1.496}$$

because $H * H$ exists and the obtained result is verified directly. Thus, as $H * H = x H(x)$, by the derivation of this product, which is allowed because the function $\mu(x) = x, x \in \mathbb{R}$, is of class $C^\infty(\mathbb{R})$, hence it is the multiplier of the space $\mathcal{D}(\mathbb{R})$, we obtain

$$(H * H)' = (x H)' = H + x H' = H + x\delta(x) = H , \tag{1.497}$$

because $x\delta(x) = 0$.

Let there be the distribution $f \in \mathcal{D}'_+$. Then, the distribution $F \in \mathcal{D}'_+$ given by the expression

$$F(x) = f(x) * H^{*m} = f(x) * \underbrace{(H * H * \cdots * H)}_{m}, \tag{1.498}$$

is a primitive of order m for the distribution f.

Indeed, $F \in \mathcal{D}'_+$ and we have

$$F^{(m)}(x) = f(x) * \underbrace{(H' * H' * \cdots * H')}_{m} = f * \underbrace{(\delta * \delta * \cdots * \delta)}_{m} = f * \delta = f. \tag{1.499}$$

It follows that the Heaviside distribution $H \in \mathcal{D}'_+$ has the role of integration operator for the distributions from $\mathcal{D}'_+$, which is of particular importance in the operational calculus.

In particular, if $f \in C^0(\mathbb{R}) \cap \mathcal{D}'_+$, then $F(x) = f(x) * H(x) \in C^1(\mathbb{R}) \cap \mathcal{D}'_+$ and $F'(x) = H'(x) * f(x) = \delta(x) * f(x) = f(x)$.

Example 1.34 If $f, g \in \mathcal{D}'(\mathbb{R})$ and $f * g$ exist, then for $P(\mathrm{D}) = \mathrm{d}^n/\mathrm{d}x^n$ we have

$$\delta^{(n)}(x) * (f * g) = (f * g)^{(n)} = f^{(n)} * g = f * g^{(n)}. \tag{1.500}$$

Hence, for the derivative of order n of a distribution from $\mathcal{D}'(\mathbb{R})$, the convolution of the respective distribution with the distribution $\delta^{(n)}(x)$ is performed.

We can say that the distribution $\delta^{(n)}(x)$ acts as an operator of derivation of order n with respect to the convolution products. This property plays an essential role in the operational calculus.

Proposition 1.47 Let there be τ_a, $a \in \mathbb{R}^n$, and ν symmetry operator with respect to the origin of the coordinates. If $f, g \in \mathcal{D}'(\mathbb{R}^n)$ and $f * g$ exists, then the following formulae take place

$$\tau_a(f * g) = \tau_a f * g = f * \tau_a g, \tag{1.501}$$

$$(f * g)^\nu = f^\nu * g^\nu. \tag{1.502}$$

Below, we will state the continuity property of the convolution product.

Proposition 1.48 Let there be the distribution $f \in \mathcal{D}'(\mathbb{R}^n)$ and the sequence of distributions $(f_i)_{i \in \mathbb{N}} \subset \mathcal{D}'(\mathbb{R}^n)$ with the property $f_i \xrightarrow{\mathcal{D}'} f$ and $\operatorname{supp}(f_i) \subset \Omega$ bounded. Then, $\forall g \in \mathcal{D}'(\mathbb{R}^n)$ we have

$$f_i * g \xrightarrow{\mathcal{D}'} f * g. \tag{1.503}$$

Proof: Because f_i has compact support $\forall \varphi \in \mathcal{D}(\mathbb{R}^n)$, we have

$$(f_i * g, \varphi) = (f_i(y), (g(x), \varphi(x+y))) = (f_i(y), h(y)(g(x), \varphi(x+y))), \tag{1.504}$$

where $h(y) \in \mathcal{D}(\mathbb{R}^n)$ and has the value 1 in a compact neighborhood of the bounded set in which are contained the supports of the distributions f_i.

Consequently, $h(y)(g(x), \varphi(x+y)) \in \mathcal{D}(\mathbb{R}^n)$ and, on the basis of the convergence of the distributions, we have

$$(f_i * g, \varphi) \to (f, h(y)(g(x), \varphi(x+y))) = (f * g, \varphi)\,, \tag{1.505}$$

namely $f_i * g \xrightarrow{\mathcal{D}'} f * g$. □

The property of continuity of the convolution product occurs in the following cases:

1. $f_i \xrightarrow{\mathcal{D}'} f,\ f_i, f \in \mathcal{D}'(\mathbb{R}^n), g \in \mathcal{E}'(\mathbb{R}^n)$ involves $f_i * g \xrightarrow{\mathcal{D}'} f * g$
2. $f_i \xrightarrow{\mathcal{D}'} f, g,\ f, f_i \in \mathcal{D}'(\mathbb{R}), \operatorname{supp}(f_i), \operatorname{supp}(g) \subset (a, \infty)$ or $(-\infty, b)$

   involves $f_i * g \xrightarrow{\mathcal{D}'} f * g$

   In particular, if $g \in \mathcal{D}'(\mathbb{R}^n)$ and $f_i \to \delta(x)$ with $\operatorname{supp}(f_i) \subset \Omega$ bounded, then $f_i * g \to \delta(x) * g = g$.
3. $f_i \xrightarrow{\mathcal{D}'} f, g_i \xrightarrow{\mathcal{D}'} g,\ f_i, g_i, f, g \in \mathcal{D}'(\mathbb{R})$ and

   $\operatorname{supp}(f_i), \operatorname{supp}(g_i), \operatorname{supp}(f), \operatorname{supp}(g) \subset (a, \infty)$ involve $f_i * g_i \xrightarrow{\mathcal{D}'} f * g$.

Proposition 1.49 Let there be $f \in \mathcal{D}'(\mathbb{R}^n), \alpha \in \mathcal{D}(\mathbb{R}^n)$ and D^p the derivation operator. Then, $f * \alpha \in C^\infty(\mathbb{R}^n), (f * \alpha)(x) = (f_t, \alpha(x-t))$ and

$$D^p(f * \alpha)(x) = (D^p f_t, \alpha(x-t)) = (f_t, D_x^p \alpha(x-t))\,. \tag{1.506}$$

Proof: Because α has compact support, then the convolution $f * \alpha$ exists and for $\varphi \in \mathcal{D}(\mathbb{R}^n)$ we have

$$\begin{aligned}
(f * \alpha, \varphi) &= (f(t) \times \alpha(u), \varphi(t+u)) = (f(t), (\alpha(u), \varphi(t+u))) \\
&= \left(f(t), \int_{\mathbb{R}^n} \alpha(u)\varphi(t+u)\mathrm{d}u\right) = \left(f(t), \int_{\mathbb{R}^n} \alpha(x-t)\varphi(x)\mathrm{d}x\right) \\
&= (f(t), (\varphi(x), \alpha(x-t))) = (f(t) \times \varphi(x), \alpha(x-t)) \\
&= (\varphi(x), (f(t), \alpha(x-t))) \\
&= \int_{\mathbb{R}^n} (f(t), \alpha(x-t))\varphi(x)\mathrm{d}x = ((f(t), \alpha(x-t)), \varphi),
\end{aligned} \tag{1.507}$$

hence $f * \alpha$ is a function of variable $x \in \mathbb{R}^n$ and $(f * \alpha)(x) = (f(t), \alpha(x-t))$.

To show that the function $\psi(x) = (f * \alpha)(x)$ is indefinitely derivable, for x fixed and $h = (0, 0, \ldots, 0, h_i, 0, 0, \ldots, 0)$, we consider

$$\frac{\psi(x+h) - \psi(x)}{h_i} = \left(f(t), \frac{\alpha(x+h-t) - \alpha(x-t)}{h_i}\right)\,. \tag{1.508}$$

For $h_i \to 0$, because f is a continuous functional on $\mathcal{D}(\mathbb{R}^n)$ and

$$\frac{\alpha(x+h-t)-\alpha(x-t)}{h_i} \to \frac{\partial\alpha(x-t)}{\partial x_i}$$

in the sense of the convergence in the space $\mathcal{D}(\mathbb{R}^n)$, we obtain

$$\begin{aligned}\frac{\partial}{\partial x_i}(f*\alpha)(x) &= \left(f(t), \frac{\partial\alpha(x-t)}{\partial x_i}\right) \\ &= \left(f(t), -\frac{\partial\alpha(x-t)}{\partial t_i}\right) = \left(\frac{\partial}{\partial t_i}f(t), \alpha(x-t)\right), \end{aligned} \tag{1.509}$$

giving

$$\mathrm{D}^p(f*\alpha)(x) = (f(t), \mathrm{D}_x^p\alpha(x-t)) = (\mathrm{D}^p f(t), \alpha(x-t)) . \tag{1.510}$$

We mention that the function $f*\alpha \in C^\infty(\mathbb{R}^n)$ is called the regularized of the distribution f.

Particularly, if $f \in \mathcal{E}'(\mathbb{R}^n)$, so that it has compact support, then the function $f*\alpha$ has compact support, because $\mathrm{supp}(f*\alpha) \subset \mathrm{supp}(f)+\mathrm{supp}(\alpha)$ and $\mathrm{supp}(f), \mathrm{supp}(\alpha)$ are compact sets, hence $f*\alpha \in \mathcal{D}(\mathbb{R}^n)$. □

Corollary 1.7 If $f \in \mathcal{E}'(\mathbb{R}^n)$ and $\alpha \in \mathcal{E}(\mathbb{R}^n)$ then

$$f*\alpha \in C^\infty(\mathbb{R}^n) \quad \text{and} \quad (f*\alpha)(x) = (f(t), \alpha(x-t)) . \tag{1.511}$$

Indeed, it follows since we have for $\varphi \in \mathcal{D}(\mathbb{R}^n)$

$$\begin{aligned}(f*\alpha, \varphi) &= (f(t)\times\alpha(u), \varphi(t+u)) \\ &= (f(t), (\alpha(u), \varphi(t+u))) = (f(t), h(t)(\alpha(u), \varphi(t+u))) , \end{aligned} \tag{1.512}$$

where $h \in \mathcal{D}(\mathbb{R}^n)$ and it has a value equal to 1 on a compact neighborhood of the distribution support $f \in \mathcal{E}'(\mathbb{R}^n)$.

Thus, if $f \in \mathcal{E}'(\mathbb{R}^n)$ and $\alpha = 1$ on $\mathbb{R}^n$, then because $\alpha \in \mathcal{E}(\mathbb{R}^n)$ we have

$$f*1 = (f(t), 1) = (f, 1) = \int_{\mathbb{R}^n} f(x)\mathrm{d}x . \tag{1.513}$$

This convolution is called the integral of the distribution f.

If D^p is a derivation operator, then we obtain $\mathrm{D}^p f*1 = \mathrm{D}^p(f*1) = 0$, hence $\int_{\mathbb{R}^n} \mathrm{D}^p f\mathrm{d}x = 0$.

Definition 1.37 We call the trace at the origin of the continuous function $f: \mathbb{R}^n \to \mathbb{C}$, the number $f(0)$ denoted by $\mathrm{Tr}\, f(x) = f(0)$.

According to the Proposition 1.49, if $f \in \mathcal{D}'(\mathbb{R}^n)$ and $\varphi \in \mathcal{D}(\mathbb{R}^n)$, then $f*\varphi \in C^\infty(\mathbb{R}^n)$ and we have

$$(f*\varphi)(x) = (f(t), \varphi(x-t)) , \tag{1.514}$$

giving

$$\mathrm{Tr}(f * \varphi) = (f * \varphi)(0) = (f(t), \varphi(-t)) = (f, \varphi^\nu) = (f^\nu, \varphi) . \tag{1.515}$$

Consequently, we have

$$\mathrm{Tr}(f^\nu * \varphi) = \mathrm{Tr}(f * \varphi^\nu) = (f^\nu, \varphi^\nu) = (f, (\varphi^\nu)^\nu) = (f, \varphi) . \tag{1.516}$$

Proposition 1.50 The necessary and sufficient condition for the distribution $f \in \mathcal{D}'(\mathbb{R}^n)$ to be null is that, for $\forall \varphi \in \mathcal{D}(\mathbb{R}^n)$, we should have $f * \varphi = 0$.
Indeed, if $f = 0$, then $\forall \varphi \in \mathcal{D}(\mathbb{R}^n)$ we can write

$$(f, \varphi) = 0 = \mathrm{Tr}(f * \varphi^\nu) \Rightarrow f * \varphi = 0 .$$

Reciprocally, if $\forall \varphi \in \mathcal{D}(\mathbb{R}^n)$ we have $f * \varphi = 0$, then $f * \varphi^\nu = 0$. Applying the formula (1.516), we have $(f, \varphi) = \mathrm{Tr}(f * \varphi^\nu) = 0$, hence $f = 0$.

Proposition 1.51 Let there be $f, g \in \mathcal{D}'(\mathbb{R}^n)$ and $\varphi \in \mathcal{D}(\mathbb{R}^n)$. If the convolution product $f * g \in \mathcal{D}'(\mathbb{R}^n)$ exists, then the formula follows:

$$(f * g, \varphi) = (f, g^\nu * \varphi) = (g, f^\nu * \varphi) . \tag{1.517}$$

Proof: Because $\varphi \in \mathcal{D}(\mathbb{R}^n)$, the product $(f * g) * \varphi$ exists, is commutative and associative. Consequently, we have

$$\begin{aligned}(f * g, \varphi) &= \mathrm{Tr}[(f * g) * \varphi^\nu] = \mathrm{Tr}[f * (g * \varphi^\nu)] \\ &= \mathrm{Tr}[f * (g^\nu * \varphi)^\nu] = (f, g^\nu * \varphi).\end{aligned} \tag{1.518}$$

We obtain the required result on the basis of the commutativity of the convolution product. □

From the formula (1.517), it follows that the convolution product $f * g$ is determined if we know one of the functions $f^\nu * \varphi$ or $g^\nu * \varphi$, $\forall \varphi \in \mathcal{D}(\mathbb{R}^n)$.
Let there be the distribution $f(x, t) \in \mathcal{D}'(\mathbb{R}^{n+m})$ and $\varphi(x) \in \mathcal{D}(\mathbb{R}^n)$. We define the distribution

$$(f(x, t), \varphi(x)) \in \mathcal{D}'(\mathbb{R}^m) \tag{1.519}$$

by the formula

$$((f(x, t), \varphi(x)), \psi(t)) = (f(x, t), \varphi(x)\psi(t)) , \tag{1.520}$$

where $\varphi \in \mathcal{D}(\mathbb{R}^n)$, $\psi \in \mathcal{D}(\mathbb{R}^m)$.

Proposition 1.52 Let there be $f(x, t) \in \mathcal{D}'(\mathbb{R}^{n+m})$ and D_t^β the derivation operator with respect to $t \in \mathbb{R}^m$. Then, $\forall \varphi \in \mathcal{D}(\mathbb{R}^n)$ we have

$$\left(\mathrm{D}_t^\beta f(x, t), \varphi(x)\right) = \mathrm{D}_t^\beta (f(x, t), \varphi(x)) , \quad \beta \in \mathbb{N}_0^m . \tag{1.521}$$

Indeed, taking into account (1.520) for $\psi(t) \in \mathcal{D}(\mathbb{R}^m)$, we can write

$$\begin{aligned}
&\left(\left(\mathrm{D}_t^\beta f(x,t), \varphi(x)\right), \psi(t)\right) = \left(\mathrm{D}_t^\beta f(x,t), \varphi(x)\psi(t)\right) \\
&= (-1)^{|\beta|}\left(f(x,t), \varphi \mathrm{D}_t^\beta \psi(t)\right) = (-1)^{|\beta|}\left((f(x,t), \varphi(x)), \mathrm{D}_t^\beta \psi\right) \\
&= \left(\mathrm{D}_t^\beta (f(x,t), \varphi(x)), \psi(t)\right) ; \qquad (1.522)
\end{aligned}$$

hence, based on the equality of two distributions, we obtain the formula (1.521).

1.3.8.3 The Convolution of Distributions Depending on a Parameter: Properties

Proposition 1.53 Let there be the distribution $f_t, g \in \mathcal{D}'(\mathbb{R}^n)$, $t \in T \subset \mathbb{R}$ being a parameter. If $\partial f_t(x)/\partial t$ and $f_t * g \in \mathcal{D}'(\mathbb{R}^n)$ exist, then we have

$$\frac{\partial}{\partial t}(f_t(x) * g(x)) = \frac{\partial f_t(x)}{\partial t} * g(x) . \qquad (1.523)$$

Remark 1.3 The formula (1.523) remains valid if $\partial f_t(x)/\partial t$, $\forall t \in T \subset \mathbb{R}$, exists and if the distributions f_t, g satisfy one of the conditions:

1. $f_t \in \mathcal{E}'(\mathbb{R}^n)$, $\quad \forall t \in T \subset \mathbb{R}$,
2. $g \in \mathcal{E}'(\mathbb{R}^n)$,
3. $f_t, g \in \mathcal{D}'(\mathbb{R})$ and $\mathrm{supp}(f_t), \mathrm{supp}(g) \subset [a, \infty)$ or $\mathrm{supp}(f_t), \mathrm{supp}(g) \subset (-\infty, b]$.

In particular, if $f_t(x) = \varphi(x,t) \in D(\mathbb{R}^{n+1})$, then $\forall g \in \mathcal{D}'(\mathbb{R}^n)$ on the basis of the formula (1.523), we obtain

$$\frac{\partial}{\partial t}\left(g(x) \underset{x}{*} \varphi(x,t)\right) = g(x) \underset{x}{*} \frac{\partial}{\partial t}\varphi(x,t) , \qquad (1.524)$$

where $t \in \mathbb{R}$ is considered parameter, and the convolution is performed with respect to the variable $x \in \mathbb{R}^n$.
Indeed, for $t \in \mathbb{R}$ fixed, $\varphi(x,t) \in \mathcal{D}(\mathbb{R}^n)$ and consequently $\forall g \in \mathcal{D}'(\mathbb{R}^n)$, the convolution product $g(x) *_x (\partial/\partial t)\varphi(x,t)$ exists and is a function of class $C^\infty(\mathbb{R}^{n+1})$, hence $g(x) *_x (\partial/\partial t)\varphi(x,t) \in C^\infty(\mathbb{R}^{n+1})$.

If $g \in \mathcal{E}'(\mathbb{R}^n)$, then $g(x) *_x (\partial/\partial t)\varphi(x,t) \in \mathcal{D}(\mathbb{R}^{n+1})$.

Proposition 1.54 Let there be the distributions $f_t, g_t \in \mathcal{D}'_+$ depending on the parameter $t \in T \subset \mathbb{R}$. If $(\partial/\partial t) f_t(x)$, $(\partial/\partial t) g_t(x) \in \mathcal{D}'_+$, then $(\partial/\partial t)(f_t(x) * g_t(x)) \in \mathcal{D}'_+$ exists and the formula follows:

$$\frac{\partial}{\partial t}(f_t * g_t) = \frac{\partial}{\partial t} f_t * g_t + f_t * \frac{\partial}{\partial t} g_t . \qquad (1.525)$$

Proof: On the basis of the definition of a derivative with respect to the parameter $t \in T \subset \mathbb{R}$ we have

$$\frac{\partial}{\partial t}(f_t * g_t) = \lim_{\Delta t \to 0} \frac{f_{t+\Delta t} * g_{t+\Delta t} - f_t * g_t}{\Delta t}$$
$$= \lim_{\Delta t \to 0} \left(\frac{f_{t+\Delta t} - f_t}{\Delta t} * g_{t+\Delta t} + \frac{g_{t+\Delta t} - g_t}{\Delta t} * f_t \right) . \quad (1.526)$$

Taking into account the Proposition 1.48, we obtain

$$\lim_{\Delta t \to 0} \frac{f_{t+\Delta t} - f_t}{\Delta t} * g_{t+\Delta t} = \frac{\partial}{\partial t} f_t * g_t \text{ and } \lim_{\Delta t \to 0} \frac{g_{t+\Delta t} - g_t}{\Delta t} * f_t = \frac{\partial}{\partial t} g_t * f_t . \quad (1.527)$$

It follows that the right-hand side of the relation (1.526) exists, which implies the existence of the derivative $(\partial/\partial t)(f_t * g_t)$ and also the formula (1.525). □

Example 1.35 Let there be the distributions $f_t(x) = \delta(x - at) \in \mathcal{D}'_+, g_t(x) = H(x - bt) \in \mathcal{D}'_+$ depending on the parameter $t \geq 0$, where $a, b \in \mathbb{R}_+$ and $H(u) \in \mathcal{D}'_+$ is the Heaviside distribution. We have

$$\frac{\partial}{\partial t}(\delta(x - at) * H(x - bt)) = -(a + b)\delta(x - (a + b)t) . \quad (1.528)$$

Indeed, applying the formula (1.525), we obtain

$$\frac{\partial}{\partial t}(\delta(x - at) * H(x - bt))$$
$$= \frac{\partial}{\partial t}\delta(x - at) * H(x - bt) + \delta(x - at) * \frac{\partial}{\partial t} H(x - bt)$$
$$= -a\delta'(x - at) * H(x - bt) - b\delta(x - at) * \delta(x - bt)$$
$$= -a\delta(x - at) * \delta(x - bt) - b\delta(x - at) * \delta(x - bt)$$
$$= -(a + b)\delta(x - (a + b)t). \quad (1.529)$$

Proposition 1.55 Let there be the distributions $f_t, g_t \in \mathcal{D}'_+$ depending on the parameter $t \in [a, b]$, continuous on $[a, b]$ and let $F(x), G(x) \in \mathcal{D}'_+$. We have

$$\int_a^b [f_t(x) * F(x) + g_t(x) * G(x)]\mathrm{d}t$$
$$= F(x) * \int_a^b f_t(x)\mathrm{d}t + G(x) * \int_a^b g_t(x)\mathrm{d}t . \quad (1.530)$$

Particularly, for $F(x) = \alpha\delta(x), G(x) = \beta\delta(x), \alpha, \beta \in \mathbb{R}$ we obtain

$$\int_a^b [\alpha f_t(x) + \beta g_t(x)]\mathrm{d}t = \alpha \int_a^b f_t(x)\mathrm{d}t + \beta \int_a^b g_t(x)\mathrm{d}t . \quad (1.531)$$

Proposition 1.56 Let there be the distributions $f_t, g_t \in \mathcal{D}'_+$ depending on the parameter $t \in [a, b]$. If $(\partial/\partial t) f_t(x)$, $(\partial/\partial t) g_t(x) \in \mathcal{D}'_+$ exist, then we have

$$\int_a^b \left(\frac{\partial}{\partial t} f_t(x) * g_t(x)\right) \mathrm{d}t = \left(f_t(x) * g_t(x)\right)\Big|_a^b - \int_a^b \left(f_t(x) * \frac{\partial}{\partial t} g_t(x)\right) \mathrm{d}t \,, \tag{1.532}$$

where

$$\left(f_t * g_t\right)\Big|_a^b = f_b(x) * g_b(x) - f_a(x) * g_a(x) \,. \tag{1.533}$$

Proof: Indeed, because $f_t, g_t \in \mathcal{D}'_+$, it follows that $\partial f_t/\partial t, \partial g_t/\partial t \in \mathcal{D}'_+$ and, as the convolution products between the distributions $f_t, g_t, \partial f_t/\partial t, \partial g_t/\partial t$ exist, according to formulae (1.525), (1.531), and (1.444), we obtain

$$\int_a^b \frac{\partial}{\partial t}(f_t * g_t)\mathrm{d}t = (f_t * g_t)\Big|_a^b = \int_a^b \left(\frac{\partial}{\partial t} f_t * g_t\right) \mathrm{d}t + \int_a^b \left(f_t * \frac{\partial}{\partial t} g_t\right) \mathrm{d}t \,, \tag{1.534}$$

that is, the formula (1.532).

The relation (1.532) represents the analogue of the integration by parts formula. □

Example 1.36 Applying the formula (1.530) to calculate the integral

$$I = \int_a^b \delta'(x) * H(x - t)\mathrm{d}t \,, \tag{1.535}$$

where $0 < a < b$ and $t \in [a, b]$ is a parameter.

We have

$$I = \delta'(x) * \int_a^b H(x - t)\mathrm{d}t = \frac{\mathrm{d}}{\mathrm{d}x} \int_a^b H(x - t)\mathrm{d}t = \frac{\mathrm{d}}{\mathrm{d}x} F_1(x) \,,$$

where, according to the Example 1.30, we obtain

$$F_1(x) = \begin{cases} 0 \,, & x \leq a < b \,, \\ b - a \,, & a < b < x \,, \\ x - a \,, & a \leq x < b \,. \end{cases} \tag{1.536}$$

Consequently, applying the differentiation formula of the functions with discontinuities of the first order we obtain

$$I = \frac{\mathrm{d}}{\mathrm{d}x} F_1(x) = \begin{cases} 0 \,, & x < a \,, \\ 1 \,, & x \in (a, b) = H(x - a) - H(x - b) \,, \\ 0 \,, & x > b \,. \end{cases} \tag{1.537}$$

Remark 1.4 By direct calculation, we have

$$I = \int_a^b \delta(x-t)\mathrm{d}t = -\int_a^b \frac{\partial}{\partial t} H(x-t)\mathrm{d}t$$
$$= -[H(x-t)]|_a^b = H(x-a) - H(x-b)\,. \tag{1.538}$$

1.3.8.4 The Partial Convolution Product for Functions and Distributions

Definition 1.38 Let there be the distributions $f(x,t) \in \mathcal{D}'(\mathbb{R}^{n+m})$ and $g(x) \in \mathcal{D}'(\mathbb{R}^n)$. We call partial convolution product of the distribution f with g, the distribution denoted $f(x,t) \otimes_x g(x) \in \mathcal{D}'(\mathbb{R}^n)$, defined by the formula

$$f(x,t) \underset{x}{\otimes} g(x) = f(x,t) * (g(x) \times \delta(t))\,, \tag{1.539}$$

where $\delta(t) \in \mathcal{D}'(\mathbb{R}^m)$ is Dirac's delta distribution.

The symbol $\otimes_x$ for the convolution product denotes that the convolution is performed only with respect to the variable $x \in \mathbb{R}^n$, common to the distributions $f(x,t) \in \mathcal{D}'(\mathbb{R}^{n+m})$ and $g(x) \in \mathcal{D}'(\mathbb{R}^n)$, considered in different spaces.

On the right-hand side of the formula (1.539), the convolution product denoted by the symbol $*$ obviously refers to the variables $(x,t) \in \mathbb{R}^n \times \mathbb{R}^m$.

In the case of existence of the partial convolution product, the latter is a distribution from $\mathcal{D}'(\mathbb{R}^{n+m})$, hence $f(x,t) \otimes_x g(x) \in \mathcal{D}'(\mathbb{R}^{n+m})$.

Taking into account the definition of the commutativity of the partial convolution product, we will not distinguish between the distributions $f(x,t) \otimes_x g(x)$ and $g(x) \otimes_x f(x,t)$.

Proposition 1.57 Let there be $f(x,t) \in \mathcal{D}'(\mathbb{R}^{n+m})$ and $g(x) \in \mathcal{D}'(\mathbb{R}^n)$. The partial convolution product $f(x,t) \otimes_x g(x) \in \mathcal{D}'(\mathbb{R}^{n+m})$ exists if one of the distributions f, g has compact support.

Indeed, if $g(x) \in \mathcal{E}'(\mathbb{R}^n)$, hence g has compact support, then $g(x) \times \delta(t)$ has compact support, hence the right-hand side of expression (1.539) exists.

Also, if $f(x,t) \in \mathcal{E}'(\mathbb{R}^{n+m})$, then the product $f(x,t) * (g(x) \times \delta(t))$ exists, wherefrom the proposition is proved.

From the above considerations, it follows that the partial convolution product denoted by the symbol $\otimes_x$ is a new law of composition for the distributions $f(x,t) \in \mathcal{D}'(\mathbb{R}^{n+m})$ and $g(x) \in \mathcal{D}'(\mathbb{R}^n)$ with respect to the common variable $x \in \mathbb{R}^n$.

This new introduced convolution product [20] has wide applications in deformable solid mechanics and in particular in viscoelasticity [21–23].

The structure relation of the partial convolution product is shown as follows:

Proposition 1.58 Representation formula Let there be the distributions $f(x,t) \in \mathcal{D}'(\mathbb{R}^{n+m})$ and $g(x) \in \mathcal{D}'(\mathbb{R}^n)$. If the partial convolution product $f(x,t) \otimes_x g(x) \in$

$\mathcal{D}'(\mathbb{R}^{n+m})$, then the relation

$$\left(f(x,t)\underset{x}{\otimes} g(x),\varphi\right) = \left(f(x,t), g^{\nu} \underset{x}{*} \varphi(x,t)\right), \quad \forall\varphi(x,t)\in\mathcal{D}(\mathbb{R}^{n+m}), \quad (1.540)$$

occurs, where g^{ν} is symmetric with respect to the origin of the distribution $g(x) \in \mathcal{D}'(\mathbb{R}^n)$.

Proof: We suppose that $g(x) \in \mathcal{E}'(\mathbb{R}^n)$. Then the convolution product exists and $\forall\varphi(x,t) \in \mathcal{D}(\mathbb{R}^{n+m})$, thus we can write

$$\begin{aligned}
\left(f(x,t)\underset{x}{\otimes} g(x),\varphi(x,t)\right) &= (f(x,t) * [g(x)\times\delta(t)],\varphi(x,t)) \\
&= (f(x,t)\times g(u)\times\delta(v),\varphi(x+u,t+v)) \\
&= (f(x,t)\times g(u),\varphi(x+u,t)) = (f(x,t),(g(u),\varphi(x+u,t)))\,. \qquad (1.541)
\end{aligned}$$

On the other hand, on the basis of the Proposition 1.49, because $g(x) \in \mathcal{E}'(\mathbb{R}^n)$, the convolution product $g^{\nu} *_x \varphi(x,t) \in \mathcal{D}'(\mathbb{R}^n)$ exists and we have

$$g^{\nu} \underset{x}{*} \varphi(x,t) = (g^{\nu}(u),\varphi(x-u,t)) = (g(u),\varphi(x+u,t))\,. \quad (1.542)$$

Taking into account (1.541), we obtain

$$(f(x,t)\underset{x}{\otimes} g(x),\varphi(x,t)) = \left(f(x,t), g^{\nu} \underset{x}{*}\varphi(x,t)\right), \quad \forall\varphi(x,t)\in\mathcal{D}(\mathbb{R}^{n+m}), \quad (1.543)$$

namely the required formula (1.540).

If $f(x,t) \in \mathcal{E}'(\mathbb{R}^{n+m})$, then on the basis of the Proposition 1.49, $\forall g \in \mathcal{D}'(\mathbb{R}^n)$ we have $g^{\nu} *_x \varphi(x,t) = (g(u),\varphi(x+u,t)) \in C^{\infty}(\mathbb{R}^{n+m})$ and the formula (1.541) becomes

$$\begin{aligned}
\left(f(x,t)\underset{x}{\otimes} g(x),\varphi(x,t)\right) &= \left(f(x,t), g^{\nu} \underset{x}{*}\varphi(x,t)\right) \\
&= \left(f(x,t), h(x,t)\left(g^{\nu} \underset{x}{*}\varphi(x,t)\right)\right), \qquad (1.544)
\end{aligned}$$

where $h(x,t) \in \mathcal{D}(\mathbb{R}^{n+m})$ and has the value 1 in a compact neighborhood of the distribution support $f(x,t) \in \mathcal{E}'(\mathbb{R}^{n+m})$.

Obviously, $h(x,t)(g^{\nu} *_x \varphi(x,t)) \in \mathcal{D}(\mathbb{R}^{n+m})$, therefore the right side of the formula (1.544) makes sense, which proves the equality (1.540). □

Comparing the formula (1.540) of the partial convolution product with the formula (1.517), that is,

$$(f*g,\varphi) = (f, g^{\nu} * \varphi) = (g, f^{\nu} * \varphi)\,, \quad f,g\in\mathcal{D}'(\mathbb{R}^n)\,, \quad \varphi\in\mathcal{D}(\mathbb{R}^n)\,, \quad (1.545)$$

we see that these two types of convolutions have the same structure, in the sense that they are expressed with respect to the common variable of both distributions which are convoluted.

From this point of view, we can say that the partial convolution product is a generalization of the ordinary convolution product.

Below, we give some properties of the partial convolution product.

Proposition 1.59 Let there be the distributions $f(x,t) \in \mathcal{D}'(\mathbb{R}^{n+m}), g(x) \in \mathcal{D}'(\mathbb{R}^n)$. If the product $f(x,t) \otimes_x g(x) \in \mathcal{D}'(\mathbb{R}^{n+m})$ exists and $\mathrm{D}_x^\alpha, \mathrm{D}_t^\beta$ are derivation operators with respect to the variables $x \in \mathbb{R}^n, t \in \mathbb{R}^m$, respectively, then the following formulae take place

$$\mathrm{D}_x^\alpha \left[f(x,t) \underset{x}{\otimes} g(x) \right] = \mathrm{D}_x^\alpha f(x,t) \underset{x}{\otimes} g(x) = f(x,t) \underset{x}{\otimes} \mathrm{D}_x^\alpha g(x) , \tag{1.546}$$

$$\mathrm{D}_t^\beta \left[f(x,t) \underset{x}{\otimes} g(x) \right] = \mathrm{D}_t^\beta f(x,t) \underset{x}{\otimes} g(x) . \tag{1.547}$$

Proposition 1.60 Let there be the distributions $f(x,t) \in \mathcal{D}'(\mathbb{R}^{n+m}), g(x) \in \mathcal{D}'(\mathbb{R}^n)$ and D_t^β the derivation operators with respect to the variable $t \in \mathbb{R}^m$. If the partial convolution product $f(x,t) \otimes_x g(x)$ exists, then the formula follows:

$$\left(\mathrm{D}_t^\beta f(x,t) \underset{x}{\otimes} g(x), \varphi(x) \right) = \mathrm{D}_t^\beta (f(x,t) \underset{x}{\otimes} g(x), \varphi(x)) , \quad \forall \varphi \in \mathcal{D}(\mathbb{R}^n) . \tag{1.548}$$

Indeed, on the basis of the Proposition 1.52 and the formula (1.547) follows the required relation.

Remark 1.5 A similar relation occurs for the distributions depending on the parameter $t \in \mathbb{R}^m$. Thus, if $f_t(x), g(x) \in \mathcal{D}'(\mathbb{R}^n)$ and $f_t(x) * g(x) \in \mathcal{D}'(\mathbb{R}^n)$ exists, then $\forall \varphi(x) \in \mathcal{D}(\mathbb{R}^n)$ and we have

$$\left(\mathrm{D}_t^\beta f_t(x) * g(x), \varphi(x) \right) = \mathrm{D}_t^\beta (f_t(x) * g(x), \varphi(x)) . \tag{1.549}$$

Proposition 1.61 Let there be the distributions $f \in \mathcal{D}'(\mathbb{R}^n), h \in \mathcal{E}'(\mathbb{R}^n), g \in \mathcal{D}'(\mathbb{R}^m)$. Then we have

$$(f(x) \times g(t)) \underset{x}{\otimes} h(x) = (f * h)(x) \times g(t) . \tag{1.550}$$

Example 1.37 Let there be D_x^α the derivation operator with respect to the variable $x \in \mathbb{R}^n$; then we have the relations

$$f(x,t) \underset{x}{\otimes} \mathrm{D}_x^\alpha \delta(x) = \mathrm{D}_x^\alpha f(x,t) , \quad \forall f(x,t) \in \mathcal{D}'(\mathbb{R}^{n+m}) , \tag{1.551}$$

$$\delta(x,t) \underset{x}{\otimes} \mathrm{D}_x^\alpha g(x) = \mathrm{D}_x^\alpha g(x) \times \delta(t), \forall g(x) \in \mathcal{D}'(\mathbb{R}^n) . \tag{1.552}$$

In particular, for $|\alpha| = 0$ we have

$$f(x,t) \underset{x}{\otimes} \delta(x) = f(x,t)\,, \quad \forall f(x,t) \in \mathcal{D}'(\mathbb{R}^{n+m})\,, \tag{1.553}$$

$$\delta(x,t) \underset{x}{\otimes} g(x) = g(x) \times \delta(t)\,, \quad \forall g(x) \in \mathcal{D}'(\mathbb{R}^{n})\,. \tag{1.554}$$

Proposition 1.62 Let there be the distributions $f(x,t) \in \mathcal{D}'(\mathbb{R}^{n+m}), g(x) \in \mathcal{D}'(\mathbb{R}^{n})$. If the product $f(x,t) \otimes_x g(x) \in \mathcal{D}'(\mathbb{R}^{n+m})$ exists, then we have

$$\left(f(x,t) \underset{x}{\otimes} g(x)\right)^{\nu} = f^{\nu} \underset{x}{\otimes} g^{\nu}\,, \tag{1.555}$$

$$\tau_a\left(f(x,t) \underset{x}{\otimes} g(x)\right) = \tau_a f(x,t) \underset{x}{\otimes} g(x)\,, \tag{1.556}$$

where τ_a is the translation operator by the vector $a \in \mathbb{R}^{n+m}$.

The partial convolution product has the property of continuity as the usual convolution product.

Proposition 1.63 Let there be the distribution $g(x) \in \mathcal{D}'(\mathbb{R}^{n})$ and the sequence of distributions $(g_i(x))_{i\in\mathbb{N}} \subset \mathcal{D}'(\mathbb{R}^{n})$ with the properties $g_i \xrightarrow{\mathcal{D}'(\mathbb{R}^{n})} g$ and $\text{supp}(g_i) \subset \Omega$ bounded. Then, $\forall f(x,t) \in \mathcal{D}'(\mathbb{R}^{n+m})$ we have

$$f(x,t) \underset{x}{\otimes} g_i(x) \xrightarrow{\mathcal{D}'(\mathbb{R}^{n+m})} f(x,t) \underset{x}{\otimes} g(x)\,. \tag{1.557}$$

As regards the support of the partial convolution product we can state [24]:

Proposition 1.64 Let there be the distributions $f(x,t) \in \mathcal{D}'(\mathbb{R}^{n+m})$ and $g(x) \in \mathcal{D}'(\mathbb{R}^{n})$. If $f \otimes_x g \in \mathcal{D}'(\mathbb{R}^{n+m})$ exists, $\text{supp}(f) = \Omega \times T$, $\text{supp}(g) = \Omega'$, $\Omega, \Omega' \subset \mathbb{R}^{n}$, $T \subset \mathbb{R}^{m}$, then

$$\text{supp}\left(f \underset{x}{\otimes} g\right) \subset \overline{(\Omega + \Omega') \times T}\,. \tag{1.558}$$

In particular, if $f(x,t) \in \mathcal{D}'(\mathbb{R} \times \mathbb{R}^{m}), g(x) \in \mathcal{D}'(\mathbb{R})$ and $\text{supp}(f) = [0,\infty) \times T$, $\text{supp}(g) \subset [0,\infty) \times T$, $T \subset \mathbb{R}^{m}$, then $\Omega + \Omega' \subset [0,\infty)$ and the formula (1.558) becomes

$$\text{supp}\left(f \underset{x}{\otimes} g\right) \subset [0,\infty) \times T\,. \tag{1.559}$$

We have seen that the partial convolution product exists if one of the factors is a distribution with compact support. Another case of existence of the partial convolution product which has particular importance in mechanics is given by [24]:

Proposition 1.65 If $f(x,t) \in \mathcal{D}'(\mathbb{R} \times \mathbb{R}^{m}), g(x) \in \mathcal{D}'(\mathbb{R})$ and $\text{supp}(f) = (a,\infty) \times T$, $\text{supp}(g) \subset (b,\infty)$, $T \subset \mathbb{R}^{m}$, then $f(x,t) \otimes_x g(x) \in \mathcal{D}'(\mathbb{R} \times \mathbb{R}^{m})$ exists.

Proposition 1.66 Let there be $f(x,t) \in \mathcal{D}'(\mathbb{R} \times \mathbb{R}^{m}), g(x) \in \mathcal{D}'(\mathbb{R})$. If $\text{supp}(f) \subset \Omega \times T$, Ω-compact, $T \subset \mathbb{R}^{m}$ and $\text{supp}(g) = \Omega'$ arbitrary, then the partial convolution product $f(x,t) \otimes_x g(x) \in \mathcal{D}'(\mathbb{R} \times \mathbb{R}^{m})$ exists.

A property that expresses a certain relation between the partial convolution product and the usual one is given by the following.

Proposition 1.67 Let there be the distributions $f(x,t) \in \mathcal{D}'(\mathbb{R}^{n+m})$ and $g_1(x)$, $g_2(x) \in \mathcal{E}'(\mathbb{R}^n)$. We have

$$f \underset{x}{\otimes} (g_1 * g_2) = \left(f \underset{x}{\otimes} g_1\right) \underset{x}{\otimes} g_2 = \left(f \underset{x}{\otimes} g_2\right) \underset{x}{\otimes} g_1 \,. \tag{1.560}$$

Remark 1.6 The formula (1.560) remains valid if $f(x,t) \in \mathcal{D}'(\mathbb{R}^{n+m})$ and $\operatorname{supp}(f) \subset [0,\infty) \times T$, $T \subset \mathbb{R}^m$, $g_1(x), g_2(x) \in \mathcal{D}'_+$, hence $\operatorname{supp}(g_1), \operatorname{supp}(g_2) \subset [0,\infty)$.

Proposition 1.68 Let there be $E(x) \in \mathcal{D}'(\mathbb{R}^n)$ a fundamental solution of the differential operator with constant coefficients $P(\mathrm{D}_x) = \sum_{|\alpha| \le l} a_\alpha \mathrm{D}_x^\alpha$, $\alpha \in \mathbb{N}_0^n$, $a_\alpha \in \mathbb{C}$. If $f(x,t) \in \mathcal{D}'(\mathbb{R}^{n+m})$; if the partial convolution product $f(x,t) \otimes_x E(x) \in \mathcal{D}'(\mathbb{R}^{n+m})$ exists, then the distribution

$$u(x,t) = f(x,t) \underset{x}{\otimes} E(x) = f(x,t) * [E(x) \times \delta(t)] \,, \tag{1.561}$$

is a solution for the equation

$$P(\mathrm{D}_x) u(x,t) = f(x,t) \,. \tag{1.562}$$

Indeed, using the formula (1.546) we obtain

$$P(\mathrm{D}_x) u(x,t) = f(x,t) \underset{x}{\otimes} P(\mathrm{D}_x) E(x) \,. \tag{1.563}$$

Observing that $P(\mathrm{D}_x) E(x) = \delta(x)$ and taking into account (1.554), the relation (1.563) becomes $P(\mathrm{D}_x) u(x,t) = f(x,t) \otimes_x \delta(x) = f(x,t)$.

Remark 1.7 It follows that a fundamental solution of the operator $P(\mathrm{D}_x)$ in $\mathcal{D}'(\mathbb{R}^{n+m})$ is the distribution $E_1(x,t) \in \mathcal{D}'(\mathbb{R}^{n+m})$ with

$$E_1(x,t) = E(x) \times \delta(t),\ E(x) \in \mathcal{D}'(\mathbb{R}^n) \,, \quad t \in \mathbb{R}^m \,. \tag{1.564}$$

Indeed, we have

$$\begin{aligned} P(\mathrm{D}_x) E_1(x,t) &= P(\mathrm{D}_x)[E(x) \times \delta(t)] \\ &= [P(\mathrm{D}_x) E(x)] \times \delta(t) = \delta(x) \times \delta(t) = \delta(x,t) \,. \end{aligned} \tag{1.565}$$

Example 1.38 Let there be $\delta_\lambda(x) = \delta(x - \lambda) \in E'(\mathbb{R}^n)$ and $f(x,t) \in \mathcal{D}'(\mathbb{R}^{n+m})$. Then, we have $f \otimes_x (\delta_a * \delta_b) = f(x - a - b, t)$. Because $f(x,t) \otimes_x \delta_a(x) = f(x,t) * [\delta_a(x) \times \delta(x)] = f(x - a, t)$, on the basis of the formula (1.560) we obtain

$$f \underset{x}{\otimes} (\delta_a * \delta_b) = \left(f \underset{x}{\otimes} \delta_a\right) \underset{x}{\otimes} \delta_b = f(x-a,t) \underset{x}{\otimes} \delta_b(x) = f(x-a-b,t) \,. \tag{1.566}$$

Example 1.39 We consider the Poisson equation

$$\Delta u(x,t) = f(x,t)\,, \tag{1.567}$$

where $\Delta = (\partial^2/\partial x_1^2) + \cdots + (\partial^2/\partial x_n^2)$ is the Laplace operator in $\mathbb{R}^n$ and $f(x,t) \in \mathcal{E}'(\mathbb{R}^{n+m}) \subset \mathcal{D}'(\mathbb{R}^{n+m})$ a distribution with compact support.

Let us show that a solution in $\mathcal{D}'(\mathbb{R}^{n+m})$ of (1.567) is the distribution $u(x,t) \in \mathcal{D}'(\mathbb{R}^{n+m})$ given by

$$u(x,t) = f(x,t) \underset{x}{\otimes} E(x) = f(x,t) * (E(x) \times \delta(t))\,, \tag{1.568}$$

where

$$E(x) = \begin{cases} \dfrac{-\Gamma[n/2]}{2(n-2)\pi^{n/2}} \cdot \dfrac{1}{r^{n-2}}\,, & n \geq 3\,, \\ \dfrac{\ln r}{2\pi}\,, & n = 2\,, \\ \dfrac{r}{2}\,, & n = 1\,, \end{cases} \tag{1.569}$$

$r = \|x\|$, $x \in \mathbb{R}^n$, while $\Gamma(p) = \int_0^\infty \mathrm{e}^{-t} t^{p-1} \mathrm{d}t$, $p > 0$ is the Euler gamma function.

Indeed, the linear differential operator with constant coefficients $P(\mathrm{D}_x)$ is the Laplace operator Δ, hence $P(\mathrm{D}_x) = \Delta$; taking into account the Example 1.19, the distribution $E(x) \in \mathcal{D}'(\mathbb{R}^n)$ given by (1.569) is the fundamental solution of the operator Δ in $\mathbb{R}^n$, $n \geq 1$.

Consequently, based on the Proposition 1.68, it follows that the distribution $u(x,t) \in \mathcal{D}'(\mathbb{R}^{n+m})$ specified by (1.569) is the solution of (1.567).

Example 1.40 We consider the equation

$$\frac{\partial^2 u(x,t)}{\partial x^2} = f(x,t)\,, \quad u, f \in \mathcal{D}'(\mathbb{R}^2)\,. \tag{1.570}$$

For $f(x,t) \in \mathcal{D}'(\mathbb{R}^2)$ where $\mathrm{supp}(f(x,t)) = [0,\infty) \times T$, $T \subset \mathbb{R}$, a solution of (1.570) is the distribution

$$u(x,t) = f(x,t) \underset{x}{\otimes} x H(x) = f(x,t) * [x H(x) \times \delta(t)]\,, \tag{1.571}$$

$H(x) \in \mathcal{D}'(\mathbb{R})$ being the Heaviside distribution.

Indeed, a fundamental solution of the operator $P(\mathrm{D}_x) = \partial^2/\partial x^2$ in $\mathcal{D}'(\mathbb{R})$ is the distribution $E(x) = x H(x) \in \mathcal{D}'(\mathbb{R})$.

Consequently, a solution of (1.570) is the distribution

$$u(x,t) = f(x,t) \underset{x}{\otimes} x H(x)\,, \tag{1.572}$$

which exists, taking into account the Proposition 1.65.

1.3.9
Partial Convolution Product of Functions

To prove the consistency of the partial convolution product introduced for distributions of different spaces, we show that in the case of functions, that is, distributions of function type, this operation coincides with the convolution operation with respect to the common variable of the two functions.

We can state the following.

Proposition 1.69 If $f(x,t) \in L^1(\mathbb{R}^{n+m})$ and $g(x) \in L^1(\mathbb{R}^n)$, then the product $f(x,t) \otimes_x g(x)$ exists and we have

$$f(x,t) \underset{x}{*} g(x) = f(x,t) \underset{x}{\otimes} g(x) \in L^1(\mathbb{R}^{n+m}) , \tag{1.573}$$

$$\int\limits_{\mathbb{R}^{n+m}} \left(f(x,t) \underset{x}{*} g(x) \right) \mathrm{d}x\mathrm{d}t = \int\limits_{\mathbb{R}^{n+m}} f(x,t)\mathrm{d}x\mathrm{d}t \cdot \int\limits_{\mathbb{R}^n} g(x)\mathrm{d}x , \tag{1.574}$$

$$\| f(x,t) \underset{x}{*} g(x) \|_1 \le \| f \|_1 \cdot \| g \|_1 . \tag{1.575}$$

From above, the symbol $*_x$ means the usual convolution with respect to $x \in \mathbb{R}^n$ considering $t \in \mathbb{R}^m$ fixed, and the symbol $\otimes_x$ represents the partial convolution product with respect to $x \in \mathbb{R}^n$ introduced by the Definition 1.38 for distribution from the spaces $\mathcal{D}'(\mathbb{R}^{n+m})$ and $\mathcal{D}'(\mathbb{R}^n)$.

Proposition 1.70 Let there be the locally integrable functions $f(x,t) \in L^1_{\text{loc}}(\mathbb{R} \times \mathbb{R}^m)$ and $g(x) \in L^1_{\text{loc}}(\mathbb{R})$. If $\operatorname{supp}(f(x,t)) \subset [0,\infty) \times T$, $\operatorname{supp}(g(x)) \subset [0,\infty)$, $T \subset \mathbb{R}^m$, then $f \otimes_x g$ is a locally integrable function on $\mathbb{R} \times \mathbb{R}^m$ and we have

$$\begin{aligned} (f \underset{x}{\otimes} g)(x,t) &= (f \underset{x}{*} g)(x,t) \\ &= \begin{cases} 0 , & (x,t) \in (-\infty, 0] \times T , \\ \displaystyle\int_0^x f(\xi,t) g(x-\xi)\mathrm{d}\xi , & (x,t) \in [0,\infty) \times T . \end{cases} \end{aligned} \tag{1.576}$$

Example 1.41 For Heaviside distributions $H(x,t) \in \mathcal{D}'(\mathbb{R} \times \mathbb{R})$ and $H(x) \in \mathcal{D}'(\mathbb{R})$ we have

$$H(x,t) \underset{x}{*} H(x) = x H(x,t) . \tag{1.577}$$

Because $H(x,t) \in L^1_{\text{loc}}(\mathbb{R}^2)$, $H(x) \in L^1_{\text{loc}}(\mathbb{R})$, the product $H(x,t) *_x H(x)$ exists, so that applying the formula (1.576) we obtain

$$H(x,t) \underset{x}{*} H(x) = \begin{cases} 0 , & (x,t) \in (-\infty, 0] \times \mathbb{R} , \\ \displaystyle\int_0^x H(\xi,t) H(x-\xi)\mathrm{d}\xi , & (x,t) \in [0,\infty) \times \mathbb{R} . \end{cases} \tag{1.578}$$

By the change of variable $x - \zeta = u$, we have

$$\int_0^x H(\zeta, t) H(x - \zeta) \mathrm{d}\zeta = \begin{cases} 0\,, & x \geq 0\,, \quad t < 0\,, \\ \int_0^x H(x - \zeta) \mathrm{d}\zeta\,, & x \geq 0\,, \quad t \geq 0\,, \end{cases}$$

$$= \begin{cases} 0\,, & (x, t) \in [0, \infty) \times (-\infty, 0)\,, \\ x\,, & (x, t) \in [0, \infty) \times [0, \infty)\,, \end{cases} \tag{1.579}$$

and thus the relation (1.578) becomes

$$H(x, t) \underset{x}{*} H(x) = \begin{cases} x\,, & \text{for } (x, t) \in [0, \infty) \times [0, \infty) \\ 0\,, & \text{for } (x, t) \in \mathbb{R} \backslash [0, \infty) \times [0, \infty) \end{cases} = x H(x, t)\,, \tag{1.580}$$

that is, the formula (1.577).

2
Integral Transforms of Distributions

2.1
Fourier Series and Series of Distributions

Of all the methods of studying and solving large classes of mathematics and physics problems, the method of functions representation by Fourier series is of major importance. The main difficulty encountered is that the Fourier series associated to a function does not always converge to the given function. Using distributions, this difficulty is overcome by showing that the Fourier series associated with a periodic distribution converges to that distribution. Since in mechanics many values are represented by singular distributions, their representation through Fourier series will play a central role in solving the boundary value problems.

2.1.1
Sequences and Series of Distributions

Definition 2.1 We say that the sequence of distributions $(f_i)_{i\in\mathbb{N}}$, $f_i \in \mathcal{D}'(\mathbb{R}^n)$ converges, in the space $\mathcal{D}'(\mathbb{R}^n)$, to the distribution $f \in \mathcal{D}'(\mathbb{R}^n)$, and we write $\lim_{i\to\infty} f_i = f$ or $f_i \xrightarrow{\mathcal{D}'(\mathbb{R})} f$, if $\forall\varphi \in \mathcal{D}(\mathbb{R}^n)$ the numerical sequence $(f_i, \varphi)_{i\in\mathbb{N}}$ converges to the number (f, φ), namely

$$\lim_{i\to\infty} (f_i, \varphi) = (f, \varphi) . \tag{2.1}$$

Proposition 2.1 Let there be the sequence of distributions $(f_i)_{i\in\mathbb{N}}$ from $\mathcal{D}'(\mathbb{R}^n)$, convergent to $f \in \mathcal{D}'(\mathbb{R}^n)$ on $\mathcal{D}'(\mathbb{R}^n)$, $f_i \xrightarrow{\mathcal{D}'(\mathbb{R})} f$; D^α is the operator of differentiation. Then, we have

$$\mathrm{D}^\alpha f_i \xrightarrow{\mathcal{D}'(\mathbb{R})} \mathrm{D}^\alpha f . \tag{2.2}$$

Indeed, because $f_i \xrightarrow{\mathcal{D}'(\mathbb{R})} f$, $\forall\varphi \in \mathcal{D}(\mathbb{R}^n)$, we have $(f_i, \varphi) \to (f, \varphi)$. On the basis of the continuity property of the derivative we obtain $(\mathrm{D}^\alpha f_i, \varphi) \to (\mathrm{D}^\alpha f, \varphi)$, namely (2.2).

Distribution Theory, First Edition. Petre P. Teodorescu, Wilhelm W. Kecs, and Antonela Toma

Definition 2.2 We say that the series of distributions $\sum_{i=1}^{\infty} f_i$, $f_i \in \mathcal{D}'(\mathbb{R}^n)$, is convergent in $\mathcal{D}'(\mathbb{R}^n)$ if the sequence of partial sums $S_p = \sum_{i=1}^{p} f_i$ is convergent.

If $S_p \xrightarrow{\mathcal{D}'(\mathbb{R}^n)} S \in \mathcal{D}'(\mathbb{R}^n)$, then we say that the distribution S is the sum of the series $\sum_{i=1}^{\infty} f_i$ and we write

$$\sum_{i=1}^{\infty} f_i = S \,. \tag{2.3}$$

We remark that the convergence of the series $\sum_{i=1}^{\infty} f_i$ is equivalent to the convergence of the numerical series $\sum_{i=1}^{\infty} (f_i, \varphi)$, $\forall \varphi \in \mathcal{D}(\mathbb{R}^n)$.

Proposition 2.2 If the distribution $f \in \mathcal{D}'(\mathbb{R}^n)$ is the sum of the series $\sum_{i=1}^{\infty} f_i$, then

$$\sum_{i=1}^{\infty} \mathrm{D}^{\alpha} f_i = \mathrm{D}^{\alpha} f \,. \tag{2.4}$$

Indeed, because $\sum_{i=1}^{\infty} f_i = f$, then $S_p = \sum_{i=1}^{p} f_i \xrightarrow{\mathcal{D}'(\mathbb{R}^n)} f$ and, on the basis of (2.2), we obtain $\sum_{i=1}^{\infty} \mathrm{D}^{\alpha} f_i = \mathrm{D}^{\alpha} f$, namely (2.4). Consequently, a series of convergent distributions can be differentiated term by term, resulting in a convergent series, a property which occurs, in a common sense, only to some uniformly convergent series of functions. This highlights the effectiveness of dealing with sequences and series of functions in the distribution space.

Proposition 2.3 If the sequence of locally integrable functions $(f_i)_{i \in \mathbb{N}} \subset L^1_{\text{loc}}(\mathbb{R}^n)$ is uniformly convergent on any compact $\Omega \subset \mathbb{R}^n$ to the function $f : \mathbb{R}^n \to \Gamma$, then $f \in L^1_{\text{loc}}(\mathbb{R}^n)$ and $T_{f_i} \xrightarrow{\mathcal{D}'} T_f$.

Proposition 2.4 Let there be the sequence $(f_i)_{i \in \mathbb{N}}$ of locally integrable functions on $\mathbb{R}^n$, $f_i \in L^1_{\text{loc}}(\mathbb{R}^n)$. If the series $\sum_{i=1}^{\infty} f_i(x)$ converges uniformly to the function $S(x)$, $x \in \mathbb{R}^n$, on any compact, then it is convergent in $\mathcal{D}'(\mathbb{R}^n)$ together with the series of the derivatives of its terms, whatever the order of differentiation.

Proof: Let there be $S_p(x) = \sum_{i=1}^{p} f_i(x)$ a partial sum of the considered series. Because $S_p \xrightarrow{u} S(x)$ for $\|x\| \leq R$, $R > 0$, and $S_p(x)$, $S(x)$ are locally integrable functions, they define regular distributions. According to the Proposition 2.3, we obtain $S_p \xrightarrow{\mathcal{D}'(\mathbb{R}^n)} S$ on $\mathcal{D}'(\mathbb{R}^n)$.

Consequently, $\mathrm{D}^{\alpha} S_p = \sum_{i=1}^{p} \mathrm{D}^{\alpha} f_i \xrightarrow{\mathcal{D}'(\mathbb{R}^n)} \mathrm{D}^{\alpha} S$ in $\mathcal{D}'(\mathbb{R}^n)$ based on the continuity of the differentiation operation from $\mathcal{D}'$ in $\mathcal{D}'$. □

Let there be the trigonometric series

$$\sum_{k=-\infty}^{\infty} a_k \mathrm{e}^{\mathrm{i}k\omega x} \,, \quad \omega = \frac{2\pi}{T} \,, \quad T > 0 \,, \quad a_k \in \mathbb{C} \,, \quad k \in \mathbb{Z} \,, \quad x \in \mathbb{R} \,. \tag{2.5}$$

Proposition 2.5 If the coefficients $a_k \in \mathbb{C}$ of the series $\sum_{k=-\infty}^{\infty} a_k e^{ik\omega x}$ satisfy the inequality $|a_k| \leq A|k|^p$, $p \in \mathbb{N}$, $A > 0$, then the trigonometric series converges in $\mathcal{D}'(\mathbb{R})$ to a certain distribution.

Proof: We consider the series

$$\sum_{\substack{k=-\infty \\ k\neq 0}}^{\infty} \frac{a_k e^{ik\omega x}}{(ik\omega)^{p+2}} + a_0 \frac{x^{p+2}}{(p+2)!} . \tag{2.6}$$

Because $|(a_k e^{ik\omega x}/(ik\omega)^{p+2})| \leq (A/(\omega^{p+2}|k|^2))$ and the numerical series $\sum_{k=1}^{\infty} (1/k^2)$ is convergent, on the basis of the Weierstrass criterion for uniform convergence of a series of functions, the series (2.6) is uniformly convergent and its sum is a continuous function $f(x)$, $x \in \mathbb{R}$. According to the Proposition 2.4, the series (2.6) converges to f in $\mathcal{D}'(\mathbb{R})$; therefore

$$\sum_{\substack{k=-\infty \\ k\neq 0}}^{\infty} \frac{a_k e^{ik\omega x}}{(ik\omega)^{p+2}} + a_0 \frac{x^{p+2}}{(p+2)!} = f(x) , \quad x \in \mathbb{R} . \tag{2.7}$$

Consequently, differentiating $p+2$ times the previous equality, we get

$$\sum_{\substack{k=-\infty \\ k\neq 0}}^{\infty} a_k e^{ik\omega x} + a_0 = \sum_{k=-\infty}^{\infty} a_k e^{ik\omega x} = f^{(p+2)}(x) . \tag{2.8}$$

Thus, the proof is complete. □

Example 2.1 We consider the sequence of functions $(f_n)_{n\in\mathbb{N}}$, $f_n(x) = (\sin nx/n)$, $x \in \mathbb{R}$. Let us state the relations

$$f_n \xrightarrow{\mathcal{D}'(\mathbb{R})} 0 , \quad \cos nx \xrightarrow{\mathcal{D}'(\mathbb{R})} 0 , \quad n \sin nx \xrightarrow{\mathcal{D}'(\mathbb{R})} 0 . \tag{2.9}$$

Indeed, $\forall \varphi \in \mathcal{D}(\mathbb{R})$ with $\operatorname{supp}(\varphi) = [a, b]$, we have

$$(f_n(x), \varphi(x)) = \frac{1}{n} \int_a^b \sin(nx)\varphi(x) dx . \tag{2.10}$$

From this relation we obtain $|(f_n, \varphi)| \leq (1/n) \int_a^b |\varphi(x)| dx$, wherefrom it results $\lim_{n\to\infty} |(f_n, \varphi)| = 0$, hence $\lim_{n\to\infty} (f_n, \varphi) = 0$, namely $\lim_{n\to\infty} f_n = \lim_{n\to\infty} (\sin nx/n) = 0$.

According to Proposition 2.1, we deduce

$$\lim_{n\to\infty} f_n'(x) = \lim_{n\to\infty} \cos nx = 0 , \quad \lim_{n\to\infty} f_n''(x) = \lim_{n\to\infty} (-n \sin nx) = 0 , \tag{2.11}$$

that is, the relations (2.9).

Example 2.2 Let there be $(f_n)_{n\in\mathbb{N}}$ the sequence of locally integrable functions, where $f_n(x) = H(x)n\mathrm{e}^{-nx}$, $x \in \mathbb{R}$, while H is the Heaviside function. Then, on $\mathcal{D}'(\mathbb{R})$ we have

$$f_n(x) \xrightarrow{\mathcal{D}'(\mathbb{R})} \delta(x)\,, \quad n^{k+1}(-1)^k\mathrm{e}^{-nx} \xrightarrow{\mathcal{D}'(\mathbb{R})} \delta^{(k)}(x)\,. \tag{2.12}$$

Indeed, we have $\int_{\mathbb{R}} f_n(x)\mathrm{d}x = \int_0^\infty n\mathrm{e}^{-nx}\mathrm{d}x = 1$. On the other hand $\forall\varphi \in \mathcal{D}(\mathbb{R})$ we have

$$(f_n(x), \varphi(x)) = \int_0^\infty n\mathrm{e}^{-nx}\varphi(x)\mathrm{d}x = -\int_0^\infty (\mathrm{e}^{-nx})'\varphi(x)\mathrm{d}x$$

$$= \varphi(0) + \int_0^\infty \mathrm{e}^{-nx}\varphi'(x)\mathrm{d}x\,. \tag{2.13}$$

Because $\operatorname{supp}(\varphi) = $ compact, then $\lim_{n\to\infty}\int_0^\infty \mathrm{e}^{-nx}\varphi'(x)\mathrm{d}x = 0$ and thus, from the relation (2.13), we obtain $\lim_{n\to\infty}(f_n(x), \varphi(x)) = \varphi(0) = (\delta(x), \varphi(x))$, namely $\lim_{n\to\infty} f_n(x) = \delta(x)$. By k times differentiation of this limit, we get the formula (2.12).

2.1.2
Expansion of Distributions into Fourier Series

Definition 2.3 We say that the function $f : \mathbb{R}^n \to \mathbb{C}$ is a periodic function if the nonzero vector $T = (T_1, T_2, \ldots, T_n) \in \mathbb{R}^n$ exists, so that $\forall x \in \mathbb{R}^n$ to have

$$\tau_T f(x) = f(x + T) = f(x)\,. \tag{2.14}$$

The symbol τ_T is the operator of translation by the vector T. The vector $T \in \mathbb{R}^n$ is called the period of the function f. It results from the definition that the periods set of the function f is $kT = (kT_1, kT_2, \ldots, kT_n)$, $k \in \mathbb{Z}$. The smallest positive period $T > 0$, $T_i > 0$, $i = \overline{1, n}$, is called the main period of the function f.

From the relation (2.14) it results $f(x) = f(x + kT)$, $k \in \mathbb{Z}$, $\forall x \in \mathbb{R}^n$. Consequently, the relation of definition (2.14) is equivalent to $f(x - T) = f(x)$, $\forall x \in \mathbb{R}^n$.

Thus, for example, the function $f : \mathbb{R} \to \mathbb{C}$, $f(x) = \mathrm{e}^{\mathrm{i}\omega x}$, $\omega \in \mathbb{R}\backslash\{0\}$ is a periodic function with the main period $T = 2\pi/|\omega|$.

Indeed, $f(x + T) = \mathrm{e}^{\mathrm{i}\omega x}\mathrm{e}^{\mathrm{i}2\pi\omega/|\omega|} = f(x)$, because $\mathrm{e}^{2\pi\mathrm{i}\omega/|\omega|} = \cos(\pm 2\pi) + \mathrm{i}\sin(\pm 2\pi) = 1$. The real number $\omega \neq 0$ is called pulsation.

Similarly, it is shown that the function $f : \mathbb{R}^2 \to \mathbb{C}$, $f(x) = f(x_1, x_2) = \exp(\mathrm{i}(x_1\omega_1 + x_2\omega_2))$, $\omega_1, \omega_2 \neq 0$ is a periodic function of period $T = (2\pi/|\omega_1|, 2\pi/|\omega_2|) \in \mathbb{R}^2$.

In particular, if the function $f : \mathbb{R}^n \to \mathbb{C}$ is a locally integrable function, then it will be considered periodic if the relation (2.14) is fulfilled a.e. (almost everywhere).

It is immediately verified that the sum, the difference, the product and the quotient of two periodic functions of period T are periodic functions with the same period.

We note that the relation (2.14) expresses the periodicity of the function f with respect to all variables $(x_1, x_2, \ldots, x_n) \in \mathbb{R}^n$, which does not imply the periodicity of the function in relation to each variable separately.

Definition 2.4 We say that the function $f : \mathbb{R}^n \to \mathbb{C}$ is periodic with respect to the variable x_i if it exists $T_i \in \mathbb{R}\backslash\{0\}$, such that $\forall x \in \mathbb{R}^n$ to have

$$f(x_1, x_2, \ldots, x_i - T_i, x_{i+1}, \ldots, x_n) = f(x_1, x_2, \ldots, x_n) . \tag{2.15}$$

The number $T_i \neq 0$ is called the period of the function f with respect to the variable x_i.

If the function $f : \mathbb{R}^n \to \mathbb{C}$ is periodic with respect to each variable of period $T_i \neq 0$, $i = \overline{1, n}$, then the function f is periodic with respect to all variables, that is, periodic of period $T = (T_1, T_2, \ldots, T_n) \in \mathbb{R}^n$.

The converse is generally not true.

Definition 2.5 Let there be the vector $T = (T_1, T_2, \ldots, T_n) \in \mathbb{R}^n$, $T \neq 0$ and the function $f : \mathbb{R}^n \to \mathbb{C}$ with $\operatorname{supp}(f)$ compact. We call the periodic transform of the function f by the vector $T \in \mathbb{R}^n\backslash\{0\}$, the function $\omega_T f : \mathbb{R}^n \to \mathbb{C}$, defined by the relation

$$\tilde{f}_T(x) = (\omega_T f)(x) = \sum_{k\in\mathbb{Z}^n} f(x - [k, T]) = \sum_{k\in\mathbb{Z}^n} f(x_1 - k_1 T_1, \ldots, x_n - k_n T_n) , \tag{2.16}$$

where $k = (k_1, k_2, \ldots, k_n) \in \mathbb{Z}^n$ and $[k, T] = (k_1 T_1, \ldots, k_n T_n) \in \mathbb{R}^n$.

Because the function f has compact support, it means that for $x \notin \operatorname{supp}(f)$, $f(x) = 0$ and thus for a fixed $x \in \mathbb{R}^n$, expression (2.16) has a finite number of terms.

Particularly, for the function $f : \mathbb{R} \to \mathbb{R}$ with $\operatorname{supp}(f) = [a, b]$, the periodic transform by the vector $T \neq 0$ is the function $\omega_T f : \mathbb{R} \to \mathbb{R}$ given by

$$\tilde{f}_T(x) = (\omega_T f)(x) = \sum_{k=-\infty}^{\infty} f(x - kT) , \quad k \in \mathbb{Z} .$$

Proposition 2.6 Let there be the function $f : \mathbb{R}^n \to \mathbb{C}$ with compact support. Then, its periodic transform $\omega_T f = \tilde{f}_T$ of vector $T \in \mathbb{R}^n\backslash\{0\}$ is periodic function of period T.
If the function f is a test function, hence $f \in \mathcal{D}(\mathbb{R}^n)$, then its periodic transform $\omega_T f$ is an indefinitely differentiable function which has not compact support. Thus, $\omega_T f \in C^\infty(\mathbb{R}^n)$ is a periodic function of period $T \in \mathbb{R}^n$, but $\omega_T f \notin \mathcal{D}(\mathbb{R}^n)$.

Particularly, let us consider the function $f : \mathbb{R} \to \mathbb{R}$, with the compact support $[-\ell, \ell]$, $\ell > 0$ (Figure 2.1).

The periodic transform $(\omega_T f)(x) = \tilde{f}_T(x) = \sum_{k=-\infty}^{\infty} f(x - kT)$, $T = 2\ell$, is a periodic function of period $T = 2\ell$ (Figure 2.2). The function $\omega_T f : \mathbb{R} \to \mathbb{R}$ represents the periodization of the function $f : \mathbb{R} \to \mathbb{R}$.

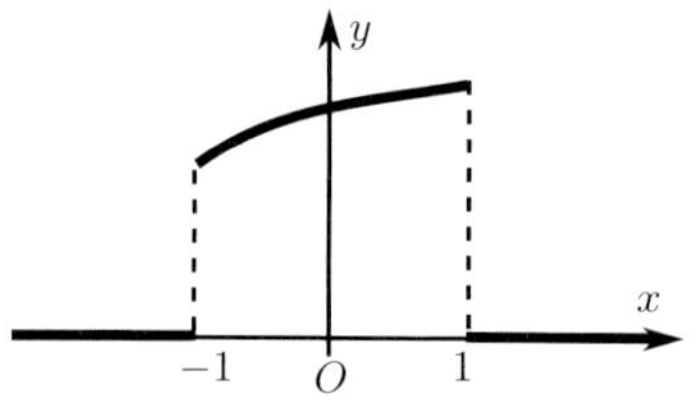

Figure 2.1

Let there be f a periodic function of period $T = 2\ell > 0$, locally integrable on $\mathbb{R}$. The Fourier series associated to the function $f \in L^1_{\text{loc}}(\mathbb{R})$ is

$$f \sim \frac{a_0}{2} + \sum_{k=1}^{\infty} \left(a_k \cos \frac{k\pi x}{\ell} + b_k \sin \frac{k\pi x}{\ell} \right) , \quad x \in \mathbb{R} , \tag{2.17}$$

where the coefficients of the series have the expressions

$$a_k = \frac{1}{\ell} \int_{-\ell}^{\ell} f(x) \cos \frac{k\pi x}{\ell} \mathrm{d}x , \quad k \in \mathbb{N}_0 ;$$

$$b_k = \frac{1}{\ell} \int_{-\ell}^{\ell} f(x) \sin \frac{k\pi x}{\ell} \mathrm{d}x , \quad k \in \mathbb{N} . \tag{2.18}$$

The integrals in these formulae can be considered on any interval $[a, b]$ for which $b - a = T = 2\ell$. The coefficients a_k and b_k have the property $\lim_{k\to\infty} a_k = 0$, $\lim_{k\to\infty} b_k = 0$.

In connection with the convergence of the Fourier series (2.17) associated to the periodic function $f : \mathbb{R} \to \mathbb{R}$, Dirichlet found sufficient conditions for the convergence, called Dirichlet's conditions.

Definition 2.6 We say that the function $f : \mathbb{R} \to \mathbb{R}$ satisfies Dirichlet's conditions on the interval $[a, b]$ if it is continuous on $[a, b]$ except for a finite number of points where it has first kind discontinuities and has a finite number of points of maximum and minimum.

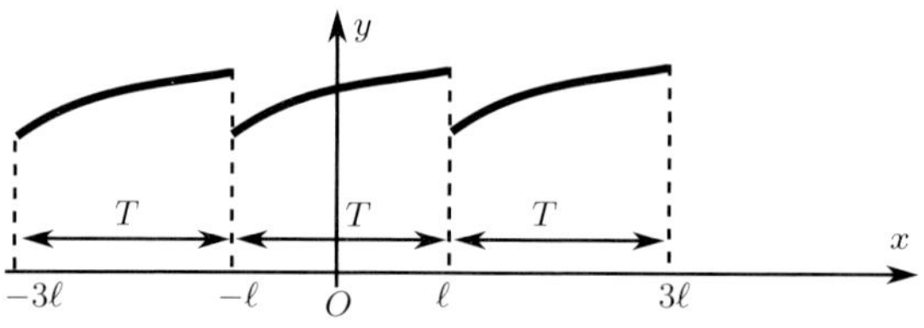

Figure 2.2

Theorem 2.1 Dirichlet If the function $f : \mathbb{R} \to \mathbb{R}$ of period $T = 2\ell > 0$ satisfies the Dirichlet conditions on the interval $[-\ell, \ell]$, then the Fourier series associated to the function f is convergent at all points. The Fourier series sum $S(x), x \in \mathbb{R}$, is equal at each point of continuity, to the value of the function f at this point. At the points of discontinuity, the value of the sum $S(x)$ is equal to the arithmetic average of the lateral limits of the point of discontinuity, that is, $S(x)|_{x=c} = (f(c-0) + f(c+0))/2$.

The distribution theory allows a unitary and general treatment of the Fourier series, achieving a remarkable result, namely that on the distribution space, any Fourier series associated to a locally integrable periodic function converges to that function.

To justify this result we establish the following:

Proposition 2.7 Let there be the sequence of the locally integrable functions $(D_n^*(x))_{n\in\mathbb{N}}$, $D_n^* \in L^1_{\text{loc}}(\mathbb{R})$, where

$$D_n^*(x) = \begin{cases} D_n(x) & \text{for } -\pi \le x \le \pi\,, \\ 0 & \text{for } |x| > \pi\,, \end{cases} \tag{2.19}$$

and

$$D_n(x) = \frac{1}{\pi}\left(\frac{1}{2} + \cos x + \cos 2x + \cdots + \cos nx\right) = \frac{\sin(n+2^{-1})x}{2\pi \sin(x/2)}\,, \qquad n \in \mathbb{N}_0\,. \tag{2.20}$$

Then, $(D_n^*)_{n\in\mathbb{N}}$ is a representative Dirac sequence, hence

$$\lim_{n\to\infty} D_n^*(x) = \delta(x)\,. \tag{2.21}$$

The sequence $(D_n^*)_{n\in\mathbb{N}}$ will be noted as a representative Dirac sequence of Dirichlet type.

Proposition 2.8 Let there be $f : \mathbb{R} \to \mathbb{R}$ a periodic function with period $T = 2\pi$, locally integrable on $\mathbb{R}$. Then, the Fourier series associated with f converges in the distribution space $\mathcal{D}'(\mathbb{R})$ to the function type distribution f, that is, we have

$$f(x) = \frac{a_0}{2} + \sum_{k=1}^{\infty} (a_k \cos kx + b_k \sin kx)\,, \quad x \in \mathbb{R}\,, \tag{2.22}$$

where

$$a_k = \frac{1}{\pi}\int_{-\pi}^{\pi} f(x)\cos kx\mathrm{d}x\,, \quad k \in \mathbb{N}_0\,; \quad b_k = \frac{1}{\pi}\int_{-\pi}^{\pi} f(x)\sin kx\mathrm{d}x\,, \quad k \in \mathbb{N}\,. \tag{2.23}$$

This proposition is also valid for periodic locally integrable functions of several variables. Finally, we note the following:

If the locally integrable function $f : \mathbb{R} \to \mathbb{R}$ on $\mathbb{R}$ is periodic and has the period $T = 2\ell > 0$, then by the substitution $x = \ell y/\pi$ one obtains the periodic function $F(y) = f(\ell y/\pi)$, $y \in \mathbb{R}$, of period 2π so that we can apply Proposition 2.8. Returning to the variable x, we obtain the Fourier series expansion

$$f(x) = \frac{a_0}{2} + \sum_{k=1}^{\infty} \left(a_k \cos \frac{k\pi x}{\ell} + b_k \sin \frac{k\pi x}{\ell} \right), \quad x \in \mathbb{R}, \tag{2.24}$$

where

$$a_k = \frac{1}{\ell} \int_{-\ell}^{\ell} f(x) \cos \frac{k\pi x}{\ell} \mathrm{d}x, \quad k \in \mathbb{N}_0,$$

$$b_k = \frac{1}{\ell} \int_{-\ell}^{\ell} f(x) \sin \frac{k\pi x}{\ell} \mathrm{d}x, \quad k \in \mathbb{N}. \tag{2.25}$$

The Fourier series (2.24) can be written in complex exponential form, namely

$$f(x) = \sum_{k=-\infty}^{\infty} C_k \mathrm{e}^{\mathrm{i}\omega k x}, \quad \omega = \frac{2\pi}{T} = \frac{\pi}{\ell}, \tag{2.26}$$

where the coefficients C_k have the expressions

$$C_k = \frac{1}{T} \int_a^b f(x) \mathrm{e}^{-\mathrm{i}\omega k x} \mathrm{d}x, \quad b - a = T = 2\ell. \tag{2.27}$$

Example 2.3 Let there be $f_T : \mathbb{R} \to \mathbb{R}$ a periodic function of period $T = 2\pi/\omega > 0$ (Figure 2.3), whose restriction to the interval $[0, T]$ is $f_0(x) = 2^{-1} - x/T$.

We shall write in complex exponential form the Fourier series of the function f_T, establishing then the formulae

$$\frac{1}{T} \sum_{k=-\infty}^{\infty} \mathrm{e}^{\mathrm{i}k\omega x} = \sum_{k=-\infty}^{\infty} \delta(x - kT), \quad k \in \mathbb{Z}, \tag{2.28}$$

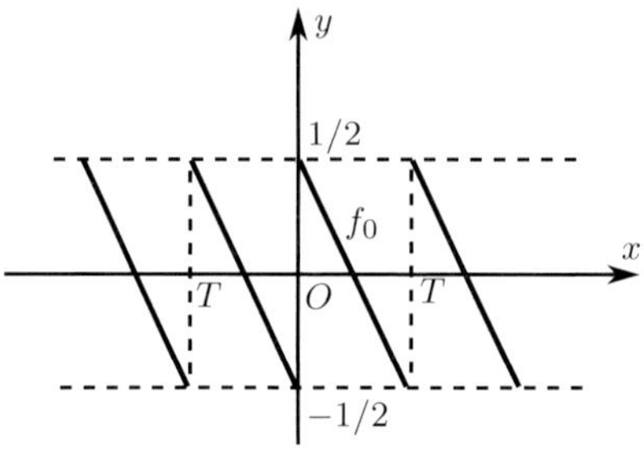

Figure 2.3

$$\sum_{k=1}^{\infty} \cos kx = -\frac{1}{2} + \pi \sum_{k=-\infty}^{\infty} \delta(x - 2\pi k) . \qquad (2.29)$$

In $\mathcal{D}'(\mathbb{R})$ the equality results:

$$f_T(x) = \sum_{k=-\infty}^{\infty} C_k e^{i\omega k x} , \quad \omega = \frac{2\pi}{T} , \qquad (2.30)$$

where

$$C_k = \frac{1}{T} \int_{-T/2}^{T/2} f_T(x) e^{-i\omega k x} dx = \frac{1}{T} \int_0^T f_0(x) e^{-i\omega k x} dx . \qquad (2.31)$$

Calculating, we obtain

$$C_0 = 0,\ C_k = \frac{-i}{2\pi k} , \quad k \neq 0 , \quad i = \sqrt{-1} . \qquad (2.32)$$

Substituting these values into (2.30), we have

$$f_T(x) = -\frac{i}{2\pi} \sum_{k=-\infty}^{\infty} \frac{e^{i\omega k x}}{k} . \qquad (2.33)$$

Differentiating this equality in the distributions sense and taking into account that the function f_T has the jumps equal to unity at the discontinuity points $x_k = kT$, $k \in \mathbb{Z}$, which are of the first kind, $f_T(kT + 0) - f_T(kT - 0) = 1$, we obtain

$$f'_T(x) = -\frac{1}{T} + \sum_{k=-\infty}^{\infty} \delta(x - kT) = \frac{1}{T} \sum_{\substack{k=-\infty \\ k\neq 0}}^{\infty} e^{i\omega k x} , \qquad (2.34)$$

giving (2.28).

We note that the first term of (2.28) can be written as

$$\frac{1}{T}\left[1 + \sum_{k=1}^{\infty} \left(e^{i\omega k x} + e^{-i\omega k x}\right)\right] = \sum_{k=-\infty}^{\infty} \delta(x - kT) . \qquad (2.35)$$

Applying the Euler formula, we obtain

$$\sum_{k=1}^{\infty} \cos \omega k x = -\frac{1}{2} + \frac{T}{2} \sum_{k=-\infty}^{\infty} \delta(x - kT) , \quad T = \frac{2\pi}{\omega} > 0 . \qquad (2.36)$$

Particularly, considering $T = 2\pi$, hence $\omega = 1$, we obtain the relation (2.29).

We note that in the usual sense the Fourier series expansion (2.33) is valid at any point of continuity of the periodic function f_T, hence for $x \neq kT$, $k \in \mathbb{Z}$.

As regards the equality (2.28), this shows that the bilateral series $(1/T) \sum_{k=-\infty}^{\infty} e^{i\omega k x}$ is divergent in the usual sense, but convergent in the sense of distributions, and the series sum represents the density of the unit mass placed at points of abscissa kT, $k \in \mathbb{Z}$.

Similarly, the series $\sum_{k=1}^{\infty} \cos kx$ is divergent in the usual sense, but convergent in the sense of distributions.

Because for $x \neq kT$ we have $\delta(x - kT) = 0$, from (2.28) and (2.29) we obtain, in the distribution theory sense, the formulae

$$\sum_{k=-\infty}^{\infty} e^{i\omega kx} = 0 \,, \quad \sum_{k=1}^{\infty} \cos kx = -\frac{1}{2} \,, \quad x \neq kT \,, \quad k \in \mathbb{Z} \,, \quad T = \frac{2\pi}{\omega} \,. \tag{2.37}$$

This means that $\forall \varphi \in \mathcal{D}(\mathbb{R})$ with the property $kT \notin \text{supp}(\varphi)$, $\forall k \in \mathbb{Z}$, we have

$$\left(\sum_{k=-\infty}^{\infty} e^{i\omega kx}, \varphi(x) \right) = 0 \,, \quad \left(\sum_{k=1}^{\infty} \cos kx, \varphi(x) \right) = 0 \,, \tag{2.38}$$

hence

$$\sum_{k=-\infty}^{\infty} (e^{i\omega kx}, \varphi(x)) = 0 \,, \quad \sum_{k=1}^{\infty} (\cos kx, \varphi(x)) = 0 \,. \tag{2.39}$$

From (2.36) for $\varphi \in \mathcal{D}(\mathbb{R})$ with the property $\text{supp}(\varphi) \subset (-T/2, T/2)$ we obtain the formulae

$$\begin{aligned} &\frac{1}{T} + \frac{2}{T} \sum_{k=1}^{\infty} \cos(\omega kx) = \delta(x) \,, \quad \frac{4\pi}{T^2} \sum_{k=1}^{\infty} kx \sin(\omega kx) = \delta(x) \,, \\ &-\frac{4\pi}{T^2} \sum_{k=1}^{\infty} k \sin(\omega kx) = \delta'(x) \,, \end{aligned} \tag{2.40}$$

valid for $\omega = 2\pi/T$ and $x \in (-T/2, T/2)$.

Definition 2.7 We say that $f \in \mathcal{D}'(\mathbb{R}^n)$ is a periodic distribution if there exists the vector $T = (T_1, T_2, \dots, T_n) \in \mathbb{R}^n \setminus \{0\}$, such that

$$f(x + T) = f(x) \,, \tag{2.41}$$

that is, $\forall \varphi \in \mathcal{D}(\mathbb{R}^n)$ we have

$$(f(x), \varphi(x)) = (f(x + T), \varphi(x)) = (f(x), \varphi(x - T)) \,. \tag{2.42}$$

Hence

$$(f(x), \varphi(x - T) - \varphi(x)) = 0 \,, \quad \forall \varphi \in \mathcal{D}(\mathbb{R}^n) \,. \tag{2.43}$$

The vector $T \in \mathbb{R}^n \setminus \{0\}$ is called the period of the distribution $f \in \mathcal{D}'(\mathbb{R}^n)$.

From the definition of the periodic distributions (2.41) it follows the relation

$$f(x-T) = f(x+T) = f(x) . \tag{2.44}$$

By successive application of the property (2.44) of the periodic distributions, it results

$$f(x+kT) = f(x) , \quad \forall k \in \mathbb{Z}^n , \tag{2.45}$$

where $kT = (kT_1, kT_2, \dots, kT_n)$, $T \in \mathbb{R}^n$.

The relation (2.45) is a generalization of the known property of periodic functions $f : \mathbb{R}^n \to \mathbb{R}$ of period $T \in \mathbb{R}^n \backslash \{0\}$.

The formula (2.45) means that $\forall \varphi \in \mathcal{D}(\mathbb{R}^n)$ we have

$$(f(x), \varphi(x)) = (f(x+kT), \varphi(x)) = (f(x), \varphi(x-kT)) , \tag{2.46}$$

hence, $(f(x), \varphi(x-kT) - \varphi(x)) = 0$.

Definition 2.8 Let there be $f \in \mathcal{E}'(\mathbb{R}^n)$ a distribution with compact support and the vector $T \in \mathbb{R}^n \backslash \{0\}$. The distribution $\omega_T f \in \mathcal{D}'(\mathbb{R}^n)$ defined by the formula

$$((\omega_T f)(x), \varphi(x)) = (f(x), (\omega_T \varphi)(x)) , \quad \forall \varphi \in \mathcal{D}(\mathbb{R}^n) , \tag{2.47}$$

is called the periodic transform by the vector $T \in \mathbb{R}^n \backslash \{0\}$ of the distribution $f \in \mathcal{E}'(\mathbb{R}^n)$, and

$$(\omega_T \varphi)(x) = \sum_{k \in \mathbb{Z}} \varphi(x - [k, T]) , \quad [k, T] = (k_1 T_1, k_2 T_2, \dots, k_n T_n) , \tag{2.48}$$

is the periodic transform by the vector $T \in \mathbb{R}^n$ of the function φ, according to the formula (2.16).

Because $\varphi \in \mathcal{D}(\mathbb{R}^n)$ has compact support, its periodic transform $\omega_T \varphi$, for any x given, has a finite number of nonzero terms, $\omega_T \varphi \in C^\infty(\mathbb{R}^n)$ and is a periodic function of period $T \in \mathbb{R}^n$.

Obviously, $\omega_T \varphi \notin \mathcal{D}(\mathbb{R}^n)$ and the set of these functions is a subset of the space $\mathcal{E}(\mathbb{R}^n)$, that is, indefinitely differentiable and periodic functions.

We note that, for the sake of simplicity, we sometimes denote $\omega_T f = \tilde{f}_T$ or $(\omega_T f)(x) = \tilde{f}_T(x)$. Therefore, the relation (2.47) can be written in the form

$$(\tilde{f}_T, \varphi) = (f, \tilde{\varphi}_T) , \quad \tilde{\varphi}_T = \omega_T \varphi .$$

Proposition 2.9 Let there be $f \in \mathcal{E}'(\mathbb{R}^n)$ a distribution with compact support. Then, the distribution $\omega_T f \in \mathcal{D}'(\mathbb{R}^n)$ is a periodic distribution of period $T \in \mathbb{R}^n$.

Example 2.4 Let there be the Dirac delta distribution $\delta(x) \in \mathcal{E}'(\mathbb{R}^n)$, then its periodic transform $\tilde{\delta}_T = \omega_T \delta \in \mathcal{D}'(\mathbb{R})$ by the vector $T \in \mathbb{R}^n \backslash \{0\}$ is

$$\begin{aligned} \tilde{\delta}_T(x) = (\omega_T \delta)(x) &= \sum_{k \in \mathbb{Z}^n} \delta(x - [k, T]) \\ &= \sum_{k \in \mathbb{Z}^n} \delta(x_1 - k_1 T_1, x_2 - k_2 T_2, \dots, x_n - k_n T_n) . \end{aligned} \tag{2.49}$$

Indeed, on the basis of the formula (2.47) we have, for any $\varphi \in \mathcal{D}(\mathbb{R}^n)$,

$$(\tilde{\delta}_T(x), \varphi(x)) = ((\omega_T\delta)(x), \varphi(x)) = (\delta(x), (\omega_T\varphi)(x))$$
$$= \left(\delta(x), \sum_{k\in\mathbb{Z}^n} \varphi(x - [k, T])\right) = \sum_{k\in\mathbb{Z}^n} \varphi(-[k, T]) = \sum_{k\in\mathbb{Z}} \varphi([k, T]) . \tag{2.50}$$

But

$$\sum_{k\in\mathbb{Z}^n} \varphi([k, T]) = \sum_{k\in\mathbb{Z}^n} (\delta(x - [k, T]), \varphi(x)) = \left(\sum_{k\in\mathbb{Z}^n} \delta(x - [k, T]), \varphi(x)\right) . \tag{2.51}$$

Consequently, we can write, $(\tilde{\delta}_T(x), \varphi(x)) = (\sum_{k\in\mathbb{Z}^n} \delta(x - [k, T]), \varphi(x))$, giving the formula (2.49).

Particularly, for $n = 1$ we obtain

$$\tilde{\delta}_T(x) = (\omega_T\delta)(x) = \sum_{k\in\mathbb{Z}} \delta(x - kT) , \quad T \in \mathbb{R}\backslash\{0\} . \tag{2.52}$$

Taking into account the formula (2.28), we find

$$\tilde{\delta}_T(x) = \sum_{k=-\infty}^{\infty} \delta(x - kT) = \frac{1}{T} \sum_{k=-\infty}^{\infty} \mathrm{e}^{\mathrm{i}k\omega x} , \quad T = \frac{2\pi}{\omega} > 0 . \tag{2.53}$$

Proposition 2.10 Let there be $f \in \mathcal{E}'(\mathbb{R})$ a distribution with compact support. Then, the periodic distribution $\omega_T f \in \mathcal{D}'(\mathbb{R})$ of period $T \in \mathbb{R}\backslash\{0\}$ has the expression

$$(\omega_T f)(x) = \tilde{f}_T(x) = \lim_{n\to\infty} \sum_{p=-n}^{n} f(x - pT) = \sum_{k\in\mathbb{Z}} f(x - kT) . \tag{2.54}$$

Definition 2.9 We call Fourier coefficients, corresponding to the periodic distribution $\tilde{f}_T = \omega_T f$, $T > 0$, the numbers

$$C_k\left(\tilde{f}_T\right) = \frac{1}{T}\left(f(x), \mathrm{e}^{-\mathrm{i}k\omega x}\right) , \quad \omega = \frac{2\pi}{T} > 0 , \quad k \in \mathbb{Z} . \tag{2.55}$$

This formula generalizes the classical formula of the Fourier coefficients corresponding to a periodic functions, locally integrable $\tilde{f}_T : \mathbb{R} \to \mathbb{C}$, namely

$$C_k\left(\tilde{f}_T\right) = \frac{1}{T}\int_0^T f(x)\mathrm{e}^{-\mathrm{i}\omega kx}\mathrm{d}x , \quad k \in \mathbb{Z} . \tag{2.56}$$

Definition 2.10 We call Fourier series associated to the periodic distribution $\tilde{f}_T = \omega_T f \in \mathcal{D}'(\mathbb{R})$, the series of distributions

$$\sum_{k=-\infty}^{\infty} C_k\left(\tilde{f}_T\right) e^{i\omega k x}, \tag{2.57}$$

where $C_k(\tilde{f}_T) \in \mathbb{C}$ are given by (2.56).

Example 2.5 Let us determine the Fourier series associated to the periodic distribution $\tilde{\delta}_T \in \mathcal{D}'(\mathbb{R})$ generated by the Dirac delta distribution $\delta(x) \in \mathcal{E}'(\mathbb{R})$.

For the Fourier coefficients, according to the formula (2.55), we have

$$C_k(\tilde{\delta}_T) = \frac{1}{T}(\delta(x), e^{-ik\omega x}) = \frac{1}{T}, \quad \omega = \frac{2\pi}{T} > 0 .$$

Consequently, the Fourier series associated to the periodic distribution $\tilde{\delta}_T = \omega_T \delta$ is

$$\sum_{k=-\infty}^{\infty} \frac{1}{T} e^{i\omega k x} = \frac{1}{T} \sum_{k\in\mathbb{Z}} e^{i\omega k x}, \quad x \in \mathbb{R}.$$

Taking into account (2.53), we can write

$$\tilde{\delta}_T = \sum_{k\in\mathbb{Z}} \delta(x - kT) = \frac{1}{T} \sum_{k\in\mathbb{Z}} e^{ik\omega x}, \quad x \in \mathbb{R}, \omega = \frac{2\pi}{T} > 0, \tag{2.58}$$

which shows that the Fourier series associated to the periodic distribution $\tilde{\delta}_T = \omega_T \delta \in \mathcal{D}'(\mathbb{R})$ converges in the distribution space $\mathcal{D}'(\mathbb{R})$ towards this distribution.

It is shown that this result holds for any periodic distribution.

Proposition 2.11 Let there be the periodic distribution $\tilde{f}_T \in \mathcal{D}'(\mathbb{R})$ of period $T > 0$. Then, the associated Fourier series converges in $\mathcal{D}'(\mathbb{R})$ towards the corresponding periodic distribution $\tilde{f}_T$, namely

$$\tilde{f}_T(x) = \sum_{k=-\infty}^{\infty} C_k\left(\tilde{f}_T\right) e^{i\omega k x}, \quad \omega = \frac{2\pi}{T} > 0 . \tag{2.59}$$

This proposition is particularly important in the study of Fourier series, since in the distribution space the convergence of the Fourier series is no longer a problem. Hence, the opportunity appears to represent by Fourier series both the locally integrable periodic functions and the periodic distributions generated by distributions with compact support.

Finally, we note that the formula (2.59) also applies in $\mathbb{R}^n$. Thus, if $f(x) \in \mathcal{E}'(\mathbb{R}^n)$ is a distribution with compact support, then the relation (2.59) will be written as

$$\tilde{f}_T(x) = \sum_{k\in\mathbb{Z}^n} C_k\left(\tilde{f}_T\right) e^{i(\omega_1 k_1 x_1 + \omega_2 k_2 x_2 + \cdots + \omega_n k_n x_n)}, \tag{2.60}$$

where $T = (T_1, T_2, \ldots, T_n) \in \mathbb{R}^n$, $T_i > 0$, $i = \overline{1, n}$, $x = (x_1, \ldots, x_n) \in \mathbb{R}^n$, $\omega_i = 2\pi/T_i > 0$, and $\tilde{f}_T(x) = (\omega_T f)(x)$ is defined by (2.47).

The numbers $C_k(\tilde{f}_T)$ are the Fourier coefficients corresponding to the periodic distribution $\tilde{f}_T$ with period $T \in \mathbb{R}^n$ and they have the expressions

$$C_k\left(\tilde{f}_T\right) = \frac{1}{T_1 T_2 \ldots T_n}\left(f(x), \mathrm{e}^{-\mathrm{i}(\omega_1 k_1 x_1 + \cdots + \omega_n k_n x_n)}\right), \tag{2.61}$$

where $k = (k_1, k_2, \ldots, k_n) \in \mathbb{R}^n$.

Example 2.6 Let there be $\delta(x, y) \in \mathcal{E}'(\mathbb{R}^2)$ the Dirac delta distribution and $\tilde{\delta}_T(x, y) = (\omega_T \delta)(x, y)$ its periodic transform of period $T = (T_1, T_2) \in \mathbb{R}^2$, $T_1, T_2 > 0$, which read

$$\tilde{\delta}_T(x, y) = \sum_{k_1, k_2 = -\infty}^{\infty} \delta(x - k_1 T_1, y - k_2 T_2)\,.$$

Next, we expand the periodic distribution $\tilde{\delta}_T \in \mathcal{D}'(\mathbb{R}^2)$ in a Fourier series.

Taking into account (2.61) for the Fourier coefficients corresponding to the distribution $\tilde{\delta}_T$, we obtain

$$C_k\left(\tilde{\delta}_T\right) = \frac{1}{T_1 T_2}\left(\delta(x, y), \mathrm{e}^{-\mathrm{i}(\omega_1 k_1 x + \omega_2 k_2 y)}\right) = \frac{1}{T_1 T_2}, \qquad k = (k_1, k_2) \in \mathbb{Z}^2,$$

where $\omega_1 = 2\pi/T_1$, $\omega_2 = 2\pi/T_2$.

The Fourier series expansion of the periodic distribution $\tilde{\delta}_T$ is

$$\begin{aligned} \tilde{\delta}_T(x, y) &= \sum_{k_1, k_2 = -\infty}^{\infty} \delta(x - k_1 T_1, y - k_2 T_2) \\ &= \frac{1}{T_1 T_2} \sum_{k_1, k_2 = -\infty}^{\infty} \exp\left(2\pi\mathrm{i}\left(\frac{k_1 x}{T_1} + \frac{k_2 y}{T_2}\right)\right). \end{aligned} \tag{2.62}$$

This last relation can be written in the form

$$\begin{aligned} \tilde{\delta}_T(x, y) &= \frac{1}{T_1 T_2} \sum_{k_1 = -\infty}^{\infty} \exp\left(\frac{2\pi\mathrm{i}k_1 x}{T_1}\right) \sum_{k_2 = -\infty}^{\infty} \exp\left(\frac{2\pi\mathrm{i}k_2 y}{T_2}\right) \\ &= \frac{1}{T_1 T_2}\left[1 + 2\sum_{k_1 = 1}^{\infty} \cos\frac{2\pi k_1 x}{T_1}\right]\left[1 + 2\sum_{k_2 = 1}^{\infty} \cos\frac{2\pi k_2 y}{T_2}\right]. \end{aligned} \tag{2.63}$$

From the mechanical point of view, the formula (2.62) represents the density of a system of material points with the same mass equal to unity and placed at the points of coordinates $x_{k_1} = k_1 T_1$, $y_{k_2} = k_2 T_2$, $(k_1, k_2) \in \mathbb{Z}^2$.

For $\varphi \in \mathcal{D}(\mathbb{R}^2)$ for which $\mathrm{supp}(\varphi) \subset (-T_1/2, T_1/2) \times (-T_2/2, T_2/2)$ from (2.62) we obtain

$$\begin{aligned} \left(\tilde{\delta}_T(x, y), \varphi(x, y)\right) &= (\delta(x, y), (\omega_T \varphi)(x, y)) = (\omega_T \varphi)(0, 0) \\ &= \sum_{k_1, k_2 = -\infty}^{\infty} \varphi(-k_1 T_1, -k_2 T_2) = \varphi(0, 0) = (\delta(x, y), \varphi(x, y))\,. \end{aligned}$$

It follows the Fourier series expansion

$$\delta(x,y) = \frac{1}{T_1 T_2}\left[1 + 2\sum_{k_1=1}^{\infty}\cos\frac{2\pi k_1 x}{T_1}\right]\left[1 + 2\sum_{k_2=1}^{\infty}\cos\frac{2\pi k_2 y}{T_2}\right],$$

where $(x,y) \in [-T_1/2, T_1/2] \times [-T_2/2, T_2/2]$.

Proposition 2.12 Let there be $f, g \in \mathcal{E}'(\mathbb{R}^n)$ distributions with compact support and $\tilde{f}_T = \omega_T f$, $\tilde{g}_T = \omega_T g$ their periodic transforms of period $T = (T_1, T_2, \ldots, T_n) \in \mathbb{R}^n$, $T_i > 0$, $i = \overline{1,n}$. Then $f * g \in \mathcal{E}'(\mathbb{R}^n)$ and the Fourier coefficients corresponding to the periodic transform $(\widetilde{f * g})_T$ are given by the formula

$$C_k\left[\left(\widetilde{f * g}\right)_T\right] = C_k[\omega_T(f * g)] = T_1 T_2 \ldots T_n C_k\left(\tilde{f}_T\right) C_k\left(\tilde{g}_T\right). \tag{2.64}$$

Proposition 2.13 Let there be the distributions with compact support $f(x) \in \mathcal{E}'(\mathbb{R})$, $g(y) \in \mathcal{E}'(\mathbb{R})$ and $\tilde{f}_{T_x} = \omega_{T_x} f$, $\tilde{g}_{T_y} = \omega_{T_y} g$ their periodic transforms of periods $T_x > 0$, $T_y > 0$. Then $f(x) \times g(y) \in \mathcal{E}'(\mathbb{R}^2)$ and for the Fourier coefficients corresponding to the periodic transform $(\widetilde{f \times g})_T$ and period $T = (T_x, T_y) \in \mathbb{R}^2$, we have the formula

$$C_k\left[\left(\widetilde{f \times g}\right)_T\right] = C_{k_1}\left(\tilde{f}_{T_x}\right) C_{k_2}\left(\tilde{g}_{T_y}\right), \tag{2.65}$$

where $k = (k_1, k_2) \in \mathbb{Z}^2$.

The Fourier series expansion of the periodic distribution $(\widetilde{f(x) \times g(y)})_T \in \mathcal{D}'(\mathbb{R}^2)$ of period $T = (T_x, T_y) \in \mathbb{R}^2$ has the expression

$$\left(\widetilde{f(x) \times g(y)}\right)_T(x,y) = \sum_{k_1,k_2=-\infty}^{\infty} C_{k_1}\left(\tilde{f}_{T_x}\right) C_{k_2}\left(\tilde{g}_{T_y}\right)\exp\left(2\pi\mathrm{i}\left(\frac{k_1 x}{T_x} + \frac{k_2 y}{T_y}\right)\right). \tag{2.66}$$

Example 2.7 We consider the Dirac delta distribution $\delta(x-\zeta_1, y-\zeta_2) \in \mathcal{E}'(\mathbb{R}^2)$ concentrated at the point $(\zeta_1, \zeta_2) \in \mathbb{R}^2$, because $\underline{\mathrm{supp}(\delta(x-\zeta_1, y-\zeta_2))} = \{(\zeta_1, \zeta_2)\}$. Let us write the Fourier series expansion of the periodic distribution $\tilde{\delta}_T(x-\zeta_1, y-\zeta_2) = \omega_T\delta(x-\zeta_1, y-\zeta_2)$ of period $T = (T_1, T_2)$, $T_1, T_2 > 0$.

Taking into account that $\delta(x-\zeta_1, y-\zeta_2) = \delta(x-\zeta_1) \times \delta(y-\zeta_2)$ we apply formulae (2.65) and (2.66). We have

$$C_k\left(\tilde{\delta}_T(x-\zeta_1, y-\zeta_2)\right) = C_{k_1}\left(\tilde{\delta}_{T_1}(x-\zeta_1)\right) C_{k_2}\left(\tilde{\delta}_{T_2}(y-\zeta_2)\right). \tag{2.67}$$

But

$$C_{k_1}\left(\tilde{\delta}_{T_1}(x-\zeta_1)\right) = \frac{1}{T_1}\left(\delta(x-\zeta_1), \exp\left(-\frac{2\pi\mathrm{i}k_1 x}{T_1}\right)\right)$$
$$= \left(\delta(x), \exp\left(-2\pi\mathrm{i}\frac{k_1}{T_1}(x+\zeta_1)\right)\right) = \exp\left(-\frac{2\pi\mathrm{i}k_1\zeta_1}{T_1}\right), \tag{2.68}$$

$$C_{k_2}\left(\tilde{\delta}_{T_2}(y-\zeta_2)\right) = \frac{1}{T_2}\left(\delta(y-\zeta_2), \exp\left(-\frac{2\pi i k_2 y}{T_2}\right)\right)$$
$$= \exp\left(-\frac{2\pi i k_2 \zeta_2}{T_2}\right) . \tag{2.69}$$

Substituting in (2.67), we obtain

$$C_k\left(\tilde{\delta}_T(x-\zeta_1, y-\zeta_2)\right) = \frac{1}{T_1 T_2}\exp\left(-\frac{2\pi i k_1 x}{T_1}\right)\exp\left(-\frac{2\pi i k_2 y}{T_2}\right) . \tag{2.70}$$

On the basis of the formula (2.66), the Fourier series corresponding to the periodic distribution $\tilde{\delta}_T(x-\zeta_1, y-\zeta_2)$ of period $T = (T_1, T_2) \in \mathbb{R}^2$ is

$$\tilde{\delta}_T(x-\zeta_1, y-\zeta_2)$$
$$= \frac{1}{T_1 T_2}\sum_{k_1,k_2=-\infty}^{\infty} \exp\left(\frac{2\pi i k_1 \zeta_1}{T_1}\right)\exp\left(\frac{2\pi i k_2 \zeta_2}{T_2}\right)\exp\left(2\pi i\left(\frac{k_1 x}{T_1}+\frac{k_2 y}{T_2}\right)\right) .$$

2.1.3
Expansion of Singular Distributions into Fourier Series

The following Fourier series expansion take place

$$\delta(x-\zeta) = \frac{2}{T} + \frac{4}{T}\sum_{k=1}^{\infty}\cos(\omega k\zeta)\cos(\omega k x) ,$$

$$\delta(x-\zeta) = \frac{4}{T}\sum_{k=1}^{\infty}\sin(\omega k\zeta)\sin(\omega k x) ,$$

$$\delta'(x-\zeta) = -\frac{4\omega}{T}\sum_{k=1}^{\infty} k\cos(\omega k\zeta)\sin(\omega k x) ,$$

$$\delta'(x-\zeta) = \frac{4\omega}{T}\sum_{k=1}^{\infty} k\sin(\omega k\zeta)\cos(\omega k x) , \tag{2.71}$$

where $x \in (0, T/2)$, $\zeta \in (0, T/2)$, $\omega = 2\pi/T > 0$.

We note that these formulae represent the Fourier cosine and sine series expansions of the distributions $\delta(x-\zeta)$, $\delta'(x-\zeta)$ for $x \in (0, T/2)$ with $\zeta \in (0, T/2)$, $T > 0$.

Finally, we notice that the cosine and sine Fourier series expansions of the distributions $\delta'(x-\zeta) \in \mathcal{E}'(\mathbb{R})$ given by the last two formulae of (2.71) can be obtained by performing the differentiation, with respect to the parameter $\zeta \in (0, T/2)$, of the first two formulas.

2.2 Fourier Transforms of Functions and Distributions

2.2.1 Fourier Transforms of Functions

Definition 2.11 Let there be $f \in L^1(\mathbb{R}^n)$ an absolutely integrable function. We call Fourier transform or Fourier image of the function f the complex function $\hat{f} : \mathbb{R}^n \to \mathbb{C}$ defined by

$$\hat{f}(\xi) = \int_{\mathbb{R}^n} f(x) e^{i\langle x,\xi\rangle} dx \, , \tag{2.72}$$

where $\langle x, \xi\rangle = \sum_{j=1}^n x_j \xi_j$ is the scalar product of the vectors $x, \xi \in \mathbb{R}^n$ and $i = \sqrt{-1}$ is the imaginary unit.

We note that the Fourier image will be denoted in the form

$$\hat{f}(\xi) = F[f](\xi) \, . \tag{2.73}$$

In this writing the symbol F is the Fourier operator and has great importance in the operational calculus based on the Fourier transform. We can therefore say that the Fourier operator F is an integral operator defined by (2.72) having as definition domain the class of absolutely integrable functions $L^1(\mathbb{R}^n)$, which will be called the original function class for the Fourier transform.

The Fourier transform $\hat{f}(\xi) = F[f](\xi)$ represents the value of the operator F corresponding to the original function $f \in L^1(\mathbb{R}^n)$.

Because $|e^{i\langle x,\xi\rangle}| = 1$ and $\forall f \in L^1(\mathbb{R}^n) \exists \int_{\mathbb{R}^n} |f| dx < \infty$, from (2.72) we obtain

$$\left|\hat{f}(\xi)\right| = |F[f](\xi)| \le \int_{\mathbb{R}^n} |f| \left|e^{i\langle x,\xi\rangle}\right| dx \le \int_{\mathbb{R}^n} |f| dx = \|f\|_1 \, , \tag{2.74}$$

wherefrom it results the existence of the Fourier image $\hat{f}(\xi)$, hence of the operator F. From the formula (2.72), due to the linearity of the integral, we deduce that the Fourier operator F is a linear operator, hence we have

$$F\left[\sum_{j=1}^m \alpha_j f_j\right] = \sum_{j=1}^m \alpha_j F\left[f_j\right] , \quad \forall \alpha_j \in \mathbb{C} \, , \quad \forall f_j \in L^1(\mathbb{R}^n) \, . \tag{2.75}$$

In particularly, if $f_j \in L^1(\mathbb{R})$, $j = \overline{1,n}$, then $f = \prod_{j=1}^n f_j \in L^1(\mathbb{R}^n)$ and we have $\hat{f} = \prod_{j=1}^n \hat{f}_j$, namely

$$F[f] = \prod_{j=1}^n F[f_j] \, . \tag{2.76}$$

Indeed, $\hat{f}(\xi) = \int_{\mathbb{R}^n} \prod_{j=1}^n f_j(x_j) e^{i\langle x,\xi\rangle} dx = \prod_{j=1}^n \int_{\mathbb{R}} f_j(x_j) e^{i\langle x_j,\xi_j\rangle} dx_j = \prod_{j=1}^n \hat{f}_j(\xi_j)$, that is, the formula (2.76).

Proposition 2.14 If $f \in L^1(\mathbb{R}^n)$, then the function $\hat{f}$ is bounded, continuous and we have

$$\left\| \hat{f} \right\|_{L^\infty} \leq \| f \|_1 , \tag{2.77}$$

$$\lim_{|\xi| \to \infty} \hat{f}(\xi) = 0 . \tag{2.78}$$

Proposition 2.15 If $f \in \mathcal{E}^k(\mathbb{R}^n)$ and $D^\alpha f \in L^1(\mathbb{R}^n)$, $\forall \alpha \in \mathbb{N}_0^n, |\alpha| \leq k$, then

$$F[D^\alpha f] = (-i\xi)^\alpha F[f] , \quad |\xi^\alpha F[f]| \leq \|D^\alpha f\|_1 . \tag{2.79}$$

From this formula it follows that from an operational point of view, to the operation of differentiation $\partial/\partial x_k$, by a Fourier transform F, corresponds the operation of multiplication by a factor $-i\xi_k$ of the Fourier image, relation which is one of the basic formulae of the operational calculus based on Fourier transform.

Corollary 2.1 Let there be $P(D) = \sum_{|\beta| \leq k} a_\beta D^\beta$ a differential operator with constant coefficients. Then, if $f \in \mathcal{E}^k(\mathbb{R}^n)$ and $D^\beta f \in L^1(\mathbb{R}^n)$, $\forall \beta \in \mathbb{N}_0^n$, we have

$$F[P(D) f] = P((-i\xi)^\beta) F[f] . \tag{2.80}$$

Indeed, because F is a linear operator, we can write

$$\begin{aligned} F[P(D) f] &= F\left[\sum_{|\beta| \leq k} a_\beta D^\beta f \right] = \sum_{|\beta| \leq k} a_\beta F\left[D^\beta f \right] \\ &= \sum_{|\beta| \leq k} a_\beta (-i\xi)^\beta F[f] = \left[\sum_{|\beta| \leq k} a_\beta (-i\xi)^\beta \right] F[f] , \end{aligned} \tag{2.81}$$

giving the formula (2.80).

Proposition 2.16 Let there be $f \in L^1(\mathbb{R}^n)$ and $x^\alpha f \in L^1(\mathbb{R}^n)$, $\forall \alpha \in \mathbb{N}_0^n$, $|\alpha| \leq k$. If D_ξ^α is a differentiation operator with respect to the variable $\xi \in \mathbb{R}^n$, then we have

$$F[f] \in \mathcal{E}^k(\mathbb{R}^n) , \quad D_\xi^\alpha F[f] = F[(ix)^\alpha f] , \quad \left| D_\xi^\alpha F[f] \right| \leq \|x^\alpha f\|_1 . \tag{2.82}$$

Proposition 2.17 Let there be $f \in L^1(\mathbb{R}^n) \cap C^0(\mathbb{R}^n)$, then $\forall x \in \mathbb{R}^n$ we have

$$f(x) = F^{-1}[F[f]] = \frac{1}{(2\pi)^n} \int_{\mathbb{R}^n} F[f](\xi) \exp(-i\langle x, \xi \rangle) d\xi . \tag{2.83}$$

The relation (2.83) is the inversion formula of the Fourier transform, hence it defines the inverse operator F^{-1} of the operator F.

Remark 2.1 The operator F will be called at times direct Fourier transform and F^{-1} the inverse Fourier operator. Their mode of action is defined by the formulae (2.72) and (2.83) and their mutual action as inverse operators arises from (2.83), namely

$$\mathrm{F}^{-1}[\mathrm{F}[f]] = f\,, \quad \mathrm{F}\left[\mathrm{F}^{-1}[\hat{f}]\right] = \hat{f}\,. \tag{2.84}$$

An immediate consequence of the formula (2.83) is that it justifies the uniqueness of the Fourier transform in the functions space $L^1(\mathbb{R}^n)$. Thus, if two functions from $L^1(\mathbb{R}^n)$ have the same Fourier image, then they coincide almost everywhere on $\mathbb{R}^n$.

From (2.83) it results that the inverse Fourier operator F^{-1}, just as the direct Fourier operator, is a linear operator, that is, we have

$$\mathrm{F}^{-1}\left[\alpha\hat{f}(\xi) + \beta\hat{g}(\xi)\right] = \alpha\mathrm{F}^{-1}\left[\hat{f}(\xi)\right] + \beta\mathrm{F}^{-1}[\hat{g}(\xi)] = \alpha f + \beta g\,,$$

where $\hat{f}(\xi) = \mathrm{F}[f](\xi)$, $\hat{g}(\xi) = \mathrm{F}[g](\xi)$, $\alpha, \beta \in \mathbb{C}$.

Also, we have

$$\mathrm{F}^{-1}\left[\hat{f}\right](x) = \frac{1}{(2\pi)^n}\mathrm{F}[\hat{f}](-x) = \frac{1}{(2\pi)^n}\mathrm{F}[\hat{f}(-\xi)]\,. \tag{2.85}$$

The properties of the Fourier operator F with respect to translation, homothety and symmetry are given by the following proposition:

Proposition 2.18 Let there be $f \in L^1(\mathbb{R}^n)$ and $a \in \mathbb{R}^n$. Then

$$\mathrm{F}[f(x-a)](\xi) = \exp(\mathrm{i}\langle a, \xi\rangle)\mathrm{F}[f(x)](\xi)\,, \tag{2.86}$$

$$\mathrm{F}[\exp(-\mathrm{i}\langle x, \xi\rangle) f(x)](\xi) = \mathrm{F}[f](\xi - a)\,. \tag{2.87}$$

Proposition 2.19 Let there be $f \in L^1(\mathbb{R}^n)$ and $a = (a_{ij}\delta_{ij})$, $\det a \neq 0$, $a_{ij} \in \mathbb{R}$, $i, j = \overline{1, n}$, a nonsingular matrix and δ_{ij} Kronecker's symbol. Then, we have

$$\mathrm{F}[f(ax)](\xi) = \frac{1}{|\det a|}\mathrm{F}[f](a^{-1}\cdot\xi)\,. \tag{2.88}$$

2.2.2
Fourier Transform and the Convolution Product

Proposition 2.20 If $f, g \in L^1(\mathbb{R}^n)$, then we have

$$\mathrm{F}[f * g] = \mathrm{F}[f]\mathrm{F}[g]\,,$$

$$\int\limits_{\mathbb{R}^n} \mathrm{F}[f](\xi)g(\xi)\mathrm{d}\xi = \int\limits_{\mathbb{R}^n} f(x)F[g](x)\mathrm{d}x\,;$$

the last relation is called Parseval's formula.

One of the remarkable properties of the space $\mathcal{S}$ is that the Fourier transform applies bijectively this space onto itself. This will allow the generalization of the Fourier transform for distributions of $\mathcal{S}'(\mathbb{R}^n)$. Because $\mathcal{S}(\mathbb{R}^n) \subset L^1(\mathbb{R}^n)$ it follows that $\forall\varphi \in \mathcal{S}, \exists \mathrm{F}[\varphi](\xi) = \hat{\varphi}$ which is a bounded continuous function on $\mathbb{R}^n$. Also, if D^β is a differentiation operator, then $\forall\varphi \in \mathcal{S}$ and $\forall\alpha, \beta \in \mathbb{N}_0^n$ we have $\mathrm{D}^\beta(x^\alpha\varphi) \in L^1(\mathbb{R}^n)$.

Proposition 2.21 The Fourier transformation is a continuous linear operator which applies bijectively the space $\mathcal{S}$ onto itself.

Proposition 2.22 If $f, g \in \mathcal{S}$, then $f * g \in \mathcal{S}$ and we have

$$\mathrm{F}[f * g] = \mathrm{F}[f]\mathrm{F}[g] , \tag{2.89}$$

$$(f * g)^\nu = f^\nu * g^\nu . \tag{2.90}$$

Proposition 2.23 Let there be $f, g \in \mathcal{S}$. Then, we have

$$\mathrm{F}^{-1}[\mathrm{F}[f]] = \mathrm{F}[\mathrm{F}^{-1}[f]] = f , \tag{2.91}$$

$$\mathrm{F}^{-1}[f] = \frac{1}{(2\pi)^n}(\mathrm{F}[f])^\nu , \tag{2.92}$$

$$\mathrm{F}[\mathrm{F}[f]] = (2\pi)^n f^\nu , \tag{2.93}$$

$$\int_{\mathbb{R}^n} \hat{f}(\xi)g(\xi)\mathrm{d}\xi = \int_{\mathbb{R}^n} f(x)\hat{g}(x)\mathrm{d}x . \tag{2.94}$$

2.2.3
Partial Fourier Transform of Functions

Definition 2.12 Let there be the absolutely integrable function $f(x, t) \in L^1(\mathbb{R}^{n+m})$, $x = (x_1, \ldots, x_n) \in \mathbb{R}^n$, $t = (t_1, \ldots, t_m) \in \mathbb{R}^m$. We call partial Fourier transform with respect to the variable $x \in \mathbb{R}^n$ of the function f, the complex function $\mathrm{F}_x[f] : \mathbb{R}^{n+m} \to \mathbb{C}$

$$\mathrm{F}_x[f(x, t)](\xi, t) = \int_{\mathbb{R}^n} f(x, t) \exp(\mathrm{i}\langle x, \xi\rangle)\mathrm{d}x , \tag{2.95}$$

where $\langle x, \xi\rangle = \sum_{j=1}^n x_j \xi_j$.

According to Fubini's theorem, $\int_{\mathbb{R}^n} f(x, t)\mathrm{d}x$ exists and $\int_{\mathbb{R}^n} f(x, t)\mathrm{d}x \in L^1(\mathbb{R}^m)$, wherefrom it results the existence of the transform $\mathrm{F}_x[f]$ and we have $\mathrm{F}_x[f](\xi, t) \in L^1(\mathbb{R}^{n+m})$, because $|\mathrm{F}_x[f](\xi, t)| \le \int_{\mathbb{R}^n} |f(x, t)|\mathrm{d}x$.

Noting by F the Fourier transform with respect to all variables $(x, t) \in \mathbb{R}^{n+m}$ and by $\mathrm{F}_x, \mathrm{F}_t$ the partial Fourier transforms with respect to the variable $x \in \mathbb{R}^n$,

$t \in \mathbb{R}^m$, respectively, then we have

$$\mathrm{F}[f(x,t)] = \mathrm{F}_x[\mathrm{F}_t[f(x,t)]] = \mathrm{F}_t[\mathrm{F}_x[f(x,t)]] \,, \quad f \in L^1(\mathbb{R}^{n+m}) \,. \tag{2.96}$$

Similarly, as for the operator F it is shown that the operators $\mathrm{F}_x^{-1}, \mathrm{F}_t^{-1}$ are continuous bijections of the space $\mathcal{S}(\mathbb{R}^{n+m})$ into itself, so we can write

$$\mathrm{F}^{-1}[\mathrm{F}[\varphi]] = \mathrm{F}[\mathrm{F}^{-1}[\varphi]] = \mathrm{F}_x^{-1}[\mathrm{F}_x[\varphi]] = \mathrm{F}_t^{-1}[\mathrm{F}_t[\varphi]] \,, \tag{2.97}$$

$$\mathrm{F}^{-1}[\varphi] = \mathrm{F}_x^{-1}\left[\mathrm{F}_t^{-1}[\varphi]\right] = \mathrm{F}_t^{-1}\left[\mathrm{F}_x^{-1}[\varphi]\right] \,, \quad \varphi \in \mathcal{S}(\mathbb{R}^{n+m}) \,. \tag{2.98}$$

Proposition 2.24 Let there be $f(x,t) \in L^1(\mathbb{R}^{n+m})$ and $g(x) \in L^1(\mathbb{R}^n)$. Then, we have

$$\mathrm{F}_x\left[f(x,t) \underset{x}{\otimes} g(x)\right] = \mathrm{F}_x[f(x,t)] \cdot \mathrm{F}_x[g(x)] \,, \tag{2.99}$$

$$\mathrm{F}_t\left[f(x,t) \underset{x}{\otimes} g(x)\right] = \mathrm{F}_t[f(x,t)] \underset{x}{\otimes} g(x) \,, \tag{2.100}$$

where the symbol $\otimes_x$ represents the partial convolution product.

2.2.4 Fourier Transform of Distributions from the Spaces $\mathcal{S}'$ and $\mathcal{D}'(\mathbb{R}^n)$

In order to extend the Fourier transform to distributions from $\mathcal{S}'$ (tempered distributions), we shall consider Parseval's formula (2.94), namely

$$\int_{\mathbb{R}^n} \mathrm{F}[f](\xi) g(\xi) \mathrm{d}\xi = \int_{\mathbb{R}^n} f(x) \mathrm{F}[g](x) \mathrm{d}x \,, \quad \forall f, g \in \mathcal{S}(\mathbb{R}^n) \,. \tag{2.101}$$

This relation can be written in the form

$$(\mathrm{F}[f], g) = (f, \mathrm{F}[g]) \,, \quad \forall f, g \in S(\mathbb{R}^n) \,, \tag{2.102}$$

as well, used to define the Fourier transform for distributions from $\mathcal{S}'$.

Definition 2.13 We call Fourier transform of the distribution $f \in \mathcal{S}'(\mathbb{R}^n)$, the distribution $\mathrm{F}[f] = \hat{f}$ defined by the relation

$$(\mathrm{F}[f], \varphi) = (f, \mathrm{F}[\varphi]) \,, \quad \forall \varphi \in \mathcal{S}(\mathbb{R}^n) \,, \tag{2.103}$$

hence $(\hat{f}, \varphi) = (f, \hat{\varphi}])$.

Proposition 2.25 Let there be $f \in \mathcal{S}'(\mathbb{R}^n)$, then $\mathrm{F}[f] = \hat{f} \in \mathcal{S}'(\mathbb{R}^n)$.

Indeed, because $\hat{\varphi} \in \mathcal{S}$, it follows that $(f, \hat{\varphi})$ defines a linear functional on $\mathcal{S}$. The continuity of this functional follows from Proposition 2.21, because if $\varphi_j \xrightarrow{\mathcal{S}} \varphi, \varphi_j, \varphi \in \mathcal{S}$, then $\hat{\varphi}_j \xrightarrow{\mathcal{S}} \hat{\varphi}$.

Hence $(f, \hat{\varphi})$ defines a linear continuous functional on $\mathcal{S}$ and thus $\hat{f} = \mathrm{F}[f] \in \mathcal{S}'(\mathbb{R}^n)$.

Definition 2.14 We call inverse Fourier transform of the distribution $f \in \mathcal{S}'(\mathbb{R}^n)$ the distribution $\mathrm{F}^{-1}[f]$, defined by the relation

$$\left(\mathrm{F}^{-1}[f], \varphi\right) = \left(f, \mathrm{F}^{-1}[\varphi]\right), \quad \forall \varphi \in \mathcal{S}, \tag{2.104}$$

where

$$\mathrm{F}^{-1}[\varphi](x) = \frac{1}{(2\pi)^n} \int_{\mathbb{R}^n} \varphi(\xi) \exp\left(-\mathrm{i}\, \langle x, \xi \rangle\right) \mathrm{d}\xi = \frac{1}{(2\pi)^n} \mathrm{F}[\varphi](-x) \in \mathcal{S}. \tag{2.105}$$

Proposition 2.26 Let there be $f \in \mathcal{S}'(\mathbb{R}^n)$. Then $\mathrm{F}^{-1}[f] \in \mathcal{S}'$ and we have

$$\mathrm{F}^{-1}[\mathrm{F}[f]] = \mathrm{F}\left[\mathrm{F}^{-1}[f]\right] = f. \tag{2.106}$$

From (2.106) it follows that the operators $\mathrm{F}, \mathrm{F}^{-1} : \mathcal{S}' \to \mathcal{S}'$ are continuous bijections of the space $\mathcal{S}'$ onto itself. Therefore, if $\mathrm{F}[f] = 0$, $f \in \mathcal{S}'$, then $f = 0$ and $f_k \xrightarrow{\mathcal{S}'} f$ implies $\mathrm{F}[f_k] \xrightarrow{\mathcal{S}'} \mathrm{F}[f]$.

This result is particularly useful in determining the Fourier image of a distribution $f \in \mathcal{S}'$, knowing that $f_k \xrightarrow{\mathcal{S}'} f$ and $\mathrm{F}_k = \mathrm{F}[f_k]$. Then $\mathrm{F}[f] = \lim_{k\to\infty} \mathrm{F}[f_k]$, the limit being considered into the distribution space $\mathcal{S}'$.

Proposition 2.27 If $f \in \mathcal{S}'$, then

$$\mathrm{F}^{-1}[f] = \frac{1}{(2\pi)^n} \mathrm{F}[f^{\nu}]. \tag{2.107}$$

Definition 2.15 Let there be the distribution $f(x, t) \in \mathcal{S}'(\mathbb{R}^{n+m})$, $x \in \mathbb{R}^n$, $t \in \mathbb{R}^m$. We call partial Fourier transform with respect to the variable $x \in \mathbb{R}^n$ of the distribution $f(x, t) \in \mathcal{S}'(\mathbb{R}^{n+m})$, the distribution $\mathrm{F}_x[f(x, t)] \in \mathcal{S}'(\mathbb{R}^{n+m})$, defined by the relation

$$(\mathrm{F}_x[f], \varphi) = (f, \mathrm{F}_x[\varphi]), \quad \forall \varphi \in \mathcal{S}(\mathbb{R}^{n+m}), \tag{2.108}$$

where

$$\mathrm{F}_x[\varphi](x, t) = \int_{\mathbb{R}^n} \varphi(\xi, t) \exp(\mathrm{i}\langle x, \xi \rangle) \mathrm{d}\xi \in \mathcal{S}(\mathbb{R}^{n+m}), \tag{2.109}$$

is the partial Fourier transform of the function $\varphi \in \mathcal{S}(\mathbb{R}^{n+m})$.

Proposition 2.28 Let there be the distributions $f(x) \in \mathcal{S}'(\mathbb{R}^n)$, $g(t) \in \mathcal{S}'(\mathbb{R}^m)$. Then we have

$$\mathrm{F}_x[f(x) \times g(t)] = \mathrm{F}_x[f(x)] \times g(t), \tag{2.110}$$

$$F_t[f(x) \times g(t)] = f(x) \times F_t[g(t)] , \tag{2.111}$$

$$F[f(x) \times g(t)] = F_x[f] \times F_t[g] = F[f(x)] \times F[g(t)] . \tag{2.112}$$

Example 2.8 Because $\delta(x,t) = \delta(x) \times \delta(t)$, $\delta(x,t) \in \mathcal{E}'(\mathbb{R}^{n+m}) \subset \mathcal{S}'(\mathbb{R}^{n+m})$ we obtain

$$F_x[\delta(x,t)](\xi,t) = F_x[\delta(x)](\xi) \times \delta(t) = 1(\xi) \times \delta(t) = \delta(t) ,$$

where $1(\xi) \times \delta(t) = \delta(t)$ is considered distribution from $\mathcal{S}'(\mathbb{R}^{n+m})$ and acts according to the formula

$$\begin{aligned}(1(\xi) \times \delta(t), \varphi(\xi,t)) &= (\delta(t), (1(\xi), \varphi(\xi,t))) \\ &= \left(\delta(t), \int_{\mathbb{R}^n} \varphi(\xi,t) d\xi\right) = \int_{\mathbb{R}^n} \varphi(x,0) dx , \quad \varphi \in \mathcal{S}(\mathbb{R}^{n+m}) .\end{aligned}$$

Conversely, if $\delta(t)$ is considered as the distribution from $\mathcal{S}'(\mathbb{R}^m)$, then for any $\psi(t) \in \mathcal{S}(\mathbb{R}^m)$ we have $(\delta(t), \psi(t)) = \psi(0)$.

We also have

$$\begin{aligned}F[\delta(x,t)](\xi,\eta) &= F_x F_t[\delta(x,t)](\xi,\eta) = F_x[\delta(x)](\xi) \times F_t[\delta(t)](\eta) \\ &= 1(\xi) \times 1(\eta) = 1(\xi,\eta) = 1, \quad (\xi,\eta) \in \mathbb{R}^{n+m} .\end{aligned}$$

Hence $F[\delta(x,t)](\xi,\eta) = 1$ is considered as the distribution from the space $\mathcal{S}'(\mathbb{R}^{n+m})$.

For distributions with compact support the Fourier transform takes a simple form, because $\mathcal{E}'(\mathbb{R}^n) \subset \mathcal{S}'(\mathbb{R}^n)$. We can assert the following.

Proposition 2.29 Let there be the distribution with compact support $f \in \mathcal{E}'(\mathbb{R}^n)$. Then, if D^α, $\alpha \in \mathbb{N}_0^n$ is the operator of differentiation, we have

$$\begin{aligned}D^\alpha(F[f](\xi)) &= D^\alpha \hat{f}(\xi) = (f(x), D^\alpha \exp(i\langle x,\xi\rangle)) \\ &= (f(x), (ix)^\alpha \exp(i\langle x,\xi\rangle)) ,\end{aligned} \tag{2.113}$$

$$\hat{f}(\xi) = F[f](\xi) = (f(x), \exp(i\langle x,\xi\rangle)) . \tag{2.114}$$

Example 2.9 Let there be the distribution $\delta(x) \in \mathcal{E}'(\mathbb{R}^n)$. Then, we have

$$F[\delta(x)] = \hat{\delta}(\xi) = 1 , \quad F^{-1}[1] = \delta(x) , \quad F[1] = (2\pi)^n \delta(\xi) .$$

Indeed, on the basis of the formula (2.114), we obtain

$$F[\delta(x)] = (\delta(x), \exp(i\langle x,\xi\rangle)) = \exp(i\langle 0,\xi\rangle) = 1 , \tag{2.115}$$

which, based on the inverse Fourier transform, we have $F^{-1}[1] = \delta(x)$.

Using the formula (2.107), we find $\mathrm{F}^{-1}[1] = \mathrm{F}[1^{\nu}]/(2\pi)^n = \mathrm{F}[1]/(2\pi)^n$, where $\mathrm{F}[1] = (2\pi)^n \mathrm{F}^{-1}[1] = (2\pi)^n \delta(\xi)$.

We note that the function $1 \notin \mathcal{S}$ but, being a slow-growing function, it generates a function type distribution from $\mathcal{S}'(\mathbb{R}^n)$. Consequently, the distribution $f = 1 \in \mathcal{S}'(\mathbb{R}^n)$ and acts according to the formula

$$(1, \varphi) = \int_{\mathbb{R}^n} \varphi(x) \mathrm{d}x \,, \quad \varphi \in \mathcal{S}(\mathbb{R}^n) \,. \tag{2.116}$$

2.2.4.1 Properties of the Fourier Transform

It can be shown that all properties of the Fourier transform of the functions from the space $\mathcal{S}$ remain valid for the distributions from $\mathcal{S}'$.

Thus, $\forall f \in \mathcal{S}'$ we have the formulae

$$\mathrm{D}^{\alpha}\mathrm{F}[f] = \mathrm{F}[(\mathrm{i}x)^{\alpha} f] \,, \quad \mathrm{D}^{\alpha} \text{ is the differentiation operator} \,, \tag{2.117}$$

$$\mathrm{F}[\mathrm{D}^{\alpha} f] = (-\mathrm{i}\xi)^{\alpha}\mathrm{F}[f] \,, \tag{2.118}$$

$$\mathrm{F}[\mathrm{F}[f]] = (2\pi)^n f^{\nu} \,, \tag{2.119}$$

$$\mathrm{F}[f(x-a)] = \exp(\mathrm{i}\langle a, \xi\rangle)\mathrm{F}[f(x)] \,, \quad a \in \mathbb{R}^n \,, \tag{2.120}$$

$$\mathrm{F}[f](\xi - a) = \mathrm{F}[\exp(-\mathrm{i}\langle a, x\rangle) f(x)] \,, \tag{2.121}$$

$$\mathrm{F}[f(x) \times g(t)] = \mathrm{F}[f(x)] \times \mathrm{F}[g(t)] \,, \quad f \in \mathcal{S}'(\mathbb{R}^n) \,, \quad g \in \mathcal{S}'(\mathbb{R}^m) \,, \tag{2.122}$$

$$\mathrm{F}^{-1}[f(x) \times g(t)] = \mathrm{F}^{-1}[f(x)] \times \mathrm{F}^{-1}[g(t)] \,, \quad f \in \mathcal{S}'(\mathbb{R}^n) \,, \quad g \in \mathcal{S}'(\mathbb{R}^m) \,. \tag{2.123}$$

Proposition 2.30 Let there be $f \in \mathcal{S}'(\mathbb{R}^n)$ and $g \in \mathcal{E}'(\mathbb{R}^n)$. Then, we have

$$\mathrm{F}[f * g] = \mathrm{F}[f] \cdot \mathrm{F}[g] \,. \tag{2.124}$$

Proposition 2.31 Let there be the distributions $f, g \in \mathcal{S}'(\mathbb{R}^n)$ and $\psi \in O_M$ a multiplier of the space $\mathcal{S}$. Then we have

$$\mathrm{F}^{-1}[f(x) * g(x)] = (2\pi)^n \mathrm{F}^{-1}[f]\mathrm{F}^{-1}[g] \,, \tag{2.125}$$

$$\mathrm{F}[f] * \mathrm{F}[\psi] = (2\pi)^n \mathrm{F}[f(x)\psi(x)] \,. \tag{2.126}$$

Indeed, on the basis of the formula (2.107) we can write $\mathrm{F}^{-1}[f(x) * g(x)] = (1/(2\pi)^n)\mathrm{F}[(f * g)^{\nu}]$, $\mathrm{F}^{-1}[f] = (1/(2\pi)^n)\mathrm{F}[f^{\nu}]$, $\mathrm{F}^{-1}[g] = (1/(2\pi)^n)\mathrm{F}[g^{\nu}]$, where we obtain $\mathrm{F}^{-1}[f * g] = (1/(2\pi)^n)\mathrm{F}[f^{\nu}]\mathrm{F}[g^{\nu}] = (2\pi)^n \mathrm{F}^{-1}[f]\mathrm{F}^{-1}[g]$.

As regards the second formula, we have $F^{-1}[F[f] * F[\psi]] = (2\pi)^n F^{-1}[F[f]] \cdot F^{-1}[F[\psi]] = (2\pi)^n f\psi$. Applying the operator F to this relation, we obtain the required relation.

Example 2.10 Let there be $\delta(x) \in \mathcal{E}'(\mathbb{R}^n)$ and D^α the differentiation operator. Then, we have

$$F[D^\alpha \delta(kx)] = (-i\xi)^\alpha \frac{1}{k^n}, \quad k \in \mathbb{R} - \{0\}, \tag{2.127}$$

$$F[\delta(x-a)] = \exp(i\langle a, \xi\rangle), \quad a \in \mathbb{R}^n, \tag{2.128}$$

$$F[x^\alpha] = (2\pi)^n (-i)^{|\alpha|} D^\alpha \delta(\xi). \tag{2.129}$$

Indeed, we have

$$F[D^\alpha \delta(kx)] = (-i\xi)^\alpha F[\delta(kx)] = (-i\xi)^\alpha F\left[\frac{\delta(x)}{k^n}\right] = (-i\xi)^\alpha \frac{1}{k^n},$$

because $F[\delta(x)] = 1$ and $\delta(kx) = (1/k^n)\delta(x)$. Applying the properties (2.117) and (2.121) we can write $F[\delta(x-a)] = \exp(i\langle a, \xi\rangle)F[\delta(x)] = \exp(i\langle a, \xi\rangle)$. Also, we have $\hat{1} = F[1] = (2\pi)^n \delta(\xi)$, hence $D^\alpha(\hat{1}) = F[(ix)^\alpha] = i^{|\alpha|}F[x^\alpha]$, where $F[x^\alpha] = (2\pi)^n i^{-|\alpha|} D^\alpha \delta(\xi) = (2\pi)^n (i^{|\alpha|}/(-1)^{|\alpha|}) D^\alpha \delta(\xi) = (2\pi)^n (-i)^{|\alpha|} D^\alpha \delta(\xi)$.

Example 2.11 Let $H(x)$, $x \in \mathbb{R}$, be the Heaviside distribution. Then, we have

$$F[H(x)](\xi) = \pi\delta(\xi) + i\,\text{p.v.}\frac{1}{\xi}, \quad F[H(-x)](\xi) = \pi\delta(\xi) - i\,\text{p.v.}\frac{1}{\xi}. \tag{2.130}$$

The distribution H is a distribution of function type in the space $\mathcal{S}'$, because it is locally integrable and with slow growth at infinity, namely $H \in L^1_{\text{loc}}$ and $|H(x)| \leq C(1+|x|)^k$, $C, k \geq 0$. The family of functions $f_a(x) = H(x)\exp(-ax)$, $a > 0$, $x \in \mathbb{R}$, consists of absolutely integrable functions, hence $f_a \in L^1(\mathbb{R})$, which generates function type distributions from $\mathcal{S}'$. Consequently, in the sense of convergence from the space $\mathcal{S}'$, we have

$$H(x) = \lim_{a \to +0} H(x)\exp(-ax). \tag{2.131}$$

Because $F[H(x)\exp(-ax)](\xi) = \int_{\mathbb{R}} \exp(i\xi x)H(x)\exp(-ax)dx = \int_0^\infty \exp(-ax) \exp(i\xi x)dx = 1/(a - i\xi) = a/(a^2+\xi^2) + i(\xi/(a^2+\xi^2))$ and $\lim_{a\to+0}(a/(a^2+\xi^2)) = \pi\delta(\xi)$, $\lim_{a\to+0}(\xi/(a^2+\xi^2)) = \text{p.v.}(1/\xi)$, from (2.131) it results

$$F[H(x)](\xi) = \lim_{a\to+0} F[H(x)\exp(-ax)](\xi) = \pi\delta(\xi) + i\,\text{p.v.}\frac{1}{\xi},$$

that is, the formula (2.130). Similarly, taking into account that $F[H(-x)\exp(-ax)](\xi) = 1/(a+i\xi)$, $a > 0$, and considering $a \to +0$ we obtain $F[H(-x)](\xi) = \pi\delta(\xi) - i\,\text{p.v.}(1/\xi)$.

We note that the formulae (2.130) can be obtained as follows:

Taking into account that $\delta(x) = H'(x)$ we can write $\mathrm{F}[\delta(x)](\xi) = 1 = \mathrm{F}[H'(x)](\xi) = (-\mathrm{i}\xi)\mathrm{F}[H]$, where $\xi\mathrm{F}[H(x)](\xi) = \mathrm{i}$.

If we denote $\mathrm{F}[H(x)](\xi) = Y(\xi)$, then we obtain the equation $\xi Y(\xi) = \mathrm{i}$ whose general solution in $\mathcal{S}'(\mathbb{R})$ is $Y(\xi) = k\delta(\xi) + \mathrm{i}\,\mathrm{p.v.}(1/\xi)$, where k is a constant. To determine the constant k, we use the obvious relation $H(x) + H(-x) = 1$, hence $\mathrm{F}[H(x)] + \mathrm{F}[H(-x)] = \mathrm{F}[1] = 2\pi\delta(\xi)$, that is, $Y(\xi) + Y(-\xi) = 2\pi\delta(\xi)$. Because $Y(\xi) = k\delta(\xi) + \mathrm{i}\,\mathrm{p.v.}(1/\xi)$ we obtain $2k\delta(\xi) = 2\pi\delta(\xi)$, hence $k = \pi$ and thus the formula (2.130) is proved. From the above considerations it follows that the distribution $Y = \pi\delta(\xi) + \mathrm{i}\,\mathrm{p.v.}(1/\xi)$ satisfies the equation $\xi Y(\xi) = \mathrm{i}$.

Because $\mathrm{F}[\mathrm{F}[H(x)]] = 2\pi H(-\xi)$, from (2.130) it results $\pi + \mathrm{i}\mathrm{F}[\mathrm{p.v.}(1/x)](\xi) = 2\pi H(-\xi)$, giving the relation

$$\mathrm{F}\left[\mathrm{p.v.}\frac{1}{x}\right](\xi) = \mathrm{i}\pi \operatorname{sgn} \xi \ . \tag{2.132}$$

Applying the operator F, to the last relation we obtain $\mathrm{F}[\mathrm{F}[\mathrm{p.v.}(1/x)]](\xi) = \mathrm{i}\pi\mathrm{F}[\operatorname{sgn}\xi]$, namely $-\pi\mathrm{p.v.}(1/x) = \mathrm{i}\pi\mathrm{F}[\operatorname{sgn}\xi](x)$, giving the relation

$$\mathrm{F}[\operatorname{sgn} x](\xi) = 2\mathrm{i}\,\mathrm{p.v.}\frac{1}{\xi} \ . \tag{2.133}$$

Example 2.12 Let there be the function $f(x) = \exp(\mathrm{i}ax)$, $x \in \mathbb{R}$, $a \in \mathbb{R}$. Then we have

$$\mathrm{F}[\exp(\mathrm{i}ax)](\xi) = 2\pi\delta(\xi + a) \ .$$

We observe that the function $f(x) = \exp(\mathrm{i}ax)$ is a locally integrable function on $\mathbb{R}$, with slow growth, because $|f| = 1$. Consequently, the function f generates a function type distribution $\mathcal{S}'(\mathbb{R})$. We can say that f is a multiplier of space $\mathcal{S}$, hence $f \in O_M$.

To determine the Fourier transform of the function $f(x) = \exp(\mathrm{i}ax) \in \mathcal{S}'(\mathbb{R})$ we apply the formula (2.107) and thus we have

$$\begin{aligned}\mathrm{F}\left[(\delta(x+a))^{\nu}\right](\xi) &= \mathrm{F}[\delta(-x+a)](\xi) \\ &= \mathrm{F}[\delta(x-a)](\xi) = \exp(\mathrm{i}a\xi) = 2\pi\mathrm{F}^{-1}[\delta(x+a)](\xi) \ ,\end{aligned} \tag{2.134}$$

which, by applying the operator F, we obtain $\mathrm{F}[\exp(\mathrm{i}ax)](\xi) = 2\pi\delta(\xi + a)$.

Taking into account the relations

$$\cos ax = \frac{\mathrm{e}^{\mathrm{i}ax} + \mathrm{e}^{-\mathrm{i}ax}}{2} \ ,$$

$$\sin ax = \frac{\mathrm{e}^{\mathrm{i}ax} - \mathrm{e}^{-\mathrm{i}ax}}{2\mathrm{i}} \ , \quad x \in \mathbb{R} \ ,$$

then, on the basis of the linearity property of the operator F, we have

$$\mathrm{F}[\cos ax](\xi) = \frac{\mathrm{F}[\mathrm{e}^{\mathrm{i}ax}](\xi) + \mathrm{F}[\mathrm{e}^{-\mathrm{i}ax}](\xi)}{2} = \pi[\delta(\xi + a) + \delta(\xi - a)].$$

Hence

$$\mathrm{F}[\cos ax](\xi) = \pi[\delta(\xi + a) + \delta(\xi - a)] . \tag{2.135}$$

Similarly, we find

$$\mathrm{F}[\sin ax](\xi) = -\mathrm{i}\pi[\delta(\xi + a) + \delta(\xi - a)] . \tag{2.136}$$

2.2.4.2 Fourier Transform of the Distributions from the Space $\mathcal{D}'(\mathbb{R}^n)$

Since $\mathcal{D} \subset \mathcal{S}$, the Fourier transform does not apply the space $\mathcal{D}$ onto itself. Thus, if $\varphi \in \mathcal{D}$, then $\mathrm{F}[\varphi] \notin \mathcal{D}$, because $\mathrm{F}[\varphi] = \hat{\varphi}$ has no compact support. On the other hand, because $\mathcal{S}' \subset \mathcal{D}'$, it follows that the Fourier transform defined by the formula (2.103) cannot be applied to any distribution from $\mathcal{D}'(\mathbb{R}^n)$.

Consider, for example, the functions $f(x) = \exp(x), g(x) = \exp(x^2), x \in \mathbb{R}$. These functions are obviously locally integrable on $\mathbb{R}$, hence $f, g \in L^1_{\text{loc}}(\mathbb{R})$, but are not with polynomial growth at infinity, because $|f|, |g|$ are not smaller than or equal to $c(1 + |x|)^k, c, k \geq 0$. Consequently, the functions $\exp(x), \exp(x^2), x \in \mathbb{R}$, do not generate function type distributions from $\mathcal{S}'(\mathbb{R})$.

Hence, $f, g \notin \mathcal{S}'(\mathbb{R})$ so that these functions do not allow the Fourier transform on $\mathcal{S}'(\mathbb{R})$.

These situations as well as other ones, require the adoption of a new definition of the Fourier transform for distributions from $\mathcal{D}'(\mathbb{R}^n)$ which constitute a generalization of the Fourier transform (2.103) given for the distributions from $\mathcal{S}'(\mathbb{R}^n)$.

Let $f \in L^1_{\text{loc}}(\mathbb{R}^n)$ be a locally integrable function with complex values and $\varphi \in \mathcal{D}(\mathbb{R}^n)$. Then, we shall associate to the function f the functional $T^*_f : \mathcal{D}(\mathbb{R}^n) \to \mathbb{C}$ defined by the formula

$$\left(T^*_f, \varphi\right) = \int_{\mathbb{R}^n} \overline{f}(x)\varphi(x)\mathrm{d}x , \tag{2.137}$$

where $\overline{f}$ is the conjugate of the function f.

Obviously, the functional T^*_f, denoted by f as well, is a function type distribution or a regular distribution from $\mathcal{D}'(\mathbb{R}^n)$.

We note that the functional T^*_f defined by (2.137) differs from the functional T_f, namely

$$(T_f, \varphi) = \int_{\mathbb{R}^n} f(x)\varphi(x)\mathrm{d}x, \varphi \in \mathcal{D}(\mathbb{R}^n) .$$

In relation to the adopted formula (2.137) on the space $\mathcal{D}'(\mathbb{R}^n)$, the operations $f + g$ and λf are defined by the formulae

$$(f + g, \varphi) = (f, \varphi) + (g, \varphi) , \quad \forall \varphi \in \mathcal{D} , \quad \forall f, g \in \mathcal{D}'(\mathbb{R}^n) , \tag{2.138}$$

$$(\lambda f, \varphi) = \overline{\lambda}(f, \varphi) = (f, \overline{\lambda}\varphi) , \quad \lambda \in \mathbb{C} , \tag{2.139}$$

where $\overline{\lambda}$ is the conjugate of λ.

The product between the function $a \in C^\infty(\mathbb{R}^n)$ and the distribution $f \in \mathcal{D}'(\mathbb{R}^n)$ is defined through the relation

$$(a f, \varphi) = (f, \overline{a}\varphi) \,. \tag{2.140}$$

The complex conjugate distribution of $f \in \mathcal{D}'(\mathbb{R}^n)$ is defined by the formula

$$(\overline{f}, \varphi) = \overline{(f, \overline{\varphi})}, \varphi \in \mathcal{D}(\mathbb{R}^n) \,. \tag{2.141}$$

Let $f \in L^1(\mathbb{R}^n)$ be an absolutely integrable function and $\varphi \in \mathcal{D}(\mathbb{R}^n)$. Then, taking into account (2.137), we can write

$$\begin{aligned}
(f, \varphi) &= \int_{\mathbb{R}^n} \overline{f}(x)\varphi(x)\mathrm{d}x = \frac{1}{(2\pi)^n}\int_{\mathbb{R}^n} \overline{f}(x)\left[\int_{\mathbb{R}^n} \hat{\varphi}(\xi)\exp(-\mathrm{i}\langle\xi, x\rangle)\mathrm{d}\xi\right]\mathrm{d}x \\
&= \frac{1}{(2\pi)^n}\int_{\mathbb{R}^n} \hat{\varphi}(\xi)\left[\int_{\mathbb{R}^n} \overline{f}(x)\overline{\exp(\mathrm{i}\langle\xi, x\rangle)}\mathrm{d}x\right]\mathrm{d}\xi \\
&= \frac{1}{(2\pi)^n}\int_{\mathbb{R}^n} \hat{\varphi}(\xi)\left[\int_{\mathbb{R}^n} \overline{f(x)\exp(\mathrm{i}\langle\xi, x\rangle)}\mathrm{d}x\right]\mathrm{d}\xi \\
&= \frac{1}{(2\pi)^n}\int_{\mathbb{R}^n} \hat{\varphi}(\xi)\overline{\hat{f}}(\xi)\mathrm{d}\xi = \frac{1}{(2\pi)^n}(\hat{f}, \hat{\varphi}) \,,
\end{aligned} \tag{2.142}$$

giving the relation

$$(\hat{f}, \hat{\varphi}) = (2\pi)^n (f, \varphi) \,, \tag{2.143}$$

named Parseval's equality.

The formula (2.143) will be adopted as definition of the Fourier transform of distributions from $\varphi \in \mathcal{D}(\mathbb{R}^n)$.

To this end, we note that the Fourier transform $\mathrm{F}[\varphi](\xi) = \hat{\varphi}(\xi) = \int_{\mathbb{R}^n} \varphi(x)\, \mathrm{e}^{\mathrm{i}\langle x, \xi\rangle}\mathrm{d}x$ of the function $\varphi \in \mathcal{D}(\mathbb{R}^n)$ may be extended from the real variable $\xi \in \mathbb{R}^n$ to the complex variables $s = \xi + \mathrm{i}\tau = (\xi_1 + \mathrm{i}\tau_1, \ldots, \xi_n + \mathrm{i}\tau_n) \in \mathbb{C}^n$, according to the formula

$$\psi(s) = \hat{\varphi}(s) = \mathrm{F}[\varphi](s) = \int_{\mathbb{R}^n} \varphi(x)\exp(\mathrm{i}\langle s, x\rangle)\mathrm{d}x \,, \tag{2.144}$$

where $(s, x) = \sum_{j=1}^n x_j s_j,\ s_j = \xi_j + \mathrm{i}\tau_j$.

The functions $\psi(s) = \hat{\varphi}(s) = \mathrm{F}[\varphi](s)$ are holomorphic, hence analytic on the whole complex plan. Consequently, the functions $\psi(s)$ can be developed in Taylor series. We shall denote by $Z(\mathbb{C}^n)$ the set of these analytical functions, that is, the Fourier transform of the space $\mathcal{D}(\mathbb{R}^n)$, hence $Z(\mathbb{C}^n) = \mathrm{F}[\mathcal{D}(\mathbb{R}^n)] = \{\hat{\varphi}(s), \hat{\varphi}(s) = \mathrm{F}[\varphi](s), \varphi \in \mathcal{D}(\mathbb{R}^n)\}$, where $\hat{\varphi}(s)$ is given by (2.144). The space Z is a complex vector space with respect to the ordinary operations of addition and scalar multiplication.

If D^α is the differentiation operator and $\psi(s) = \hat{\varphi}(s) \in Z(\mathbb{C}^n)$, then we have

$$\mathrm{F}[\mathrm{D}^\alpha \varphi(x)](s) = (-\mathrm{i}s)^\alpha \mathrm{F}[\varphi(x)](s) = (-\mathrm{i}s)^\alpha \psi(s) . \tag{2.145}$$

Acknowledging that the support of the function $\varphi \in \mathcal{D}(\mathbb{R}^n)$ is included in the n-dimensional parallelepiped $|x_k| \leq a_k, k = \overline{1, n}$, then from (2.145) we obtain the evaluation

$$\begin{aligned} |s^\alpha \psi(s)| &\leq \left| \int_{-a_1}^{a_1} \int_{-a_2}^{a_2} \dots \int_{-a_n}^{a_n} \mathrm{D}^\alpha \varphi(x) \exp(\mathrm{i}\langle s, x\rangle) \mathrm{d}x_1 \dots \mathrm{d}x_n \right| \\ &\leq c_\alpha \exp(a_1|\tau_1| + \dots + a_n|\tau_n|) , \end{aligned} \tag{2.146}$$

where $s = \xi + \mathrm{i}\tau \in \mathbb{C}^n$ and c_α, a_i are constants that depend only on $\psi(s)$, $\alpha \in \mathbb{N}_0^n$. The relation (2.146) characterizes the analytical functions of the space $Z = \mathrm{F}[\mathcal{D}]$.

It can be shown, using Cauchy's theorem for holomorphic functions, that any analytic function $\psi(s)$, which verifies the inequality (2.146) for any $\alpha \in \mathbb{N}_0^n$, is the Fourier transform of a function $\varphi \in \mathcal{D}(\mathbb{R}^n)$ that is canceled for $|x_k| > a_k$, $k = \overline{1, n}$.

Thus the function $\varphi \in \mathcal{D}(\mathbb{R}^n)$ is given by

$$\begin{aligned} \mathrm{F}^{-1}[\psi] = \varphi(x) &= \frac{1}{(2\pi)^n} \int_{\mathbb{R}^n} \psi(\xi) \exp(-\mathrm{i}\langle \xi, x\rangle) \mathrm{d}\xi \\ &= \frac{1}{(2\pi)^n} \int_{\mathbb{R}^n} \psi(s) \exp(-\mathrm{i}\langle s, x\rangle) \mathrm{d}s \end{aligned} \tag{2.147}$$

and thus the Fourier transformation establishes an isomorphism between the vector spaces $\mathcal{D}$ and $Z = \mathrm{F}[\mathcal{D}]$.

Using this isomorphism we can define the convergence to zero of the sequence $(\psi_j) \subset Z(\mathbb{C}^n)$ through the convergence to zero in the space $\mathcal{D}(\mathbb{R}^n)$ of the sequence $\varphi_j = \mathrm{F}^{-1}[\psi_j]$, where

$$\mathrm{F}^{-1}[\psi_j] = \frac{1}{(2\pi)^n} \int_{\mathbb{R}^n} \psi_j(s) \exp(-\mathrm{i}\langle s, x\rangle) \mathrm{d}\xi .$$

Hence, $\psi_j \to 0$ on $Z(\mathbb{C}^n)$ if $\varphi_j \to 0$ on $\mathcal{D}(\mathbb{R}^n)$.

Definition 2.16 Let there be the distribution $f \in \mathcal{D}'(\mathbb{R}^n)$. We call Fourier transform of the distribution f the functional $\hat{f}(s) = \mathrm{F}[f](s)$ defined on the space $Z(\mathbb{C}^n) = \mathrm{F}[\mathcal{D}(\mathbb{R}^n)]$ by the equality

$$(\hat{f}(s), \hat{\varphi}(s)) = (2\pi)^n (f(x), \varphi(x)) , \quad \varphi \in \mathcal{D}(\mathbb{R}^n) , \quad \hat{\varphi} \in Z = \mathrm{F}[\mathcal{D}] . \tag{2.148}$$

The relation (2.148) is a generalization of Parseval's equality (2.143).

Due to the isomorphism established by the Fourier transform between the spaces $\mathcal{D}$ and $Z = \mathrm{F}[\mathcal{D}]$, as well as to the convergence defined on Z, it follows that the Fourier transform $\hat{f} = \mathrm{F}[f]$ is a linear continuous functional defined on the space Z.

The set of the linear and continuous functionals defined on Z is denoted Z' and $\hat{f} \in Z'$ is a distribution defined on Z. The distributions from Z' are called ultradistributions.

The formula (2.148) takes a simple form for distributions with compact support. Let there be $f(x) \in \mathcal{E}'(\mathbb{R}^n)$, then its Fourier transform is given by

$$\mathrm{F}[f(x)](s) = \hat{f}(s) = (\overline{f}(x), \exp(\mathrm{i}\langle s, x\rangle)) = (\overline{f}(x), \eta(x)\exp(\mathrm{i}\langle s, x\rangle)),$$

where $\overline{f}$ is the complex conjugate of f and $\eta(x) \in \mathcal{D}'(\mathbb{R}^n)$, $\eta(x) = 1$ on a compact neighborhood of the support of the distribution $f \in \mathcal{E}'(\mathbb{R}^n)$.

The common property of the distributions from $\mathcal{D}'(\mathbb{R}^n)$ and $Z'(\mathbb{C}^n)$ is that both types of distributions are indefinitely differentiable, and the distinctive property is that the distributions from $Z'(\mathbb{C}^n)$ are analytical, therefore they can be developed in Taylor series.

Thus, if $g(s) \in Z'(\mathbb{C}^n)$, then for any $h \in \mathbb{C}^n$ we can write the Taylor series expansion

$$g(s+h) = \sum_{\alpha \geq 0} \frac{h^\alpha}{\alpha!} \frac{\partial^{|\alpha|} g(s)}{\partial s_1^{\alpha_1} \ldots \partial s_n^{\alpha_n}}, \tag{2.149}$$

where $s = (s_1, \ldots s_n) \in \mathbb{C}^n$, $s_j = \xi_j + \mathrm{i}\tau_j$, $h = (h_1, \ldots h_n) \in \mathbb{C}^n$, $\alpha \in \mathbb{N}_0^n$, $\alpha = (\alpha_1, \ldots, \alpha_n)$, $\alpha! = \alpha_1! \ldots \alpha_n!$, $\alpha \geq 0$, $\alpha_j \geq 0$, $|\alpha| = \sum_{i=1}^n \alpha_i$, $h^\alpha = h_1^{\alpha_1} \ldots h_n^{\alpha_n}$.

The convergence of the series (2.149) is considered in the sense of the space $Z'(\mathbb{C}^n)$. Thus $g_i \xrightarrow{Z'} g$, $g_i \in Z'(\mathbb{C}^n)$; if $\forall \psi(s) \in Z(\mathbb{C}^n)$ we have

$$\lim_i (g_i(s), \psi(s)) = (g(s), \psi(s)) .$$

Note that, by the formula (2.148), each distribution $f \in \mathcal{D}'(\mathbb{R}^n)$ is associated with the distribution $\hat{f} = \mathrm{F}[f] \in Z'$. This means that the Fourier transform is a linear operator defined on $\mathcal{D}'$ with values on Z', that is, $\mathrm{F} : \mathcal{D}' \to Z'$.

The inverse operator F^{-1} which applies Z' on $\mathcal{D}'$ is also defined with the formula (2.148). We have thus

$$(\mathrm{F}^{-1}[g], \varphi) = \frac{1}{(2\pi)^n}(g, \mathrm{F}[\varphi]), \quad g \in Z', \quad f \in \mathcal{D}' \tag{2.150}$$

$$\mathrm{F}^{-1}[\mathrm{F}[f]] = f, \quad \mathrm{F}[\mathrm{F}^{-1}[g]] = g . \tag{2.151}$$

Obviously, the inverse operator $\mathrm{F}^{-1} : Z' \to \mathcal{D}'$ as well as the direct operator $\mathrm{F} : \mathcal{D}' \to Z'$ are bijective linear operators, which results from the formulae (2.148), (2.150), and (2.151). The bijective correspondence between the spaces $\mathcal{D}$ and Z can be written using the operators F and F^{-1} in the form $Z = \mathrm{F}[\mathcal{D}]$, $\mathcal{D} = \mathrm{F}^{-1}[Z]$.

We note that the formulae regarding the Fourier transforms established for the distributions from $\mathcal{S}'$ will also apply to distributions from $\mathcal{D}'(\mathbb{R}^n)$.

The differentiation operation in the space Z' is defined similarly as in $\mathcal{D}'$. Thus, if D^α is a differentiation operator, then we have

$$(D^\alpha g(s), \psi(s)) = (-1)^{|\alpha|}(g(s), D^\alpha \psi(s)) , \tag{2.152}$$

where $g \in Z'(\mathbb{C}^n), \forall \psi \in Z(\mathbb{C}^n)$.

Example 2.13 We use the formula (2.148) to determine the Fourier images of the distributions $1, \delta(x) \in \mathcal{D}'(\mathbb{R}^n)$. Thus, we have

$$(F[\delta(x)](s), \psi(s)) = (2\pi)^n(\delta(x), \varphi(x)) = (2\pi)^n\varphi(0) = \int_{\mathbb{R}^n} \psi(\xi)d\xi = (1, \psi) ,$$

where we obtain $F[\delta(x)] = 1$.

Proceeding analogously, we can write

$$\begin{aligned}(F[1], \psi(s)) &= (2\pi)^n(1, \varphi(x)) = (2\pi)^n \int_{\mathbb{R}^n} \varphi(x)dx \\ &= (2\pi)^n \int_{\mathbb{R}^n} \varphi(x)\exp(i\langle 0, x\rangle)dx = (2\pi)^n\psi(0) = ((2\pi)^n\delta(s), \psi(s)) ,\end{aligned}$$

hence $F[1](s) = (2\pi)^n\delta(s)$.

Proposition 2.32 Let there be the distribution $f \in \mathcal{D}'(\mathbb{R}^n)$. Then, we have

$$\frac{d}{ds}F[f](s) = F[ixf](s) . \tag{2.153}$$

Proof: Taking into account the formula (2.148), we have

$$\begin{aligned}(F[ixf](s), \psi(s)) &= 2\pi(ixf(x), \varphi(x)) \\ &= 2\pi(f(x), \overline{ix}\varphi(x)) = 2\pi(f(x), -ix\varphi(x)) .\end{aligned} \tag{2.154}$$

Because $\hat{\varphi} = \psi(s) = \int_{\mathbb{R}} \varphi(x)\exp(isx)dx$, we get

$$\frac{d}{ds}\hat{\varphi} = \frac{d\psi(s)}{ds} = \int_{\mathbb{R}} ix\varphi(x)dx = F[ix\varphi(x)] ,$$

thus the formula (2.154) becomes

$$(F[ixf], \psi) = \left(F[f], -\frac{d}{ds}\hat{\varphi}\right) = \left(\frac{d}{ds}F[f], \hat{\varphi}\right) ,$$

giving the formula (2.153). □

Example 2.14 For the distribution $x^n \in \mathcal{D}(\mathbb{R})$, $n \in \mathbb{N}$, we have

$$\mathrm{F}[x^n](s) = 2\pi(-\mathrm{i})^n \delta^{(n)}(s) . \tag{2.155}$$

Indeed, observing that $\mathrm{F}[1] = 2\pi\delta(s)$ and applying the formula (2.153), we have

$$\frac{\mathrm{d}^n}{\mathrm{d}s^n}\mathrm{F}[1] = 2\pi\delta^{(n)}(s) = \mathrm{F}[(\mathrm{i}x)^n] ,$$

giving the relation (2.155).

Example 2.15 The Fourier transform of the distribution $f(x) = \mathrm{e}^{ax} \in \mathcal{D}'(\mathbb{R})$, where $a \in \mathbb{C}$, is given by the expression

$$\mathrm{F}[\mathrm{e}^{ax}](s) = 2\pi\delta(s - \mathrm{i}a) .$$

On the basis of the above formula we get

$$\mathrm{F}[\sin(ax)] = \mathrm{F}\left[\frac{\mathrm{e}^{\mathrm{i}ax} - \mathrm{e}^{-\mathrm{i}ax}}{2\mathrm{i}}\right] = \mathrm{i}\pi[\delta(s-a) - \delta(s+a)] , \tag{2.156}$$

$$\mathrm{F}[\cos(ax)] = \mathrm{F}\left[\frac{\mathrm{e}^{\mathrm{i}ax} + \mathrm{e}^{-\mathrm{i}ax}}{2}\right] = \pi[\delta(s-a) + \delta(s+a)] , \tag{2.157}$$

$$\mathrm{F}[\cosh(ax)] = \mathrm{F}\left[\frac{\mathrm{e}^{ax} + \mathrm{e}^{-ax}}{2}\right] = \pi[\delta(s-\mathrm{i}a) + \delta(s+\mathrm{i}a)] , \tag{2.158}$$

$$\mathrm{F}[\sinh(ax)] = F\left[\frac{\mathrm{e}^{ax} - \mathrm{e}^{-ax}}{2}\right] = \pi[\delta(s-\mathrm{i}a) - \delta(s+\mathrm{i}a)] . \tag{2.159}$$

2.2.4.3 Fourier Transform and the Partial Convolution Product

Taking into account the definition of the partial convolution product we will determine the properties of the Fourier transform with respect to this new type of convolution product.

Proposition 2.33 Let there be the distributions $f(x,t) \in \mathcal{S}'(\mathbb{R}^{n+m})$, $g(x) \in \mathcal{E}'(\mathbb{R}^n)$. Then we have

$$\mathrm{F}[f \underset{x}{\otimes} g] = \mathrm{F}[f(x,t)]\mathrm{F}[g(x)] , \tag{2.160}$$

$$\mathrm{F}^{-1}[f \underset{x}{\otimes} g] = (2\pi)^n \mathrm{F}^{-1}[f]\mathrm{F}^{-1}[g] . \tag{2.161}$$

In the Section 2.2.3 we have introduced the operators $\mathrm{F}, \mathrm{F}_x, \mathrm{F}_t$, representing the Fourier transforms with respect to the variables $(x,t) \in \mathbb{R}^{n+m}$, $x \in \mathbb{R}^n$ and $t \in \mathbb{R}^m$, respectively. Denoting $\mathrm{F} = \mathrm{F}_{x,t}$ the relation $\mathrm{F} = \mathrm{F}_{x,t} = \mathrm{F}_x\mathrm{F}_t = \mathrm{F}_t\mathrm{F}_x$, which shows that the Fourier transform with respect to all variables (x,t) is equal to the successive application of two partial transforms with respect to the variables $x \in \mathbb{R}^n$ and $t \in \mathbb{R}^m$ was stated.

Using this property, we can state the following.

Proposition 2.34 Let the distributions be $f(x,t) \in \mathcal{S}'(\mathbb{R}^{n+m})$ and $g(x) \in \mathcal{E}'(\mathbb{R}^n)$. Then we have

$$\mathrm{F}_t[f(x,t) \underset{x}{\otimes} g(x)] = \mathrm{F}_t[f(x,t)] \underset{x}{\otimes} g(x)\,, \tag{2.162}$$

$$\mathrm{F}_x[f(x,t) \underset{x}{\otimes} g(x)] = \mathrm{F}_x[f(x,t)]\mathrm{F}_x[g(x)]\,. \tag{2.163}$$

Also, if $f(x,t) \in \mathcal{S}'(\mathbb{R}^{n+m})$ and $g(x) \in \mathcal{E}'(\mathbb{R}^n)$ then we have

$$\mathrm{F}_t^{-1}[f(x,t) \underset{x}{\otimes} g(x)] = \mathrm{F}_t^{-1}[f(x,t)] \underset{x}{\otimes} g(x)\,, \tag{2.164}$$

$$\mathrm{F}_x^{-1}[f(x,t) \underset{x}{\otimes} g(x)] = (2\pi)^n \mathrm{F}_x^{-1}[f(x,t)]\mathrm{F}_x^{-1}[g(x)]\,. \tag{2.165}$$

Example 2.16 Let $P(\partial/\partial t)$ be a linear differential operator with constant coefficients, $t \in \mathbb{R}$, $u(x,t)$, $f(x,t) \in S'(\mathbb{R}^n \times \mathbb{R})$. We consider the equation

$$P\left(\frac{\partial}{\partial t}\right) u(x,t) = f(x,t)\,,$$

where f is a given distribution.

The equation can be written as a convolution equation

$$u(x,t) \underset{t}{\otimes} P\left(\frac{\partial}{\partial t}\right) \delta(t) = f(x,t)\,.$$

Applying the Fourier transformation with respect to the variables $x \in \mathbb{R}^n$, $t \in \mathbb{R}$, we obtain

$$\tilde{u}(\xi,\alpha) P(-\mathrm{i}\alpha) = \tilde{f}(\xi,\alpha),$$

where $\tilde{u} = \mathrm{F}[u(x,t)] = \mathrm{F}_t\mathrm{F}_x[u]$, $\tilde{f} = \mathrm{F}_t\mathrm{F}_x[f]$, $\mathrm{F}_t[\delta^{(k)}(t)] = (-\mathrm{i}\alpha)^k$.

Consequently, we have $\tilde{u}(\xi,\alpha) = \tilde{f}/P(-\mathrm{i}\alpha)$.

Let $E(t) \in \mathcal{S}'(\mathbb{R})$ be the fundamental solution of the operator $P(\partial/\partial t)$. Then $P(\partial/\partial t)E(t) = P(\partial/\partial t)\delta(t) * E(t) = \delta(t)$. By applying the Fourier transform F_t we find $\mathrm{F}_t[E] = \tilde{E}(\alpha) = 1/P(-\mathrm{i}\alpha)$.

In conclusion we can write $\tilde{u}(\xi,\alpha) = \tilde{f}(\xi,\alpha)\tilde{E}(\alpha)$, wherefrom, by applying the inverse Fourier transforms $\mathrm{F}_t^{-1}, \mathrm{F}_x^{-1}$, we obtain the solution of the differential equation in the form $u(x,t) = f(x,t) \otimes_t E(t)$.

2.3 Laplace Transforms of Functions and Distributions

The Laplace transform is a variant of the Fourier transform, hence the two are closely related. In many cases, the Laplace transform is easier to apply than the Fourier transform, due to the difficulties of Fourier images calculation.

2.3.1
Laplace Transforms of Functions

Definition 2.17 The Laplace transform of the function $f : \mathbb{R} \to \mathbb{C}$, which is null for $t < 0$, is the complex function $\mathrm{L}[f]$ of complex variable $p = \sigma + \mathrm{i}\tau$, given by

$$\mathrm{L}[f](p) = \int_0^\infty f(t)\mathrm{e}^{-pt}\mathrm{d}t \,. \tag{2.166}$$

Proposition 2.35 If $f(t)\mathrm{e}^{-pt} \in L^1(\mathbb{R})$ for $\mathrm{Re}\, p = \sigma_0$, then its Laplace transform $\mathrm{L}[f]$ exists $\forall p \in \mathbb{C}$, $\mathrm{Re}\, p \geq \sigma_0$.

Indeed, if $\mathrm{Re}\, p = \sigma \geq \sigma_0$, then $\mathrm{e}^{-\sigma t} \leq \mathrm{e}^{-\sigma_0 t}$ and, since $f(t)\mathrm{e}^{-\sigma_0 t} \in L^1(\mathbb{R})$, it follows that $|f(t)\mathrm{e}^{-pt}| = |f(t)\mathrm{e}^{-\sigma t}| \leq |f(t)|\mathrm{e}^{-\sigma_0 t}$; hence, $f(t)\mathrm{e}^{-\sigma t} \in L^1(\mathbb{R})$ and, consequently, $\mathrm{L}[f]$ exists for $\mathrm{Re}\, p \geq \sigma_0$. The lower bound of the real numbers $\mathrm{Re}\, p = \sigma_0$ for which $f(t)\mathrm{e}^{-pt} \in L^1(\mathbb{R})$ is called the convergence abscissa of *f*. If $\inf \sigma_0 = \alpha$, then $\mathrm{L}[f]$ exists for $\mathrm{Re}\, p > \alpha$.

Definition 2.18 The functions $f : \mathbb{R} \to \mathbb{C}$ such that

1. $f(t) = 0, t < 0$;
2. $f(t)$ is continuous except for a finite number of points in which it has discontinuities of the first kind;
3. $|f(t)| \leq M\mathrm{e}^{\sigma_0 t}$, $M > 0$, $\sigma_0 \geq 0$,

are called original functions. The number σ_0 is called the rise index of the original function. Notice that the convergence abscissa α of the original function *f* is at most σ_0, $\alpha \leq \sigma_0$.

The evaluation

$$|f(t)\mathrm{e}^{-pt}| = |f(t)|\mathrm{e}^{-\sigma t} \leq M\mathrm{e}^{-t(\sigma-\sigma_0)} \tag{2.167}$$

holds for original functions, indicating that, for $\mathrm{Re}\, p = \sigma > \sigma_0$, the right-hand side of the inequality is integrable. Hence, the Laplace transforms of the original functions exist in the right-hand half-plane $\mathrm{Re}\, p > \sigma_0$; they are analytical in this domain and $|\mathrm{L}[f]| \leq M/(\sigma - \sigma_0)$. Notice that the analytical property of the original functions in the half-plane $\mathrm{Re}\, p > \sigma_0$ implies that they are indefinitely differentiable in this half-plane.

Proposition 2.36 Let f_1 and f_2 be two original functions having convergence abscissae s_1 and s_2, respectively. Then the sum $f_1 + f_2$ and the convolution product $f_1 * f_2$ are original functions, having the convergence abscissa $s = \max(s_1, s_2)$; also, the product of the above functions is an original function having the convergence abscissa $s = s_1 + s_2$.

Proposition 2.37 Mellin–Fourier formula If f is an original function having the convergence abscissa σ_0, then at each point of continuity of the function f we have

$$f(t) = \mathrm{L}^{-1}[\mathrm{L}[f]] = \frac{1}{2\pi \mathrm{i}} \int_{\sigma-\mathrm{i}\infty}^{\sigma+\mathrm{i}\infty} \mathrm{L}[f]\mathrm{e}^{pt}\mathrm{d}p, \mathrm{Re}\, p > \sigma_0 \,. \tag{2.168}$$

This integral is sometimes called *the Bromwich integral.*

From the definition of the Laplace transform and the formula (2.168) it results that the direct and inverse Laplace transforms, L and L^{-1}, are linear operators, namely $\forall \alpha, \beta, A, B \in \mathbb{C}$, f, g original functions and f_1, g_1 the Laplace images of some original functions, we have

$$\mathrm{L}[\alpha f + \beta g] = \alpha \mathrm{L}[f] + \beta \mathrm{L}[g] \,, \tag{2.169}$$

$$\mathrm{L}^{-1}[A f_1 + B g_1] = A\mathrm{L}^{-1}[f_1] + B\mathrm{L}^{-1}[g_1] \,. \tag{2.170}$$

Let f be an original function having the convergence abscissa σ_0, the Laplace transform has the following properties:

$$\forall q \in \mathbb{C} \,, \quad \mathrm{L}[\mathrm{e}^{qt} f(t)] = \mathrm{L}[f(t)](p-q) \,, \quad \mathrm{Re}\, p > \sigma_0 + \mathrm{Re}\, q \,, \tag{2.171}$$

$$\forall a \in \mathbb{R} - \{0\} \,, \quad \mathrm{L}[f(t-a)](p) = \mathrm{e}^{-ap}\mathrm{L}[f](p) \,, \tag{2.172}$$

$$\forall a > 0 \,, \quad \mathrm{L}[f(at)](p) = \frac{1}{a}\mathrm{L}[f(t)]\left(\frac{p}{a}\right) \,, \tag{2.173}$$

$$\mathrm{L}[f'(t)](p) = p\mathrm{L}[f] - f(0+0) \,, \tag{2.174}$$

$$\begin{aligned} \mathrm{L}[f^{(n)}(t)](p) = p^n \mathrm{L}[f] - \big(p^{n-1} f(0+0) + p^{n-2} f'(0+0) + \ldots \\ + f^{(n-1)}(0+0)\big) \,, \end{aligned} \tag{2.175}$$

$$\mathrm{L}\left[\int_0^t f(\tau)\mathrm{d}\tau\right] = \frac{1}{p}\mathrm{L}[f(t)] \,, \quad \mathrm{Re}\, p > \sigma_0 \,, \tag{2.176}$$

$$\mathrm{L}[f * g] = \mathrm{L}[f]\mathrm{L}[g] \,, \tag{2.177}$$

$$\frac{\mathrm{d}^n}{\mathrm{d}p^n}\mathrm{L}[f](p) = \mathrm{L}[(-t)^n f(t)](p) \,. \tag{2.178}$$

We shall extend the definition of the Laplace transform for the function of one variable to the case of a function of several variables. Thus, if $f(x_1, x_2, \ldots, x_n)$ is a complex function of n variables and satisfies the conditions:

1. $f(x_1, x_2, \ldots, x_n) = 0$, for $x_1 < 0$, or $x_2 < 0$, or ..., or $x_n < 0$,
2. $f(x_1, x_2, \ldots, x_n)$ has partial derivatives of the first order,
3. $|f(x_1, x_2, \ldots, x_n)| \le M\mathrm{e}^{a_1 x_1 + a_2 x_2 + \cdots + a_n x_n}$, $M > 0$, $a_1, a_2, \ldots, a_n \ge 0$,

then the Laplace transform of the function $f(x_1, x_2, \ldots, x_n)$ is a complex function of n complex variables $p_j = \sigma_j + i\tau_j$, $j = \overline{1, n}$, defined by the expression

$$\mathrm{L}[f(x_1, x_2, \ldots, x_n)](p) = \int_0^\infty \int_0^\infty \ldots \int_0^\infty f(x_1, x_2, \ldots x_n) \mathrm{e}^{-(p_1 x_1 + p_2 x_2 + \cdots + p_n x_n)} \mathrm{d}x_1 \mathrm{d}x_2 \ldots \mathrm{d}x_n \,, \tag{2.179}$$

we can denote

$$\mathrm{L}[f](p) = \int_{\mathbb{R}^n} f(x) \mathrm{e}^{-\langle p, x \rangle} \mathrm{d}x \,, \quad x \in \mathbb{R}^n \,, \tag{2.180}$$

where

$$\langle p, x \rangle = p_1 x_1 + p_2 x_2 + \cdots + p_n x_n, \mathrm{d}x = \mathrm{d}x_1 \mathrm{d}x_2 \ldots \mathrm{d}x_n \,. \tag{2.181}$$

The image function $\mathrm{L}[f](p)$ is an analytical function in the domain $\operatorname{Re} p_j > a_j$, $j = \overline{1, n}$. The inverse Laplace transform is defined by the relation

$$f(x_1, x_2, \ldots, x_n) = \mathrm{L}^{-1}[\mathrm{L}[f(x_1, x_2, \ldots, x_n)]] = \frac{1}{(2\pi i)^n} \int_{\sigma_1 - i\infty}^{\sigma_1 + i\infty} \int_{\sigma_2 - i\infty}^{\sigma_2 + i\infty} \ldots \int_{\sigma_n - i\infty}^{\sigma_n + i\infty} \mathrm{L}[f] \mathrm{e}^{\langle p, x \rangle} \mathrm{d}p_1 \mathrm{d}p_2 \ldots \mathrm{d}p_n \,, \tag{2.182}$$

for $\sigma_j > a_j$, $j = \overline{1, n}$.

Particularly, for $f(x_1, x_2, \ldots, x_n) = f_1(x_1) f_2(x_2) \ldots f_n(x_n)$ we can write

$$\mathrm{L}[f(x_1, x_2, \ldots, x_n)] = \mathrm{L}[f_1(x_1)] \mathrm{L}[f_2(x_2)] \ldots \mathrm{L}[f_n(x_n)] \,. \tag{2.183}$$

Thus, for the Heaviside function $H(x_1, x_2, \ldots, x_n) = H(x_1) H(x_2) \ldots H(x_n)$ we obtain

$$\mathrm{L}[H(x_1, x_2, \ldots, x_n)] = \mathrm{L}[H(x_1)] \mathrm{L}[H(x_2)] \ldots \mathrm{L}[H(x_n)] = \frac{1}{p_1 p_2 \ldots p_n} \,, \quad \operatorname{Re} p_j > 0 \,, \quad j = \overline{1, n} \,. \tag{2.184}$$

Analogously, we can show

$$\begin{aligned} &\mathrm{L}\left[H(x_1, x_2, \ldots, x_n) \mathrm{e}^{\lambda_1 x_1 + \lambda_2 x_2 + \cdots + \lambda_n x_n} \right] \\ &= \mathrm{L}[H(x_1) \mathrm{e}^{\lambda_1 x_1}] \mathrm{L}[H(x_2) \mathrm{e}^{\lambda_2 x_2}] \ldots \mathrm{L}[H(x_n) \mathrm{e}^{\lambda_n x_n}] \\ &= \frac{1}{(p_1 - \lambda_1)(p_2 - \lambda_2) \ldots (p_n - \lambda_n)} \,, \end{aligned} \tag{2.185}$$

for $\operatorname{Re} p_j > \operatorname{Re} \lambda_j$, $j = \overline{1, n}$.

The theorem of the differentiation of the image becomes

$$\frac{\partial^{k_1+k_2+\ldots k_n}}{\partial p_1^{k_1}\partial p_2^{k_2}\ldots\partial p_n^{k_n}}\mathrm{L}[f](p)$$
$$= \mathrm{L}\left[(-x_1)^{k_1}(-x_2)^{k_2}\ldots(-x_n)^{k_n} f(x_1, x_2, \ldots, x_n)\right] . \tag{2.186}$$

As regards the image of the derivative of a function of several variables, we can write

$$\mathrm{L}\left[\frac{\partial}{\partial x_i} f(x_1, x_2, \ldots, x_n)\right] = p_i \mathrm{L}[f(x_1, x_2, \ldots, x_n)]$$
$$- \mathrm{L}[f(x_1, x_2, \ldots, x_{i-1}, 0, x_{i+1}, \ldots, x_n)] , \tag{2.187}$$

where $\mathrm{L}[f(x_1, x_2, \ldots, x_n)]$ is the Laplace transform in $\mathbb{R}^n$ analogously, the transform $\mathrm{L}[f(x_1, x_2, \ldots, x_{i-1}, 0, x_{i+1}, \ldots, x_n)]$ is the Laplace transform in $\mathbb{R}^{n-1}$.

For example,

$$\mathrm{L}\left[\frac{\partial}{\partial x_2} H(x_1, x_2)\mathrm{e}^{\lambda_1 x_1+\lambda_2 x_2}\right] = p_2 \mathrm{L}[H(x_1, x_2)\mathrm{e}^{\lambda_1 x_1+\lambda_2 x_2}]$$
$$- [H(x_1)\mathrm{e}^{\lambda_1 x_1}] = \frac{p_2}{(p_1-\lambda_1)(p_2-\lambda_2)} - \frac{1}{p_1-\lambda_1}$$
$$= \frac{\lambda_2}{(p_1-\lambda_1)(p_2-\lambda_2)} , \quad \mathrm{Re}\, p_1 > \mathrm{Re}\, \lambda_1, \mathrm{Re}\, p_2 > \mathrm{Re}\, \lambda_2 . \tag{2.188}$$

2.3.2
Laplace Transforms of Distributions

Definition 2.17 can be extended to distributions. To this purpose, we state the following.

Definition 2.19 Let f be a distribution from $\mathcal{D}'(\mathbb{R})$ with its support in $[0, \infty)$, hence $f \in \mathcal{D}'_+(\mathbb{R})$. If there exists a $\sigma_0 \in \mathbb{R}$ such that $\mathrm{e}^{-\sigma t} f \in \mathcal{S}'(\mathbb{R})$ for $\mathrm{Re}\, p > \sigma_0$, then

$$\mathrm{L}[f](p) = (f, \mathrm{e}^{-pt}) , \tag{2.189}$$

which represents the Laplace transform of the distribution f defined for $\mathrm{Re}\, p = \sigma > \sigma_0$.

We notice that the function e^{-pt} is indefinitely derivable with respect to t and is not a function from $\mathcal{S}$. For this reason the formula (2.189) cannot be applied to all distributions, but only to the distributions with the supports on $[0, \infty)$, for which $f(t)\mathrm{e}^{-pt}$ is a distribution from $\mathcal{S}'$.

The Laplace transform thus defined is an analytical complex function of complex variable p on the half-plane $\mathrm{Re}\, p > \sigma_0$, where σ_0 is the lower real number σ such that $\mathrm{e}^{-\sigma t} f(t) \in \mathcal{S}'(\mathbb{R})$. Performing the distribution differentiation of the relation (2.189), we obtain

$$\frac{\mathrm{d}^n}{\mathrm{d}p^n}\mathrm{L}[f](p) = (f, (-t)^n \mathrm{e}^{-pt}) , \tag{2.190}$$

hence

$$L[(-t)^n f(t)](p) = L^{(n)}[f](p) . \tag{2.191}$$

We have the following formulae:

$$\forall a \in \mathbb{R} - \{0\}, L[f(t-a)](p) = e^{-ap} L[f](p) , \tag{2.192}$$

$$\forall a > 0 , \quad L[f(at)](p) = \frac{1}{a} L[f(t)]\left(\frac{p}{a}\right) , \tag{2.193}$$

$$L[f * g] = L[f]L[g] , \tag{2.194}$$

$$L[f'(t)](p) = p L[f] , \tag{2.195}$$

$$L[f^{(n)}(t)](p) = p^n L[f] , \tag{2.196}$$

$$L[e^{qt} f(t)] = L[f](p-q) , \tag{2.197}$$

$$L[\delta(kt)] = \frac{1}{k} , \quad k > 0 , \quad L[\delta^{(n)}(t)] = p^n , \tag{2.198}$$

$$L[\delta(t-a)] = e^{-ap} . \tag{2.199}$$

We notice that, unlike the classical formula, the image of the derivative is expressed by the product of the image by p.

It should be observed that the defining relation (2.179) may be extended to distributions of several variables. Thus, we give the following.

Definition 2.20 If $f \in \mathcal{D}'(\mathbb{R}^n)$ is a distribution with the support on the domain $x_1 > 0, x_2 > 0, \ldots, x_n > 0$, so that the following distribution $f(x_1, x_2, \ldots, x_n)$ $e^{-\langle p,x\rangle}$ is a temperate distribution, then

$$L[f(x_1, x_2, \ldots, x_n)] = \left(f(x_1, x_2, \ldots, x_n), e^{-(p_1x_1+p_2x_2+\cdots+p_nx_n)}\right) \tag{2.200}$$

represents the Laplace transform of that distribution.

As in the case of the Laplace transform of a distribution of one variable, we have

$$L[f(k_1x_1, k_2x_2, \ldots, k_nx_n)] = \frac{1}{k_1k_2\ldots k_n} L[f(x_1, x_2, \ldots, x_n)]\left(\frac{p_1}{k_1}, \frac{p_2}{k_2}, \ldots, \frac{p_n}{k_n}\right) , \quad k_j > 0 , \quad j = \overline{1, n} , \tag{2.201}$$

$$L[f(x_1 - a_1, x_2 - a_2, \ldots, x_n - a_n)] = e^{-(p_1a_1+p_2a_2+\cdots+p_na_n)} L[f(x_1, x_2, \ldots, x_n)] , \tag{2.202}$$

$$L\left[\frac{\partial}{\partial x_j} f(x_1, x_2, \ldots, x_n)\right] = p_j L[f(x_1, x_2, \ldots, x_n)] . \tag{2.203}$$

3
Variational Calculus and Differential Equations in Distributions

3.1
Variational Calculus in Distributions

In order to broaden the applicability of results obtained in classical variational calculus and the possible treatment of the variational calculation problems in which the admissible lines have first-order discontinuities, we define the notion of variation of a functional in the distributions space.

Let there be the functional

$$I[y] = \int_a^b F(x, y, y')\mathrm{d}x \,, \tag{3.1}$$

where F is a function of class $C^2(\Delta_3)$ in its domain of definition $\Delta_3 \subset \mathbb{R}^3$ with respect to the variables $(x, y, y') \in \Delta_3$.

The set of the admissible lines D for the functional (3.1), is the set of functions

$$D = \{y \in C^1[a, b], y(a) = y_1, y(b) = y_2\} \,. \tag{3.2}$$

The first-order variation of the functional (3.1) is

$$\delta I(y; \eta) = \int_a^b \left(F_y \eta + F_{y'} \eta'\right) \mathrm{d}x \,, \tag{3.3}$$

where $\eta \in C^1[a, b]$ is an arbitrary function, verifying the conditions $\eta(a) = \eta(b) = 0$. Instead of the function η we consider a test function $\varphi \in \mathcal{D}(\mathbb{R})$, with the support included in the interval $[a, b]$, hence $\mathrm{supp}(\varphi) \subset [a, b]$. Thus, (3.3) becomes

$$\delta I(y; \varphi) = \int_{\mathbb{R}} \left(F_y \varphi + F_{y'} \varphi'\right) \mathrm{d}x \,. \tag{3.4}$$

On the other hand, the Lagrangian F with null values may be extended beyond its domain of definition $\Delta_3 \subset \mathbb{R}^3$, although this is not absolutely necessary, because in the integral (3.4) values only from $\Delta_3 \subset \mathbb{R}^3$ occur.

Distribution Theory, First Edition. Petre P. Teodorescu, Wilhelm W. Kecs, and Antonela Toma

Analogously, we make an extension of the admissible line $y \in D$, out of the interval $[a, b]$, in order to be of class C^2 on $\mathbb{R}$, which is always possible. Therefore, both the Lagrangian F and the admissible lines are defined on $\mathbb{R}^3$ and $\mathbb{R}$, respectively. We shall denote by $\mathcal{D}_{[a,b]} \subset \mathcal{D}$ the test functions set $\varphi \in \mathcal{D}(\mathbb{R})$, with the property $\operatorname{supp}(\varphi) \subset [a, b]$; hence, we have

$$\varphi^{(k)}(a) = \varphi^{(k)}(b) = 0\,, \quad k = 0, 1, 2, \ldots \tag{3.5}$$

Thus, the first-order variation δI reads

$$\delta I(y; \varphi) = (\delta I, \varphi) = (F_y, \varphi) + (F_{y'}, \varphi')\,, \tag{3.6}$$

which shows that the variation of the first order is a distribution defined on the subspace $\mathcal{D}_{[a,b]} \subset \mathcal{D}$ of indefinitely differentiable functions with supports in $[a, b]$. As regards the definition of a neighborhood of a certain order of the admissible line y, defined by the extension on $\mathbb{R}$, it remains the same, that is,

$$\overline{y} = y + t\varphi\,, \quad t \in \mathbb{R}$$

where $|t\varphi^{(k)}(x)| < \varepsilon$, $k = 0, 1$.

In connection with the conditions for which a distribution $f \in \mathcal{D}'(\mathbb{R})$ is zero on the interval $[a, b]$, we have the following.

Proposition 3.1 The necessary and sufficient condition for the distribution $f \in \mathcal{D}'(\mathbb{R})$ to be zero on $[a, b]$ is

$$(f(x), \varphi(x)) = 0 \tag{3.7}$$

for any $\varphi \in \mathcal{D}_{[a,b]} \subset \mathcal{D}$, hence $\operatorname{supp}(\varphi) \subset [a, b]$.

The Proposition 3.1 contains as special cases the Lagrange and the DuBois–Reymond lemmas from the variational calculus. Indeed, if $f \in C^0$, hence if it is a function type distribution generated by a continuous function, then from (3.7) it results

$$(f, \varphi) = \int_a^b f(x)\varphi(x)\mathrm{d}x = 0\,,$$

which occurs only if $f \equiv 0$, $x \in [a, b]$. This is the Lagrange lemmas. If, instead of the function φ, we consider the function $\varphi' \in \mathcal{D}_{[a,b]}$, then, from (3.7), we have

$$(f, \varphi') = -(f', \varphi) = -\int_a^b f'\varphi\mathrm{d}x = 0\,,$$

wherefrom $f' \equiv 0$; hence $f = \text{const}$, which is the lemma of DuBois–Reymond.

The Proposition 3.1 is the fundamental lemma of the variational calculus in the case in which the admissible lines are distributions from the space $\mathcal{D}'(\mathbb{R})$.

We note that the definition (3.6) of the first-order variation in distributions also has meaning in the case when the Lagrangian F is a continuous function with respect to the variables x, y, y', and the admissible line y is a distribution from $\mathcal{D}'$, assuming that the operations indicated by F have a meaning. Sometimes, we can consider F as a continuous function, while the admissible lines are function type distributions, generated by functions of class C^1.

Taking into account the rule of differentiation in distributions, the expression (3.6) can be written in the form

$$(\delta I, \varphi) = (F_y, \varphi) - \left(\frac{\mathrm{d}}{\mathrm{d}x} F_{y'}, \varphi \right) = \left(F_y - \frac{\mathrm{d}}{\mathrm{d}x} F_{y'}, \varphi \right) , \tag{3.8}$$

wherefrom, on the basis of the Proposition 3.1, we obtain the Euler equation in distributions

$$F_y - \frac{\mathrm{d}}{\mathrm{d}x} F_{y'} = 0 ; \tag{3.9}$$

the differentiation operations are considered in the distributions space.

In relation to the functional (3.1), if $F_{y'y'} \neq 0$, then the extremal line $y \in \mathcal{D}$ allows for a continuous derivative of the second order, which simultaneously verifies the equations

$$\frac{\mathrm{d}}{\mathrm{d}x} F_{y'} = F_y , \quad \frac{\mathrm{d}}{\mathrm{d}x} \left(F - y' F_{y'} \right) = F_x . \tag{3.10}$$

Because, in the meaning of distributions, an admissible line allows for derivatives of any order, it follows that, in the distributions space, an admissible line must satisfy (3.10), which obviously are identical with the Euler equation (3.9) considered in the distributions space. Therefore, any extremal of the functional (3.1), considered as distribution, satisfies the system of (3.10) in the distributions space.

Let us assume now that, at a point $x_0 \in (a, b)$, the extremal has a discontinuity of the first order. This means that the admissible lines for the functional (3.1) will be functions of class C^1 only on the intervals $[a, x_0)$ and $(x_0, b]$. Thus, for the functional (3.1), the class of the admissible lines changes, now becoming the set of piecewise smooth functions which verifies the conditions $y(a) = y_1$ and $y(b) = y_2$.

Let us assume that $x_0 \in (a, b)$ is a point of discontinuity of the first order for the extremal of the functional (3.1); hence, y is of class $C^1(\mathbb{R})$, except x_0. The admissible lines, extended on $\mathbb{R}$, have no discontinuities outside the interval $[a, b]$, because we assumed them to be of the class C^2 outside the interval $[a, b]$. Hence, the function type distributions $F - y' F_{y'}$ and $F_{y'}$ have the point x_0 as discontinuity of the first order.

Denoting with $\mathrm{d}/\mathrm{d}x$, $\tilde{\mathrm{d}}/\mathrm{d}x$, the derivative in the distributions meaning and the derivative in classical sense, respectively, from (3.10) we get

$$\frac{\mathrm{d}}{\mathrm{d}x} \left(F - y' F_{y'} \right) = \frac{\tilde{\mathrm{d}}}{\mathrm{d}x} \left(F - y' F_{y'} \right) + S_{x_0} \left(F - y' F_{y'} \right) \delta(x - x_0) = F_x , \tag{3.11}$$

$$\frac{\mathrm{d}}{\mathrm{d}x}F_{y'} = \frac{\tilde{\mathrm{d}}}{\mathrm{d}x}F_{y'} + S_{x_0}\left(F_{y'}\right)\delta(x - x_0) = F_y \,, \tag{3.12}$$

where $S_{x_0}(F - y'F_{y'})$, $S_{x_0}(F_{y'})$ are the jumps of the functions $F - y'F_{y'}$ and $F_{y'}$ at the point x_0, that is,

$$\begin{aligned} S_{x_0}\left(F - y'F_{y'}\right) &= \lim_{\substack{x \to x_0 \\ x > x_0}} \left(F - y'F_{y'}\right) - \lim_{\substack{x \to x_0 \\ x < x_0}} \left(F - y'F_{y'}\right) \\ &= \left(F - y'F_{y'}\right)\big|_{x_0+0} - \left(F - y'F_{y'}\right)\big|_{x_0-0} \,, \\ S_{x_0}(F_{y'}) &= \lim_{\substack{x \to x_0 \\ x > x_0}} F_{y'} - \lim_{\substack{x \to x_0 \\ x < x_0}} F_{y'} \,. \end{aligned} \tag{3.13}$$

Proposition 3.2 If f is a continuous function on $\mathbb{R}$ and A is a constant, then the equality

$$f(x) + A\delta(x - x_0) = 0 \,, \tag{3.14}$$

where $\delta(x - x_0)$ is the Dirac delta distribution concentrated at the point $x_0 \in \mathbb{R}$, occurs only if $f = 0$ and $A = 0$.

Indeed, let $\varphi \in \mathcal{D}(\mathbb{R})$ be a test function whose support does not contain the point x_0, hence $x_0 \notin \operatorname{supp}(\varphi)$. Then $(\delta(x - x_0), \varphi(x)) = 0$, therefore

$$(f(x) + A\delta(x - x_0), \varphi) = (f(x), \varphi) = 0 \,, \tag{3.15}$$

which means that $f = 0$ on $\mathbb{R}$, except possibly the point x_0. On the basis of the continuity of the function f, resulting in $f = 0$ on $\mathbb{R}$. It follows, obviously, $A = 0$ and the proof is complete.

On the basis of the Proposition 3.2, the equalities (3.11) and (3.12) take place only if

$$\frac{\tilde{\mathrm{d}}}{\mathrm{d}x}\left(F - y'F_{y'}\right) = F_x \,, \quad \frac{\tilde{\mathrm{d}}}{\mathrm{d}x}F_{y'} = F_y \,, \tag{3.16}$$

$$S_{x_0}\left(F - y'F_y\right) = 0 \,, \quad S_{x_0}(F_{y'}) = 0 \,. \tag{3.17}$$

We note that (3.16) are the Euler equations which the extremal line verifies on the intervals $[a, x_0)$, $(x_0, b]$, and that the relations (3.17) express the additional conditions which the extremal line must meet at the point of discontinuity $x_0 \in (a, b)$.

The additional conditions (3.17) are called the Erdmann–Weierstrass conditions.

In conclusion, if an extremal line has a discontinuity of the first order at the point $x_0 \in (a, b)$, then it satisfies the Euler equation on the intervals $[a, x_0)$, $(x_0, b]$; at the point of discontinuity x_0, it must verify the Erdmann–Weierstrass conditions.

Proceeding as in the case of the Proposition 3.2, we obtain an extension of it in the following form.

Proposition 3.3 If A is a continuous function and B_j^k, $j = \overline{1, m}$, $k = 0, 1, \ldots, n$, are constants, the equality

$$A(x) + \sum_{k=0}^{n} \sum_{j=1}^{m} B_j^k \delta^{(k)}(x - x_j) = 0 \tag{3.18}$$

implies $A(x) = 0$, $B_j^k = 0$.

Example 3.1 Let us consider the functional

$$I[y] = \int_{-1}^{1} x^2 y'^2 \mathrm{d}x \,. \tag{3.19}$$

The question is how to determine the curve $y \in C^1[-1, 1]$, which achieves the minimum of the functional (3.19) and passes through the points $A(-1, -1)$, $B(1, 1)$ (Figure 3.1). Because $F = x^2 y'^2 \geq 0$, it follows that $I[y] \geq 0$.

Because $\inf I[y] = 0$, it follows that the minimum value of the functional is $I[y] = 0$. This implies $F = 0$, hence $y' = 0$, that is, $y = \text{const}$. This is a function of class $C^1[-1, 1]$, but which does not pass through the points A and B. Therefore, the functional (3.19) does not attain its minimum in the set of the admissible lines of the class $C^1[-1, 1]$. Thus, the problem has no solution in C^1.

The Euler equation corresponding to the functional (3.19) is

$$\frac{\mathrm{d}}{\mathrm{d}x}(x^2 y') = 0 \,, \tag{3.20}$$

wherefrom we get $x^2 y' = 0$, the equation being considered in distributions. The solution of this equation is the function type distribution

$$y(x) = 2H(x) - 1 = \begin{cases} 1 \,, & x > 0 \,, \\ -1 \,, & x \leq 0 \,. \end{cases} \tag{3.21}$$

Differentiating in the distributions sense, we have $y' = 2\delta(x)$, hence $x^2 y' = 2x^2 \delta(x) = 0$, which shows that (3.21) is the solution of the Euler equation (3.20) in distributions.

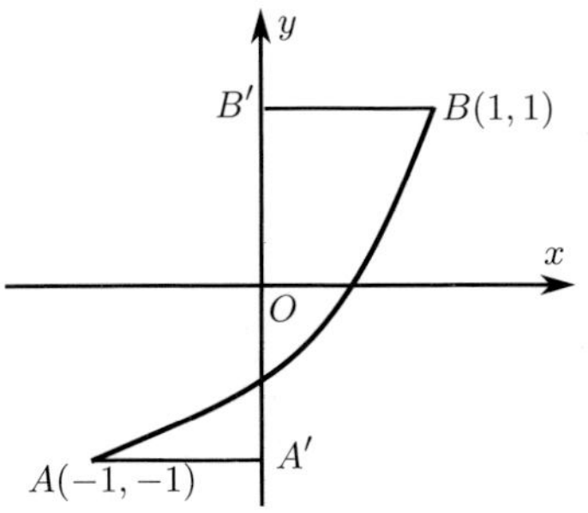

Figure 3.1

Therefore, the curve that performs the minimum of the functional (3.19) consists of segments parallel to the Ox-axis, that is, AA' and BB' passing through the given points A and B. The discontinuity point of the solution (3.21) is the origin $x_0 = 0$. At this point, the two conditions Erdmann–Weierstrass are met, because $S_0(F - y'F_{y'}) = S_0(-x^2y'^2) = 0$, since $y' = 0$, for $x \neq 0$. Analogously, $S_0(F_{y'}) = S_0(2x^2y') = 0$, based on the previous results. The problem formulated for the functional (3.19) was put forth by K. Weierstrass.

What has been presented in relation to the functional (3.1) can be applied to more general functionals as well. Thus, if we consider the functional

$$I[y] = \int_a^b F(x, y, y', \dots, y^{(n)}) \mathrm{d}x \,, \tag{3.22}$$

then the first-order variation will be the distribution δI, acting as

$$(\delta I, \varphi) = (F_y, \varphi) + \left(F_{y'}, \varphi'\right) + \cdots + \left(F_{y^{(n)}}, \varphi^{(n)}\right) \,, \tag{3.23}$$

where $\varphi \in \mathcal{D}_{[a,b]} \subset \mathcal{D}$, hence $\operatorname{supp}(\varphi) \subset [a, b]$.

From (3.23), we obtain

$$(\delta I, \varphi) = \left(F_y - \frac{\mathrm{d}}{\mathrm{d}x} F_{y'} + \frac{\mathrm{d}^2}{\mathrm{d}x^2} F_{y''} + \cdots + (-1)^n \frac{\mathrm{d}^n}{\mathrm{d}x^n} F_{y^{(n)}}, \varphi\right) \tag{3.24}$$

and, on the basis of the Proposition 3.1, we obtain the Euler–Poisson equation in distributions

$$F_y - \frac{\mathrm{d}}{\mathrm{d}x} F_{y'} + \cdots + (-1)^n \frac{\mathrm{d}^n}{\mathrm{d}x^n} F_{y^{(n)}} = 0 \,. \tag{3.25}$$

Let

$$I\left[y_1(x), y_2(x), \dots, y_m(x)\right] = \int_a^b F\left(x, y_1, \dots, y_m, y'_1, \dots y'_m, \dots, y_1^{(n)}, \dots, y_m^{(n)}\right) \mathrm{d}x \tag{3.26}$$

be a functional where F is a function of class C^0 with respect to all its arguments.

We shall consider as admissible lines the curves expressed by distributions in the case where the operations indicated by F have a meaning, or, if they have not, the curves of class C^n for which the derivatives of order n may have discontinuities of the first species at a finite number of points x_j, $j = \overline{1, p}$.

If y_j is an admissible line expressed by a distribution, in order that F have a meaning, then, by definition, the neighboring admissible lines are expressed by

$$\overline{y}_j(x) = y_j(x) + \varphi_j(x) \,, \tag{3.27}$$

with $\varphi_i(x) \in \mathcal{D}_{[a,b]} \subset \mathcal{D}$.

Obviously, in the case in which y_j is a distribution, the neighboring admissible line will be expressed by a distribution as well.

In this way, the notion of neighborhood is retained; the curve $\overline{y}_j$ will belong to the neighborhood of order r of the curve y_j if the conditions

$$\left|\overline{y}_j^{(k)}(x) - y_j^{(k)}(x)\right| < \varepsilon \,, \quad k = 0, 1, 2, \ldots, r \,, \tag{3.28}$$

are satisfied. This is equivalent to $\left|t\varphi_j^{(k)}(x)\right| < \varepsilon$, where $\varepsilon > 0$ is arbitrary.

Proposition 3.4 The necessary and sufficient condition that the distribution f be zero over the interval (a, b) is

$$(f(x), \varphi(x)) = 0 \,, \tag{3.29}$$

for any $\varphi(x) \in \mathcal{D}_{[a,b]} \subset \mathcal{D}$.

Definition 3.1 The distribution $\delta I = \delta I[y_j]$, defined by

$$\left(\delta I[y_j], \sum_{j=1}^{m} \varphi_j\right) = \sum_{j=1}^{m} \sum_{k=0}^{n} \left(F_{y_j^{(k)}}, \varphi_j^{(k)}\right) , \tag{3.30}$$

where $\varphi \in \mathcal{D}_{[a,b]} \subset \mathcal{D}$ and

$$F_{y_j^{(k)}} = \frac{\partial F}{\partial y_j^{(k)}} \tag{3.31}$$

representing derivatives in the sense of the theory of distributions, is called a variation of the first order of the functional (3.26).

It is obvious that the expression (3.30) of the first variation of the functional (3.26) includes, as a particular case, the expression of the variation $\delta I[y_j]$ known in the classical variational calculus.

We remark that

$$\left(F_{y_j^{(k)}}, \varphi_j^{(k)}\right) = (-1)^k \left(\frac{\mathrm{d}^k}{\mathrm{d}x^k} F_{y_j^{(k)}}, \varphi_i\right) , \tag{3.32}$$

so that the expression (3.30) becomes

$$\left(\delta I[y_j], \sum_{j=1}^{m} \varphi_j\right) = \sum_{j=1}^{m} \sum_{k=0}^{n} (-1)^k \left(\frac{\mathrm{d}^k}{\mathrm{d}x^k} F_{y_j^{(k)}}, \varphi_j\right) , \tag{3.33}$$

where, as it has been mentioned before, the differentiations are performed in the sense of the distribution theory.

Expanding the relation (3.33), we may write

$$\left(\delta I[y_j], \sum_{j=1}^{m} \varphi_j\right)$$
$$= \sum_{j=1}^{m} \left(F_{y_j} - \frac{\mathrm{d}}{\mathrm{d}x} F_{y'_j} + \frac{\mathrm{d}^2}{\mathrm{d}x^2} F_{y''_j} + \cdots + (-1)^n \frac{\mathrm{d}^n}{\mathrm{d}x^n} F^{(n)}_{y_j^{(n)}}, \varphi_j \right) . \qquad (3.34)$$

3.1.1
Equations of the Euler–Poisson Type

The extremals y_j, $j = \overline{1, m}$, of the functional (3.26) are obtained by imposing the condition

$$\delta I[y_j] = 0 .$$

Using the expression (3.34) for the distribution $\delta I[y_j]$ and taking into account the Proposition 3.4, we find the following necessary equations

$$F_{y_j} - \frac{\mathrm{d}}{\mathrm{d}x} F_{y'_j} + \frac{\mathrm{d}^2}{\mathrm{d}x^2} F_{y''_j} + \cdots + (-1)^n \frac{\mathrm{d}^n}{\mathrm{d}x^n} F^{(n)}_{y_j^{(n)}} = 0 , \quad j = \overline{1, m} .$$

We remark that these equations are of the same form as the classical Euler–Poisson equation; however, they differ substantially from the latter one, since the function F has been assumed to be of class C^0 with respect to all arguments, while the differentiation is performed in the sense of the theory of distributions. For these reasons, we shall say that such equations are of the Euler–Poisson type.

The extremum equation for the functional (3.26) may be written in the form

$$F_{y_j} - \frac{\tilde{\mathrm{d}}}{\mathrm{d}x} F_{y'_j} + \frac{\tilde{\mathrm{d}}^2}{\mathrm{d}x^2} F_{y''_j} + \cdots + (-1)^n \frac{\tilde{\mathrm{d}}^n}{\mathrm{d}x^n} F^{(n)}_{y_j^{(n)}} = 0 ,$$

where $j = \overline{1, m}$; also, in the case in which the extremal curves y_j, $j = \overline{1, m}$, are of class C^n everywhere except at the points x_ℓ, $\ell = \overline{1, p}$, where the derivatives of the nth order may have discontinuities of the first species, the conditions at these points will be

$$S_\ell \left(F_{y_j^{(k+1)}} \right) - S_\ell \left(\frac{\tilde{\mathrm{d}}}{\mathrm{d}x} F_{y_j^{(k+2)}} \right) + \cdots + (-1)^{n-(k+1)} S_\ell \left(\frac{\tilde{\mathrm{d}}^{n-(k+1)}}{\mathrm{d}x^{n-(k+1)}} F_{y_j^{(n)}} \right) = 0 ,$$

where $j = \overline{1, m}$, $\ell = \overline{1, p}$, $k = 0, 1, \ldots, n-1$.

Thus, in the case of the functional

$$I[y] = \int_a^b F(x, y, y', y'') \mathrm{d}x ,$$

the necessary equation for an extremum is

$$F_y - \frac{\tilde{\mathrm{d}}}{\mathrm{d}x} F_{y'} + \frac{\tilde{\mathrm{d}}^2}{\mathrm{d}x^2} F_{y''} = 0$$

and the natural limiting conditions may be written in the form

$$S_\ell\left(F_{y'}\right) - S_\ell\left(\frac{\tilde{\mathrm{d}}}{\mathrm{d}x} F_{y''}\right) = 0,\ S_\ell\left(F_{y''}\right) = 0\,, \quad \ell = \overline{1, p}\,.$$

For example, the functional

$$I[y] = \int_{-1}^{1} x^3 y y' \mathrm{d}x \tag{3.35}$$

has a meaning if we take as admissible line $y = \delta(x)$. In that case, we obtain

$$F(x, y, y') = x^3 y y' = 0\,,$$

regardless of the order of the operations $(x^3 y')y$ or $(x^3 y)y'$; we remark that the operation $(y y')x^3$ is meaningless for $y = \delta(x)$.

Let us determine the function (distribution) which realizes the maximum of the functional (3.35) such that

$$y(-1) = y(1) = 0\,. \tag{3.36}$$

We remark that the functional may be written in the form

$$I[y] = \frac{1}{2} \int_{-1}^{1} x^3 \mathrm{d}y^2\,,$$

from which, integrating by parts and taking into account the conditions (3.36), we obtain

$$I[y] = -\frac{3}{2} \int_{-1}^{1} x^2 y^2 \mathrm{d}x\,.$$

Obviously, the maximum of the functional occurs when the integral vanishes; in this case, we obtain

$$y = 0\,. \tag{3.37}$$

However, by writing the equation of the Euler–Poisson type in distributions, we have $3x^2 y = 0$, from which, besides the solution (3.37), we also obtain the solutions

$$y = \delta(x)\,, \quad y = \delta'(x)\,.$$

Indeed, we have $I[\delta(x)] = 0$, $I[\delta'(x)] = 0$, if the operations in the function F are performed in the order $(x^3 y')y$ or $(x^3 y)y'$.

It should be noted that $\delta(x)$ and $\delta'(x)$ constitute extremals of the functional (3.35), since they vanish in the neighborhood of the points -1 and 1, as specified in conditions (3.36).

The existence of three solutions is due to the fact that $y = 0$ belongs to the class of admissible lines of the class C^1 (classical case), whereas $y = \delta(x)$ and $y = \delta'(x)$ are distributions, hence admissible lines considered as distributions.

3.2
Ordinary Differential Equations

The study of differential equations in the distributions space differs from the study of the same equations for which operators and functions are used in the classical sense. Establishing these equations and finding their solutions is particularly important for applications; indeed, solving differential equations in distributions allows us to determine new solutions that cannot be found by conventional means.

It should be noted that the deduction of equations of mathematical physics cannot always be done directly in the distribution space, because of the difficulties that are encountered in modeling physical phenomena.

In general, the equations which describe such phenomena are obtained first by classical methods. Next, an extension of the unknown functions is made with null values, so they are defined on the whole space; the derivatives, considered in the ordinary sense, are replaced by other expressions given by relations which connect derivatives in the sense of the theory of distributions to derivatives in the ordinary sense of a function continuous almost everywhere and having a finite number of discontinuities of the first order. In this way, the unknowns of the problem will be regular distributions; then, it will be assumed that these unknowns may be arbitrary distributions.

Another possibility which is frequently used is to assume from the very beginning that the unknowns of the problem are arbitrary distributions, assuming the same form in distributions, for the differential equations obtained by classical methods (obviously these are no longer valid for the whole space). However, there is not a general method for passing to differential equations in distributions.

Integration of differential equations leads naturally to the notion of primitive of a function. Because we have defined in distributions the operations of addition, multiplication with functions (multipliers) and the operation of differentiation, we can now build differential equations in the distributions space.

Definition 3.2 We say that the distribution $g \in \mathcal{D}'(\mathbb{R})$ is a primitive of order $m \in \mathbb{N}$ of the distribution $f \in \mathcal{D}'(\mathbb{R})$ if we have satisfied the relation $g^{(m)}(x) = f(x)$, hence if

$$(g^{(m)}(x), \varphi(x)) = (f(x), \varphi(x)) \,, \quad \forall \varphi \in \mathcal{D}(\mathbb{R}) \,.$$

In particular, if $m = 1$, then $g \in \mathcal{D}'(\mathbb{R})$ is a primitive of first order or the primitive of the distribution *f*. The primitive of a distribution is called the indefinite integral of that distribution.

Thus, for example, a primitive of Dirac's delta distribution $\delta(x) \in \mathcal{D}'(\mathbb{R})$ is the regular distribution of Heaviside $H(x)$, because $H'(x) = \delta(x)$.

Also, if we consider the distribution $f(x) = H(x)\cos x$, then the distribution $g(x) = H(x)(1 - \cos x)$ is a primitive of second order for $f(x)$, because we have $g''(x) = f$.

Proposition 3.5 Any distribution $f \in \mathcal{D}'(\mathbb{R})$ allows for a primitive $g \in \mathcal{D}'(\mathbb{R})$. Two primitives g_1, g_2, differ from each other by a constant.

From this proposition, which ensures the existence of the primitive of any distribution, it follows that the equation $g' = f$ has the solution in $\mathcal{D}'(\mathbb{R})$ and, in particular, the equation $g' = 0$, allowing for only the classical solution $g = c = \text{const}$.

If *f* is a continuous function, then its primitive coincides with the classical one.

Using the induction method, we can state:

Proposition 3.6 Any distribution $f \in \mathcal{D}'(\mathbb{R})$ allows for a primitive $g \in \mathcal{D}'(\mathbb{R})$ of a certain order m. Two primitives g_1, g_2 of the same order m differ from each other by a polynomial of degree $\leq m - 1$.

Let us now consider the system of homogeneous first-order differential equations in the distributions space $\mathcal{D}'(\mathbb{R})$

$$\frac{\mathrm{d}y_i(x)}{\mathrm{d}x} = \sum_{k=1}^{n} a_{ik}(x) y_k, \; i = \overline{1, n}\,, \tag{3.38}$$

where $a_{ik}(x) \in C^\infty(\mathbb{R})$ and $y_k(x) \in \mathcal{D}'(\mathbb{R})$ are unknown distributions.

Using the matrices

$$A = (a_{ij}) = \begin{pmatrix} a_{11} & \dots & a_{1n} \\ a_{21} & \dots & a_{2n} \\ \dots & \dots & \dots \\ a_{n1} & \dots & a_{nn} \end{pmatrix}, \quad y = \begin{pmatrix} y_1 \\ y_2 \\ \vdots \\ y_n \end{pmatrix},$$

the system (3.38) is written in the form

$$\frac{\mathrm{d}y(x)}{\mathrm{d}x} = A(x) y(x)\,. \tag{3.39}$$

We know that there is a square matrix $U(x)$, nonsingular, ($\det U \neq 0$), called the fundamental matrix of the system of (3.39), which is the classical solution of this system, hence for which we can write

$$\frac{\mathrm{d}U(x)}{\mathrm{d}x} = A(x) U(x)\,. \tag{3.40}$$

Proposition 3.7 The matrix differential equation (3.39), for which elements of the matrix $A(x)$ are of class $C^\infty(\mathbb{R})$, does not allow for any other solution besides the classical one.

Proof: Let y be the new solution in distributions of (3.39). Then, we consider a new unknown matrix given by the relation $y(x) = U(x)z(x)$. Differentiating and taking into account (3.40), we obtain

$$\frac{dy}{dx} = \frac{dU}{dx}z + U\frac{dz}{dx} = AUz + U\frac{dz}{dx} = AUz\,,$$

giving $U dz/dx = 0$.

Because $\det U \neq 0$ means that U^{-1} exists and by left multiplication of the above equation we find $dz/dx = 0$, thus $z = c$ the proposition is proved. □

Let us consider now the nonhomogeneous matrix equation

$$\frac{dy}{dx} = A(x)y(x) + f\,, \tag{3.41}$$

where f is a column matrix whose elements are distributions of $\mathcal{D}'(\mathbb{R})$.

Substituting $y = Uz$ and taking into account (3.40) and (3.41), we obtain the equation $U dz/dx = f$, which is equivalent to $dz/dx = U^{-1} \cdot f$.

The latter matrix equation, according to the Proposition 3.5 regarding the existence of the primitives in the distributions space, always allows for a solution in $\mathcal{D}'(\mathbb{R})$. In conclusion, (3.41) has solutions in distributions.

Let there be the linear and homogeneous differential equation of order n, with coefficients of class $C^\infty(\mathbb{R})$,

$$P\left(\frac{d}{dx}\right)y = y^{(n)} + a_1(x)y^{(n-1)} + \cdots + a_n(x)y(x) = 0\,. \tag{3.42}$$

Using the substitutions $y_i(x) = y^{(i-1)}(x)$, $i = \overline{1, n-1}$, (3.42) is equivalent to a matrix equation (3.39), wherefrom we have the following.

Proposition 3.8 The homogeneous linear differential equation $P(d/dx)\,y = 0$ with variable coefficients $a_i(x) \in C^\infty(\mathbb{R})$, $i = \overline{1, n}$, does not allow any other solution besides the classical solution.

Let there be the nonhomogeneous linear differential equation of order n

$$P\left(\frac{d}{dx}\right)y(x) = f(x)\,, \quad y, f \in \mathcal{D}'(\mathbb{R})\,, \tag{3.43}$$

with the coefficients $a_i(x) \in C^\infty(\mathbb{R})$, $i = \overline{1, n}$.

Definition 3.3 We call the generalized solution (solution in distributions) of the differential equation (3.43) on the interval (a, b) a distribution $y \in \mathcal{D}'(\mathbb{R})$, which verifies the equation on the interval (a, b), in the sense of distributions, that is,

$$\left(P\left(\frac{d}{dx}\right)y, \varphi\right) = (f, \varphi)\,, \quad \forall \varphi \in \mathcal{D}(\mathbb{R})\,, \quad \text{supp}(\varphi) \subset (a, b)\,.$$

In connection with the integration of differential equations, the following is of importance.

Proposition 3.9 The solution in distributions, of the equation $x^n T = 0, T \in \mathcal{D}'(\mathbb{R})$, is

$$T = \sum_{i=0}^{n-1} c_i \delta^{(i)}(x) , \quad c_i \in \mathbb{R} .$$

We can also show that the equation

$$(x-a)^n T = 0 , \quad T \in \mathcal{D}'(\mathbb{R}) \tag{3.44}$$

has the solution

$$T = \sum_{i=0}^{n-1} c_i \delta^{(i)}(x-a) .$$

This result can be obtained if we use the properties of the translation operator τ_a. Thus, because $\tau_{-a} T(x) = T(x-a)$, $\tau_{-a} x^n = (x-a)^n$, (3.44) can be written in the form

$$(\tau_{-a} x^n) \cdot T = \tau_{-a} (x^n \tau_a T) = 0 ,$$

giving $x^n \tau_a T = 0$.

Applying the Proposition 3.9, we find $\tau_a T = \sum_{i=0}^{n-1} c_i \delta^{(i)}(x)$ and, after applying the operator τ_{-a} to this expression, we obtain

$$\tau_{-a}(\tau_a T) = T = \sum_{i=0}^{n-1} c_i \tau_{-a} \delta^{(i)}(x) = \sum_{i=0}^{n-1} c_i \delta^{(i)}(x-a) ,$$

that is, which had to be demonstrated.

Let us now consider the equation

$$a(x) T = f(x) , \tag{3.45}$$

where $a(x) \in C^\infty(\mathbb{R})$, $f \in \mathcal{D}'(\mathbb{R})$, while the unknown distribution is $T \in \mathcal{D}'(\mathbb{R})$.

If T_0 is the general solution of the homogeneous equation $a(x)T = 0$ and T_1 is a particular solution of the inhomogeneous equation (3.45), then the general solution of (3.45) is

$$T = T_0 + T_1 ,$$

because we have $a(x)(T_0 + T_1) = a(x)T_0 + a(x)T_1 = 0 + f(x) = f(x)$.

Let us solve, in this manner, the equation

$$(x-a)(x-b)T = 0 , \quad a \neq b . \tag{3.46}$$

This can be written in the form $(x-a)[(x-b)T] = 0$, from which we obtain

$$(x-b)T = c_1\delta(x-a)\,, \quad c_1 \in \mathbb{R}\,. \tag{3.47}$$

We observe that $T_0 = c_2\delta(x-b)$, $c_2 \in \mathbb{R}$, is the general solution of the homogeneous equation $(x-b)T = 0$ and $T_1 = c_1\delta(x-a)/(a-b)$ is a particular solution of the inhomogeneous equation (3.47).

Thus, the solution of (3.46) is given by the distribution $T = c_1^*\delta(x-a)+c_2\delta(x-b)$, where $c_1^* = c_1/(a-b)$.

One of the classical methods of linear homogeneous differential equations integration with variable coefficients is the constants variation method.

This method is applicable to linear differential equations considered in distributions as well.

Example 3.2 We consider the differential equation

$$xy'' - y' = \delta(x)\,. \tag{3.48}$$

A fundamental system of solutions for the homogeneous equation $xy'' - y' = 0$ consists of the linearly independent particular solutions $y_1 = 1$ and $y_2 = x^2$, $x \in \mathbb{R}$.

We will search the solution in the distributions of (3.48) in the form

$$y = c_1(x) + x^2 c_2(x),\, x \in \mathbb{R}\,,$$

where c_1, c_2 are distributions from $\mathcal{D}'(\mathbb{R})$.

According to the variation of constants method to determine the distributions c_1', c_2', we impose the condition $c_1' + x^2 c_2' = 0$.

In view of this equation, from the expression of y' it follows $y' = 2xc_2$, wherefrom, we get $y'' = 2c_2 + 2xc_2'$.

Substituting the expressions of the derivatives y' and y'' in (3.48), we obtain the equation $2x^2c_2' = \delta(x)$.

From the system $c_1' + x^2c_2' = 0, 2x^2c_2' = \delta(x)$, it follows $c_1' = -\delta(x)/2$, hence $c_1(x) = -H(x)/2 + h_1$, where h_1 is a constant.

Regarding the second equation, by multiplying it by x, it gives $x^3c_2' = 0$, from which we find $c_2(x) = \delta'(x)/4 + h_2\delta(x) + h_3H(x) + h_4$, $h_1, h_2, h_3, h_4 = \text{const}$.

Consequently, the solution in distributions of (3.48) is

$$y = -\frac{H(x)}{2} + \alpha_1 x^2 + \alpha_2 x^2 H(x) + \alpha_3\,, \quad \alpha_1, \alpha_2, \alpha_3 = \text{const}\,.$$

To verify the correctness of the solution, the relations $x\delta(x) = 0$, $x^2\delta(x) = 0$, $x^3\delta(x) = 0$ are used.

From the first relation, by differentiation, we obtain $\delta(x) + x\delta'(x) = 0$, that is, $x\delta'(x) = -\delta(x)$. From the second relation, it follows $x^2\delta'(x) + x\delta(x) = 0 \Leftrightarrow x^2\delta'(x) = 0$. From the third relation, by differentiation, we obtain $x^3\delta'(x) + 3x^2\delta(x) = 0 \Leftrightarrow x^3\delta'(x) = 0$. From this, by differentiation, it follows $x^3\delta''(x) + 3x^2\delta'(x) = 0 \Leftrightarrow x^3\delta''(x) = 0$.

Therefore, we have

$$x\delta'(x) = -\delta(x) \,, \quad x^2\delta'(x) = 0 \,, \quad x^3\delta''(x) = 0 \,.$$

Definition 3.4 We say that the distribution $E(x) \in \mathcal{D}'(\mathbb{R}^n)$ is the fundamental solution of the operator $P(\mathrm{D})$ (1.296) if it satisfies the relation

$$P(\mathrm{D})\,E(x) = \delta(x) \,.$$

The fundamental solution of an operator is generally not unique. Thus, if $f \in \mathcal{D}'(\mathbb{R}^n)$ satisfies the equation $P(\mathrm{D})\,f = 0$ and E is the fundamental solution for $P(\mathrm{D})$, then $E_1 = f + E$ is also a fundamental solution, because, on the linearity basis of $P(\mathrm{D})$, we can write

$$P(\mathrm{D})\,E_1 = P(\mathrm{D})(f + E) = P(\mathrm{D})\,f + P(\mathrm{D})\,E = \delta \,.$$

Thus, the regular distribution

$$E(x) = H(x)\frac{\sin \omega x}{\omega} \,, \qquad \omega > 0$$

is the fundamental solution of the operator $\mathrm{d}^2/\mathrm{d}x^2 + \omega^2$; indeed, we have

$$E'(x) = H(x)\cos \omega x \,, \qquad E''(x) = \delta(x) - H(x)\sin \omega x \,,$$

and the verification is immediate.

Also, the function type distribution

$$E(x) = -\frac{1}{2\omega}\exp(-\omega|x|) \,, \qquad \omega > 0$$

is the fundamental solution of the operator $\mathrm{d}^2/\mathrm{d}x^2 - \omega^2$.

These operators have many applications in mechanics, particularly in vibration problems.

The Fourier and the Laplace integral transformations can be particularly useful for determining the solutions of the differential equations in distributions, even if they are not with constant coefficients.

Let there be such a linear equation with variable coefficients

$$xy'' + 2y' = \delta(x) \,.$$

Applying the Laplace transform, we obtain

$$-\frac{\mathrm{d}}{\mathrm{d}p}(p^2\mathrm{L}[y]) + 2p\mathrm{L}[y] = 1 \,,$$

where

$$-p^2\frac{\mathrm{d}}{\mathrm{d}p}\mathrm{L}[y] = 1 \,,$$

thus $\mathrm{L}[y] = 1/p + C_1$, the inverse Laplace transform allows us to write

$$y(x) = H(x) + C_1\delta(x) + C_2 \,.$$

3.3 Convolution Equations

3.3.1 Convolution Algebras

Definition 3.5 The vector space (X, Γ) is called an algebra if we can define the mapping $(x, y) \to x \circ y$ of $X \times X$ into X, satisfying:

1. $\forall x, y, z \in X, (x \circ y) \circ z = x \circ (y \circ z)$, associativity;
2. $\forall x, y, z \in X, x \circ (y + z) = x \circ y + x \circ z, (y + z) \circ x = y \circ x + z \circ x$, distributivity with respect to addition;
3. $\forall \lambda, \mu \in \Gamma, \forall x, y \in X, (\lambda x) \circ (\mu y) = (\lambda\mu)(x \circ y)$.

If $\forall x, y \in X, x \circ y = y \circ x$, then we can say that the algebra $(X, \Gamma, \circ)$ is commutative.

The algebra $(X, \Gamma, \circ)$ is said to have unity elements if $\forall x \in X, \exists u \in X$ such that $u \circ x = x \circ u = x$.

We say that the vector subspace $\mathcal{K}' \subset \mathcal{D}'(\mathbb{R}^n)$ forms a convolution algebra if it satisfies the following conditions:

1. $\forall f, g \in \mathcal{K}'(\mathbb{R}^n), \exists f * g \in \mathcal{K}'(\mathbb{R}^n)$;
2. $\forall f, g, h \in \mathcal{K}'(\mathbb{R}^n), (f * g) * h = f * (g * h)$.

We note that the existence of the convolution product implies its commutativeness, hence the subspace $\mathcal{K}'(\mathbb{R}^n) \subset \mathcal{D}'(\mathbb{R}^n)$ forms a commutative algebra. If the Dirac delta distribution $\delta(x)$ also belongs to $\mathcal{K}'(\mathbb{R}^n)$, then it represents the unity element of $\mathcal{K}'(\mathbb{R}^n)$, which, in this case, is a convolution algebra with unity. Thus, the space $\mathcal{E}'(\mathbb{R}^n)$ of distributions with compact support forms a convolution algebra with unity.

In the symbolic calculus [25, 26], an outstanding role is played by the space $\mathcal{D}'_+$. It may be easily checked that the space $\mathcal{D}'_+$ forms a convolution algebra with the unity $\delta(x) \in \mathcal{D}'_+$.

In contrast, the spaces $L^1(\mathbb{R}^n)$, $\mathcal{D}(\mathbb{R}^n)$, $\mathcal{S}(\mathbb{R}^n)$ form convolution algebras without unity. The distribution space $\mathcal{D}'(\mathbb{R}^n)$ does not form a convolution algebra, since the convolution product is not necessarily defined for every distribution pair.

Let $\mathcal{K}'(\mathbb{R}^n)$ be a convolution algebra with unity. Denote by $(\mathcal{K}')^{m\times m}$ the set of square matrices of order $m \geq 1$, having as elements distributions from the algebra with unity $\mathcal{K}'$.

Thus, if $(a) \in (\mathcal{K}')^{m\times m}$, then its elements are $a_{ij} \in \mathcal{K}'(\mathbb{R}^n)$, $i, j = \overline{1, m}$.

Since the set of square matrices has a vector space structure with respect to the sum and scalar multiplication, we shall define the convolution product in the space $(\mathcal{K}')^{m\times m}$ by the formula

$$\forall (a), (b) \in \left(\mathcal{K}'\right)^{m\times m}, \quad (a) * (b) = (c)\,,$$

where $c_{ij} = \sum_{k=1}^{m} a_{ik} * b_{kj}, (a) = (a_{ij}), (b) = (b_{ij}), i, j = \overline{1, m}$.

The unity element of $(\mathcal{K}')^{m\times m}$, with respect to the convolution product, is the matrix $(\delta) \in (\mathcal{K}')^{m\times m}$ which is defined by

$$(\delta) = \begin{cases} \delta(x)\,, & i = j\,, \\ 0\,, & i \neq j\,, \end{cases} \qquad i, j = \overline{1, m}\,.$$

Since the matrix product is associative but, generally, noncommutative, it follows that the space $(\mathcal{K}')^{m\times m}$ forms a noncommutative algebra with unity element.

If $\mathcal{K}'(\mathbb{R}) = \mathcal{D}'_+(\mathbb{R}) \subset \mathcal{D}'(\mathbb{R})$, then $\mathcal{D}'_+(\mathbb{R})$ is a convolution algebra with unity and thus forms a noncommutative algebra with the unity element

$$(\delta) = \begin{cases} \delta(x)\,, & i = j\,, \\ 0\,, & i \neq j\,, \end{cases} \qquad i, j = \overline{1, n}\,, \quad \delta(x) \in \mathcal{D}'_+\,.$$

Definition 3.6 Let $\mathcal{K}'(\mathbb{R}^n) \subset \mathcal{D}'(\mathbb{R}^n)$ be a convolution algebra with unity. We call a convolution equation the equation of the form

$$A * X = B\,,$$

where $A \in \mathcal{K}'(\mathbb{R}^n)$, $B \in \mathcal{K}'(\mathbb{R}^n)$ are given distributions and $X \in \mathcal{K}'(\mathbb{R}^n)$ is an unknown distribution.

Similarly, the equation

$$(a) * (X) = (b)\,, \tag{3.49}$$

where $(a) \in (\mathcal{K}')^{m\times m}$, $(b) \in (\mathcal{K}')^{m\times m}$ and $(X) \in (\mathcal{K}')^{m\times 1}$ (column matrix of m elements from $\mathcal{K}'(\mathbb{R}^n)$) is the unknown matrix, is a convolution matrix equation.

The matrix equation (3.49) is equivalent to a system of m equations of convolution with m unknown distributions X_j, namely

$$\sum_{j=1}^{m} a_{ij} * X_j = b_i\,, \quad i = \overline{1, m}\,, \tag{3.50}$$

where $a_{ij} \in \mathcal{K}'(\mathbb{R}^n)$, $b_i, X_j \in \mathcal{K}'(\mathbb{R}^n)$.

The matrix $(a) = (a_{ij})$ is the matrix of the system coefficients, whereas (b) is the matrix of the right-hand side of the system (3.50).

In particular, if $(b) = 0$, then (3.49) and (3.50) are called homogeneous.

Definition 3.7 We call a fundamental solution corresponding to the matrix equation (3.49), the matrix $(E) \in (\mathcal{K}')^{m\times m}$, satisfying the relation

$$(a) * (E) = (\delta)\,,$$

where $(\delta) \in (\mathcal{K}')^{m\times m}$ is the unity element in the convolution algebra with unity $(\mathcal{K}')^{m\times m}$.

Definition 3.8 We say that the matrix $(a) \in (\mathcal{K}')^{m\times m}$ is invertible in $(\mathcal{K}')^{m\times m}$ if there exists a matrix $(a)^{-1} \in (\mathcal{K}')^{m\times m}$ such that

$$(a) * (a)^{-1} = (a)^{-1} * (a) = (\delta) .$$

The matrix $(a)^{-1}$ is called the inverse of the matrix (a) in $(\mathcal{K}')^{m\times m}$. It follows that any inverse of the matrix $(a) \in (\mathcal{K}')^{m\times m}$ represents a fundamental solution of (3.49). The reciprocal is not generally true.

Proposition 3.10 Let $(E) \in (\mathcal{K}')^{m\times m}$ be the fundamental solution of (3.49) and let $(X_1) \in (\mathcal{K}')^{m\times 1}$ be a solution of its associate homogeneous equation. Then, any solution of (3.49) has the expression

$$(X) = (E) * (b) + (X_1) .$$

Indeed, because $(a) * (X_1) = (0)$ and $(E) * (b)$ exists, we have

$$(a) * (X) = ((a) * (E)) * (b) + (a) * (X_1) = (\delta) * (b) = (b) .$$

Let there be the convolution equation

$$A * X = B . \tag{3.51}$$

Proposition 3.11 If $A \in \mathcal{D}(\mathbb{R}^n)$, $B \in \mathcal{D}'(\mathbb{R}^n)$ and $X \in \mathcal{D}'(\mathbb{R}^n)$ is an unknown distribution, then (3.51) has no fundamental solution.
Indeed, if there would exist a fundamental solution $E \in \mathcal{D}'(\mathbb{R}^n)$ then, the function $A * E$ would be indefinitely differentiable and hence it could not be equal to $\delta(x)$.

Proposition 3.12 Let $\mathcal{K}'(\mathbb{R}^n) \subset \mathcal{D}'(\mathbb{R}^n)$ be a convolution algebra with unity. The necessary and sufficient condition for the equation

$$(a) * (X) = (b) , \quad (a) \in (\mathcal{K}')^{m\times m} , \quad (X), (b) \in (\mathcal{K}')^{m\times 1} \tag{3.52}$$

to have a solution for any (b) is that the determinant of the matrix (a) be invertible on $\mathcal{K}'(\mathbb{R}^n)$. Then, the fundamental solution exists, is unique and has the expression

$$(E) = (a)^{-1}$$

where $(a)^{-1}$ is the inverse of the matrix (a).

The solution of (3.52) is also unique and reads

$$(X) = (E) * (b) = (a)^{-1} * (b) . \tag{3.53}$$

Proof: To prove the necessity of the condition, let us suppose that (3.52) allows for a solution for any (b). Then, in particular, for (b) successively taking the values

$$(b_1) = \begin{pmatrix} \delta \\ 0 \\ \vdots \\ \vdots \\ 0 \end{pmatrix}, \quad (b_2) = \begin{pmatrix} 0 \\ \delta \\ 0 \\ \vdots \\ 0 \end{pmatrix}, \quad \ldots, \quad (b_m) = \begin{pmatrix} 0 \\ 0 \\ \vdots \\ \vdots \\ \delta \end{pmatrix},$$

we get the corresponding solutions

$$(X_1) = \begin{pmatrix} X_{11} \\ X_{21} \\ \vdots \\ X_{m1} \end{pmatrix}, \quad \ldots, \quad (X_m) = \begin{pmatrix} X_{1m} \\ X_{2m} \\ \vdots \\ X_{mm} \end{pmatrix} \in (\mathcal{K}')^{m\times 1},$$

which satisfy the equation $(a) * (X_i) = (b_i)$, $i = \overline{1, m}$.

Based on these equations it follows that the matrix

$$(C) = \begin{pmatrix} X_{11} & X_{12} & \ldots & X_{1m} \\ X_{21} & X_{22} & \ldots & X_{2m} \\ \vdots & \vdots & \vdots & \vdots \\ X_{m1} & X_{m2} & \ldots & X_{mm} \end{pmatrix} \in (\mathcal{K}')^{m\times m}$$

satisfies the relation $(a) * (C) = (\delta)$.

Consequently, we have

$$\det(a) * \det(C) = \det(\delta) = \delta(x), \quad \delta(x) \in \mathcal{K}'(\mathbb{R}^n),$$

hence $\det(a)$ is invertible in $\mathcal{K}'(\mathbb{R}^n)$; thus, the necessity of the condition is proved.

In order to prove the sufficiency, we shall suppose that $\det(a)$ is invertible in $\mathcal{K}'(\mathbb{R}^n)$. Let us consider the matrix $(E) \in (\mathcal{K}')^{m\times m}$, whose elements $E_{ij} \in \mathcal{K}'(\mathbb{R}^n)$ have the expression

$$E_{ij} = (\det(a))^{-1} * \alpha_{ji}, \quad i, j = \overline{1, m},$$

where α_{ji} is the algebraic complement of the element a_{ij} in $\det(a)$.

Performing the products $(a) * (E), (E) * (a)$, we obtain

$$(a) * (E) = (E) * (a) = (\delta).$$

Hence, the matrix (E) thus constructed is a fundamental solution of (3.52) and it clearly represents the inverse of the matrix (a) in $(\mathcal{K}')^{m\times m}$ that is, $(E) = (a)^{-1}$.

Under these assumptions, the matrix $(X) = (E) * (b)$ exists for any (b) and represents a solution of (3.52), because we have

$$(a) * (X) = [(a) * (E)] * (b) = (\delta) * (b) = (b).$$

Thus the relations are sufficiently proved.

The uniqueness of the solution given by (3.53) follows from the equivalence of (3.52) and (3.53), since by multiplying the left-hand side of (3.53) with (a) we obtain (3.52).

To prove the uniqueness of the fundamental solution (E), suppose that there also exists another solution, namely (E'). Then, because $(E') * (a) = (\delta)$, we get $(E') * (a) * (E) = (E)$ but, since $(a) * (E) = (\delta)$ we obtain $(E') = (E)$, so that the proposition is completely proved. □

Example 3.3 Let the Poisson equation be

$$\Delta\delta(x) * X = B \,, \tag{3.54}$$

where Δ is the Laplace operator in $\mathbb{R}^3$ and $\delta(x)$ is the Dirac delta distribution.

Because $\Delta\delta(x) \in \mathcal{E}'(\mathbb{R}^n)$, we shall look for solutions of the convolution equation (3.54) in $\mathcal{D}'(\mathbb{R}^3)$. The function type distribution $E = -1/4\pi r \in \mathcal{D}'(\mathbb{R}^3)$, where $r = \sqrt{x_1^2 + x_2^2 + x_3^2}$ is a fundamental solution of (3.54), because $\Delta 1/r = -4\pi\delta(x)$, hence $\Delta\delta(x) * E = \delta(x)$.

If we denote by Y a harmonic distribution, then $\Delta Y = 0$ and the distribution $\tilde{E} = E + Y$ is also a fundamental solution of (3.54).

To construct the general solution of (3.54) using the fundamental solution E, we notice that the support of the distribution E is the whole space $\mathbb{R}^3$. Consequently, taking $B \in \mathcal{E}'(\mathbb{R}^3)$, according to the Proposition 3.10, we obtain for the general solution of (3.54) the expression

$$X = -\frac{1}{4\pi r} * B + Y \,. \tag{3.55}$$

We mention that the fundamental solution E does not represent the inverse of the element $\Delta\delta(x) \in \mathcal{E}'(\mathbb{R}^3)$ in $\mathcal{E}'(\mathbb{R}^3)$, because $E \notin \mathcal{E}'(\mathbb{R}^3)$.

In conclusion, the expression (3.55) is the general solution of the Poisson equation $\Delta X = B$, that is, (3.54).

3.3.2
Convolution Algebra $\mathcal{D}'_+$: Convolution Equations in $\mathcal{D}'_+$

The set $C(\mathbb{R}_+)$ of complex functions defined and continuous on $\mathbb{R}_+ = [0, \infty)$, together with the convolution product $*$, form a commutative algebra, without unity element, called convolution algebra $C(\mathbb{R}_+)$. According to Titchmarsch's theorem, $C(\mathbb{R}_+)$ does not have divisors of zero.

A generalization of the convolution algebra $C(\mathbb{R}_+)$ is the convolution algebra $\mathcal{D}'_+$. This algebra as well as $C(\mathbb{R}_+)$ does not have divisors of zero either. Hence, if $f, g \in \mathcal{D}'_+$ and $f * g = 0$, then $f = 0$ or $g = 0$.

Let there be the convolution equation in $\mathcal{D}'_+$

$$A * X = B \,, \tag{3.56}$$

where $A, B, X \in \mathcal{D}'_+$.

Proposition 3.13 If $A \in \mathcal{D}'_+$ is invertible in $\mathcal{D}'_+$, then its inverse $A^{-1} \in \mathcal{D}'_+$ is unique and represents a fundamental solution of (3.56) which has only one solution, namely,

$$X = A^{-1} * B\,. \tag{3.57}$$

Proof: Obviously, any inverse of the element $A \in \mathcal{D}'_+$, if it exists, is a fundamental solution of (3.56).

Suppose that (3.56) has two fundamental solutions, E and E'; then the relations $A * E = \delta$, $A * E' = \delta$ imply that $A * (E - E') = 0$ and since $A \neq 0$, whereas the algebra $\mathcal{D}'_+$ has no divisors of zero, it follows that $E = E'$. Hence the uniqueness of the inverse of the fundamental solution is proved.

Multiplying (3.56) by $E = A^{-1}$, we obtain (3.57), which reinforces the uniqueness of the solution of (3.56). □

Proposition 3.14 Let there be $A, B \in \mathcal{D}'_+$. If A and B are invertible in $\mathcal{D}'_+$, then $A * B$ is invertible in $\mathcal{D}'_+$ and we have $(A * B)^{-1} = A^{-1} * B^{-1}$.

Indeed, because $A * A^{-1} = \delta$, $B * B^{-1} = \delta$, we have

$$(A * B) * (A^{-1} * B^{-1}) = (A * A^{-1}) * (B * B^{-1}) = \delta * \delta = \delta\,,$$

that is $(A * B)^{-1} = A^{-1} * B^{-1}$.

Remark 3.1 If $A, B \in (\mathcal{D}'+)^{m\times m}$ and A, B are invertible in $(\mathcal{D}'+)^{m\times m}$, then $A * B \in (\mathcal{D}'+)^{m\times m}$ is invertible in $(\mathcal{D}'+)^{m\times m}$ and we have $(A * B)^{-1} = B^{-1} * A^{-1}$.

Indeed, we have

$$(A * B) * (B^{-1} * A^{-1}) = A * (B * B^{-1}) * A^{-1} = A * (\delta) * A^{-1} = A * A^{-1} = (\delta)\,,$$

because $A * A^{-1} = A^{-1} * A = (\delta)$, $B * B^{-1} = B^{-1} * B = (\delta)$.

We note that the convolution algebra $(\mathcal{D}'+)^{m\times m}$ is noncommutative. For solving the convolution equations in algebra $\mathcal{D}'_+$, it is useful to apply the Laplace transform [27]. Thus, if the distributions A, B, X which appear in (3.56) allow for the Laplace transform then, we obtain

$$\mathrm{L}[X] = \frac{\mathrm{L}[B]}{\mathrm{L}[A]} = \mathrm{L}(p)\,, \quad p = \sigma + i\tau\,.$$

If $\mathrm{L}(p)$ allows for an original belonging to the algebra $\mathcal{D}'_+$, then this is the required solution.

We emphasize that the convolution algebras $\mathcal{D}'_+$ and $(\mathcal{D}'+)^{3\times 3}$ have important applications in the study of the bending of an elastic bar on elastic foundation, as well as in the study of linear vibration of mechanical systems and of inductively coupled circuits.

Example 3.4 Let there be the convolution equation

$$H(x)\cos x * y(x) = \frac{H(x)(\sin x + x\cos x)}{2}$$

in $\mathcal{D}'_+$. Applying the Laplace transform, we obtain

$$\frac{p}{p^2+1}\mathrm{L}[y] = \frac{1}{2}\left[\frac{1}{p^2+1} + \frac{p^2-1}{(p^2+1)^2}\right] = \frac{p^2}{(p^2+1)^2} ,$$

where $\mathrm{L}[y] = p/(p^2+1)$.

Consequently, the solution of the equation is

$$y(x) = H(x)\cos x , \quad x \in [0,\infty) .$$

Example 3.5 The fundamental matrix of the complete system of equations of bending of elastic bars on elastic foundation Let $v, q, T, M \in \mathcal{D}'_+$, $\operatorname{supp}(v, q, T, M) \subset [a, b]$, be distributions which represent the deflection, the intensity of distributed forces on an elastic bar $[a, b] \subset [0, \infty)$, the shear force and the bending moment, respectively. The complete system of equations of bending of the bounded elastic bars on elastic foundation satisfies in $\mathcal{D}'_+$ [28] the equations

$$\begin{aligned} &EI\partial_x^2 v(x) + M(x) = B_1 , \\ &\partial_x T(x) - kv(x) = B_2 , \\ &\partial_x M(x) - T(x) = B_3 , \end{aligned} \tag{3.58}$$

where $B_1, B_2, B_3 \in \mathcal{D}'_+$, EI is the bar rigidity, E the modulus of elasticity, I the moment of inertia with respect to the neutral axis of the cross-section and $k > 0$ the coefficient of elasticity of the foundation.

We introduce the matrices $(A) \in (\mathcal{D}'+)^{3\times 3}$, $(X) = (\mathcal{D}'+)^{3\times 1}$, $(B) = (\mathcal{D}'+)^{3\times 1}$ by the expressions

$$(A) = \begin{pmatrix} EI\delta''(x) & 0 & \delta(x) \\ -k\delta(x) & \delta'(x) & 0 \\ 0 & -\delta(x) & \delta'(x) \end{pmatrix} ,$$

$$(X) = \begin{pmatrix} v(x) \\ T(x) \\ M(x) \end{pmatrix} , \quad (B) = \begin{pmatrix} B_1(x) \\ B_2(x) \\ B_3(x) \end{pmatrix} .$$

Consequently, the system of (3.58) can be written in the form of the matrix convolution equation in $(\mathcal{D}'+)^{3\times 3}$

$$(A) * (X) = (B) . \tag{3.59}$$

The determinant $\Delta = \det(A)$ associated to the matrix $(A) \in (\mathcal{D}'+)^{3\times 3}$ is

$$\Delta = \det(A) = EI\delta^{(4)}(x) + k\delta(x) \in \mathcal{D}'_+ .$$

Noting $\omega = \sqrt[4]{k/4EI}$, we obtain $\Delta = EI(\delta^{(4)}(x) + 4\omega^4\delta(x)) \in \mathcal{D}'_+$.

From this we obtain the following proposition. The determinant $\Delta = \det(A) \in \mathcal{D}'_+$ is invertible in the convolution algebra $\mathcal{D}'_+$, hence the inverse $\Delta^{-1} \in \mathcal{D}'_+$ is expressed by

$$\Delta^{-1} = \frac{1}{4EI\omega^3} H(x)(\cosh \omega x \sin \omega x - \sinh \omega x \cos \omega x) \in \mathcal{D}'_+ \,, \qquad (3.60)$$

where H is the Heaviside distribution.

Indeed, we have $\mathrm{L}[\Delta](p) = EI(p^4 + 4\omega^4)$ because $\mathrm{L}[\delta(x)] = 1$, $\mathrm{L}[\delta^{(4)}(x)] = p^4$. Taking into account that $\Delta * \Delta^{-1} = \delta(x)$, by applying the Laplace transform, we obtain

$$\mathrm{L}[\Delta^{-1}] = \frac{1}{\mathrm{L}[\Delta]} = \frac{1}{EI}\frac{1}{p^4 + 4\omega^4} \,.$$

Applying the inverse Laplace transform L^{-1}, we get

$$\mathrm{L}^{-1}\left[\frac{1}{p^4 + 4\omega^4}\right] = \frac{H(x)}{4\omega^3}[\cosh \omega x \sin \omega x - \sinh \omega x \cos \omega x] \,. \qquad (3.61)$$

Consequently, we obtain the formula (3.60).

The matrix $(A) \in (\mathcal{D}'+)^{3\times 3}$ being invertible in $(\mathcal{D}'+)^{3\times 3}$, we obtain

$$(E) = (A)^{-1} \in (\mathcal{D}'+)^{3\times 3}$$

where $E_{ij} \in \mathcal{D}'_+$, $E_{ij} = \Delta^{-1} \times \alpha_{ji}$, $i, j = 1, 2, 3$ and α_{ji} is the algebraic complement of the element α_{ij} from $\det(A)$.

The solution of (3.59) is unique and reads

$$(X) = (E) * (B) = (A^{-1}) * (B) \,, \qquad (3.62)$$

where the fundamental matrix $(E) = (A)^{-1}$ has the expression

$$(E) = (A)^{-1} = \begin{pmatrix} \dfrac{H(x)u_2(x)}{4EI\omega^3} & -\dfrac{H(x)u(x)}{4EI\omega^3} & -\dfrac{H(x)u_1(x)}{4EI\omega^3} \\ \dfrac{kH(x)u_1(x)}{4EI\omega^3} & \dfrac{H(x)u_3(x)}{4\omega^3} & -\dfrac{kH(x)u(x)}{4EI\omega^3} \\ \dfrac{kH(x)u(x)}{4EI\omega^3} & \dfrac{H(x)u_2(x)}{4\omega^3} & \dfrac{H(x)u_3(x)}{4\omega^3} \end{pmatrix} = \begin{pmatrix} E_{11} & E_{12} & E_{13} \\ E_{21} & E_{22} & E_{23} \\ E_{31} & E_{32} & E_{33} \end{pmatrix}$$

where

$$\left.\begin{aligned} u(x) &= \cosh \omega x \sin \omega x - \sinh \omega x \cos \omega x \,, \\ u_1(x) &= u'(x) = 2\omega \sinh \omega x \sin \omega x \,, \\ u_2(x) &= u''(x) = 2\omega^2(\cosh \omega x \sin \omega x + \sinh \omega x \cos \omega x) \,, \\ u_3(x) &= u'''(x) = 4\omega^3 \cosh \omega x \cos \omega x \,, \end{aligned}\right. \quad x \in \mathbb{R} \,.$$

Specifying (3.62), we obtain

$$\begin{aligned} v(x) &= E_{11} * B_1 + E_{12} * B_2 + E_{13} * B_3 \,, \\ T(x) &= E_{21} * B_1 + E_{22} * B_2 + E_{23} * B_3 \,, \\ M(x) &= E_{31} * B_1 + E_{32} * B_2 + E_{33} * B_3 \,. \end{aligned} \tag{3.63}$$

3.4 The Cauchy Problem for Linear Differential Equations with Constant Coefficients

Let there be the differential equation

$$P(\mathrm{D})X = f \,, \quad x > 0 \,, \tag{3.64}$$

where f is a locally integrable function on $[0, \infty)$ and $P(\mathrm{D})$ the differential operator with constant coefficients

$$P(\mathrm{D}) = \frac{\mathrm{d}^n}{\mathrm{d}x^n} + a_1 \frac{\mathrm{d}^{n-1}}{\mathrm{d}x^{n-1}} + \cdots + a_{n-1}\frac{\mathrm{d}}{\mathrm{d}x} + a_n \,.$$

Definition 3.9 We call Cauchy problem corresponding to (3.64), the determination of the function $X \in C^n[0, \infty)$, which, for $x > 0$, verifies the differential equation (3.64) and, for $x = 0$, verifies the initial conditions

$$X(0) = X_0 \,, \quad X'(0) = X_1, \ldots, X^{(n-1)}(0) = X_{n-1} \,.$$

This problem can be reduced to an equation of convolution in $\mathcal{D}'_+$ by extending with null values for $x < 0$ both the function f, and the solution X. For this purpose we introduce the function type distributions $\overline{f}(x) = H(x) f(x)$, $\overline{X} = H(x)X(x)$.

We note that $\overline{X}$ is a function of class $C^n(\mathbb{R})$, except the origin where it has a discontinuity of the first order. The initial conditions represent the jumps of the function $\overline{X}$ and its derivatives up to order $n-1$ are inclusive.

Taking into account the rule of differentiation of functions with discontinuities of first order, we obtain

$$\overline{X}^{(p)}(x) = \tilde{\overline{X}}^{(p)}(x) + \sum_{k=1}^{p} X_{k-1}\delta^{(p-k)}(x) \,, \quad p = \overline{1, n} \,,$$

where $\overline{X}^{(p)}$ and $\tilde{\overline{X}}^{(p)}$ are the derivatives of order p in the sense of distributions and of ordinary sense, respectively.

Denoting

$$F(x) = \overline{f}(x) + \sum_{i=1}^{n} b_i \delta^{(n-i)}(x) \,, \tag{3.65}$$

where $b_i = X_{i-1} + a_1 X_{i-2} + \cdots + a_{i-2}X_1 + a_{i-1}X_0$, we obtain the differential equation in the distribution space

$$P(\mathrm{D})\overline{X}(x) = F(x)\,, \quad \overline{X}\,, \quad F \in \mathcal{D}'_+\,. \tag{3.66}$$

This equation has the advantage to include the initial conditions of the problem; for $x > 0$ it coincides with the initial equation (3.64).

Therefore, solving the corresponding Cauchy problem of (3.64) is equivalent to solving (3.66) in $\mathcal{D}'_+$, that is, the convolution equation in $\mathcal{D}'_+$

$$P(\mathrm{D})\delta(x) * X(x) = F(x)\,. \tag{3.67}$$

Let $E(x) = H(x)Y(x) \in \mathcal{D}'_+$ be the fundamental solution corresponding to the operator $P(\mathrm{D})$. Then, the unique solution of (3.66), respectively (3.67), will be $\overline{X} = E * F$.

Taking into account the expression (3.65), we obtain the solution in an explicit form

$$\overline{X} = E * f + \sum_{i=1}^{n} b_i E^{(n-i)}\,.$$

For $x > 0$ we shall obtain the solution of the Cauchy problem corresponding to (3.64), that is,

$$X(x) = \int_0^x Y(x-t)\,f(t)\mathrm{d}t + \sum_{i=1}^{n} b_i Y^{(n-i)}(x)\,. \tag{3.68}$$

In order to verify that this is a solution, it is sufficient to differentiate the equation under the integral sign. Thus, because $Y(0) = 0$, $Y'(0) = 0, \ldots, Y^{(n-2)}(0) = 0$, $Y^{(n-1)}(0) = 1$, we obtain

$$X'(x) = \int_0^x Y'(x-t)\,f(t)\mathrm{d}t + Y(0)\,f(x) + \sum_{i=1}^{n} b_i \frac{\mathrm{d}}{\mathrm{d}x} Y^{(n-i)}(x)$$
$$= \int_0^x Y'(x-t)\,f(t)\mathrm{d}t + \sum_{i=1}^{n} b_i \frac{\mathrm{d}}{\mathrm{d}x} Y^{(n-i)}(x)\,.$$

Similarly, we have

$$X^{(p)} = \int_0^x Y^{(p)}(x-t)\,f(t)\mathrm{d}t + \sum_{i=1}^{n} b_i \frac{\mathrm{d}^p}{\mathrm{d}x^p} Y^{(n-i)}(x)\,, \quad p = \overline{1, n-1}\,,$$
$$X^{(n)} = \int_0^x Y^{(n)}(x-t)\,f(t)\mathrm{d}t + f(x) + \sum_{i=1}^{n} b_i \frac{\mathrm{d}^n}{\mathrm{d}x^n} Y^{(n-i)}(x)\,.$$

Consequently, we obtain

$$P(\mathrm{D})X(x) = \int_0^x P(\mathrm{D})Y(x-t)f(t)\mathrm{d}t + f(x) + \sum_{i=1}^{n} b_i P(\mathrm{D})Y^{(n-i)}(x) = f(x) ,$$

because $P(\mathrm{D})Y(x-t) = 0$ and $P(\mathrm{D})Y^{(n-i)}(x) = 0$.

Using the above relations, it may be verified that the function X satisfies the initial conditions.

Example 3.6 Consider the equation

$$X''(x) + \omega^2 X(x) = f(x) , \quad x > 0 \tag{3.69}$$

with the initial conditions $X(0) = X_0$, $X'(0) = X_1$; $\omega \in \mathbb{R}$, where f is a locally integrable function on $[0, \infty)$.

Because

$$\delta''(x) + \omega^2\delta(x) = (\delta'(x) + \mathrm{i}\omega\delta(x)) * (\delta'(x) - \mathrm{i}\omega\delta(x)) ,$$

we have

$$\begin{aligned} E(x) &= (\delta''(x) + \omega^2\delta(x))^{-1} = (\delta'(x) + \mathrm{i}\omega\delta(x))^{-1} * (\delta'(x) - \mathrm{i}\omega\delta(x))^{-1} \\ &= H(x)\mathrm{e}^{-\mathrm{i}\omega x} * H(x)\mathrm{e}^{\mathrm{i}\omega x} . \end{aligned}$$

In an explicit form, we have

$$E(x) = H(x)\int_0^x \mathrm{e}^{-\mathrm{i}\omega t}\mathrm{e}^{\mathrm{i}\omega(x-t)}\mathrm{d}t = H(x)\mathrm{e}^{\mathrm{i}\omega x}\int_0^x \mathrm{e}^{-2\mathrm{i}\omega t}\mathrm{d}t = \frac{H(x)}{\omega}\sin\omega x ,$$

hence $Y(x) = \sin\omega x/\omega$.

According to the formula (3.68), the Cauchy problem solution will be

$$\begin{aligned} X(x) &= Y(x) * f(x) + X_0 Y'(x) + X_1 \\ &= \frac{1}{\omega}\int_0^x \sin\omega(x-t) f(t)\mathrm{d}t + X_0\cos\omega x + \frac{X_1}{\omega}\sin\omega x . \end{aligned}$$

By direct calculation, we have

$$\begin{aligned} \overline{X} &= H(x)X(x) , \quad \overline{X}' = X_0\delta(x) + H(x)\tilde{X}' , \\ \overline{X}'' &= X_0\delta'(x) + X_1\delta(x) + H(x)\tilde{X}'' . \end{aligned}$$

Thus, (3.69) transposed into the distribution space will have the expression $\overline{X}'' + \omega^2\overline{X} = H(x)f(x) + X_0\delta'(x) + X_1\delta(x)$. Taking into account that the fundamental solution is $E = H(x)Y(x) = (H(x)\sin\omega x)/\omega$, we obtain, for the solution of the equation, the expression $\overline{X}(x) = E(x) * H(x)f(x) + X_0 E' + X_1 E$. For $x > 0$, we get for the Cauchy problem, the solution $X(x) = Y(x) * f(x) + X_0 Y'(x) + X_1$, which is identical to that previously found.

3.5 Partial Differential Equations: Fundamental Solutions and Solving the Cauchy Problem

Let

$$P(\mathrm{D}) = \sum_{|\alpha| \le m} a_\alpha \mathrm{D}^\alpha , \quad a_\alpha \in \mathbb{C} , \quad \mathrm{D}^\alpha = \frac{\partial^{|\alpha|}}{\partial x_1^{\alpha_1} \partial x_2^{\alpha_2} \dots \partial x_n^{\alpha_n}} \tag{3.70}$$

be a linear differential operator with constant coefficients. Then, the equation

$$P(\mathrm{D}) X = f , \quad X, f \in \mathcal{D}'(\mathbb{R}^n) \tag{3.71}$$

is equivalent to the convolution equation

$$P(\mathrm{D})\delta(x) * X(x) = f(x) . \tag{3.72}$$

The distribution $E(x) \in \mathcal{D}'(\mathbb{R}^n)$ which verifies the equation $P(\mathrm{D})E(x) = \delta(x)$ is called the fundamental solution of the operator $P(\mathrm{D})$.

The existence and uniqueness of the fundamental solution in the convolution algebra $\mathcal{D}'_+$ has already been shown for ordinary linear differential equations with constant coefficients. Using the Fourier transform, it is proved in [4, 6, 29] that any linear differential operator with constant coefficients has a fundamental solution in the space $\mathcal{D}'(\mathbb{R}^n)$.

Proposition 3.15 The necessary and sufficient condition for a distribution $E \in \mathcal{S}'(\mathbb{R}^n)$ be fundamental solution for the operator $P(\mathrm{D})$ defined by (3.70) is that the Fourier $\hat{E}(\xi)$ does satisfy the equation

$$P(-\mathrm{i}\xi)\hat{E} = 1 , \tag{3.73}$$

where $P(\xi) = \sum_{|\alpha| \le m} a_\alpha \xi^\alpha$, $\xi^\alpha = \xi_1^{\alpha_1} \dots \xi_n^{\alpha_n}$.

Indeed, let us suppose that $E \in \mathcal{S}'(\mathbb{R}^n)$ is a fundamental solution of the operator $P(\mathrm{D})$. Then, because $P(\mathrm{D})E = P(\mathrm{D})\delta * E = \delta$, by the Fourier transform, we get $\widehat{P(\mathrm{D})E} = 1$, that is, $P(-\mathrm{i}\xi)\hat{E} = 1$; thus the necessity of the condition is proved. Regarding the sufficiency of the condition, we notice that, if $\hat{E}$ satisfies (3.73), then, by a converse procedure, we obtain $\widehat{P(\mathrm{D})E} = P(-\mathrm{i}\xi)\hat{E} = 1$, where $P(\mathrm{D})E = \delta$ and thus the proof is complete.

Proposition 3.16 Let $E \in \mathcal{D}'(\mathbb{R}^n)$ be a fundamental solution of the operator $P(\mathrm{D})$ and let $f \in \mathcal{D}'(\mathbb{R}^n)$. If the convolution product $E * f$ exists, then the expression

$$X = E * f \tag{3.74}$$

represents a solution of (3.71) and (3.72), respectively.

Indeed, using the formula for convolution differentiation, we have

$$P(\mathrm{D})X = P(\mathrm{D})(E * f) = (P(\mathrm{D})E) * f = \delta * f = f .$$

Similarly, let there be $(a) \in (\mathcal{D}')^{m\times m}$ the matrix with elements $a_{ij} = P_{ij}(\mathrm{D})\delta$, where $P_{ij}(\mathrm{D})$, $i, j = \overline{1, m}$ are linear differential operators with constant coefficients.

The system of partial differential equations

$$\sum_{j=1}^{m} P_{ij}(\mathrm{D})X_j = f_i\,, \quad X_i, f_i \in \mathcal{D}'(\mathbb{R}^n)\,, \quad i, j = \overline{1, m}$$

is equivalent to the convolution matrix equation $(a) * (X) = (f)$, where

$$(X) = \begin{pmatrix} X_1 \\ \vdots \\ X_m \end{pmatrix}, \quad (f) = \begin{pmatrix} f_1 \\ \vdots \\ f_m \end{pmatrix}.$$

Definition 3.10 We call the fundamental solution corresponding to the matrix of operators $(a) = (P_{ij}(\mathrm{D})\delta)$ a matrix distribution $(E) \in (\mathcal{D}')^{m\times m}$ which satisfies the relation

$$(E) * (P_{ij}(\mathrm{D})\delta) = (P_{ij}(\mathrm{D})\delta) * (E) = (\delta)\,. \tag{3.75}$$

Consider the partial differential equation with constant coefficients

$$P\left(\frac{\partial}{\partial x}, \frac{\partial}{\partial t}\right) u(x, t) = 0\,, \tag{3.76}$$

where P is a polynomial of $n + 1$ variables and degree m with respect to the variable t, whereas $u(x, t) = u_t(x)$ is a distribution in $\mathcal{D}'(\mathbb{R}^n)$ depending on the real parameter t.

The fundamental solution corresponding to the operator $P(\partial/\partial x, \partial/\partial t)$ is the distribution $E(x, t) \in \mathcal{D}'(\mathbb{R}^{n+1})$, which satisfies the relation

$$P\left(\frac{\partial}{\partial x}, \frac{\partial}{\partial t}\right) E(x, t) = \delta(x, t)\,,$$

where $\delta(x, t) \in \mathcal{D}'(\mathbb{R}^{n+1})$ represents the Dirac delta distribution concentrated at the origin.

Definition 3.11 We call the Cauchy problem for (3.76) the determination of the distribution $u(x, t) = u_t(x) \in \mathcal{D}'(\mathbb{R}^n)$, which is the solution of (3.76), also satisfying the initial conditions

$$u_t(x)|_{t=0} = u_0(x),\ \frac{\partial}{\partial t}u_t(x)\bigg|_{t=0} = u_1(x), \ldots, \frac{\partial^{m-1}}{\partial t^{m-1}}u_t(x)\bigg|_{t=0} = u_{m-1}(x)\,, \tag{3.77}$$

where $u_0, u_1, \ldots, u_{m-1} \in \mathcal{D}'(\mathbb{R}^n)$ are given distributions.

The notion of fundamental solution of the Cauchy problem is extremely important for the solution of this problem.

Definition 3.12 The distribution $E_t(x) \in \mathcal{D}'(\mathbb{R}^n)$, which depends on the real parameter t, is called the fundamental solution of the Cauchy problem corresponding to (3.76) if, for $t > 0$, it represents a solution of this equation and also satisfies the following initial conditions:

$$E_t(x)|_{t=0} = 0, \; \frac{\partial}{\partial t} E_t(x)\bigg|_{t=0} = 0, \ldots,$$
$$\frac{\partial^{m-2}}{\partial t^{m-2}} E_t(x)\bigg|_{t=0} = 0, \; \frac{\partial^{m-1}}{\partial t^{m-1}} E_t(x)\bigg|_{t=0} = \delta(x) .$$

Proposition 3.17 Let $E_t \in \mathcal{D}'(\mathbb{R}^n)$ be the fundamental solution of the Cauchy problem for (3.76) and let $u_{m-1} \in \mathcal{D}'(\mathbb{R}^n)$ be a distribution. If the convolution product $E_t * u_{m-1}$ exists, then the distribution

$$u(x,t) = u_t(x) = E_t(x) * u_{m-1}(x)$$

represents the solution of the Cauchy problem corresponding to (3.76) with the initial conditions

$$u_t(x)|_{t=0} = u_0(x), \; \frac{\partial}{\partial t} u_t(x)\bigg|_{t=0} = u_1(x), \ldots,$$
$$\frac{\partial^{m-2}}{\partial t^{m-2}} u_t(x)\bigg|_{t=0} = 0, \; \frac{\partial^{m-1}}{\partial t^{m-1}} u_t(x)\bigg|_{t=0} = u_{m-1}(x) . \tag{3.78}$$

Indeed, we have

$$P\left(\frac{\partial}{\partial x}, \frac{\partial}{\partial t}\right) u_t(x)$$
$$= P\left(\frac{\partial}{\partial x}, \frac{\partial}{\partial t}\right) E_t(x) * u_{m-1}(x) = P\left(\frac{\partial}{\partial x}, \frac{\partial}{\partial t}\right) E_t(x) * u_{m-1}(x) = 0 ,$$

hence u_t is a solution of (3.76).

Similarly, we can write

$$\frac{\partial^k}{\partial t^k} u_t = \frac{\partial^k}{\partial t^k}(E_t * u_{m-1}) = \frac{\partial^k}{\partial t^k} E_t * u_{m-1} , \quad k = 0, 1, 2, \ldots, m-1 .$$

Taking into account the definition of the fundamental solution of the Cauchy problem, the initial conditions (3.78) are also satisfied.

Proposition 3.18 Let $E_t \in \mathcal{D}'(\mathbb{R}^n)$ be the fundamental solution of the Cauchy problem corresponding to (3.76) and the operator

$$P\left(\frac{\partial}{\partial x}, \frac{\partial}{\partial t}\right) = \frac{\partial^m}{\partial t^m} - L\left(\frac{\partial}{\partial x}, \frac{\partial}{\partial t}\right) ,$$

where $L(\partial/\partial x, \partial/\partial t)$ is a linear differential operator with constant coefficients of maximum order $m-2$ with respect to t.

If the convolution products between E_t and $u_0, u_1, \ldots, u_{m-1} \in \mathcal{D}'(\mathbb{R}^n)$ exist, then the solution of (3.76) with the initial conditions (3.77) is given by the distribution

$$u_t = E_t * u_{m-1} + \frac{\partial}{\partial t} E_t * u_{m-2} + \cdots + \frac{\partial^{m-1}}{\partial t^{m-1}} E_t * u_0 \,.$$

Let's consider the equation

$$P\left(\frac{\partial}{\partial x}, \frac{\partial}{\partial t}\right) u(x,t) = 0\,, \tag{3.79}$$

where P is a polynomial of degree m in $n+1$ variables with respect to t; this may be expressed as

$$P\left(\frac{\partial}{\partial x}, \frac{\partial}{\partial t}\right) = \frac{\partial^m}{\partial t^m} - L\left(\frac{\partial}{\partial x}, \frac{\partial}{\partial t}\right), \tag{3.80}$$

$L(\partial/\partial x, \partial/\partial t)$ being a polynomial of maximum degree $m-1$ in $n+1$ variables with respect to t.

Proposition 3.19 Let $E_t \in \mathcal{S}'(\mathbb{R}^n)$ be the fundamental solution of the Cauchy problem for (3.79). Then, there exists a fundamental solution $E(x,t) \in \mathcal{S}'(\mathbb{R}^{n+1})$ of the operator (3.80) of the form

$$E(x,t) = H(x)\,E_t(x) = \begin{cases} 0\,, & t < 0\,, \\ E_t(x)\,, & t \geq 0\,. \end{cases} \tag{3.81}$$

Proposition 3.20 Let $E(x,t) \in \mathcal{S}'(\mathbb{R}^{n+1})$ be a fundamental solution of the operator (3.80). If $E(x,t)$ is of the form (3.81), where $E_t \in \mathcal{S}'(\mathbb{R}^n)$, t being a parameter, and if it satisfies the condition

$$\left.\frac{\partial^k}{\partial t^k} E\right|_{t=0} = 0, \quad \left.\frac{\partial^{m-1}}{\partial t^{m-1}} E\right|_{t=0} = \delta(x)\,, \quad k = 0, 1, 2, \ldots, m-2\,,$$

then the distribution E_t represents the fundamental solution of the Cauchy problem for (3.79).

3.5.1 Fundamental Solution for the Longitudinal Vibrations of Viscoelastic Bars of Maxwell Type

In the study of the longitudinal vibrations in viscoelastic bars of Maxwell type appears the operator

$$P\left(\frac{\partial}{\partial t}, \frac{\partial}{\partial x}\right) = \frac{\partial^2}{\partial t^2} + \lambda \frac{\partial}{\partial t} - c^2 \frac{\partial^2}{\partial x^2}\,, \quad \lambda > 0\,, \quad c > 0\,. \tag{3.82}$$

We shall determine the fundamental solution of the operator [30], as well as the fundamental solution of the corresponding Cauchy problem.

Let $E(x,t) \in \mathcal{D}'(\mathbb{R}^2)$ be the fundamental solution of the operator (3.82). Applying the Laplace transform with respect to the temporal variable $t \in \mathbb{R}$ to the equation

$$P\left(\frac{\partial}{\partial t}, \frac{\partial}{\partial x}\right) E(x,t) = \delta(x,t) = \delta(x) \times \delta(t)\,,$$

we obtain

$$c^2 \frac{\mathrm{d}^2}{\mathrm{d}x^2} \hat{E}(x,p) - \omega^2(p)\hat{E}(x,p) = -\delta(x)\,, \tag{3.83}$$

where $\hat{E}(x,p) = \mathrm{L}[E(x,t)](p)$, $\omega^2(p) = p^2 + \lambda p$.

The solution of (3.83) is the function

$$\hat{E}(x,p) = \frac{1}{2\omega c} \exp\left(-\frac{\omega|x|}{c}\right) = \frac{\exp\left(-|x|/c\sqrt{(p+(\lambda/2))^2 - (\lambda/2)^2}\right)}{2c\sqrt{(p+(\lambda/2))^2-(\lambda/2)^2}}\,. \tag{3.84}$$

Taking into account [31] we have

$$\mathrm{L}^{-1}\left[\frac{\exp\left(-b\sqrt{p^2-k^2}\right)}{\sqrt{p^2-k^2}}\right] = \begin{cases} 0\,, & 0 < t < b\,, \\ I_0\left(k\sqrt{t^2-b^2}\right)\,, & t \geq b\,, \end{cases} \quad b > 0\,, \tag{3.85}$$

where I_0 is the modified Bessel function of zero order.

Using the formula $\mathrm{L}[\mathrm{e}^{qt} f(t)] = \mathrm{L}[f(t)](p-q)$, and taking into account (3.85) and (3.84) we shall obtain for the fundamental solution $E(x,t) \in \mathcal{D}'(\mathbb{R}^2)$ the expression

$$E(x,t) = \mathrm{L}^{-1}[\hat{E}(x,p)] = \frac{1}{2c}\exp\left(-\frac{\lambda t}{2}\right) I_0\left(\frac{\lambda}{2}\sqrt{t^2 - \frac{x^2}{c^2}}\right) H\left(t - \frac{|x|}{c}\right)\,. \tag{3.86}$$

Because we have $H(t - |x|/c) = H((ct - |x|)/c) = H(ct - |x|)$, the expression (3.86) becomes

$$E(x,t) = \frac{1}{2c}\exp\left(-\frac{\lambda t}{2}\right) I_0\left(\frac{\lambda}{2}\sqrt{t^2-\frac{x^2}{c^2}}\right) H(ct-|x|)$$

$$= \begin{cases} 0\,, & t < 0\,, \\ \dfrac{1}{2c}\exp\left(-\dfrac{\lambda t}{2}\right) I_0\left(\dfrac{\lambda}{2}\sqrt{t^2-\dfrac{x^2}{c^2}}\right)\,, & t \geq \dfrac{|x|}{c} \geq 0\,, \\ 0\,, & 0 \leq t \leq \dfrac{|x|}{c}\,. \end{cases} \tag{3.87}$$

We mention that

$$I_0(x) = \sum_{k=0}^{\infty} \frac{1}{k!\Gamma(k+1)} \left(\frac{x}{2}\right)^{2k} , \quad I_0(0) = 1 , \quad I_0'(0) = 0 ,$$

where $\Gamma(k)$ is the gamma function.

Particularly, for $\lambda = 0$ the operator $P(\partial/\partial t, \partial/\partial x)$ reads

$$P\left(\frac{\partial}{\partial t}, \frac{\partial}{\partial x}\right) = \frac{\partial^2}{\partial t^2} - c^2 \frac{\partial^2}{\partial x^2} ,$$

which represents the operator of the wave equation.

From (3.87), in this particular case, we obtain for the fundamental solution

$$E^*(x,t) = \frac{H(ct - |x|)}{2c} .$$

Let $E_t(x) \in \mathcal{S}'(\mathbb{R})$ be the solution of the Cauchy problem for the operator (3.82). Then, between the fundamental solution $E(x,t) \in \mathcal{S}'(\mathbb{R}^2)$ and $E_t(x) \in \mathcal{S}'(\mathbb{R})$ follows the relation

$$E(x,t) = H(t) E_t(x) .$$

Hence, we have

$$E_t(x) = \begin{cases} 0 , & |x| \geq ct , \\ \dfrac{1}{2c} \exp\left(-\dfrac{\lambda}{2} t\right) I_0\left[\dfrac{\lambda}{2}\sqrt{t^2 - \dfrac{x^2}{c^2}}\right] , & x \in [-ct, ct] , \end{cases} \tag{3.88}$$

for $t > 0$.

Consequently, we have

$$E_t(x)|_{t=0} = 0, \ \frac{\partial}{\partial t} E_t(x)\bigg|_{t=0} = \delta(x), \ \frac{\partial^2}{\partial t^2} E_t(x)\bigg|_{t=0} = -\lambda\delta(x)$$

and

$$\frac{\partial^2 E_t(x)}{\partial t^2} + \lambda \frac{\partial E_t(x)}{\partial t} - c^2 \frac{\partial^2 E_t(x)}{\partial x^2} = 0 .$$

Example 3.7 Let us determine the solution of the Cauchy problem for the wave equation

$$\frac{\partial^2 u_t(x)}{\partial t^2} = a^2 \frac{\partial^2 u_t(x)}{\partial x^2} , \quad u_t(x) \in \mathcal{D}'(\mathbb{R}) , \quad t \geq 0 , \tag{3.89}$$

with the initial conditions

$$u_t(x)|_{t=0} = u_0(x), \ \frac{\partial u_t(x)}{\partial t}\bigg|_{t=0} = u_1(x) , \tag{3.90}$$

where $u_0, u_1 \in \mathcal{D}'(\mathbb{R})$.

Let $E_t(x) \in \mathcal{D}'(\mathbb{R})$, $t \geq 0$, be the solution of the Cauchy problem for (3.89). We consider the solution of the problem in the form

$$u_t(x) = E_t(x) * f_0(x) + \frac{\partial}{\partial t} E_t(x) * f_1(x) . \tag{3.91}$$

It is shown that the distribution (3.91) verifies (3.89) for any distribution $f_0, f_1 \in \mathcal{D}'(\mathbb{R})$.

Indeed, we have

$$\frac{\partial^2}{\partial t^2} u_t(x) = \frac{\partial^2 E_t(x)}{\partial t^2} * f_0(x) + \frac{\partial^3 E_t(x)}{\partial t^3} * f_1(x) ,$$
$$a^2 \frac{\partial^2 u_t(x)}{\partial x^2} = a^2 \frac{\partial^2 E_t(x)}{\partial x^2} * f_0(x) + a^2 \frac{\partial}{\partial t} \frac{\partial^2 E_t(x)}{\partial x^2} * f_1(x) .$$

Taking into account (3.89), we obtain

$$\frac{\partial^2}{\partial t^2} u_t(x) - a^2 \frac{\partial^2 u_t(x)}{\partial x^2} = \left(\frac{\partial^2 E_t(x)}{\partial t^2} - a^2 \frac{\partial^2 E_t(x)}{\partial x^2} \right) * f_0(x)$$
$$+ \frac{\partial}{\partial t} \left(\frac{\partial^2 E_t(x)}{\partial t^2} - a^2 \frac{\partial^2 E_t(x)}{\partial x^2} \right) * f_1(x) .$$

Because $(\partial^2 E_t(x)/\partial t^2) = a^2(\partial^2 E_t(x)/\partial x^2)$, we get $(\partial^2 u_t(x)/\partial t^2) = a^2(\partial^2 u_t(x)/\partial x^2)$, hence $u_t(x)$ given by (3.91) verifies (3.89).

The distributions $f_0, f_1 \in \mathcal{D}'(\mathbb{R})$ are determined from the initial condition (3.90).

Thus, because $E_t(x)|_{t=0} = 0$ and $(\partial E_t(x)/\partial t)|_{t=0} = \delta(x)$, from (3.91), the following results:

$$u_t(x)|_{t=0} = u_0(x) = f_1(x) ,$$
$$\left. \frac{\partial u_t(x)}{\partial t} \right|_{t=0} = u_1(x) = \left. \frac{\partial E_t(x)}{\partial t} \right|_{t=0} * f_0(x) + \left. \frac{\partial^2 E_t(x)}{\partial t^2} \right|_{t=0} * f_1(x) = f_0(x) .$$

But, from the relation $\partial^2 E_t(x)/\partial t^2 = a^2(\partial^2 E_t(x)/\partial x^2)$ for $t \to +0$, we obtain

$$\left. \frac{\partial^2 E_t(x)}{\partial t^2} \right|_{t=0} = a^2 \frac{\partial^2}{\partial x^2} (E_t(x)|_{t=0}) = 0 .$$

Consequently, we have $u_0(x) = f_1(x)$ and $u_1(x) = f_0(x)$.

Thus, the solution of the Cauchy problem for the wave equation (3.89) with initial conditions (3.90) becomes

$$u_t(x) = E_t(x) * u_1(x) + \frac{\partial E_t(x)}{\partial t} * u_0(x) .$$

From (3.88), considering $\lambda = 0$ and $c = a$, we obtain the fundamental solution of the Cauchy problem

$$E_t(x) = \begin{cases} 0 , & |x| > at \\ \dfrac{1}{2a} , & |x| \leq at \end{cases} = \frac{1}{2a} [H(x + at) - H(x - at)] . \tag{3.92}$$

From (3.92), it results $\partial E_t(x)/\partial t = [\delta(x+at)+\delta(x-at)]/2$. If u_0, u_1 are function type distributions, then we obtain the d'Alembert formula

$$u_t(x) = \frac{1}{2a}\int_{-at}^{at} u_1(x-\xi)\mathrm{d}\xi + u_0(x) * \frac{1}{2}[\delta(x+at)+\delta(x-at)]$$

$$= \frac{u_0(x+at)+u_0(x-at)}{2} + \frac{1}{2a}\int_{x-at}^{x+at} u_1(y)\mathrm{d}y\,.$$

3.6
Wave Equation and the Solution of the Cauchy Problem

In the study of various physical phenomena, such as sound propagation, electromagnetic waves, vibrations of strings and membranes, there appears the differential equation of hyperbolic type

$$\Box_a u = 0\,, \tag{3.93}$$

which is called the wave equation in $\mathbb{R}^n$, where $u(x,t)$ is a distribution from $\mathcal{D}'(\mathbb{R}^n)$, depending on the parameter $t \geq 0$, while $\Box_a = \partial^2/\partial t^2 - a^2\Delta$ is d'Alembert's operator.

For $n = 1, 2, 3$ we obtain the vibrating string equation, the cylindrical wave equation and the spherical wave equation, respectively.

Let $E(x,t) \in \mathcal{S}'(\mathbb{R}^{n+1})$ be the fundamental solution corresponding to d'Alembert's operator.

Then, we have

$$\Box_a E(x,t) = \delta(x,t)\,.$$

Performing the Fourier transform with respect to the variable x, we obtain the differential equation

$$\frac{\mathrm{d}^2}{\mathrm{d}t^2}\hat{E}(\xi,t) + a^2\|\xi\|^2\hat{E}(\xi,t) = \delta(t)\,.$$

The fundamental solution of this equation is the distribution

$$\hat{E}(\xi,t) = H(t)\frac{\sin(a\|\xi\|t)}{a\|\xi\|}\,.$$

Consequently, we obtain the following expression for the fundamental solution of the wave equation and of d'Alembert's operator

$$E(x,t) = H(t)\mathrm{F}_\xi^{-1}\left[\frac{\sin(a\,\|\xi\|\,t)}{a\,\|\xi\|}\right] \in \mathcal{S}'(\mathbb{R}^{n+1})\,. \tag{3.94}$$

In evaluating the inverse Fourier transform in the spaces having dimensions $2m+3$, $m \in \mathbb{N}_0$, the role played by Dirac's delta distribution δ_S concentrated on a sphere having the radius R and its center at the origin is paramount.

Proposition 3.21 Let S be a bounded, piecewise smooth surface. If g and f are piecewise continuous functions in $\mathbb{R}^n$ and on S, respectively, then the following formula holds

$$g * f\delta_S = \int_S f(y)g(x-y)\mathrm{d}S_y \,. \tag{3.95}$$

Accordingly with the Fourier transform formula for distributions with compact supports, we have

$$\mathrm{F}[\delta_{S_R}] = (\delta_{S_R}, \exp(\mathrm{i}\langle x, \xi \rangle)) = \int_{S_R} \exp(\mathrm{i}\langle x, \xi \rangle)\mathrm{d}S \,, \tag{3.96}$$

where S_R is a sphere of radius R with the center at the origin.

Let us denote by θ the angle formed by the vector $x = (x_1, \ldots, x_n) \in S_R$ with the Ox_n-axis. Hence, the differential element of area $\mathrm{d}S$ on the sphere S_R is

$$\mathrm{d}S = \frac{2\pi^{(n-1)/2}}{\Gamma((n-1)/2)}(R\sin\theta)^{n-2}R\mathrm{d}\theta \,, \quad n \geq 2 \,.$$

Since the Ox_n-axis is arbitrary, we shall select it so as to coincide in direction and sign with the vector $\xi = (\xi_1, \xi_2, \ldots, \xi_n) \in \mathbb{R}^n$; hence, $\langle x, \xi \rangle = R\|\xi\|\cos\theta$ so that the formula (3.96) reads

$$\mathrm{F}[\delta_{S_R}] = \frac{2R^{n-1}\pi^{(n-1)/2}}{\Gamma((n-1)/2)} \int_0^{\pi} \exp(\mathrm{i}R\|\xi\|\cos\theta)\sin^{n-2}\theta\,\mathrm{d}\theta \,.$$

This formula allows us to determine the fundamental solution $E(x,t)$ given by the expression (3.94) for n-dimensional spaces with $n = 2m+3$, $m \in \mathbb{N}_0$. Thus, for $n = 3$, since $\Gamma(1) = 1$, we obtain

$$\mathrm{F}\left[\delta_{S_R}\right] = 4\pi R \frac{\sin(R\|\xi\|)}{\|\xi\|} \,. \tag{3.97}$$

By comparing the formulae (3.94) and (3.97), we find the following expression for the fundamental solution corresponding to the wave equation in $\mathbb{R}^3$, that is,

$$E(x,t) = \frac{H(t)}{4\pi a^2 t}\delta_{S_{at}} \in \mathcal{S}'(\mathbb{R}^4) \,. \tag{3.98}$$

Taking into account the expression (3.98) of the fundamental solution $E(x,t)$, as well as its properties

$$\forall \varphi \in \mathcal{S}(\mathbb{R}^3), (E(x,t), \varphi(x)) = \frac{H(t)}{4\pi a^2 t}(\delta_{S_{at}}, \varphi(x)) = \frac{H(t)}{4\pi a^2 t}\int_{S_{at}} \varphi(x)\mathrm{d}S \,,$$

$$E(x,t)|_{t=0} = 0 \,,$$

according to the Proposition 3.18, we deduce that the solution of the Cauchy problem for the wave equation (3.93) with initial conditions

$$u_t(x)|_{t=0} = u_0(x)\,, \ \frac{\partial}{\partial t} u_t(x)\bigg|_{t=0} = u_1(x)$$

is the distribution

$$u_t(x) = E_t * u_1 + \frac{\partial}{\partial t} E_t * u_0\,, \tag{3.99}$$

where the distributions u_0 and u_1 are with compact support.

Particularly, if u_0 and u_1 are function type distributions with compact support, generated by piecewise continuous functions, then, on the basis of the formula (3.95), the solution of the Cauchy problem (3.99) reads

$$u_t(x) = \frac{1}{4\pi a^2 t} \int\limits_{\|y\|=at} u_1(x-y)\mathrm{d}S_y + \frac{1}{4\pi a^2}\frac{\partial}{\partial t}\frac{1}{t} \int\limits_{\|y\|=at} u_0(x-y)\mathrm{d}S_y$$

or, making the change of variable,

$$u_t(x) = \frac{1}{4\pi a^2} \int\limits_{\|x-\xi\|=at} \frac{u_1(\xi)}{t}\mathrm{d}S + \frac{1}{4\pi a^2}\frac{\partial}{\partial t} \int\limits_{\|x-\xi\|=at} \frac{u_0(\xi)}{t}\mathrm{d}S\,.$$

It can be shown, taking into account (3.94), that for $n = 1$ the fundamental solution corresponding to the vibrating string is the distribution $E(x,t) \in \mathcal{S}'(\mathbb{R}^2)$, given by

$$E(x,t) = \frac{1}{2a} H(at - |x|) = \begin{cases} 0\,, & t < 0\,, \\ \dfrac{1}{2a}\,, & at \geq x \geq 0\,, \\ 0\,, & |x| > at \geq 0\,. \end{cases}$$

Thus, for $\varphi \in \mathcal{D}(\mathbb{R})$ and $t > 0$ we obtain

$$(E(x,t), \varphi(x)) = \frac{1}{2a}\int\limits_{-at}^{at} \varphi(x)\mathrm{d}x = \frac{t}{2}\int\limits_{-1}^{1} \varphi(atu)\mathrm{d}u\,,$$

$$\left(\frac{\partial E}{\partial t}, \varphi\right) = \frac{1}{2}\int\limits_{-1}^{1} \varphi(atu)\mathrm{d}u + \frac{ta}{2}\int\limits_{-1}^{1} u\varphi'(atu)\mathrm{d}u\,.$$

For $t \to +0$ we have

$$E(x,t)|_{t=0} = 0,\ \frac{\partial}{\partial t} E(x,t)\bigg|_{t=0} = \delta(x)\,.$$

According to Proposition 3.20, we deduce that the fundamental solution of the Cauchy problem for the vibrating string equation is the distribution

$$E_t(x) = \begin{cases} \dfrac{1}{2a}\,, & |x| \le at\,, \\ 0\,, & |x| > at\,, \end{cases} \qquad t > 0\,.$$

Consequently, if $u_0, u_1 \in \mathcal{E}'(\mathbb{R})$ are distributions with compact support, then the Cauchy problem solution with initial conditions

$$u_t(x)|_{t=0} = u_0(x)\,, \qquad \left.\frac{\partial}{\partial t} u_t(x)\right|_{t=0} = u_1(x)$$

is given by the distribution

$$u_t(x) = E_t(x) * u_1(x) + \frac{\partial}{\partial t}\left(E_t(x) * u_0(x)\right)\,.$$

If u_0, u_1 are function type distributions, we obtain the formula of d'Alembert

$$\begin{aligned} u_t(x) &= \frac{1}{2a}\int_{-at}^{at} u_1(x-\xi)\mathrm{d}\xi + \frac{1}{2a}\frac{\partial}{\partial t}\int_{-at}^{at} u_0(x-\xi)\mathrm{d}\xi \\ &= \frac{u_0(x+at)+u_0(x-at)}{2} + \frac{1}{2a}\int_{x-at}^{x+at} u_1(y)\mathrm{d}y\,. \end{aligned}$$

For $n = 2$, the fundamental solution corresponding to the cylindrical wave equation is

$$E(x,t) = \frac{H(at - \|x\|)}{2\pi a\sqrt{a^2t^2 - \|x\|^2}}\,, \qquad \|x\| = \sqrt{x_1^2 + x_2^2}\,,$$

and the corresponding fundamental solution of the Cauchy problem reads

$$E_t(x) = \begin{cases} \dfrac{1}{2\pi a\sqrt{a^2t^2 - \|x\|^2}}\,, & \|x\| < at\,, \\ 0\,, & \|x\| \ge at\,, \end{cases} \qquad t > 0\,.$$

3.7
Heat Equation and Cauchy Problem Solution

The study of thermal conductivity and diffusion phenomena leads to the parabolic type equation

$$\frac{\partial u(x,t)}{\partial t} - a^2\Delta u(x,t) = 0\,, \tag{3.100}$$

called the heat equation in $\mathbb{R}^n$.

The Cauchy problem for (3.100) consists of determining the distribution $u(x,t) = u_t(x) \in \mathcal{D}'(\mathbb{R}^n)$, depending on the real parameter t, which is the solution of (3.100) with the initial condition

$$u_t(x)|_{t=0} = u_0(x) , \quad u_0 \in \mathcal{D}'(\mathbb{R}^n) . \tag{3.101}$$

Let $E(x,t) \in \mathcal{S}'(\mathbb{R}^{n+1})$ be the fundamental solution of the heat propagation operator $\partial/\partial t - a^2\Delta$.

The fundamental solution $E(x,t)$ satisfies the equation

$$\frac{\partial E}{\partial t} - a^2 \Delta E = \delta(x,t) .$$

Applying the Fourier transform with respect to the variable x to this equation, we get

$$\frac{\mathrm{d}}{\mathrm{d}t} \hat{E}(\xi,t) + a^2 \|\xi\|^2 \hat{E}(\xi,t) = 1(\xi) \times \delta(t) = \delta(t) .$$

The fundamental solution of this differential equation is the function type distribution

$$\hat{E}(\xi,t) = H(t) \exp(-a^2\|\xi\|^2 t) \in \mathcal{S}'(\mathbb{R}^{n+1}) .$$

Taking into account the formula

$$\mathrm{F}[\exp(-a^2\|x\|^2)](\xi) = \left(\frac{\sqrt{\pi}}{|a|}\right)^n \exp\left(\frac{-\|\xi\|^2}{4a^2}\right) ,$$

we have for the fundamental solution the expression

$$E(x,t) = \begin{cases} 0 , & t < 0 , \\ \dfrac{\exp\left(-\|x\|^2/4a^2 t\right)}{\left(2a\sqrt{\pi t}\right)^n} , & t > 0 . \end{cases}$$

It follows that the fundamental solution of the Cauchy problem corresponding to the heat equation (3.100) is the distribution $E_t(x) \in \mathcal{S}'(\mathbb{R}^n)$, depending on the parameter $t > 0$, namely

$$E_t(x) = \frac{\exp\left(-\|x\|^2/4a^2 t\right)}{\left(2a\sqrt{\pi t}\right)^n} , \quad t > 0 .$$

Because the distribution of E_t converges to $\delta(x)$ in the sense of distributions, this is a representative Dirac sequence.

Using E_t, the Cauchy problem solution corresponding to (3.100) and to the initial condition (3.101) will be the distribution

$$u_t(x) = E_t(x) * u_0(x) ,$$

where $u_0 \in \mathcal{E}'(\mathbb{R}^n)$.

In particular, if u_0 is a bounded function from $\mathbb{R}^n$, then we obtain the Poisson formula

$$u_t(x) = \frac{1}{\left(2a\sqrt{\pi t}\right)^n} \int\limits_{\mathbb{R}^n} u_0(\xi) \exp\left(\frac{-\|x-\xi\|^2}{4a^2 t}\right) \mathrm{d}\xi \,,$$

which is the bounded solution of the heat equation (3.100).

3.8
Poisson Equation: Fundamental Solutions

We consider the Poisson equation

$$\Delta u(x) = f(x)\,, \quad x \in \mathbb{R}^n\,, \tag{3.102}$$

where f is a given distribution from $\mathcal{D}'(\mathbb{R}^n)$.

This equation is of elliptic type and for, $f = 0$, we obtain the Laplace equation.

The function type distributions

$$E_2(x) = \frac{1}{2\pi} \ln r\,, \quad x \in \mathbb{R}^2 \backslash \{0\}\,, \quad r = \|x\|\,, \tag{3.103}$$

$$E_n(x) = -\frac{1}{r^{n-2}} \frac{1}{(n-2)\,S_1}\,, \quad x \in \mathbb{R}^n \backslash \{0\}\,, \quad r = \|x\|\,, \quad n \geq 3\,, \tag{3.104}$$

are the fundamental solutions of the Laplace operator in $\mathbb{R}^2$ and $\mathbb{R}^n$, $n \geq 3$, where $S_1 = 2\pi^{n/2}/\Gamma(n/2)$ is the area of the unit radius sphere from $\mathbb{R}^n$, $n \geq 3$.

For the deduction of the formula (3.104), we apply the rule of differentiation of homogeneous functions of degree $\lambda = -n + 2$.

We consider thus, the local integrable homogeneous function $f(x) = r^{2-n}$, $x \in \mathbb{R}^n \backslash \{0\}$, $r = \|x\|$, $n \geq 3$. We have

$$\frac{\partial f}{\partial x_i} = \frac{\tilde{\partial} f}{\partial x_i} + (-n+2)\frac{x_i}{r^n}\,,$$

$$\frac{\partial^2 f}{\partial x_i^2} = \frac{\tilde{\partial}^2 f}{\partial x_i^2} + (-n+2)\delta(x) \int\limits_{S_1} \frac{x_i}{r^n} \cos\alpha_i \mathrm{d}S_1$$

$$= (-n+2)\frac{r^2 - n x_i^2}{r^{n+2}} + (-n+2)\delta(x) \int\limits_{S_1} x_i^2 \mathrm{d}S_1\,.$$

By summing, we get

$$\Delta \frac{1}{r^{n-2}} = \tilde{\Delta} \frac{1}{r^{n-2}} + (-n+2)\delta(x) \int\limits_{S_1} \mathrm{d}S_1 = (-n+2)\,S_1 \delta(x)\,,$$

because $\tilde{\Delta} r^{2-n} = 0$, $\tilde{\Delta} = \tilde{\partial}^2/\partial x_1^2 + \cdots + \tilde{\partial}^2/\partial x_n^2$.

With this, the formula (3.104) is proved.

Analogously, formula (3.103) is deduced based on the locally integrable function $h(x) = \ln r$, $r = \|x\|$.

It is noted that the function

$$F(x, y) = \frac{x}{x^2 + y^2} = \frac{\partial}{\partial x} \ln(x^2 + y^2) = \frac{\tilde{\partial}}{\partial x} \ln(x^2 + y^2)$$

is a homogeneous function of degree $\lambda = -1$.

With this, the solution of the Poisson equation (3.102) is given by the distribution

$$u(x) = E_n(x) * f(x) ,$$

where f is a distribution with compact support.

3.9 Green's Functions: Methods of Calculation

The solution of many problems of mathematical physics is related to the construction of Green's function with the help of which the solutions of the boundary value problems may be determined explicitly and presented in an integral form. Thus, the Green functions are closely connected to the fundamental solution of the linear differential operator corresponding to a given problem.

In the case in which the fundamental solution is a distribution, we have to deal with Green's distributions.

To illustrate the construction of a Green function as well as its connection to the fundamental solution of the operator corresponding to the given boundary value problem, we shall consider the heat conduction equation for an infinite bar, the generalized Poisson equation and the vibrating string equation.

3.9.1 Heat Conduction Equation

We shall consider a cylindrical homogeneous and isotropic bar whose lateral surface is isolated from the rest of the medium. We take the symmetry axis of the bar as Ox-axis. We shall denote by $u(x, t)$, $x \in \mathbb{R}$, $t \geq 0$, the temperature of the bar at the point x and at the moment t, and by ρ, c, k the density, the specific heat and the thermal conductivity, respectively.

We use the Fourier–Newton law regarding the amount of heat flowing in the unit time across the unit cross-section of the bar; then, in the absence of internal sources of heat, *the heat conduction equation* is

$$\frac{\partial u(x, t)}{\partial t} = a^2 \frac{\partial^2 u(x, t)}{\partial x^2} , \quad x \in \mathbb{R} , \quad t > 0 , \tag{3.105}$$

where the constant a^2 has the expression $a^2 = k/\rho c$.

We consider the parabolic equation (3.105) with the initial condition

$$u(x,0) = \varphi(x)\,, \quad \varphi \in C^0(\mathbb{R})\,. \tag{3.106}$$

Equation (3.105) with the condition (3.106) represents the boundary value problem of the heat conduction in an infinite homogeneous bar.

The boundary value problem has a unique solution for the given function φ. The problem solution can be represented in the form

$$u(x,t) = \mathcal{L}(\varphi(x)) = \int_{-\infty}^{\infty} G(x,\xi,t)\varphi(\xi)\mathrm{d}\xi = (\varphi(\xi), G(x,\xi,t))\,, \tag{3.107}$$

where the kernel of the integral operator $\mathcal{L}$ is $G(x,\xi,t)$ and is called the Green function.

To obtain the Green function, we will consider $\varphi(x) = \delta(x - x_0)$, where $x_0 \in \mathbb{R}$ is a parameter. From (3.107), the following equation results

$$u(x,t) = (\delta(\xi - x_0), G(x,\xi,t)) = (\delta(\xi), G(x,\xi + x_0,t)) = G(x,x_0,t)\,,$$

hence

$$u(x,t) = G(x,x_0,t)\,.$$

Consequently, the Green function $G(x,x_0,t)$ represents the solution of (3.105) with the condition $u(x,0) = \delta(x - x_0)$.

Hence, the Green function $G(x,\xi,t)$ satisfies (3.105)

$$\frac{\partial G(x,\xi,t)}{\partial t} = \frac{\partial^2 G(x,\xi,t)}{\partial x^2} \tag{3.108}$$

and the initial condition

$$G(x,\xi,t)|_{t=0} = \delta(x - \xi)\,. \tag{3.109}$$

Let $E(x,t) \in \mathcal{D}'(\mathbb{R} \times \mathbb{R})$ be the fundamental solution of (3.105), hence

$$\frac{\partial E}{\partial t} - a^2\frac{\partial^2 E}{\partial x^2} = \delta(x,t) = \delta(x) \times \delta(t)\,. \tag{3.110}$$

Applying the Fourier transform with respect to x to (3.110), we obtain

$$\frac{\mathrm{d}}{\mathrm{d}t}\hat{E}(\alpha,t) - a^2(-\mathrm{i}\alpha)^2\hat{E}(\alpha,t) = \delta(t)\,, \tag{3.111}$$

where $\hat{E}(\alpha,t) = \mathrm{F}_x[E(x,t)]$.

Equation (3.111) shows that the distribution $\hat{E}(\alpha,t)$ is the fundamental solution of the operator

$$\frac{\mathrm{d}}{\mathrm{d}t} + a^2\alpha^2\,. \tag{3.112}$$

To determine the distribution $\hat{E}(\alpha, t)$, we consider the homogeneous equation

$$\frac{\mathrm{d}V(t)}{\mathrm{d}t} + a^2\alpha^2 V(t) = 0 ,$$

the general solution of which is $V(t) = C\exp(-a^2\alpha^2 t)$.

Imposing the condition $V(0) = 0$, we obtain $C = 1$ and, consequently, the fundamental solution of the operator (3.112) is

$$V(t) = H(t)\exp(-a^2\alpha^2 t) = \hat{E}(\alpha, t) . \tag{3.113}$$

Applying the inverse Fourier transform, we yield

$$E(x, t) = \mathrm{F}^{-1}[\hat{E}(\alpha, t)] = \frac{H(t)}{2a\sqrt{\pi t}}\exp\left(-\frac{x^2}{4a^2 t}\right) , \quad x \in \mathbb{R} , \quad t \in \mathbb{R} ,$$

because $\mathrm{F}^{-1}[\exp(-b^2x^2)] = \sqrt{\pi}\exp(-x^2/4b^2)/\sqrt{|b|}$.

Considering t as parameter, from (3.113), we obtain, for $t > 0$,

$$E_t(x) = \frac{1}{2a\sqrt{\pi t}}\exp\left(-\frac{x^2}{4a^2 t}\right) . \tag{3.114}$$

We remark that $E_t(x), t > 0$, is an element of the test space $\mathcal{S}(\mathbb{R})$ and that it satisfies the relation $\int_{\mathbb{R}} E_t(x)\mathrm{d}x = 1$, because $\int_{\mathbb{R}} \exp(-u^2)\mathrm{d}u = \sqrt{\pi}$.

By direct calculation, it is verified that E_t is the solution of (3.105) and satisfies the initial condition $E_t(x)|_{t=+0} = \delta(x)$.

Indeed, taking into account (3.113), we have

$$\lim_{t\to+0} \mathrm{F}_x[E(x, t)] = \mathrm{F}_x[\lim_{t\to+0} E_t(x)] = \lim_{t\to+0} \hat{E}(\alpha, t) = 1 = \mathrm{F}_x[\delta(x)] ,$$

hence $\lim_{t\to+0} E_t(x) = \delta(x)$.

Taking account of (3.108), (3.109), and (3.114), we obtain

$$G(x, \xi, t) = E_t(x - \xi) = \frac{1}{2a\sqrt{\pi t}}\exp\left(-\frac{(x-\xi)^2}{4a^2 t}\right) , \quad t > 0 ,$$

$$E(x - \xi) = H(t)G(x, \xi, t) .$$

These relations show the dependence between the Green function and the fundamental solution of the heat conduction equation in an infinite homogeneous bar.

Consequently, the solution of the boundary value problem (3.105), (3.106) is given by the convolution with respect to $x \in \mathbb{R}$

$$u(x, t) = E_t(x) * \varphi(x) = \int_{-\infty}^{\infty} E_t(x - \xi)\varphi(\xi)\mathrm{d}\xi = \int_{-\infty}^{\infty} G(x, \xi, t)\varphi(\xi)\mathrm{d}\xi ,$$

where φ is considered to be a bounded function and where $\lim_{t\to+0} u(x, t) = \varphi(x)$ in all continuity points of the function φ.

All the results can be generalized to $\mathbb{R}^n$. Thus, the heat conduction equation in $\mathbb{R}^n$ is

$$\frac{\partial u(x,t)}{\partial t} = a^2 \Delta u(x,t)\,, \quad t > 0\,, \quad a > 0\,, \quad x \in \mathbb{R}^n\,, \tag{3.115}$$

where Δ represents the Laplace operator.

The initial condition is

$$u(x,0) = \varphi(x)\,, \quad \varphi \in C^0(\mathbb{R}^n)\,. \tag{3.116}$$

The Green function $G(x,\xi,t), x,\xi \in \mathbb{R}^n, t > 0$, for the boundary value problem (3.115), (3.116) represents the solution of (3.115) with the initial condition

$$G(x,\xi,t)|_{t=+0} = \delta(x-\xi)\,.$$

Let $E_t(x), x \in \mathbb{R}^n, t > 0$, be the solution of (3.115) with the initial condition

$$E_t(x)|_{t=+0} = \delta(x)\,. \tag{3.117}$$

Then, the Green function $G(x,\xi,t)$ is given by the relation

$$G(x,\xi,t) = E_t(x-\xi)\,.$$

The boundary value problem (3.115), (3.116) is

$$u(x,t) = E_t(x) * \varphi(x) = \int\limits_{\mathbb{R}^n} E_t(x-\xi)\varphi(\xi)\mathrm{d}\xi = \int\limits_{\mathbb{R}^n} G(x,\xi,t)\varphi(\xi)\mathrm{d}\xi\,, \tag{3.118}$$

where φ is considered a bounded function and where $\lim_{t\to+0} u(x,t) = \varphi(x)$ in all continuity points of the function φ.

We remark that the solution $E_t(x)$ of (3.115) with the condition (3.117) can be obtained applying the Fourier transform with respect to the variable $x \in \mathbb{R}^n$. Thus, applying the Fourier transform to the equation

$$\frac{\partial E_t(x)}{\partial t} = a^2 \Delta E_t(x)\,,$$

as well the condition (3.117), we obtain

$$\frac{\mathrm{d}}{\mathrm{d}t}\hat{E}_t(\alpha) + a^2\|\alpha\|^2 \hat{E}_t(\alpha) = 0\,, \quad \left.\frac{\mathrm{d}}{\mathrm{d}t}\hat{E}_t(\alpha)\right|_{t=+0} = 1\,, \tag{3.119}$$

where $\hat{E}_t(\alpha) = \mathrm{F}[E_t(x)](\alpha), \alpha \in \mathbb{R}^n$.

The solution of the problem (3.119) is the function

$$\hat{E}_t(\alpha) = \exp(-a^2\|\alpha\|^2 t)\,, \quad t > 0\,.$$

According to the formula

$$\mathrm{F}[\exp(-b^2\|x\|^2)] = \left(\frac{\sqrt{\pi}}{b}\right)^n \exp\left(-\frac{\|\alpha\|^2}{4b^2}\right)\,, \quad b > 0\,,$$

we obtain

$$E_t(x) = \mathrm{F}^{-1}[\hat{E}_t(\alpha)](x) = \mathrm{F}^{-1}[\exp(-a^2\|\alpha\|^2 t)]$$
$$= \frac{1}{(2a\sqrt{\pi t})^n} \exp\left(-\frac{\|x\|^2}{4a^2 t}\right) .$$

Hence, the Green function corresponding to the boundary value problem of the heat conduction in $\mathbb{R}^n$ is

$$G(x,\xi,t) = E_t(x-\xi) = \frac{1}{(2a\sqrt{\pi t})^n} \exp\left(-\frac{\|x-\xi\|^2}{4a^2 t}\right) . \tag{3.120}$$

The formulae (3.118), (3.120) generalize those obtained for the case $n = 1$.

Let $E(x,t) \in \mathcal{D}'(\mathbb{R}^{n+1})$ be the fundamental solution of (3.115), hence of the operator $\partial/\partial t - a^2\Delta$ of heat conduction in $\mathbb{R}^n$. Proceeding as in the case $n = 1$, we can establish the relation $E(x,t) = H(t)\,E_t(x)$.

Thus, we can write

$$\frac{\partial E(x,t)}{\partial t} - a^2\Delta E(x,t) = \delta(x,t) = \delta(x) \times \delta(t) .$$

3.9.1.1 Generalized Poisson Equation

The generalized Poisson equation is

$$\Delta u(x,y,z) - k^2 u(x,y,z) = f(x,y,z) , \tag{3.121}$$

where Δ is the Laplace operator, $k = \text{const}$, and f is a distribution from $\mathcal{D}'(\mathbb{R}^3)$.

For $k = 0$ we obtain the Poisson equation

$$\Delta u(x,y,z) = f(x,y,z) . \tag{3.122}$$

By definition, the Green function corresponding to (3.121) represents the solution of the equation

$$\Delta u(x,y,z) - k^2 u(x,y,z) = \delta(x-\xi, y-\eta, z-\tau) ,$$

where ξ, η, τ are parameters.

Let $E(x,y,z) \in \mathcal{D}'(\mathbb{R}^3)$ be the fundamental solution of the operator $\Delta - k^2$; hence, we have

$$\Delta E(x,y,z) - k^2 E(x,y,z) = \delta(x,y,z) . \tag{3.123}$$

The Green function $G(x,y,z;\xi,\eta,\tau)$ satisfies the equation

$$\Delta G(x,y,z;\xi,\eta,\tau) - k^2 G(x,y,z;\xi,\eta,\tau) = \delta(x-\xi, y-\eta, z-\tau)$$

and, consequently, the dependence between G and E is

$$G(x,y,z;\xi,\eta,\tau) = E(x-\xi, y-\eta, z-\tau) . \tag{3.124}$$

This relation shows that the Green function is obtained by a translation of the distribution $E(x, y, z)$ to the point (ξ, η, τ).

Applying the Fourier transform to (3.123) in $\mathbb{R}^3$, we obtain

$$-(\alpha^2 + \beta^2 + \gamma^2)\hat{E}(\alpha, \beta, \gamma) - k^2 \hat{E}(\alpha, \beta, \gamma) = 1 \tag{3.125}$$

where $\mathrm{F}[E(x, y, z)] = \hat{E}(\alpha, \beta, \gamma)$ and α, β, γ are real variables.

From (3.125) it results

$$\hat{E}(\alpha, \beta, \gamma) = -\frac{1}{k^2 + \alpha^2 + \beta^2 + \gamma^2} . \tag{3.126}$$

We shall denote by $\mathrm{F}_x[]$, $\mathrm{F}_y[]$, $\mathrm{F}_z[]$ the Fourier transforms in $\mathbb{R}$, with respect to the variables x, y, z, respectively; we have

$$\mathrm{F}_x\left[\frac{1}{r}\exp(-kr)\right] = 2K_0\left(\sqrt{\alpha^2 + k^2}\sqrt{y^2 + z^2}\right) ,$$

where K_0 represents the modified Bessel function of second kind and zero order and $r = \sqrt{x^2 + y^2 + z^2}$ is the radius vector.

Also, we can write

$$\mathrm{F}_y\left[K_0\left(\sqrt{\alpha^2 + k^2}\sqrt{y^2 + z^2}\right)\right] = \frac{\pi \exp\left(-|z|\sqrt{\alpha^2 + \beta^2 + k^2}\right)}{\sqrt{\alpha^2 + \beta^2 + k^2}} ,$$

$$\mathrm{F}_z\left[\frac{\exp\left(-|z|\sqrt{\alpha^2 + \beta^2 + k^2}\right)}{\sqrt{\alpha^2 + \beta^2 + k^2}}\right] = \frac{2}{\alpha^2 + \beta^2 + \gamma^2 + k^2} .$$

Because $\mathrm{F}[]$ represents the Fourier transform in $\mathbb{R}^3$, with respect to all variables, we have

$$\mathrm{F}\left[\frac{1}{r}\exp(-kr)\right] = \frac{4\pi}{\alpha^2 + \beta^2 + \gamma^2 + k^2} ,$$

where

$$\mathrm{F}^{-1}\left[\frac{1}{\alpha^2 + \beta^2 + \gamma^2 + k^2}\right] = \frac{1}{4\pi r}\exp(-kr) ,$$

r being the radius vector.

Taking into account (3.126), we obtain

$$E(x, y, z) = \mathrm{F}^{-1}\left[\hat{E}(\alpha, \beta, \gamma)\right] = -\frac{1}{4\pi r}\exp(-kr) ; \tag{3.127}$$

in accordance with the relation (3.124), the Green function corresponding to the generalized Poisson equation reads as

$$G(x, y, z; \xi, \eta, \tau) = -\frac{1}{4\pi\rho}\exp(-k\rho) , \tag{3.128}$$

where $\rho = \sqrt{(x - \xi)^2 + (y - \eta)^2 + (z - \tau)^2}$.

Particularly, for $k = 0$, we obtain the Green function corresponding to (3.122) in the form

$$G(x, y, z; \xi, \eta, \tau) = E_1(x - \xi, y - \eta, z - \tau) = -\frac{1}{4\pi\rho} ,$$

where $E_1(x, y, z) = -1/(4\pi r)$ represents the fundamental solution of the Laplace operator Δ in $\mathbb{R}^3$, hence $\Delta E_1(x, y, z) = \delta(x, y, z)$.

With the help of the Green function (3.128), the solution of the generalized Poisson equation is

$$u(x, y, z) = E(x, y, z) * f(x, y, z) , \tag{3.129}$$

where $f \in \mathcal{D}'(\mathbb{R}^3)$ is a distribution with compact support.

If f is a distribution of function type for which the convolution product $E(x, y, z)*$ $f(x, y, z)$ exists, then, developing the convolution product (3.129), we obtain

$$\begin{aligned} u(x, y, z) &= \int_{\mathbb{R}^3} E(x - \xi, y - \eta, z - \tau) f(\xi, \eta, \tau) \mathrm{d}\xi \mathrm{d}\eta \mathrm{d}\tau \\ &= \int_{\mathbb{R}^3} G(x, y, z; \xi, \eta, \tau) f(\xi, \eta, \tau) \mathrm{d}\xi \mathrm{d}\eta \mathrm{d}\tau = \mathcal{L}(f(x, y, z)) , \end{aligned} \tag{3.130}$$

where the kernel of the integral operator $\mathcal{L}$ is the Green function.

Remark 3.2 The fundamental solution $E(x, y, z) \in \mathcal{D}'(\mathbb{R}^3)$ of the operator $\Delta + k^2$ can be obtained using the differentiation rule of the homogeneous functions having at the origin, $x = 0$, a singular point. Then, according with [21, p. 126], the function $h(x) = \exp(\mathrm{i}kr)r^{2-n}$, $x \in \mathbb{R}^n - \{0\}$, $r = \|x\|$, $k \in \mathbb{C}$, $n \geq 3$, is locally integrable and the functions $x_j r^{1-n}$, $x_j r^{-n}$ are homogeneous functions. It is shown that

$$\Delta h(x) + k^2 h(x) = -\mathrm{i}k(n-3)\frac{h(x)}{r} - (n-2) S_1 \delta(x) ,$$

where $S_1 = 2\pi^{n/2}/\Gamma(n/2)$ represents the area of the unit sphere from $\mathbb{R}^n$.

Particularly, for $n = 3$, $S_1 = 4\pi$ and considering $k = \mathrm{i}k_1$, we obtain the relation

$$\Delta h_1(x, y, z) - k_1^2 h_1(x, y, z) = -4\pi\delta(x, y, z) , \tag{3.131}$$

where $h_1(x, y, z) = \exp(-k_1 r)/r$, $r = \sqrt{x^2 + y^2 + z^2}$.

From (3.131), it results

$$\Delta\left(-\frac{h_1}{4\pi}\right) - k_1^2\left(-\frac{h_1}{4\pi}\right) = \delta(x, y, z) ;$$

hence, the fundamental solution of the operator $\Delta - k_1^2$ is

$$E(x, y, z) = -\frac{h_1(x, y, z)}{4\pi} = -\frac{1}{4\pi r}\exp(-k_1 r) .$$

3.9.2
Green's Function for the Vibrating String

For small vibrations of a homogeneous string, in the absence of the external forces, the motion equation of hyperbolic type is

$$\frac{\partial^2 u(x,t)}{\partial t^2} - a^2 \frac{\partial^2 u(x,t)}{\partial x^2} = 0 \,, \tag{3.132}$$

where the constant a^2 has the expression $a^2 = T/\rho$, where T is the tension in the string and ρ is the density of the string per unit length.

In the case of an infinite string, the Cauchy problem consists in the determination of the function $u(x,t) \in C^2(\mathbb{R} \times \mathbb{R}_+)$ which satisfies (3.132), as well as the initial conditions

$$u(x,t)|_{t=0} = \varphi(x) \,, \quad \left.\frac{\partial u(x,t)}{\partial t}\right|_{t=0} = \psi(x) \,, \tag{3.133}$$

where $\varphi, \psi \in C^0(\mathbb{R})$.

We remark that the vibrating string (3.132) as well as the initial condition (3.133) can be considered in the distributions space.

The solution of the Cauchy problem will be thus the distribution $u(x,t) \in \mathcal{D}'(\mathbb{R})$ depending on the parameter $t \geq 0$; φ and ψ are distributions from $\mathcal{D}'(\mathbb{R})$.

The Green function corresponding to the Cauchy problem (3.132) and (3.133) for the vibrating string is $G(x,\xi,t) \in C^2(\mathbb{R} \times \mathbb{R}_+)$, $x \in \mathbb{R}$, $t \geq 0$, where $\xi \in \mathbb{R}$ is a parameter, satisfying the conditions

$$\frac{\partial^2 G}{\partial t^2} - \frac{\partial^2 G}{\partial x^2} = 0 \,, \quad x \in \mathbb{R} \,, \quad t > 0 \,,$$

$$G(x,\xi,t)|_{t=+0} = 0 \,, \quad \left.\frac{\partial}{\partial t} G(x,\xi,t)\right|_{t=+0} = \delta(x-\xi) \,.$$

Let $E_t(x) = E(x,t) \in C^2(\mathbb{R} \times \mathbb{R}_+)$ be the solution of (3.132), hence

$$\frac{\partial^2}{\partial t^2} E_t(x) - a^2 \frac{\partial^2}{\partial x^2} E_t(x) = 0 \,,$$

which satisfies the conditions

$$E_t(x)|_{t=+0} = 0 \,, \quad \left.\frac{\partial}{\partial t} E_t(x)\right|_{t=+0} = \delta(x) \,. \tag{3.134}$$

In this way, between the functions $G(x,\xi,t)$ and E_t the relation follows

$$g(x,\xi,t) = \mathcal{E}_t(x-\xi) \,. \tag{3.135}$$

With the help of Green's function, the solution of the Cauchy problem (3.132), (3.133) for the vibrating string can be written in the form

$$\begin{aligned} u(x,t) &= E_t(x) * \psi(x) + \frac{\partial}{\partial t}(E_t(x) * \varphi(x)) \\ &= \int\limits_{\mathbb{R}} E_t(x-\xi)\psi(\xi)\mathrm{d}\xi + \frac{\partial}{\partial t}\int\limits_{\mathbb{R}} E_t(x-\xi)\varphi(\xi)\mathrm{d}\xi \,. \end{aligned} \tag{3.136}$$

Taking into account (3.135), we have

$$u(x,t) = \int_{\mathbb{R}} G(x,\xi,t)\psi(\xi)\mathrm{d}\xi + \frac{\partial}{\partial t}\int_{\mathbb{R}} G(x,\xi,t)\varphi(\xi)\mathrm{d}\xi \,. \tag{3.137}$$

We shall denote by $P(\mathrm{D}) = P(\partial_t, \partial_x)$ the operator corresponding to the vibrating string, namely

$$P(\mathrm{D}) = \frac{\partial^2}{\partial t^2} - a^2\frac{\partial^2}{\partial x^2} \,. \tag{3.138}$$

Because

$$P(\mathrm{D})\,E_t(x) = \frac{\partial^2 E_t(x)}{\partial t^2} - a^2\frac{\partial^2 E_t(x)}{\partial x^2} = 0 \,,$$

on the basis of the differentiation rule of the convolution product we obtain

$$P(\mathrm{D})u(x,t) = P(\mathrm{D})\,E_t(x) * \psi(x) + \frac{\partial}{\partial t}[P(\mathrm{D})\,E_t(x) * \varphi(x)] = 0 \,.$$

It results thus that the two terms of the solution (3.136), that is, $E_t(x) * \psi(x)$ and $\partial\,(E_t(x) * \varphi(x))/\partial t$, are solutions of the vibrating string equation (3.132).

Taking into account the conditions (3.134), we obtain

$$\begin{aligned} u(x,t)|_{t=+0} &= \mathcal{E}_t(x)|_{t=+0} * \psi(x) + \left.\frac{\partial \mathcal{E}_t(x)}{\partial t}\right|_{t=+0} * \varphi(x) \\ &= 0 * \psi(x) + \delta(x) * \varphi(x) = \varphi(x) \,, \\ \left.\frac{\partial u(x,t)}{\partial t}\right|_{t=+0} &= \left.\frac{\partial E_t(x)}{\partial t}\right|_{t=+0} * \psi(x) + \left.\frac{\partial^2 E_t(x)}{\partial t^2}\right|_{t=+0} * \varphi(x) \\ &= \delta(x) * \psi(x) + 0 * \varphi(x) = \psi(x) \,, \end{aligned}$$

because

$$\left.\frac{\partial^2 E_t(x)}{\partial t^2}\right|_{t=+0} = 0 \,.$$

Indeed, from the equation

$$\frac{\partial^2 E_t(x)}{\partial t^2} - a^2\frac{\partial^2 E_t(x)}{\partial x^2} = 0 \,, \tag{3.139}$$

by passing to the limit, we obtain

$$\lim_{t\to+0}\frac{\partial^2 E_t(x)}{\partial t^2} = a^2\frac{\partial^2}{\partial x^2}\left[\lim_{t\to+0} E_t(x)\right] = a^2\frac{\partial^2}{\partial x^2}(0) = 0 \,.$$

We shall determine now the expression of the function $E_t(x)$, $x \in \mathbb{R}$, $t > 0$.

Applying the Fourier transform with respect to the variable $x \in \mathbb{R}$ to (3.139) and to the conditions (3.134), we obtain

$$\frac{d^2}{dt^2}\hat{E}_t(\alpha) + a^2\alpha^2\hat{E}_t(\alpha) = 0\,, \quad \frac{d}{dt}\hat{E}_t(\alpha)\bigg|_{t=+0} = 1\,, \quad \hat{E}_t(\alpha)|_{t=+0} = 0\,.$$

The solution of this Cauchy problem is

$$\hat{E}_t(\alpha) = \frac{\sin(a\alpha t)}{a\alpha}\,.$$

Applying the inverse Fourier transform, we obtain

$$E_t(x) = \mathrm{F}_x^{-1}[\hat{E}_t(\alpha)] = \begin{cases} \dfrac{1}{2a}\,, & |x| \le at\,, \\ 0\,, & |x| > at\,. \end{cases}$$

Indeed, we can write

$$\mathrm{F}_x[E_t(x)] = \int_{\mathbb{R}} e^{i\alpha x} E_t(x)dx = \frac{1}{2a}\int_{-at}^{at} e^{i\alpha x}dx = \frac{1}{2i\alpha x}e^{i\alpha x}|_{-at}^{at}$$
$$= \frac{1}{2i\alpha a}(e^{i\alpha t a} - e^{-i\alpha t a}) = \frac{1}{a\alpha}\sin(at\alpha)\,.$$

Taking into account (3.135), we shall obtain for Green's function corresponding to the Cauchy problem of the vibrating spring the expression

$$G(x,\xi,t) = \mathcal{E}_t(x-\xi) = \begin{cases} \dfrac{1}{2a}\,, & |x-\xi| \le at \\ 0\,, & |x-\xi| > at \end{cases}$$
$$= \begin{cases} \dfrac{1}{2a}\,, & \xi \in [x-at, x+at]\,, \\ 0\,, & \xi \notin [x-at, x+at]\,. \end{cases}$$

Thus, (3.137) becomes

$$u(x,t) = \frac{1}{2a}\int_{x-at}^{x+at} \psi(\xi)d\xi + \frac{1}{2a}\frac{\partial}{\partial t}\int_{x-at}^{x+at} \varphi(\xi)d\xi\,.$$

Applying the Leibniz formula of differentiation of integrals depending on a parameter we obtain

$$u(x,t) = \frac{\varphi(x+at)+\varphi(x-at)}{2} + \frac{1}{2a}\int_{x-at}^{x+at} \psi(\xi)d\xi\,,$$

which represents d'Alembert's formula.

We remark that the formula (3.136) occurs in the case in which (3.132) is considered in the distributions space $\mathcal{D}'(\mathbb{R})$, where $u(x,t)$ is a distribution from $\mathcal{D}'(\mathbb{R})$, $t > 0$ being the parameter and φ, ψ from the initial conditions (3.133) being distributions with compact support.

Let $E(x,t) \in \mathcal{D}'(\mathbb{R} \times \mathbb{R})$ be the fundamental solution of the operator (3.138), hence

$$\frac{\partial^2 E(x,t)}{\partial t^2} - a^2 \frac{\partial^2 E(x,t)}{\partial x^2} = \delta(x,t) = \delta(x) \times \delta(t) .$$

The expression of $E(x,t)$ is

$$E(x,t) = \frac{1}{2a} H(at - |x|) = \begin{cases} 0\,, & t < 0\,, \\ \dfrac{1}{2a}\,, & 0 \le |x| \le at\,, \\ 0\,, & |x| > at \ge 0\,, \end{cases}$$

or

$$E(x,t) = H(t)\, E_t(x) . \tag{3.140}$$

Indeed, applying the Fourier transform with respect to the variable $x \in \mathbb{R}$, we obtain

$$\frac{\mathrm{d}^2}{\mathrm{d}t^2} \hat{E}(\alpha,t) + a^2\alpha^2 \hat{E}(\alpha,t) = \delta(t) ,$$

where $\hat{E}(\alpha,t) = \mathrm{F}_x[E(x,t)]$.

This relation shows that the distribution $\hat{E}(\alpha,t)$ is the fundamental solution of the operator $\mathrm{d}^2/\mathrm{d}t^2 + a^2\alpha^2$.

But the fundamental solution of this operator is the distribution

$$\hat{E}(\alpha,t) = H(t) \frac{\sin(a\alpha t)}{a\alpha} .$$

Applying the inverse Fourier transform we get

$$E(x,t) = \mathrm{F}_x^{-1}[\hat{E}(\alpha,t)] = \frac{1}{2a} H\left(at - |x|\right) ,$$

because

$$\mathrm{F}[H(R - |x|)] = \int_{-R}^{R} \mathrm{e}^{\mathrm{i}\alpha x} \mathrm{d}x = 2\frac{\sin(R\alpha)}{\alpha} , \qquad H(R - |x|) = \begin{cases} 0\,, & R < |x|\,, \\ 1\,, & |x| \le R\,. \end{cases}$$

From (3.140) and (3.135), it results

$$E(x - \xi, t) = H(t)\mathcal{E}_t(x - \xi) = H(t) G(x, \xi, t) , \tag{3.141}$$

which shows the dependence between the fundamental solution $E(x,t)$ of the vibrating spring equation (3.132) and Green's function corresponding to the Cauchy problem (3.132), (3.133).

4
Representation in Distributions of Mechanical and Physical Quantities

4.1
Representation of Concentrated Forces

In mechanics, the force is a fundamental concept which is modeled mathematically by a vector characterized by following elements: direction, sense, magnitude and point of application [32].

In deformable solids mechanics, the force is considered a bound vector, unlike in the rigid solids mechanics, where the force is modeled by a sliding vector. Geometrically, the bound vectors are represented by an oriented segment applied at a point. Because the transmission area of the action of a force may be considered theoretically a point, the point of application of the force leads to a singularity that will be represented mathematically using Dirac delta distributions.

Let $\boldsymbol{F}(F_x, F_y)$ be a force reported to the orthogonal frame Oxy and applied at the point O (Figure 4.1), where $\boldsymbol{i}$ and $\boldsymbol{j}$ are the unit vectors of the coordinate axes.

We assign to the force $\boldsymbol{F}$, which acts on the axis Ox at the point O, the force field $\boldsymbol{Q}_a(x) = \boldsymbol{F} q_a(x), x \in \mathbb{R}, a > 0$, which act on the segment $[-a, a]$ (Figure 4.2).

The intensity of the force field $\boldsymbol{Q}_a(x)$ per unit length is $Q_a(x) = F q_a(x)$, $F = |\boldsymbol{F}|$, where the repartition function $q_a(x)$, $x \in \mathbb{R}$ (Figure 4.3), of this field has the

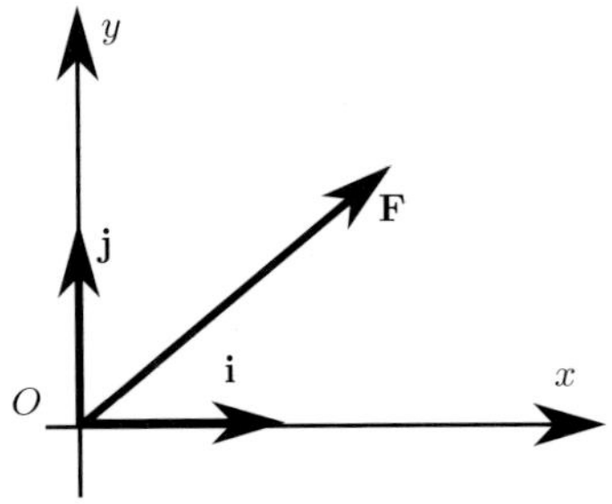

Figure 4.1

Distribution Theory, First Edition. Petre P. Teodorescu, Wilhelm W. Kecs, and Antonela Toma

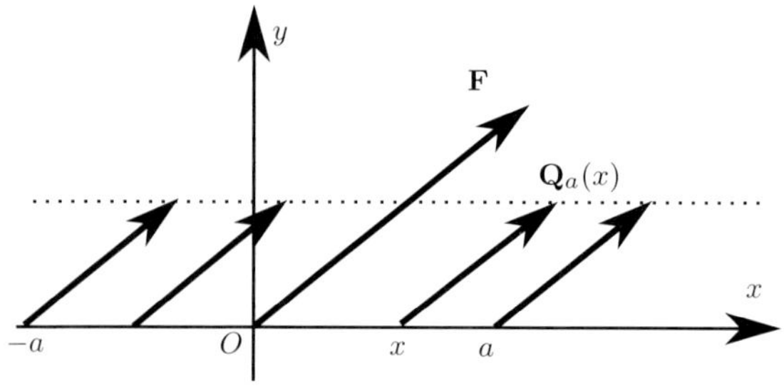

Figure 4.2

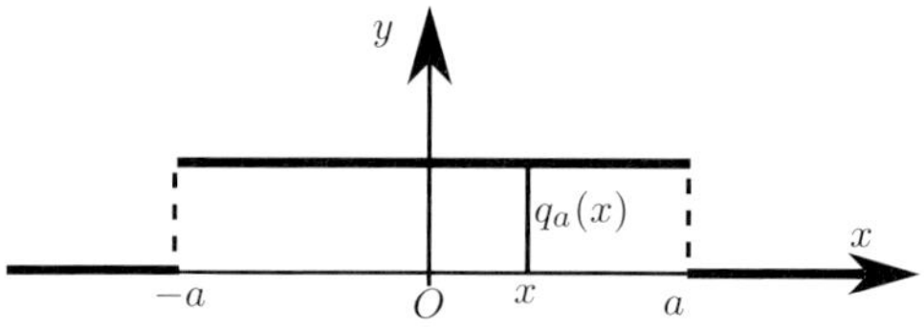

Figure 4.3

expression

$$q_a(x) = \begin{cases} \dfrac{1}{2a}\,, & x \in [-a, a]\,, \\ 0\,, & x \notin [-a, a]\,. \end{cases} \tag{4.1}$$

We note that on the interval $[-a, a]$ the force field $\boldsymbol{Q}_a(x)$, $x \in \mathbb{R}$, has a uniform distribution and the resulting vector is $\boldsymbol{F}$.

Indeed, we have

$$\boldsymbol{R}_a(x) = \int\limits_{\mathbb{R}} \boldsymbol{Q}_a(x)\mathrm{d}x = \boldsymbol{F}\int\limits_{\mathbb{R}} q_a(x)\mathrm{d}x = \boldsymbol{F}\int\limits_{-a}^{a} \frac{\mathrm{d}x}{2a} = \boldsymbol{F}\,.$$

The resultant moment $\boldsymbol{M}_a$ with respect to the point O of the force field $\boldsymbol{Q}_a$, has the expression

$$\boldsymbol{M}_a = \int\limits_{\mathbb{R}} (x\boldsymbol{i} \times \boldsymbol{Q}_a(x))\mathrm{d}x = (\boldsymbol{i} \times \boldsymbol{F})\int\limits_{\mathbb{R}} x q_a(x)\mathrm{d}x = \frac{1}{2a}(\boldsymbol{i} \times \boldsymbol{F})\int\limits_{-a}^{a} x\mathrm{d}x = 0\,.$$

Consequently, the resulting vector $\boldsymbol{R}_a$ of the force field $\boldsymbol{Q}_a$ is $\boldsymbol{F}$ and is not dependent on $a > 0$, while the resultant moment $\boldsymbol{M}_a$ vanishes. It follows that the action of the force $\boldsymbol{F}$ applied at the point O can be approximated by the action of the force field $\boldsymbol{Q}_a(x) = \boldsymbol{F}q_a(x)$, $x \in \mathbb{R}$, $a > 0$, uniformly distributed on the interval $[-a, a]$. The smaller the length $2a > 0$, the better the approximation in terms of mechanical effect will be. Because, for $a \to +0$ the quantity $1/2a$ tends to infinity in the usual sense, it means that $\lim_{a\to+0} \mathbf{Q}_a(x) = \mathbf{F} \lim_{a\to+0} q_a(x)$ does not exist, which leads us to consider the limit in the sense of the distribution theory.

On the other hand, we see that the distribution function $q_a(x)$, $x \in \mathbb{R}$ of the force field $\mathbf{Q}_a = \mathbf{F} q_a$ is a Dirac sequence, that is, we have

$$\lim_{a\to+0} q_a(x) = \delta(x) . \tag{4.2}$$

Taking into account (4.2) we obtain

$$\mathbf{Q}(x) = \lim_{a\to+0} \mathbf{Q}_a(x) = \mathbf{F} \lim_{a\to+0} q_a(x) = \mathbf{F}\delta(x) , \tag{4.3}$$

where obviously, the limit is considered in the distributions space $\mathcal{D}'(\mathbb{R})$.

Therefore, the distribution $\mathbf{Q}(x) = \mathbf{F}\delta(x)$ is the mathematical representation of the force $\mathbf{F} = F_x \mathbf{i} + F_y \mathbf{j}$ applied at the origin.

We can say that the vectorial repartition $\mathbf{Q}(x)$ is the vector field equivalent to the force $\mathbf{F}$ which acts at the point O and can be interpreted as a linear density, hence, per unit length of the force $\mathbf{F}$.

If force $\mathbf{F} = F_x \mathbf{i} + F_y \mathbf{j}$ is applied at a point $x_0 \in \mathbb{R}$, then the vector field $\mathbf{Q}(x)$, equivalent to the action of the force $\mathbf{F}$, is expressed by

$$\mathbf{Q}(x) = \mathbf{F}\delta(x - x_0) = F_x \delta(x - x_0)\mathbf{i} + F_y \delta(x - x_0)\mathbf{j} . \tag{4.4}$$

The force field $\mathbf{Q}_a(x) = \mathbf{F} q_a(x)$, $x \in \mathbb{R}$ has the dimension of a force per length, so that its dimensional equation is $[\mathbf{Q}_a] = \mathrm{MT}^{-2}$, where M is mass and T is time.

Because

$$\mathbf{Q}(x) = \mathbf{F}\delta(x) = \lim_{a\to+0} \mathbf{Q}_a(x) ,$$

it follows that $\mathbf{F}\delta(x)$ has the same dimensional equation as $\mathbf{Q}_a(x)$, that is,

$$[\mathbf{Q}] = [\mathbf{F}\delta(x)] = [\mathbf{Q}_a] = \mathrm{MT}^{-2} .$$

Taking into account that $\mathbf{F}$ has the dimension of a force, that is, $[\mathbf{F}] = \mathrm{MLT}^{-2}$, where L is the length, it follows from the previous relation that the Dirac delta distribution $\delta(x)$ must be interpreted as a physical quantity of dimension L^{-1}, that is,

$$[\delta(x)] = \mathrm{L}^{-1} .$$

On the basis of the above relation, it follows that the dimensional equation of the distribution $\delta'(x) \in \mathcal{D}'(\mathbb{R})$ is L^{-2}, that is,

$$[\delta'(x)] = \mathrm{L}^{-2} . \tag{4.5}$$

Indeed, for $\delta'(x)$ we can write

$$\delta'(x) = \lim_{h\to 0} \frac{\delta(x+h) - \delta(x)}{h} .$$

Because $[\delta(x+h) - \delta(x)] = \mathrm{L}^{-1}$ and $[h] = \mathrm{L}$, from the foregoing relation it thus results in $[\delta'] = \mathrm{L}^{-1}/\mathrm{L} = \mathrm{L}^{-2}$, that is (4.5).

We note that the repartition function $q_a(x)$ given by (4.1) and representing a Dirac sequence, has an auxiliary role in deriving the expression (4.3). Instead of it, any family of functions which form a Dirac sequence can be considered.

Proceeding similarly and using Dirac sequences of two and three variables, we obtain a representation in distributions of a concentrated force $\boldsymbol{F}(F_x, F_y, F_z)$ which is applied at a point A of $\mathbb{R}^2$ or $\mathbb{R}^3$, respectively.

Let $\boldsymbol{F}(F_x, F_y, F_z)$ be a force reported to the orthogonal frame $Oxyz$, acting on the Oxy-plane at the point O (Figure 4.4), where $\boldsymbol{i}, \boldsymbol{j}, \boldsymbol{k}$ are the unit vectors of the coordinate axes.

Then, the vector field $\boldsymbol{Q}(x, y)$, equivalent to the force $\boldsymbol{F} = F_x\boldsymbol{i} + F_y\boldsymbol{j} + F_z\boldsymbol{k}$, has the expression

$$\boldsymbol{Q}(x, y) = \boldsymbol{F}\delta(x, y) . \tag{4.6}$$

This field may be interpreted as the limit in the distributions space $\mathcal{D}'(\mathbb{R}^2)$ of the force field $\boldsymbol{Q}_a(x, y) = \boldsymbol{F} q_a(x, y), (x, y) \in \mathbb{R}^2$, for $a \to +0$ that acts on the square $[-a, a] \times [-a, a] \subset \mathbb{R}^2$ (Figure 4.5), where $q_a(x, y), (x, y) \in \mathbb{R}^2$, is the repartition function of the field $\boldsymbol{Q}_a$ and has the expression

$$q_a(x, y) = \begin{cases} \dfrac{1}{a^2}, & (x, y) \in [-a, a] \times [-a, a] , \quad a > 0 , \\ 0 , & \text{otherwise} . \end{cases} \tag{4.7}$$

The family of functions $\{q_a(x, y)\}$, $a > 0$, $q_a \in L^1_{\mathrm{loc}}(\mathbb{R}^2)$, is a Dirac sequence, because

$$\lim_{a \to +0} q_a(x, y) = \delta(x, y) . \tag{4.8}$$

Consequently, because $\boldsymbol{Q}_a(x, y) = \boldsymbol{F} q_a(x, y)$ we obtain

$$\boldsymbol{Q}(x, y) = \lim_{a \to +0} \boldsymbol{Q}_a(x, y) = \boldsymbol{F} \lim_{a \to +0} q_a(x, y) = \boldsymbol{F}\delta(x, y) ,$$

hence, the expression (4.6).

Obviously, instead of the partition function given by (4.7), any other Dirac sequence can be considered, because it does not interfere with the final result (4.6) by its concrete form, but only through its property (4.8).

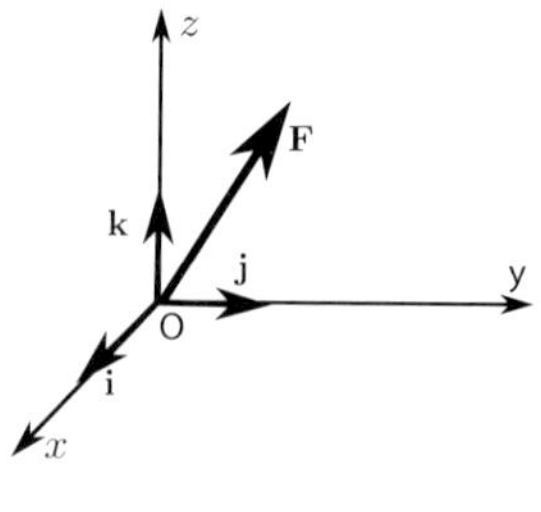

Figure 4.4

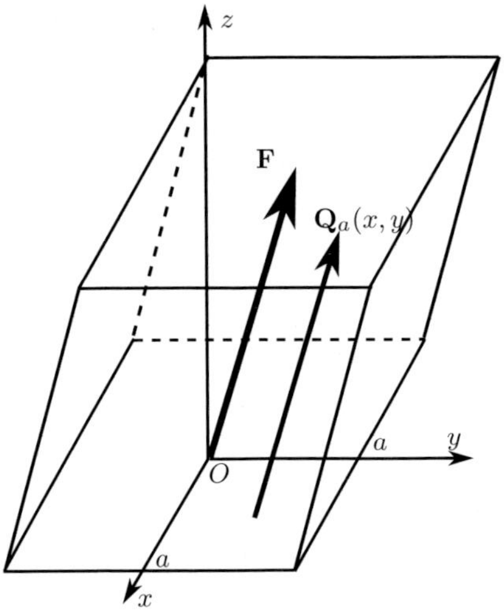

Figure 4.5

We note that the force field intensity $Q_a(x, y)$, $(x, y) \in \mathbb{R}^2$, per unit area is $Q_a(x, y) = F q_a(x, y)$, $F = |\boldsymbol{F}|$.

For this reason, the vector field $\boldsymbol{Q}(x, y) = \boldsymbol{F}\delta(x, y)$ can be interpreted as the planar density of the force $\boldsymbol{F}$ applied at the point O.

Indeed, because $[\delta(x)] = \mathrm{L}^{-1}$ we have

$$[\boldsymbol{Q}(x, y)] = [\boldsymbol{F}\delta(x) \times \delta(y)] = [\boldsymbol{F}][\delta(x)][\delta(y)] = \mathrm{MLT}^{-2}\mathrm{L}^{-2} = \mathrm{ML}^{-1}\mathrm{T}^{-2} .$$

If the force $\boldsymbol{F}(F_x, F_y, F_z)$ is applied at the point $A(x_0, y_0) \in \mathbb{R}^2$ in the Oxy-plane, then the equivalent vector field represents the vectorial distribution $\boldsymbol{Q}(x, y)$, which has the expression

$$\boldsymbol{Q}(x, y) = \boldsymbol{F}\delta(x - x_0, y - y_0) .$$

Let us now assume that the force $\boldsymbol{F}(F_x, F_y, F_z)$ is applied at the point $M_0(x_0, y_0, z_0) \in \mathbb{R}^3$. Then, the equivalent vector field $\boldsymbol{Q}(x, y, z)$ has the expression

$$\boldsymbol{Q}(x, y, z) = \boldsymbol{F}\delta(x - x_0, y - y_0, z - z_0)$$

and can be interpreted as the volume density (per unit volume) of the force $\boldsymbol{F}$ applied at M_0.

Indeed, we have

$$[\boldsymbol{Q}(x, y, z)] = [\boldsymbol{F}\delta(x-x_0)\times\delta(y-y_0)\times\delta(z-z_0)] = \mathrm{MLT}^{-2}\mathrm{L}^{-3} = \mathrm{ML}^{-2}\mathrm{T}^{-2} ,$$

because $[\delta(x - x_0)] = \mathrm{L}^{-1}$.

4.2 Representation of Concentrated Moments

Another important notion in mechanics is that of a couple, which means the set $(\boldsymbol{F}, -\boldsymbol{F})$ of two parallel forces, equal in magnitude and of opposite directions, whose lines of action do not coincide (Figure 4.6), [13, 16, 17].

The two forces $\boldsymbol{F}$ and $-\boldsymbol{F}$ that make up the couple act at the points A and B, respectively. The arm of the couple $(\boldsymbol{F}, -\boldsymbol{F})$ is the distance d between the action lines of the two forces, and the magnitude of the couple moment is $M = Fd$, where $F = |\boldsymbol{F}|$.

Because in rigid solids mechanics the forces are considered to be sliding vectors, the couple $(\boldsymbol{F}, -\boldsymbol{F})$ is represented by a free vector $\boldsymbol{M}$ (Figure 4.6), called the moment of the couple and whose magnitude is $|\boldsymbol{M}| = M = Fd$ and is perpendicular to the (P)-plane, determined by the parallel forces $\boldsymbol{F}$ and $-\boldsymbol{F}$.

The sense of the moment is indicated by the right-hand rule.

Consequently, in rigid solids mechanics, a couple $(\boldsymbol{F}, -\boldsymbol{F})$ is fully characterized by its moment $\boldsymbol{M}$, which is a free vector.

In deformable solids mechanics the forces are modeled by bounded vectors, hence the action of the couple $(\boldsymbol{F}, -\boldsymbol{F})$ depends on the points of application of the component forces, as well as on the direction of the parallel forces.

As a result, in the deformable solids mechanics, a couple $(\boldsymbol{F}, -\boldsymbol{F})$ cannot be characterized only by the moment $\boldsymbol{M}$ as in the rigid solids mechanics. These considerations determine the introduction of the concept of directed concentrated moment (couple) that takes into account all the bounded forces.

Let $(\boldsymbol{F}, -\boldsymbol{F})$ be a couple, where $\boldsymbol{F}$ has the point of application at the variable point $A(\xi, \eta, \tau)$ (Figure 4.7) with the position vector $\boldsymbol{\rho} = \overrightarrow{OA}$ such that the unit vector $\boldsymbol{u} = \boldsymbol{\rho}/\|\boldsymbol{\rho}\|$ is constant.

The force $-\boldsymbol{F}$ acts in the origin O of the orthogonal frame $Oxyz$. We denote by $\boldsymbol{F}^0$ the unit vector of $\boldsymbol{F}$, which will be considered constant, and by d the arm of the couple $(\boldsymbol{F}, -\boldsymbol{F})$.

For the magnitude M of the moment of this couple we have the expression

$$M = Fd = Fh\cos\alpha = Fh\sin(\boldsymbol{u}, \boldsymbol{F}^0) = Fh|\boldsymbol{u} \times \boldsymbol{F}^0| , \tag{4.9}$$

where $h = |\boldsymbol{\rho}| = |\overrightarrow{OA}|$.

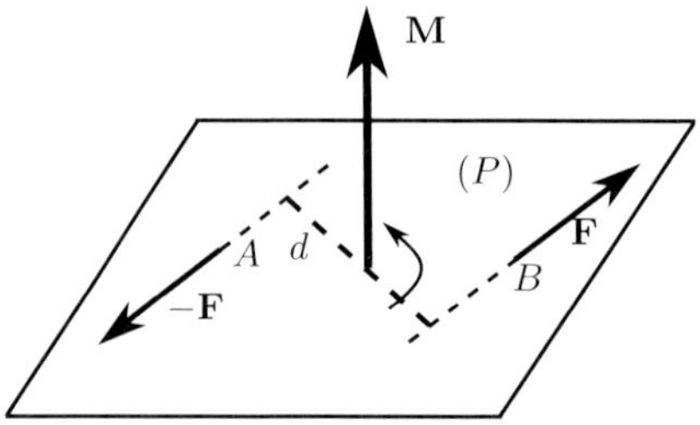

Figure 4.6

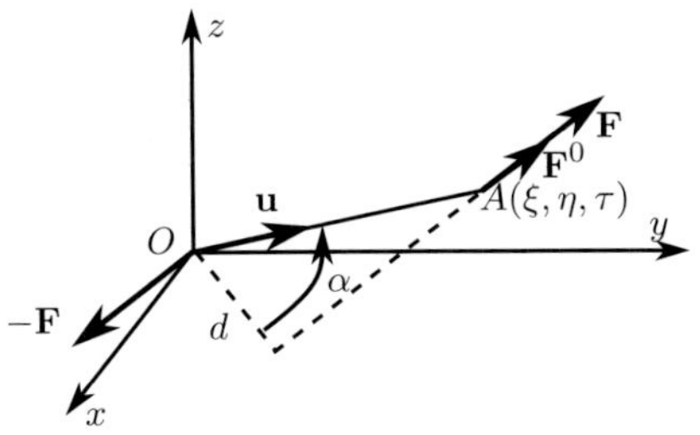

Figure 4.7

Definition 4.1 We call directed concentrated moment at the origin the limit, in the sense of distributions, of the set of concentrated forces $\boldsymbol{F}$ and $-\boldsymbol{F}$, when the arm of the couple d tends to zero. The point A is considered variable, the unit vectors $\boldsymbol{F}^0$ and $\boldsymbol{u}$ are constant and the magnitude $M d$ of the moment is constant.

To obtain the mathematical expression of the directed concentrated moment at the origin, we observe that the action of the two forces is equivalent to the vectors field action:

$$\begin{aligned}\boldsymbol{q}_d(x, y, z) &= \boldsymbol{F}\delta(x-\xi, y-\eta, z-\tau) - \boldsymbol{F}\delta(x, y, z)\\ &= F\boldsymbol{F}^0[\delta(\boldsymbol{r}-h\boldsymbol{u}) - \delta(\boldsymbol{r})]\,,\end{aligned}$$

where $h = |\overrightarrow{OA}|$, hence $h\boldsymbol{u}(\xi, \eta, \tau) = \overrightarrow{OA}$.

Taking into account (4.9) we obtain

$$\boldsymbol{q}_d(x, y, z) = \frac{M\boldsymbol{F}^0}{|\boldsymbol{u}\times\boldsymbol{F}^0|}\left(\frac{\delta(\boldsymbol{r}-h\boldsymbol{u}) - \delta(\boldsymbol{r})}{h}\right),$$

where $|\boldsymbol{u}\times\boldsymbol{F}^0| = \sin(\boldsymbol{u}, \boldsymbol{F}^0) \neq 0$.

Proposition 4.1 The vector distribution $\boldsymbol{q}(x, y, z) \in \mathcal{D}'(\mathbb{R}^3)$ corresponding to the directed concentrated moment $\lim_{d\to+0}\boldsymbol{q}_d$ is

$$\boldsymbol{q}(x, y, z) = \lim_{d\to+0}\boldsymbol{q}_d = -\frac{M\boldsymbol{F}^0}{|\boldsymbol{u}\times\boldsymbol{F}^0|}\frac{\partial}{\partial u}\delta(\boldsymbol{r})\,, \quad |\boldsymbol{u}\times\boldsymbol{F}^0| = \sin(\boldsymbol{u}, \boldsymbol{F}^0) \neq 0\,, \tag{4.10}$$

where $\partial\delta(\boldsymbol{r})/\partial u$ is the directional derivative of the Dirac delta distribution $\delta(x, y, z)$ in the direction of the unit vector $\boldsymbol{u}(l_1, l_2, l_3) = \overrightarrow{OA}/|\overrightarrow{OA}|$ and has the expression

$$\begin{aligned}\lim_{h\to+0}\frac{\delta(\boldsymbol{r}-h\boldsymbol{u}) - \delta(\boldsymbol{r})}{-h} &= \frac{\partial}{\partial u}\delta(\boldsymbol{r}) = l_1\frac{\partial\delta(x, y, z)}{\partial x}\\ &+ l_2\frac{\partial\delta(x, y, z)}{\partial y} + l_3\frac{\partial\delta(x, y, z)}{\partial z} = \boldsymbol{u}\cdot\mathrm{grad}\delta(x, y, z)\,.\end{aligned} \tag{4.11}$$

In particular, if the angle $(\boldsymbol{u}, \boldsymbol{F}^0) = \pi/2$, hence $|\boldsymbol{u}\times\boldsymbol{F}^0| = \sin(\boldsymbol{u}\times\boldsymbol{F}^0) = 1$, which results in

$$\boldsymbol{q}(x, y, z) = -M\boldsymbol{F}^0\frac{\partial}{\partial u}\delta(\boldsymbol{r})\,. \tag{4.12}$$

To determine the magnitude of $\boldsymbol{q}$ given by (4.10), we see that $\boldsymbol{F}^0$ and $|\boldsymbol{u} \times \boldsymbol{F}^0|$ are dimensionless quantities and the magnitude of the directed concentrated moment M has the dimension $[M] = \mathrm{ML^2T^{-2}}$.

Consequently, we have

$$[\boldsymbol{q}] = [M]\left[\frac{\partial}{\partial u}\delta(\boldsymbol{r})\right] = \mathrm{ML^2T^{-2} \cdot L^{-1} \cdot L^{-3} = ML^{-2}T^{-2}}\,,$$

that is, the vector distribution $\boldsymbol{q} \in \mathcal{D}'(\mathbb{R}^3)$ has the dimension of a force per volume and must be interpreted mechanically as a force density per unit volume.

It follows that the quantities

$$-\frac{M\boldsymbol{F}^0}{|\boldsymbol{u} \times \boldsymbol{F}^0|}\frac{\partial}{\partial u}\delta(\boldsymbol{r})$$

and $\boldsymbol{P}\delta(\boldsymbol{r})$, the directed concentrated moment and the force $\boldsymbol{P}$ applied at the origin, respectively, have the same dimensional equation (force per unit volume), which is why we can perform, from a mechanical point of view, the addition operation with these quantities.

If the directed concentrated moment is applied at a point different from the origin, namely $M_0(x_0, y_0, z_0)$, then the corresponding vector distribution has the expression

$$\boldsymbol{q}(x, y, z) = -\frac{M\boldsymbol{F}^0}{|\boldsymbol{u} \times \boldsymbol{F}^0|}\frac{\partial}{\partial u}\delta(\boldsymbol{r} - \boldsymbol{r}_0)\,, \quad |\boldsymbol{u} \times \boldsymbol{F}^0| = \sin(\boldsymbol{u}, \boldsymbol{F}^0) \neq 0\,, \qquad (4.13)$$

where $\boldsymbol{r}_0 = x_0\boldsymbol{i} + y_0\boldsymbol{j} + z_0\boldsymbol{k}$ is the position vector of point M_0, and $\boldsymbol{i}, \boldsymbol{j}, \boldsymbol{k}$ are the unit vectors of the coordinate axes.

From the expressions (4.13), (4.10) it follows that the directed concentrated moment is characterized by the following:

1. point of application $M_0(x_0, y_0, z_0)$;
2. the unit vector $\boldsymbol{F}^0$ of the forces which have generated this moment;
3. the unit vector $\boldsymbol{u}$ of the direction in which the limit is performed;
4. the magnitude M of the directed concentrated moment.

4.2.1
Concentrate Moment of Linear Dipole Type

We consider the case in which the action lines of the forces $\boldsymbol{F}$ and $-\boldsymbol{F}$ applied at A and B, respectively, coincide (Figure 4.8).

We note that in the case of the rigid solid the considered system of forces is equivalent to zero, but this does not take place for a deformable solid.

In this case, the formulas (4.10) and (4.12) are inapplicable, $(\boldsymbol{u} = \boldsymbol{F}^0)$, which will allow us to introduce a new type of directed concentrated moment, that of dipole type.

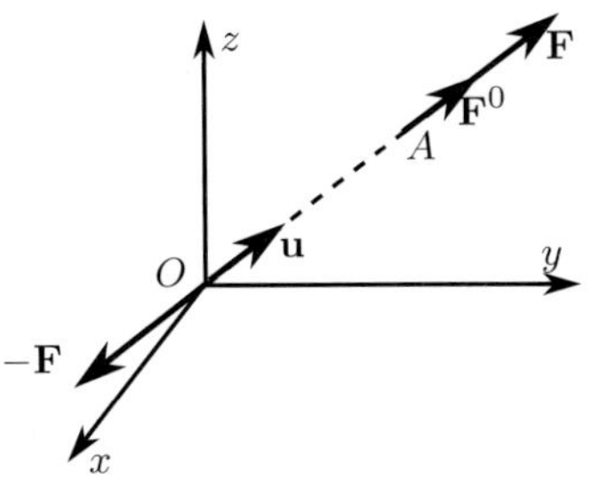

Figure 4.8

By definition, the magnitude of the dipole moment $(\boldsymbol{F}, -\boldsymbol{F})$ is considered $D = Fh$, where $F = |\boldsymbol{F}|$ and $h = |\overrightarrow{OA}|$.

Definition 4.2 We call the concentrated moment of linear dipole type at the origin, the limit, in the sense of distributions, of the set of forces $(\boldsymbol{F}, -\boldsymbol{F})$, if the distance $h \to 0$; the point A is considered variable, and the magnitude $D = Fh$ of the dipole moment is constant.

To obtain the mathematical expression of the concentrated moment of dipole type at the point O, we notice that the equivalent vector field with the action of the two concentrated forces $\boldsymbol{F}$ and $-\boldsymbol{F}$ is

$$\boldsymbol{q}_h(x, y, z) = \boldsymbol{F}\delta(x-\xi, y-\eta, z-\tau) - \boldsymbol{F}\delta(x, y, z) = F\boldsymbol{F}^0\left[\delta(\boldsymbol{r} - h\boldsymbol{u}) - \delta(\boldsymbol{r})\right] ,$$

where $h = |\overrightarrow{OA}|$ and $h\boldsymbol{u}(\xi, \eta, \tau) = \overrightarrow{OA}$.

Because $D = Fh$ is constant we can write

$$\boldsymbol{q}_h(x, y, z) = -D\boldsymbol{F}^0 \frac{\delta(\boldsymbol{r} - h\boldsymbol{u}) - \delta(\boldsymbol{r})}{-h} .$$

Taking into account (4.11), we obtain

$$\boldsymbol{q}(x, y, z) = \lim_{h \to +0} \boldsymbol{q}_h(x, y, z) = -D\boldsymbol{F}^0 \frac{\partial}{\partial u}\delta(\boldsymbol{r}) = -D\boldsymbol{F}^0(\boldsymbol{u}^0 \cdot \operatorname{grad}\delta(x, y, z)) .$$

The vector distribution $\boldsymbol{q} \in \mathcal{D}'(\mathbb{R}^3)$ is the mathematical expression of the concentrated moment of linear type at the origin.

If, instead of the origin, another point $M_0(x_0, y_0, z_0)$ is chosen, then the concentrated moment of dipole type applied at M_0 has the expression

$$\boldsymbol{q}(x, y, z) = -D\boldsymbol{F}^0 \frac{\partial}{\partial u}\delta(\boldsymbol{r} - \boldsymbol{r}_0) = -D\boldsymbol{F}^0(\boldsymbol{u} \cdot \operatorname{grad}\delta(x - x_0, y - y_0, z - z_0)) , \tag{4.14}$$

where $\boldsymbol{r}_0 = x_0\boldsymbol{i} + y_0\boldsymbol{j} + z_0\boldsymbol{k}$ is the position vector of the point M_0.

This moment of dipole type concentrated at the point M_0 and of magnitude D is called concentrated dipole of dilatation and it will be represented in Figure 4.9.

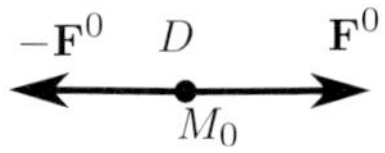

Figure 4.9

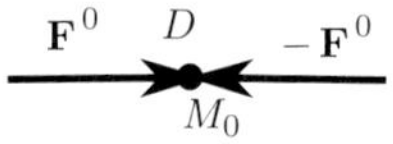

Figure 4.10

If we change the sense of the unit vector $\boldsymbol{F}^0$, then, taking into account (4.14), the expression of the moment of dipole type concentrated at the point M_0 and magnitude D is

$$q(x, y, z) = D\boldsymbol{F}^0 \frac{\partial}{\partial u} \delta(\boldsymbol{r} - \boldsymbol{r}_0) . \tag{4.15}$$

We say that the expression (4.14) corresponds to the concentrated dipole of contraction and will be represented in Figure 4.10.

4.2.2
Rotational Concentrated Moment (Center of Rotation)

Using the Dirac delta distribution $\delta_{C_r} \in \mathcal{D}'(\mathbb{R}^2)$ having as support the circle of radius r, centered at the origin of the Oxy-plane, we introduce the notion of rotational concentrated moment (or center of rotation), corresponding to the origin O and to the Oxy-plane.

Let $C_r = \{(x, y) \in \mathbb{R}^2, x^2 + y^2 - r^2 = 0\}$ be the circle with center at the origin and of radius r (Figure 4.11). We acknowledge that along the tangent to the circle acts a field of forces uniformly distributed with the intensity per unit length $p_r = M/2\pi r^2$ (Figure 4.11).

If we denote by $\boldsymbol{u}$ the unit vector tangent to the circle at the current point $A(x, y)$, oriented into the direct sense, then the vector field of forces per unit length has the expression

$$\boldsymbol{p}_r(x, y) = p_r \boldsymbol{u} = \frac{M}{2\pi r^2} \boldsymbol{u} .$$

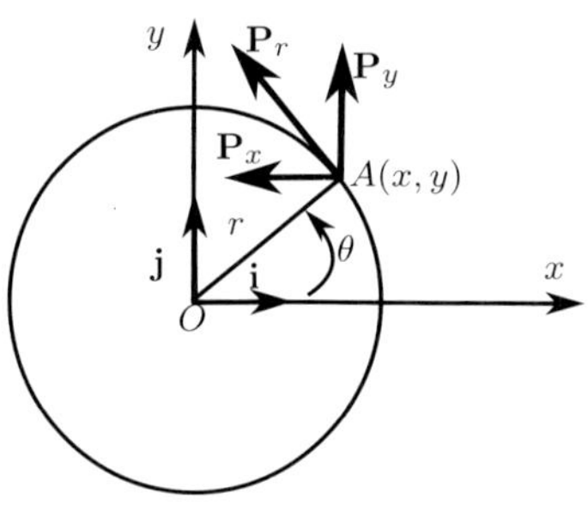

Figure 4.11

The components of field of forces $\boldsymbol{p}_r$ on the two coordinate axes are

$$\boldsymbol{p}_x = -\frac{M\boldsymbol{i}}{2\pi r^2}\sin\theta\,, \quad \boldsymbol{p}_y = \frac{M\boldsymbol{j}}{2\pi r^2}\cos\theta\,, \quad \theta \in [0, 2\pi]\,,$$

where $\boldsymbol{i}$, $\boldsymbol{j}$ are the unit vectors of the Ox-axis and Oy-axis, respectively.

On the basis of the definition of the Dirac delta distribution δ_{C_r}, concentrated on the circle $C_k \subset \mathbb{R}^2$, to the field of forces $\boldsymbol{p}_r$, uniformly distributed on the circle C_r, corresponds the vector distribution $\boldsymbol{q}_r(x, y) \in \mathcal{D}'(\mathbb{R}^2)$ with the expression

$$\boldsymbol{q}_r(x, y) = \boldsymbol{p}_r\delta_{C_r} = \frac{M}{2\pi r^2}\boldsymbol{u}\delta_{C_r}\,. \tag{4.16}$$

Therefore $\forall\varphi \in \mathcal{D}(\mathbb{R}^2)$ we have

$$(\boldsymbol{q}_r(x, y), \varphi(x, y)) = (\boldsymbol{p}_r\delta_{C_r}, \varphi) = \oint_{C_r} \boldsymbol{p}_r(x, y)\varphi(x, y)\mathrm{d}s\,,$$

where $\mathrm{d}s$ is the arc element corresponding to the circle C_r.

The resultant moment of the field of forces $\boldsymbol{p}_r$, per unit length, with respect to the origin O is

$$\boldsymbol{M}_0(\boldsymbol{p}_r) = \oint_{C_r} (\overrightarrow{OA} \times \boldsymbol{p}_r)\mathrm{d}s = \boldsymbol{k}\oint_{C_r} p_r r\mathrm{d}s = p_r r\boldsymbol{k}\oint_{C_r} \mathrm{d}s = \boldsymbol{k}M\,, \tag{4.17}$$

where $\boldsymbol{k}$ is the unit vector of the Oz-axis, hence $\boldsymbol{k} = \boldsymbol{i} \times \boldsymbol{j}$.

As regards the resultant vector $\boldsymbol{R}(\boldsymbol{p}_r)$ of this field of forces, it is equal to zero, hence $\boldsymbol{R}(\boldsymbol{p}_r) = 0$.

Indeed, we have

$$\boldsymbol{R}(\boldsymbol{p}_r) = \oint_{C_r} \boldsymbol{p}_r\mathrm{d}s = \int_0^{2\pi}\left(-\frac{M}{2\pi r^2}\sin\theta\,\boldsymbol{i} + \frac{M}{2\pi r^2}\cos\theta\,\boldsymbol{j}\right) r\mathrm{d}\theta = \boldsymbol{0}\,.$$

This result is obtained directly because $\boldsymbol{u} = \mathrm{d}\boldsymbol{\rho}/\mathrm{d}s$, $\boldsymbol{\rho} = \overrightarrow{OA}$, hence

$$\oint_{C_r} \boldsymbol{p}_r\mathrm{d}s = \frac{M}{2\pi r^2}\oint_{C_r} \boldsymbol{u}\mathrm{d}s = \frac{M}{2\pi r^2}\oint_{C_r} \mathrm{d}\boldsymbol{\rho} = \boldsymbol{0}.$$

Definition 4.3 We call rotational concentrated moment (or center of rotation), corresponding to the origin O and to the Oxy-plane, the limit, in the sense of distributions, of the vector field $\boldsymbol{q}_r$ given by (4.16) when $r \to 0$.

Proposition 4.2 The vector distribution $\boldsymbol{q}(x, y) \in \mathcal{D}'(\mathbb{R}^2)$, corresponding to the rotational concentrated moment, $\lim_{r\to+0} \boldsymbol{q}_r$ is

$$\boldsymbol{q}(x, y) = \lim_{r\to+0} \boldsymbol{q}_r(x, y) = -\frac{M}{2}\boldsymbol{k} \times \operatorname{grad}\delta(x, y)\,, \tag{4.18}$$

where $\boldsymbol{k} = \boldsymbol{i} \times \boldsymbol{j}$ is the unit vector of the Oz-axis.

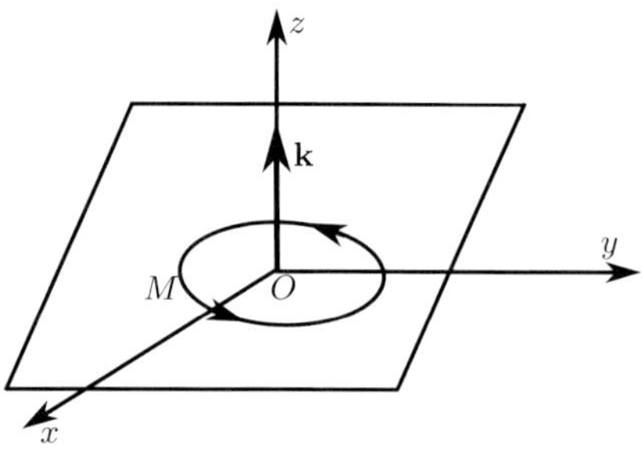

Figure 4.12

The vector distribution $\boldsymbol{q} \in \mathcal{D}'(\mathbb{R}^2)$ given by (4.18) for the rotational concentrated moment of magnitude M corresponding to the origin O and to the Oxy-plane, with normal unit vector $\boldsymbol{k}$, will be represented in Figure 4.12.

If the concentrated rotational moment is applied at a point $M_0(x_0, y_0)$, then the corresponding vector distribution has the expression

$$\boldsymbol{q}(x, y) = -\frac{M}{2}\boldsymbol{k} \times \operatorname{grad} \delta(x - x_0, y - y_0) .$$

4.2.3
Concentrated Moment of Plane Dipole Type (Center of Dilatation or Contraction)

This concentrated moment represents the correspondent of the concentrated moment of linear dipole type. As in the case of the rotational moment, we consider a plane (P) defined by the point $M_0(x_0, y_0, z_0) \in \mathbb{R}^3$ and by the normal unit vector $\boldsymbol{n}(\cos\alpha, \cos\beta, \cos\gamma)$.

Let $C_a \in (P)$ be a closed curve in the plane (P) depending on the parameter $a > 0$ and containing the point M_0 (Figure 4.13), such that, when $a \to +0$, it is continuously deformed to the point M_0, hence $\lim_{a \to +0} C_a = M_0$.

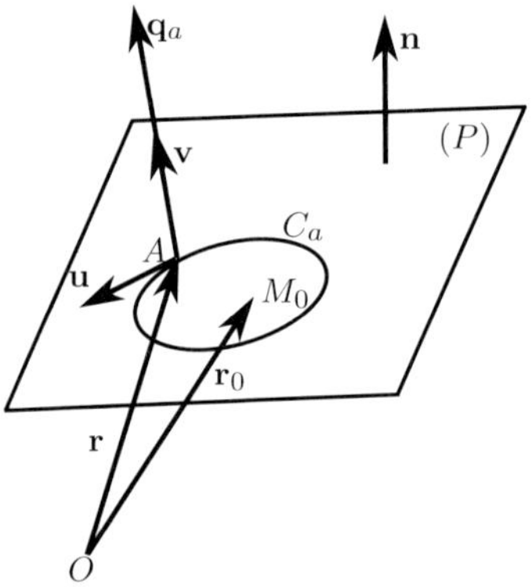

Figure 4.13

We shall denote by $\boldsymbol{u}$ the unit vector tangent to the curve at the point $A(x, y, z)$ and by $\boldsymbol{v}$ the unit vector normal to the curve at A in the plane (P). We acknowledge that, along the curve C_a, the normal $\boldsymbol{v}$ acts a field of forces $\boldsymbol{q}_a$, whose intensity per unit length of the arc of the curve C_a is

$$q_a(x, y, z) = |\boldsymbol{q}_a| = \frac{D}{2\omega_a},$$

where ω_a is the area of the domain D_a from the plane (P) having as border the curve C_a.

To the vector field $\boldsymbol{q}_a$, uniformly distributed along the curve C_a, corresponds the vector distribution $\boldsymbol{h}_a \in \mathcal{D}'(\mathbb{R}^3)$

$$\boldsymbol{h}_a(x, y, z) = \boldsymbol{q}_a \delta_{C_a} = \frac{-D}{2\omega_a}(\boldsymbol{n} \times \boldsymbol{u})\delta_{C_a}, \tag{4.19}$$

where δ_{C_a} is the Dirac delta distribution concentrated on the curve C_a.

Definition 4.4 We call concentrated moment of plane dipole type (center of plane dilatation) corresponding to the point $M_0(x_0, y_0, z_0) \in \mathbb{R}^3$ and to the oriented plane (P), defined by the point M_0 and by the normal unit vector $\boldsymbol{n}(\cos\alpha, \cos\beta, \cos\gamma)$, the limit, in the distributions sense, of the field vector given by (4.19) when $a \to +0$, $\lim_{a\to+0} C_a = M_0$, hence

$$\boldsymbol{h}(x, y, z) = \lim_{a\to+0} \boldsymbol{h}_a = \lim_{a\to+0}(q_a \delta_{C_a}).$$

Proposition 4.3 The vector distribution $\boldsymbol{h}_a \in \mathcal{D}'(\mathbb{R}^3)$ corresponding to the concentrated moment of planar dipole type $\lim_{a\to+0} \boldsymbol{h}_a$ is

$$\boldsymbol{h}(\boldsymbol{r}) = \frac{D}{2}\boldsymbol{n} \times [\boldsymbol{n} \times \operatorname{grad}\delta(\boldsymbol{r} - \boldsymbol{r}_0)]$$

where $\boldsymbol{r}_0 = x_0\boldsymbol{i} + y_0\boldsymbol{j} + z_0\boldsymbol{k}$ is the position vector of the point M_0 and $\boldsymbol{r} = x\boldsymbol{i} + y\boldsymbol{j} + z\boldsymbol{k}$ is the position vector of the current point $(x, y, z) \in \mathbb{R}^3$.

4.3
Representation in Distributions of the Shear Forces and the Bending Moments

The shear force and the bending moment are important quantities that occur in the study of bending and of transverse vibration of elastic and viscoelastic bars.

We consider an elastic straight bar of finite length $[a, b]$ and we take along Ox the bar axis and along Ov the deflection of it (Figure 4.14).

We acknowledge the Bernoulli–Euler hypothesis of plane cross-sections, according to which the cross-sections before bending remain cross-section after bending as well.

Following this hypothesis, the axis of the bar is its elastic line, because its points do not have longitudinal displacement, but only displacements along the Ov-axis.

We will also acknowledge that the bending of the bar of length $l = b - a$ is due to the action of distributed forces $q \in L^1_{\text{loc}}([a, b])$ per unit length and due to concentrated forces P_i and concentrated moments m_i, applied at the points $c_1 = a, c_2, c_3, \ldots, c_i, \ldots, c_{n-1}, c_n = b$. These forces act along the normal to the Ox-axis and in the Oxv-plane.

Concentrated forces act perpendicular to the Ox-axis and in the Oxv-plane; they are considered positive if they are oriented in the positive direction of the Ov-axis (Figure 4.14). The concentrated moments m_i are considered positive if they have the direct sense of rotation represented in Figure 4.14.

Given concentrated loads, which represent concentrated forces P_i and moments m_i, can act at the points $x = c_i$.

As noted previously, these concentrated loads, in terms of the distribution theory, will be represented by the distributions

$$q_{1i}(x) = P_i \delta(x - c_i), \quad x \in \mathbb{R}, \quad q_{2i}(x) = m_i \delta'(x - c_i), \quad x \in \mathbb{R}.$$

Corresponding to the concentrated force P_i and to the concentrated moment m_i, in the vector sense, are the vector distributions $\boldsymbol{q}_{1i}, \boldsymbol{q}_{2i} \in \mathcal{D}'(\mathbb{R})$, with the expressions

$$\boldsymbol{q}_{1i}(x) = \boldsymbol{j} P_i \delta(x - c_i), \boldsymbol{q}_{2i}(x) = \boldsymbol{j} m_i \delta'(x - c_i),$$

where $\boldsymbol{j}$ is the unit vector of Ov-axis.

These representations are consistent with the formulas (4.4) and (4.12), where $\boldsymbol{F}^0 = -\boldsymbol{j}$, and $\boldsymbol{u} = \boldsymbol{i}$ are the unit vectors of the Ox and Ov axes, respectively.

By definition, the shear force corresponding to the forces which act on the elastic bar $[a, b]$ means the function $T : \mathbb{R} \to \mathbb{R}$, the value $T(x)$ of which at a point $x \in \mathbb{R}$, is the resultant of the given and constraint forces on the interval (x, ∞).

Thus, for example, if the force P acts at the point $x = c \in [a, b]$ (Figure 4.15) then the shear force is the function $T : \mathbb{R} \to \mathbb{R}$, where

$$T(x) = \begin{cases} P, & x \in (-\infty, c), \\ 0, & \text{otherwise}. \end{cases} \tag{4.20}$$

The graph of this function is drawn in Figure 4.15.

The bending moment corresponding to forces which act on the bar $[a, b]$ is given by the function $M : \mathbb{R} \to \mathbb{R}$, where the value of the bending moment $M(x)$ at a point $x \in \mathbb{R}$ is the algebraic sum of the moments forces on the interval (x, ∞), with

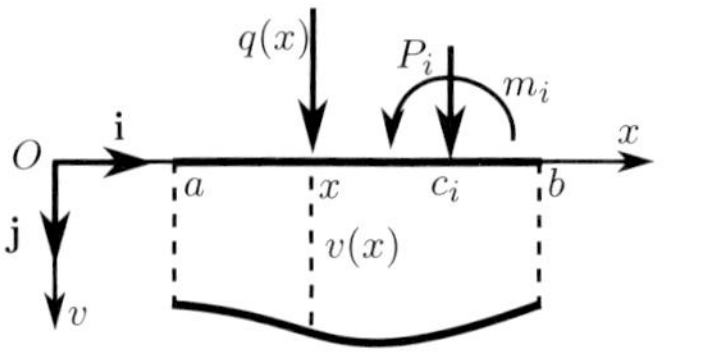

Figure 4.14

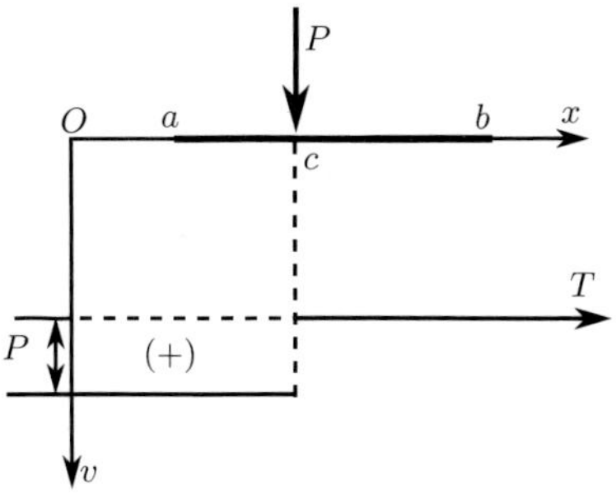

Figure 4.15

respect to $x \in \mathbb{R}$, and the concentrated moments on (x, ∞), taking into account the sign convention adopted for the moments.

Thus, for the force P acting at the point $x = c \in [a, b]$, the bending moment is the function $M : \mathbb{R} \to \mathbb{R}$, where

$$M(x) = \begin{cases} (x - c)P\,, & x \in (-\infty, c)\,, \\ 0\,, & \text{otherwise}\,. \end{cases} \tag{4.21}$$

The graph of this function is given in Figure 4.16. We note that on the interval $(-\infty, c)$ the bending moment $M(x)$ is negative due to the convention signs adopted. We note that the point of action of a concentrated force is a discontinuity point of the first order for the shear force T.

We shall generalize the concepts of shear force and bending moment for any load expressed by a distribution from $\mathcal{D}'(\mathbb{R})$ [14].

Definition 4.5 The application $T^* : \mathcal{E}'(\mathbb{R}) \to \mathcal{D}'(\mathbb{R})$ defined by formula

$$T^*(q(x)) = q(x) * (1 - H(x)) \tag{4.22}$$

will be referred to as the shear force operator, H is the Heaviside distribution and $*$ the convolution product.

The distribution $T(x) = T^*(q(x)) \in \mathcal{D}'(\mathbb{R})$ will be referred to as the generalized shear force corresponding to the load with compact support $q \in \mathcal{E}'(\mathbb{R})$, normal to the elastic line of the bar.

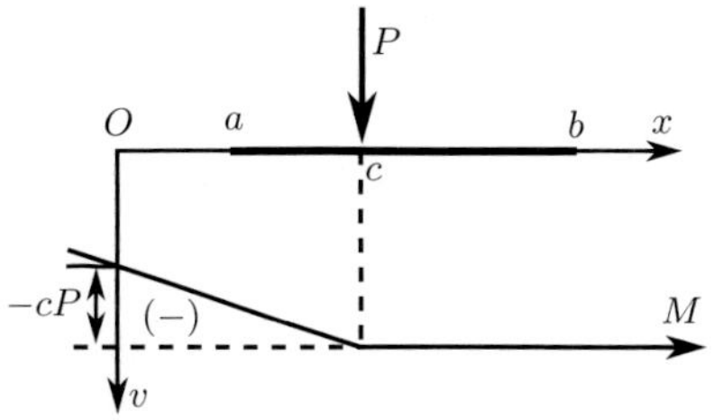

Figure 4.16

Definition 4.6 The application $M^* : \mathcal{E}'(\mathbb{R}) \to \mathcal{D}'(\mathbb{R})$ defined by the formula

$$M^*(q(x)) = q(x) * x(1 - H(x)) , \tag{4.23}$$

will be called the bending moment operator of the load $q \in \mathcal{E}'(\mathbb{R})$ and the distribution $M^*(q)$ will be called the generalized bending moment.

Due to the distributivity of the convolution product with respect to addition, it results that the operators T^* and M^* are linear operators, hence we can write

$$\begin{aligned} T^*(\alpha f + \beta g) &= \alpha T^*(f) + \beta T^*(g) , \\ M^*(\alpha f + \beta g) &= \alpha M^*(f) + \beta M^*(g) , \end{aligned} \qquad f, g \in E'(\mathbb{R}) , \quad \alpha, \beta \in \mathbb{R} .$$

Because the distributions $f, g \in \mathcal{E}'(\mathbb{R})$ have compact support, the operators T^* and M^* exist.

The dependence between the generalized shear force and the generalized bending moment is given by the following.

Proposition 4.4 For any distribution $q \in \mathcal{E}'(\mathbb{R})$ we have

$$T^*(q) = \frac{\mathrm{d}}{\mathrm{d}x} M^*(q) . \tag{4.24}$$

Proof: Taking into account the expression of $M^*(q)$, we obtain

$$\frac{\mathrm{d}}{\mathrm{d}x} M^*(q) = \frac{\mathrm{d}}{\mathrm{d}x} \left[q(x) * x\,(1 - H(x)) \right] = q(x) * \frac{\mathrm{d}}{\mathrm{d}x} \left[x\,(1 - H(x)) \right] .$$ □

Because $(\mathrm{d}/\mathrm{d}x)[x(1 - H(x))] = 1 - H(x) + x(-\delta(x))$ and $x\delta(x) = 0$, the previous relation becomes

$$\frac{\mathrm{d}}{\mathrm{d}x} M^*(q) = q(x) * (1 - H(x)) = T^*(q) .$$

The relation (4.24) is a generalization, in the distributions space $\mathcal{D}'(\mathbb{R})$, of the known dependence between the shear force and the bending moment as stated in the strength of materials.

The formula (4.24) allows us to express the dependence between the operators T^* and M^* under the form

$$T^* = \frac{\mathrm{d}}{\mathrm{d}x} M^* .$$

Further, we will show that the formulas (4.22) and (4.23) are a generalization of the concepts of shear force and bending moment. On the other hand, the linearity property of the operators T^* and M^* is consistent with the known properties of the shear force and the bending moment regarding the simultaneous action of several loads.

For this reason, it is sufficient to verify formulas (4.22) and (4.23) in the following cases: concentrated force, concentrated couple and distributed forces on an interval $[a_1, b_1] \subset [a, b]$.

4.3.1
Concentrated Force of Magnitude P Applied at the Point $c \in [a, b]$

To this force (Figure 4.16) corresponds the distribution $q \in \mathcal{D}'(\mathbb{R})$ having the expression

$$q(x) = P\delta(x - c) ,$$

because the force acts in the positive direction of the Ov-axis.

According to the formula (4.22), we obtain

$$T^*(q) = P\delta(x - c) * (1 - H(x)) = P(1 - H(x - c)) ,$$

namely

$$T^*(q) = \begin{cases} P , & x \in (-\infty, c) , \\ 0 , & \text{otherwise} . \end{cases}$$

Comparing the generalized shear force $T^*(q)$ with the shear force $T(x)$, $x \in \mathbb{R}$, given by (4.20), we see that they coincide, hence $T^*(q) = T(x)$.

Applying the formula (4.23), we obtain the generalized bending moment $M^*(q)$, namely

$$M^*(q) = P\delta(x - c) * [x(1 - H(x))] = P(x - c)[1 - H(x - c)] .$$

Specifying, we get

$$M^*(q) = \begin{cases} (x - c)P , & x \in (-\infty, c) , \\ 0 , & \text{otherwise} . \end{cases}$$

We note that $M^*(q)$ coincides with the bending moment $M(x)$, $x \in \mathbb{R}$, having the expression (4.21), which shows the correctness of the formulas (4.22) and (4.23) in the case of the concentrated forces.

4.3.2
Concentrated Moment of Magnitude m Applied at the Point $c \in [a, b]$

The concentrated moment is considered positive oriented (Figure 4.17) and in the distribution space $\mathcal{D}'(\mathbb{R})$ it has the expression

$$q(x) = m\delta'(x - c) = m\frac{\mathrm{d}}{\mathrm{d}x}\delta(x - c) .$$

For the generalized bending moment we obtain

$$\begin{aligned} M^*(q) &= m\delta'(x - c) * [x(1 - H(x))] \\ &= m(\delta(x - c) * [x(1 - H(x))]') = m[1 - H(x - c)] , \end{aligned}$$

because $x\delta(x) = 0$.

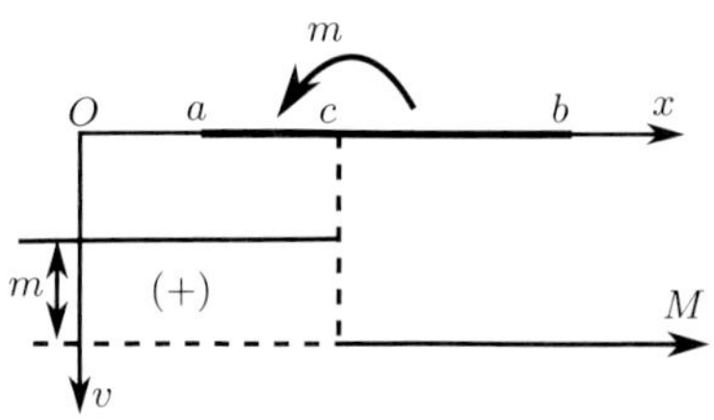

Figure 4.17

Consequently, we can write

$$M^*(q) = \begin{cases} m\,, & x \in (-\infty, c)\,, \\ 0\,, & \text{otherwise}\,, \end{cases}$$

which is identical to the bending moment $M(x), x \in \mathbb{R}$, whose graph is given in Figure 4.17.

For $q \in \mathcal{D}'(\mathbb{R})$ we can calculate the generalized shear force $T^*(q) \in \mathcal{D}'(\mathbb{R})$, but it has no concrete mechanical meaning.

Thus, we obtain $T^*(q) = m\delta'(x-c) * [1 - H(x)] = -m\delta(x-c)$, because $H'(x) = \delta(x)$.

It follows that formally the generalized shear force $T^*(q)$ can be represented as a concentrated force applied at the point $x = c$ and of intensity $-m$.

4.3.3 Distributed Forces of Intensity $q \in L^1_{\text{loc}}([a, b])$

We acknowledge that on the bar $[a, b]$ act the normal distributed forces whose intensity per unit length is expressed by a locally integrable function $q \in L^1_{\text{loc}}([a, b])$ (Figure 4.18).

We shall denote by $q_1 \in L^1_{\text{loc}}(\mathbb{R})$ the extension of the load q on $\mathbb{R}$ with null values, hence

$$q_1(x) = \begin{cases} q(x)\,, & x \in [a, b]\,, \\ 0\,, & x \notin [a, b]\,. \end{cases}$$

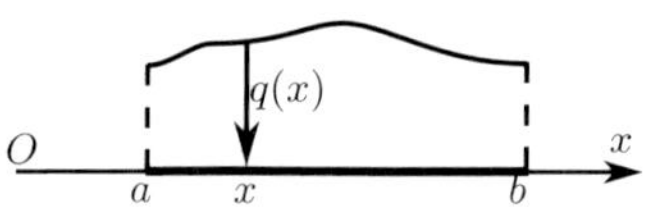

Figure 4.18

Consequently, the shear force corresponding to the load q_1 is the function $T : \mathbb{R} \to \mathbb{R}$, where

$$T(x) = \int_x^\infty q_1(\zeta)\mathrm{d}\zeta = \begin{cases} \int_a^b q(x)\mathrm{d}x\,, & x \le a\,, \\ \int_x^b q(x)\mathrm{d}x\,, & a \le x \le b\,, \\ 0\,, & x > b \end{cases} . \tag{4.25}$$

We note that in the strength of materials the shear force T is considered suitable only to the segment $[a, b]$.

According to the formula (4.22), the generalized shear force corresponding to the load $q \in \mathcal{D}'(\mathbb{R})$, which is a function type distribution, has the expression

$$T^*(q_1) = q_1(x) * (1 - H(x)) = q_1(x) * 1(x) - q_1(x) * H(x)\,.$$

Specifying, we obtain

$$\begin{aligned} T^*(q_1) &= \int_{-\infty}^{\infty} q_1(\zeta)1(x-\zeta)\mathrm{d}\zeta - \int_{-\infty}^{\infty} q_1(x-\zeta)H(\zeta)\mathrm{d}\zeta \\ &= \int_a^b q(\zeta)\mathrm{d}\zeta - \int_0^\infty q_1(x-\zeta)\mathrm{d}\zeta\,. \end{aligned}$$

Making the change of variable $x - \zeta = u$, the previous relation becomes

$$T^*(q_1) = \int_a^b q(\zeta)\mathrm{d}\zeta - \int_{-\infty}^{x} q_1(u)\mathrm{d}u\,,$$

namely

$$T^*(q_1) = \int_a^b q(\zeta)\mathrm{d}\zeta - \int_{-\infty}^{x} q_1(\zeta)\mathrm{d}\zeta\,.$$

It follows

$$T^*(q_1) = \begin{cases} \int_a^b q(\zeta)\mathrm{d}\zeta\,, & x \le a\,, \\ \int_a^b q(\zeta)\mathrm{d}\zeta - \int_a^x q(\zeta)\mathrm{d}\zeta = \int_x^b q(\zeta)\mathrm{d}\zeta\,, & a \le x \le b\,, \\ 0\,, & x > b\,. \end{cases} \tag{4.26}$$

Comparing (4.25) with (4.26), it follows $T^*(q_1) = T(x)$, which certifies the accuracy of the formula (4.22).

4.4
Representation by Distributions of the Moments of a Material System

In the dynamics of discrete and continuous material systems (in particular, in rigid solids dynamics) the laws of motion are expressed by means of geometrical and mechanical quantities called moments. This has allowed the development of a geometry of the masses, where the static moments and the inertia moments play an essential role. These quantities characterize the mass distribution in space of a system of material points.

Let $A_i(x_i, y_i, z_i)$ be a discrete system (S) of n material points of masses m_i, specified by the position vectors $\boldsymbol{r}_i$, $i = \overline{1, n}$ (Figure 4.19).

The expression

$$I_p = \sum_{i=1}^{n} m_i x_i^{\alpha} y_i^{\beta} z_i^{\gamma} ,$$

where $\alpha, \beta, \gamma \in \mathbb{N}$, is called the moment of order $p = \alpha + \beta + \gamma$ of the points system.

For $\alpha = \beta = \gamma = 0$, $p = 0$, we obtain the zero-order moment,

$$M = \sum_{i=1}^{n} m_i ,$$

which is the total mass of the system of material points.

For $p = 1$ we obtain the first-order moments (*static moment*) and for $p = 2$ we obtain the second-order moments (*moments of inertia*).

The expressions

$$S_{Oyz} = \sum_{i=1}^{n} m_i x_i, \; S_{Ozx} = \sum_{i=1}^{n} m_i y_i, \; S_{Oxy} = \sum_{i=1}^{n} m_i z_i$$

represent the static moments of the material points system with respect to the planes Oyz, Ozx and Oxy, respectively.

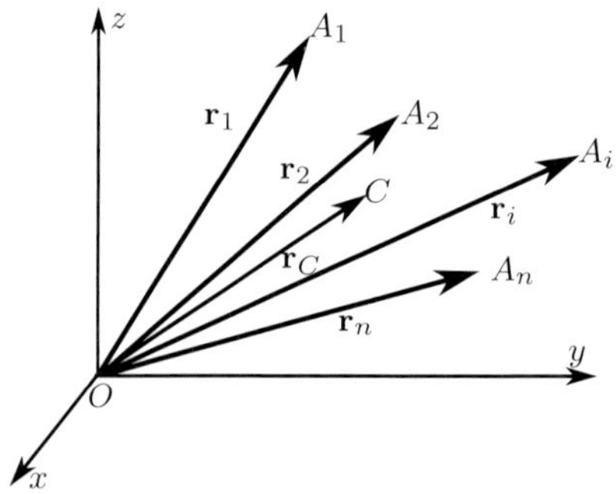

Figure 4.19

The center of mass $C(x_C, y_C, z_C)$, indicated by the position vector $\boldsymbol{r}_C$ will be given, by definition, by the relations

$$x_C = \frac{S_{Oyz}}{M}, \quad y_C = \frac{S_{Ozx}}{M}, \quad z_C = \frac{S_{Oxy}}{M}.$$

In addition to the planar static moments, we shall define the polar static moment, which is a vector, under the form

$$\boldsymbol{S}_O = \sum_{i=1}^{n} m_i \boldsymbol{r}_i ;$$

in this case, the position of the center of mass will be given by $M\boldsymbol{r}_C = \boldsymbol{S}_O$.

Let us now focus on the material point A_i; from the point of view of the distribution theory, it will be characterized by the volume density

$$\rho_i(\boldsymbol{r}) = m_i \delta(\boldsymbol{r} - \boldsymbol{r}_i) = m_i \delta(x - x_i, y - y_i, z - z_i) .$$

The mass of the point is obtained by applying this distribution to the test function $\varphi(\boldsymbol{r}) = 1 \in \mathcal{E}(\mathbb{R}^3)$, thus we have

$$(\rho_i(\boldsymbol{r}), 1) = (m_i \delta(\boldsymbol{r} - \boldsymbol{r}_i), 1) = (\delta(\boldsymbol{r} - \boldsymbol{r}_i), m_i) = m_i|_{\boldsymbol{r}=\boldsymbol{r}_i} = m_i .$$

The system (S) of material points will be characterized by the volume density

$$\rho(\boldsymbol{r}) = \rho(x, y, z) = \sum_{i=1}^{n} \rho_i(\boldsymbol{r}) = \sum_{i=1}^{n} m_i \delta(\boldsymbol{r} - \boldsymbol{r}_i)$$
$$= \sum_{i=1}^{n} m_i \delta(x - x_i, y - y_i, z - z_i) ;$$

the total mass of the points system is obtained by applying this density to the test function $\varphi(\boldsymbol{r}) = 1 \in \mathcal{E}(\mathbb{R}^3)$, thus we have

$$M = (\rho(\boldsymbol{r}), 1) = \left(\sum_{i=1}^{n} m_i \delta(\boldsymbol{r} - \boldsymbol{r}_i), 1 \right) = \sum_{i=1}^{n} (m_i \delta(\boldsymbol{r} - \boldsymbol{r}_i), 1) = \sum_{i=1}^{n} m_i .$$

With the help of the above results, we can calculate the moments of any order of the given system (S) of material points.

Thus, the static moment of the system (S) with respect to the Oyz-plane will be calculated as follows

$$S_{Oyz} = (\rho(\boldsymbol{r}), x) = \left(\sum_{i=1}^{n} \rho_i(\boldsymbol{r}), x \right) = \sum_{i=1}^{n} (\rho_i(\boldsymbol{r}), x) = \sum_{i=1}^{n} (m_i \delta(\boldsymbol{r} - \boldsymbol{r}_i), x)$$
$$= \sum_{i=1}^{n} (\delta(\boldsymbol{r} - \boldsymbol{r}_i), m_i x) = \sum_{i=1}^{n} m_i x|_{\boldsymbol{r}=\boldsymbol{r}_i} = \sum_{i=1}^{n} m_i x_i ,$$

where $\varphi(x, y, z) = x = x \times 1(y, z)$ is a test function from $\mathcal{E}(\mathbb{R}^3)$.

Similarly, for static moments with respect to the planes Ozx, Oxy, respectively, we can write

$$S_{Ozx} = (\rho(\boldsymbol{r}), y) = \left(\sum_{i=1}^{n} m_i \delta(\boldsymbol{r} - \boldsymbol{r}_i), y\right) = \sum_{i=1}^{n} m_i y_i ,$$

$$S_{Oxy} = (\rho(\boldsymbol{r}), z) = \left(\sum_{i=1}^{n} m_i \delta(\boldsymbol{r} - \boldsymbol{r}_i), z\right) = \sum_{i=1}^{n} m_i z_i , \tag{4.27}$$

where $x, y, z \in \mathcal{E}(\mathbb{R}^3)$.

The position vector $\boldsymbol{r}_C$ of the mass center of the material points system will be

$$\boldsymbol{r}_C = \left(\rho(\boldsymbol{r}), \frac{\boldsymbol{r}}{M}\right) = \left(\sum_{i=1}^{n} m_i \delta(\boldsymbol{r} - \boldsymbol{r}_i), \frac{\boldsymbol{r}}{M}\right) = \frac{1}{M} \sum_{i=1}^{n} m_i \boldsymbol{r}_i ,$$

where M is the total mass of the system of material points.

The expression

$$I_O = \sum_{i=1}^{n} m_i r_i^2 = \sum m_i \left(x_i^2 + y_i^2 + z_i^2\right)$$

is the polar moment of inertia of the system (S) of material points with respect to the origin O.

Similarly, the expressions

$$I_{Oyz} = \sum_{i=1}^{n} m_i x_i^2 ,\ I_{Ozx} = \sum_{i=1}^{n} m_i y_i^2 ,\ I_{Oxy} = \sum_{i=1}^{n} m_i z_i^2$$

are the planar moments of inertia, about the planes Oyz, Ozx, and Oxy, respectively; the sum of these moments of inertia is equal to the polar moment of inertia

$$I_O = I_{Oyz} + I_{Ozx} + I_{Oxy} .$$

The axial moments of inertia of the system (S) of material points with respect to the axes Ox, Oy and Oz are

$$I_x = \sum_{i=1}^{n} m_i \left(y_i^2 + z_i^2\right) , \quad I_y = \sum_{i=1}^{n} m_i \left(z_i^2 + x_i^2\right) , \quad I_z = \sum_{i=1}^{n} m_i \left(x_i^2 + y_i^2\right) ;$$

also, we introduce the centrifugal moments of inertia

$$I_{yz} = \sum_{i=1}^{n} m_i y_i z_i ,\ I_{zx} = \sum_{i=1}^{n} m_i z_i x_i ,\ I_{xy} = \sum_{i=1}^{n} m_i x_i y_i .$$

Similar to the polar and planar moments of inertia, the axial and centrifugal moments of inertia are second order moments.

In distributions sense, we obtain the polar moment of inertia

$$I_O = (\rho(\mathbf{r}), \mathbf{r}^2) = (\rho(\mathbf{r}), x^2 + y^2 + z^2)$$
$$= \left(\sum_{i=1}^{n} m_i \delta(\mathbf{r} - \mathbf{r}_i), \mathbf{r}^2\right) = \sum_{i=1}^{n} m_i \mathbf{r}_i^2 ,$$

where $\varphi(x, y, z) = \mathbf{r}^2 = x^2 + y^2 + z^2$ is a test function from $\mathcal{E}(\mathbb{R}^3)$.

The axial moments of inertia are

$$I_x = (\rho(\mathbf{r}), y^2 + z^2) = \sum_{i=1}^{n} m_i (y_i^2 + z_i^2) ,$$

$$I_y = (\rho(\mathbf{r}), z^2 + x^2) = \sum_{i=1}^{n} m_i (z_i^2 + x_i^2) ,$$

$$I_z = (\rho(\mathbf{r}), x^2 + y^2) = \sum_{i=1}^{n} m_i (x_i^2 + y_i^2) ,$$

and the centrifugal moments of inertia read

$$I_{yz} = (\rho(\mathbf{r}), yz) = \sum_{i=1}^{n} m_i y_i z_i, \; I_{zx} = (\rho(\mathbf{r}), zx) = \sum_{i=1}^{n} m_i z_i x_i ,$$

$$I_{xy} = (\rho(\mathbf{r}), xy) = \sum_{i=1}^{n} m_i x_i y_i .$$

As in the case of static moments, we introduce some distributions, which completely characterize a certain moment of inertia.

Thus

$$\mathcal{J}_O = \mathbf{r}^2 \rho(\mathbf{r})$$

is the distribution corresponding to the polar moment of inertia with respect to the pole O; indeed, we can write

$$I_O = (\mathcal{J}_O, 1) = (\mathbf{r}^2 \rho(\mathbf{r}), 1) = \left(\sum_{i=1}^{n} m_i \mathbf{r}_i^2 \delta(\mathbf{r} - \mathbf{r}_i), 1\right) = \sum_{i=1}^{n} m_i \mathbf{r}_i^2 .$$

Similarly, the planar moments of inertia distributions read

$$\mathcal{J}_{Oyz} = x^2 \rho(\mathbf{r}) , \quad \mathcal{J}_{Ozx} = y^2 \rho(\mathbf{r}) , \quad \mathcal{J}_{Oxy} = z^2 \rho(\mathbf{r}) ,$$

the axial moments of inertia distributions will be expressed in the form

$$\mathcal{J}_x = (y^2 + z^2) \rho(\mathbf{r}) , \quad \mathcal{J}_y = (z^2 + x^2) \rho(\mathbf{r}) , \quad \mathcal{J}_z = (x^2 + y^2) \rho(\mathbf{r}) ,$$

and the centrifugal moments of inertia distributions can be written

$$\mathcal{J}_{yz} = yz \rho(\mathbf{r}) , \quad \mathcal{J}_{zx} = zx \rho(\mathbf{r}) , \quad \mathcal{J}_{xy} = xy \rho(\mathbf{r}) .$$

Hereafter, we give two examples of calculation in which appear both distributed masses and concentrated masses; such problems are encountered in the case of vibrations of mechanical structure elements.

Example 4.1 Let there be a straight bar AB of length ℓ; on the bar acts a uniformly distributed mass of linear density m_0, on AA_1 (segment of length a) and the concentrated masses m_1 and m_2 at the points B_1, B_2, placed at the distances b_1 and b_2 from the bar end A, respectively (Figure 4.20a). We choose the coordinate axis Ax to coincide with the bar axis.

The law of mass distribution will be expressed in the form

$$m(x) = \overline{m}(x) + \overline{\overline{m}}(x) ,$$

where

$$\overline{m}(x) = m_0(x(H(x) - H(x-a)) + aH(x-a))$$

corresponds to the distributed mass (Figure 4.20b), and

$$\overline{\overline{m}}(x) = m_1 H(x-b_1) + m_2 H(x-b_2)$$

corresponds to concentrated masses (Figure 4.20c).

The linear density will be calculated as follows:

$$\overline{\rho}(x) = \frac{\mathrm{d}\overline{m}(x)}{\mathrm{d}x} = m_0(H(x) - H(x-a) + x\delta(x) - (x-a)\delta(x-a)) ,$$

$$\overline{\overline{\rho}}(x) = \frac{\mathrm{d}\overline{\overline{m}}(x)}{\mathrm{d}x} = m_1\delta(x-b_1) + m_2\delta(x-b_2) ;$$

so, we can write

$$\rho(x) = m_0(H(x) - H(x-a)) + m_1\delta(x-b_1) + m_2\delta(x-b_2) .$$

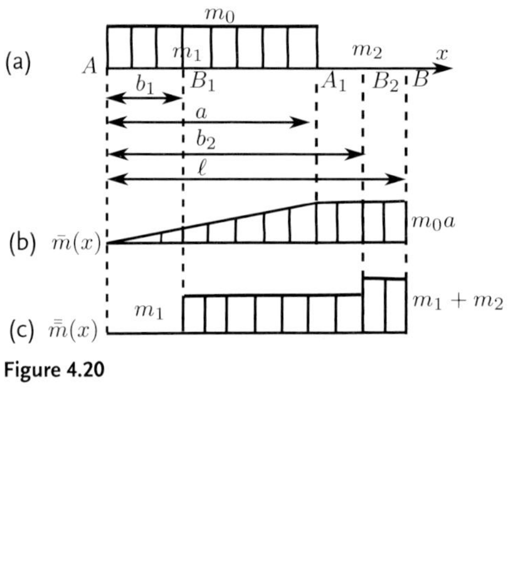

Figure 4.20

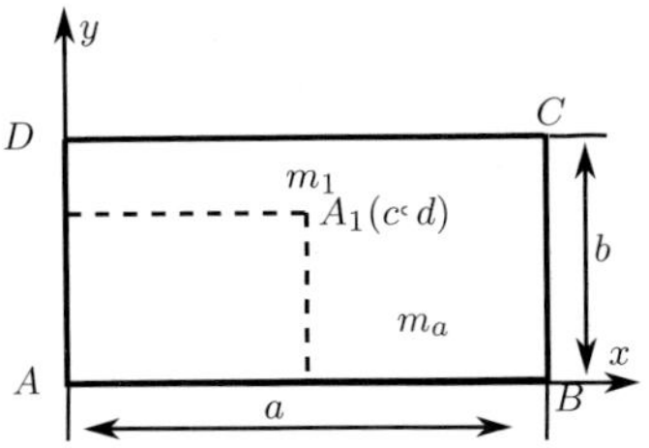

Figure 4.21

We calculate, for example, the moment of inertia with respect to the end A of the bar in the form

$$\begin{aligned} I_A &= (\rho(x), x^2) = (m_0 (H(x) - H(x-a)), x^2) \\ &+ (m_1 \delta(x - b_1) + m_2 \delta(x - b_2), x^2) \\ &= \int_0^a m_0 x^2 \mathrm{d}x + m_1 b_1^2 + m_2 b_2^2 = \frac{1}{3} m_0 a^3 + m_1 b_1^2 + m_2 b_2^2 . \end{aligned}$$

Example 4.2 Let there be a rectangular plate $ABCD$ of dimensions a and b; on it acts a uniformly distributed mass of superficial mass m_0 and a concentrated mass m_1 at the point A_1 (Figure 4.21). The mass density will be

$$\rho(x, y) = m_0 \delta_S + m_1 \delta(x - c, y - d) ,$$

where we used the Dirac delta distribution δ_S concentrated on the rectangle $ABCD$.

The total mass of the plate will be

$$\begin{aligned} M &= (\rho(x, y), 1) = (m_0 \delta_S + m_1 \delta(x - c, y - d), 1) \\ &= \iint_{ABCD} m_0 \mathrm{d}x \mathrm{d}y + m_1 = m_0 ab + m_1 \end{aligned}$$

and the moment of inertia about the Ox-axis, for example, reads as

$$\begin{aligned} I_x &= (\rho(x, y), y^2) = (m_0 \delta_S + m_1 \delta(x - c, y - d), y^2) \\ &= \iint_{ABCD} m_0 y^2 \mathrm{d}x \mathrm{d}y + m_1 y^2 |_{\substack{x=c \\ y=d}} = \frac{1}{3} m_0 a b^3 + m_1 d^2 . \end{aligned}$$

4.5 Representation in Distributions of Electrical Quantities

4.5.1 Volume and Surface Potential of the Electrostatic Field

In physics and in particular, in electrostatics, Coulomb's law of interaction of electric charges is of great importance.

According to this law, two idealized point electric charges A and B, are repulsive or attractive each other as the charges have the same sign or an opposite one, and the interaction force $\boldsymbol{F}$ has the expression

$$F = \frac{1}{4\pi\varepsilon_0}\frac{|q_1 q_2|}{r^2}\,, \qquad \boldsymbol{F} = \frac{1}{4\pi\varepsilon_0} q_1 q_2 \frac{\boldsymbol{r}}{r^3}\,, \tag{4.28}$$

where r is the separation distance between the electric charges and ε_0 is the permittivity of space.

We can notice that, besides the continuous distributed density of electrical charges we must consider the densities of the point charges. This can be done using the theory of distributions. In this way, problems with continuous or discontinuous quantities will be treated in a uniform manner.

For example, if the load $q \in C^1(\mathbb{R})$ is distributed continuously on Ox-axis, then its linear density is $\rho(x) = \mathrm{d}q/\mathrm{d}x$.

Obviously, if the point charge is placed at the origin of the Ox-axis, then this formula is inapplicable and we observe that the density in this case will be zero everywhere except the point $x = 0$ at which it tends to infinity. We recognize from this the property of the distribution $\delta(x) \in D'(\mathbb{R})$, that is, $\delta(x) = 0$, $x \neq 0$.

Thus, an electric point charge q placed at the origin of the Ox-axis will have the density

$$\rho(x) = q\delta(x)\,, \tag{4.29}$$

where $\delta \in \mathcal{D}'(\mathbb{R})$ is the Dirac delta distribution.

We note that the unit of measurement of electric charges is the coulomb and from (4.28) it results that the dimension of the electric charge q is

$$[q] = \left[r\sqrt{F}\right] = L\sqrt{\mathrm{LMT}^{-2}} = \mathrm{M}^{1/2}\mathrm{L}^{3/2}\mathrm{T}^{-1}\,.$$

Consequently, the linear density ρ of the charge q given by (4.29) is expressed in coulombs per meter and has the dimension

$$[\rho] = [q][\delta(x)] = \mathrm{M}^{1/2}\mathrm{L}^{3/2}\mathrm{T}^{-1} \cdot \mathrm{L}^{-1} = \mathrm{M}^{1/2}\mathrm{L}^{1/2}\mathrm{T}^{-1}\,.$$

Similarly, the surface density of a point electric charge q, located at the origin in the Oxy-plane is the distribution $q \in \mathcal{D}'(\mathbb{R}^2)$ with the expression

$$\rho(x, y) = q\delta(x, y)\,.$$

Its dimensional equation is

$$[\rho] = [q][\delta(x, y)] = \mathrm{M}^{1/2}\mathrm{L}^{3/2}\mathrm{T}^{-1} \cdot \mathrm{L}^{-2} = \mathrm{M}^{1/2}\mathrm{L}^{-1/2}\mathrm{T}^{-1}\,.$$

An electric point charge q located at the origin of the orthogonal system $Oxyz$ will have the volume density expressed by the distribution

$$\rho(x, y, z) = q\delta(x, y, z)\,;$$

and its dimensional equation is

$$[\rho] = [q][\delta(x, y, z)] = [q]\mathrm{L}^{-3} = \mathrm{M}^{1/2}\mathrm{L}^{-3/2}\mathrm{T}^{-1} .$$

This volume density can be expressed in coulombs per m^3.

Let us now consider the set of two-point electric charges $-q$ and q, $q > 0$, located at the points $M_0(x_0, y_0, z_0)$ and $M(x', y', z')$ (Figure 4.22), referred to as the electric dipole. The magnitude $D = qd$, $d = \overline{M_0 M}$, is called the dipole moment.

The unit vector $\boldsymbol{u}$ directed from the negative charge to the positive one determines the dipole axis. The densities of the two charges have the expressions

$$\rho_-(x, y, z) = -q\delta(\boldsymbol{r} - \boldsymbol{r}_0) , \quad \rho_+(x, y, z) = q\delta(\boldsymbol{r} - \boldsymbol{r}') ,$$

where $\boldsymbol{r} = x\boldsymbol{i} + y\boldsymbol{j} + z\boldsymbol{k}$, $\boldsymbol{r}_0 = x_0\boldsymbol{i} + y_0\boldsymbol{j} + z_0\boldsymbol{k}$, $\boldsymbol{r}' = x'\boldsymbol{i} + y'\boldsymbol{j} + z'\boldsymbol{k} = \boldsymbol{r}_0 + d\boldsymbol{u}$, and $\boldsymbol{i}, \boldsymbol{j}, \boldsymbol{k}$ are the unit vectors of the coordinate axes.

Consequently, for the volume density corresponding to the electric dipole we obtain the distribution

$$\rho_d(x, y, z) = \rho_- + \rho_+ = D\frac{\delta(\boldsymbol{r} - \boldsymbol{r}') - \delta(\boldsymbol{r} - \boldsymbol{r}_0)}{d} , \tag{4.30}$$

because $D = qd$.

Definition 4.7 We call the electric dipole concentrated at the point $M_0(x_0, y_0, z_0)$, the limit, in the sense of the distribution theory, of the volume density ρ_d, for $d \to 0$, $M \to M_0$, considering that both the unit vector $\boldsymbol{u}$ and the dipole moment $D = qd$ are constant.

From (4.30), we obtain

$$\rho(\boldsymbol{r}) = \lim_{d \to +0} \rho_d = -D\frac{\partial\delta(\boldsymbol{r} - \boldsymbol{r}_0)}{\partial u} = -D\boldsymbol{u} \cdot \operatorname{grad}\delta(\boldsymbol{r} - \boldsymbol{r}_0) ,$$

which is the volume density corresponding to the electric dipole moment D concentrated at the point $M_0(x_0, y_0, z_0)$.

In particular, the linear density of an electric dipole of magnitude D concentrated at the origin of the Ox-axis is

$$\rho(x) = -D\delta'(x) , \tag{4.31}$$

because $\boldsymbol{u} = \boldsymbol{i}$, the unit vector of the Ox-axis.

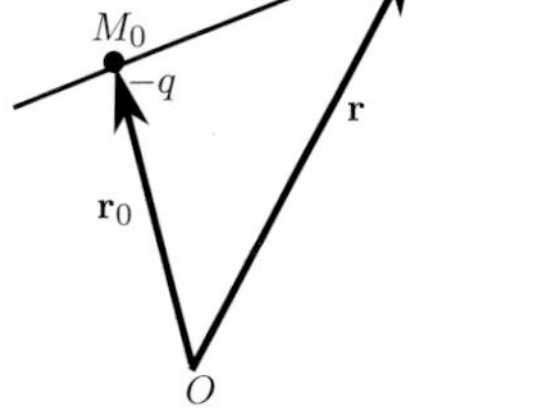

Figure 4.22

4.5.2
Electrostatic Field

Let q be a point electric charge located at the origin of orthogonal frame $Oxyz$. If at any point $M \neq O$ (Figure 4.23) is an electric charge +1, then, according to Coulomb's law, the force acting on it is

$$E = \frac{q}{r^2}\,, \quad r = \overline{OM} = \sqrt{x^2 + y^2 + z^2}\,, \quad \boldsymbol{E} = q\frac{\boldsymbol{r}}{r^3}\,, \quad r \neq 0\,. \tag{4.32}$$

By definition, the vector function $\boldsymbol{E} : \mathbb{R}^3\backslash\{0\} \to \mathbb{R}^3$ given by (4.32) is called the electrostatic field intensity created by the electric point charge q located at the origin of the frame $Oxyz$.

We observe that the vector function $\boldsymbol{E}$ defined by (4.32) is a locally integrable function, because $\forall \Omega \subset \mathbb{R}^3$ exists and the integral $\iiint_\Omega \left(x^2 + y^2 + z^2\right)^{-1} \mathrm{d}x\mathrm{d}y\mathrm{d}z$ is finite, which can be proved by passing to spherical coordinates.

Consequently, the intensity of the electrostatic field $\boldsymbol{E}$ created by charge q located at the origin O generates a function type distribution from $\mathcal{D}'(\mathbb{R}^3)$.

By definition, the function $V : \mathbb{R}^3\backslash\{0\} \to \mathbb{R}$ given by

$$V(x, y, z) = \frac{q}{r}\,, \quad r = \sqrt{x^2 + y^2 + z^2}$$

is called the potential of the electrostatic field created by the charge q located at the point O.

The dimension of this quantity is

$$[V] = \frac{[q]}{\mathrm{L}} = \mathrm{M}^{1/2}\mathrm{L}^{1/2}\mathrm{T}^{-1}\,.$$

The potential function V is also a locally integrable function which generates a function type distribution from $\mathcal{D}'(\mathbb{R})$.

The functions E and V are homogeneous functions of degrees -2 and -1, respectively.

By direct calculation, it follows that between the electrostatic field intensity $\boldsymbol{E}$ and the potential V of the field we obtain the relation

$$\boldsymbol{E} = -\operatorname{grad} V\,. \tag{4.33}$$

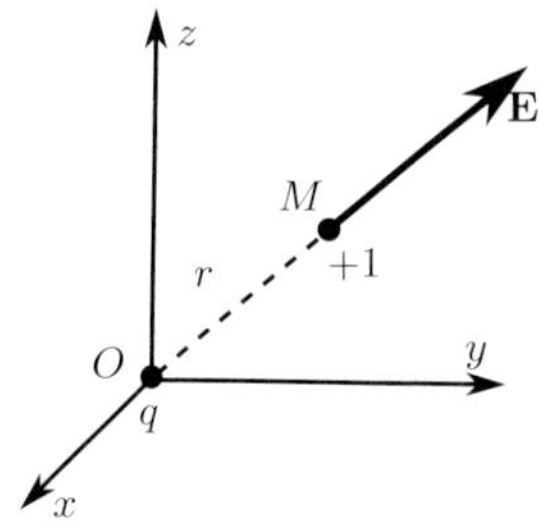

Figure 4.23

In general, if we denote by ρ the volume density of electric charges, expressed as a distribution from $\mathcal{D}'(\mathbb{R}^3)$, and by V the potential field created by these charges, then the quantities $\boldsymbol{E}$, V and ρ, considered as distributions from $\mathcal{D}'(\mathbb{R}^3)$, verify the Poisson equation

$$\Delta V = -4\pi\rho \, , \tag{4.34}$$

namely

$$\operatorname{div} \boldsymbol{E} = -\operatorname{div}\operatorname{grad} V = -\Delta V = 4\pi\rho \, .$$

Equations 4.33 and 4.34 considered in the distribution space are the fundamental equations of electrostatics.

The Poisson equation allows the determination of the field potential of some electric charges, knowing the density, that is, the repartition in space of these charges.

We note that (4.33) and (4.34) can be applied in particular on $\mathbb{R}$ or on $\mathbb{R}^2$. In these cases, ρ is the linear density and the surface density of electric charges, respectively.

Using the fundamental solution of the Laplace operator Δ, we can write the expression of the potential V of an electrostatic field.

As it is well known, the fundamental solution $U \in \mathcal{D}'(\mathbb{R}^3)$ of the operator Δ in $\mathbb{R}^3$ is the function type distribution

$$U(\boldsymbol{r}) = U(x, y, z) = -\frac{1}{4\pi r} \, , \quad r = \sqrt{x^2 + y^2 + z^2} \, ,$$

namely

$$\Delta\left(\frac{-1}{4\pi r}\right) = \delta(\boldsymbol{r}) = \delta(x, y, z), \Delta = \frac{\partial^2}{\partial x^2} + \frac{\partial^2}{\partial y^2} + \frac{\partial^2}{\partial z^2} \, .$$

In $\mathbb{R}^2$ and $\mathbb{R}$ the fundamental solutions of the operator Δ are

$$U(\boldsymbol{r}) = U(x, y) = \frac{1}{2\pi} \ln r \, , \quad r = \sqrt{x^2 + y^2} \, , \quad \Delta = \frac{\partial^2}{\partial x^2} + \frac{\partial^2}{\partial y^2} \, , \tag{4.35}$$

$$U(x) = \frac{1}{2}|x| \, , \quad x \in \mathbb{R} \, , \quad \Delta = \frac{\mathrm{d}^2}{\mathrm{d}x^2} \, , \tag{4.36}$$

which, obviously, are function type distributions from $\mathcal{D}'(\mathbb{R}^2)$ and $\mathcal{D}'(\mathbb{R})$, respectively.

If the volume density $\rho(\boldsymbol{r})$, $\boldsymbol{r} \in \mathbb{R}^3$, of the electric charge is expressed by a distribution with compact support, hence $\rho \in \mathcal{E}'(\mathbb{R}^3)$, then, by definition, the distribution

$$V(\boldsymbol{r}) = V(x, y, z) = \frac{1}{r} * \rho \tag{4.37}$$

is called the volume potential with density ρ.

If the density ρ is an integrable function with compact support, then the formula (4.37) becomes

$$V(x, y, z) = \int\limits_{\mathbb{R}^3} \frac{\rho(\zeta, \eta, \tau)}{\sqrt{(x-\zeta)^2 + (y-\eta)^2 + (z-\tau)^2}} \mathrm{d}\zeta \mathrm{d}\eta \mathrm{d}\tau \,.$$

Since the fundamental solution $U \in \mathcal{D}'(\mathbb{R}^3)$ of the Laplace operator Δ in $\mathbb{R}^3$ is $U = -1/4\pi r$, the volume potential (4.37) is a particular solution of the Poisson equation (4.34).

Indeed, the general solution of this equation is

$$V(\boldsymbol{r}) = U(\boldsymbol{r}) * (-4\pi\rho) = \frac{1}{r} * \rho(\boldsymbol{r}) + \psi \,,$$

where ψ is a harmonic function, $\Delta\psi = 0$.

For $\psi = 0$, we obtain the expression (4.37) of the volume potential.

In particular, for a point charge q located at the origin of the frame $Oxyz$, the volume density is the distribution $\rho(\boldsymbol{r}) \in \mathcal{E}'(\mathbb{R}^3)$, namely

$$\rho(\boldsymbol{r}) = q\delta(\boldsymbol{r}) = q\delta(x, y, z) \,.$$

Consequently, for the volume potential V we obtain, from the (4.37), the known formula

$$V(\boldsymbol{r}) = \frac{1}{r} * q\delta(\boldsymbol{r}) = \frac{q}{r} \,, \qquad r = \sqrt{x^2 + y^2 + z^2} \,, \qquad r > 0 \,.$$

If the charge q is located at the point $M_0(x_0, y_0, z_0)$, then, for the volume density of this electric charge, we have

$$\rho(\boldsymbol{r}) = q\delta(\boldsymbol{r} - \boldsymbol{r}_0) \,, \qquad \boldsymbol{r}_0 = x_0\boldsymbol{i} + y_0\boldsymbol{j} + z_0\boldsymbol{k} \,,$$

where $\boldsymbol{i}, \boldsymbol{j}, \boldsymbol{k}$ are the unit vectors of the coordinate axes.

Consequently, the potential of volume corresponding to this charge is

$$V(\boldsymbol{r}) = \frac{1}{r} * q\delta(\boldsymbol{r} - \boldsymbol{r}_0) = \frac{q}{R} \,,$$
$$R = \overline{M_0 M} = \sqrt{(x - x_0)^2 + (y - y_0)^2 + (z - z_0)^2} \,.$$

Let us consider the case of an electric charge q located at the origin of the frame Oxy; the surface density of this charge in the Oxy-plane has the expression

$$\rho(x, y) = q\delta(x, y) \,, \tag{4.38}$$

namely $\rho(\boldsymbol{r}) = q\delta(\boldsymbol{r})$, $\boldsymbol{r} = x\boldsymbol{i} + y\boldsymbol{j}$.

Let $\rho \in \mathcal{E}'(\mathbb{R}^2)$ be a distribution with compact support representing the surface density of electric charge in the Oxy-plane. Since the fundamental solution of the Laplace operator in $\mathbb{R}^2$ is given by (4.35), from (4.34) we obtain for the potential $V \in \mathcal{D}'(\mathbb{R}^2)$ of the electrostatic field the expression

$$V(x,y) = U(x,y) * [-4\pi\rho(x,y)] = -2\ln\sqrt{x^2+y^2} * \rho(x,y) . \tag{4.39}$$

The distribution V exists, because ρ has compact support.

In particular, if the surface density has the expression (4.38), then we get

$$V(x,y) = -2\ln\sqrt{x^2+y^2} * q\rho(x,y) = -2q\ln\sqrt{x^2+y^2} .$$

This potential is called the logarithmic potential and it corresponds to the electrostatic field created by a charge q located at the origin of the frame Oxy.

We note that this potential is a function type distribution expressed as a homogeneous function of degree one, locally integrable.

If the surface density ρ is an integrable function with compact support, then the formula (4.39) becomes

$$V(x,y) = -2\iint\limits_{\text{supp}(\rho)} \rho(\xi,\eta)\ln\sqrt{(x-\xi)^2+(y-\eta)^2}\,\mathrm{d}\xi\,\mathrm{d}\eta .$$

Further, we will consider an electric charge q located at the origin of the Ox-axis. The linear density of this charge on the Ox-axis is the distribution $\rho \in \mathcal{D}'(\mathbb{R})$, namely $\rho(x) = q\delta(x)$.

Taking into account (4.34) and (4.36) for the created field potential, we obtain

$$V(x) = U(x) * [-4\pi\rho(x)] = -2\pi|x| * q\delta(x) = -2\pi q|x| , \quad x \in \mathbb{R} .$$

In general, if the linear density ρ is a distribution with compact support, hence $\rho \in \mathcal{E}'(\mathbb{R})$, then the corresponding potential of the electrostatic field has the expression

$$V(x) = \frac{1}{2}|x| * [-4\pi\rho(x)] = -2\pi|x| * \rho(x) . \tag{4.40}$$

If ρ is an integrable function with compact support, then this formula becomes

$$V(x) = -2\pi\int\limits_{\text{supp}(\rho)} \rho(\xi)|x-\xi|\,\mathrm{d}\xi .$$

Let $\rho(x) = -D\delta(x)$ (4.31) be the linear density of an electric dipole concentrated at the origin of the Ox-axis.

According to the formula (4.40), the potential of this electrostatic field is

$$V(x) = -2|x| * [-D\delta'(x)] = 2D|x|' = \begin{cases} -2D , & x < 0 , \\ 2D , & x > 0 . \end{cases}$$

4.5.3
Electric Potential of Single and Double Layers

Of great importance in solving various boundary value problems in electrostatics are the cases in which the electric field is created by electric charges distributed on a bounded area $S \subset \mathbb{R}^3$, piecewise smooth, which is called the single electric layer.

Let $\sigma, \sigma = \mathrm{d}q/\mathrm{d}S$, be the single layer surface density expressed by a piecewise continuous function defined on S. With the help of the Dirac delta distribution δ_S concentrated on the surface S, the volume density $\rho \in \mathcal{D}'(\mathbb{R}^3)$ of electric charges distributed on the surface S has the expression

$$\rho(x, y, z) = \sigma(x, y, z)\delta_S .$$

Because $\mathrm{supp}(\sigma\delta_S) \subset S$, it follows that the distribution $\sigma\delta_S$ has compact support, hence $\sigma\delta_S \in \mathcal{E}'(\mathbb{R}^3) \subset \mathcal{D}'(\mathbb{R}^3)$.

Consequently, if $(x, y, z) \notin S$, then $\delta_S = 0$ and $\forall\varphi \in \mathcal{D}(\mathbb{R}^3)$ we have

$$(\sigma\delta_S, \varphi) = \int\limits_S \sigma(x, y, z)\varphi(x, y, z)\mathrm{d}S ,$$

where $\mathrm{d}S$ is the area element corresponding to the surface S.

Taking into account the formula (4.40), we call single layer potential of density σ, the distribution $V \in \mathcal{D}'(\mathbb{R}^3)$, namely

$$V(x, y, z) = \frac{1}{r} * \sigma\delta_S, r = \sqrt{x^2 + y^2 + z^2} .$$

The convolution product exists, because the distribution $\sigma\delta_S$ has compact support.

Proposition 4.5 Let S be a bounded surface, piecewise smooth. If g and f are piecewise continuous functions on $\mathbb{R}^n$ and on S, respectively, then we may write the formula

$$g * f\delta_S = \int\limits_S f(y)g(x - y)\mathrm{d}S(y) . \tag{4.41}$$

Indeed, because $\mathrm{supp}(f\delta_S) \subset S$, the distribution $f\delta_S$ has compact support; therefore, the convolution product (4.41) exists.

For any $\varphi \in \mathcal{D}(\mathbb{R}^n)$ we can write

$$(g * f\delta_S, \varphi) = (f(y)\delta_S(y) \times g(z), \varphi(y + z))$$

$$= (f(y)\delta_S(y), (g(z), \varphi(y + z))) = \int\limits_S f(y)\left[\int\limits_{\mathbb{R}^n} g(z)\varphi(y + z)\mathrm{d}z\right]\mathrm{d}S(y)$$

$$= \int\limits_S f(y)\left[\int\limits_{\mathbb{R}^n} g(x - y)\varphi(x)\mathrm{d}x\right]\mathrm{d}S(y) = \int\limits_{\mathbb{R}^n} \varphi(x)\left[\int\limits_S f(y)g(x - y)\mathrm{d}S(y)\right]\mathrm{d}x .$$

This last relation can be written in the form

$$(g * f\delta_S, \varphi) = \left(\int_S f(y) g(x-y) \mathrm{d}S(y), \varphi \right), \quad \forall \varphi \in \mathcal{D}(\mathbb{R}^n),$$

which results in the formula (4.41).

Using the formula (4.41), the single-layer potential takes the form

$$V(x, y, z) = \int_S \frac{\sigma(\zeta, \eta, \tau)}{R} \mathrm{d}S(\zeta, \eta, \tau),$$

where $R = \sqrt{(x-\zeta)^2 + (y-\eta)^2 + (z-\tau)^2}$.

Next, we will define the notion of electric double layer.

Let S be a bounded surface piecewise smooth and $\boldsymbol{r}$ be the current position vector of the point $M(x, y, z) \in S$ (Figure 4.24).

At each point $M \in S$ we consider an oriented segment h along the interior normal unit vector $\boldsymbol{n}$ to the surface of the S.

In this way, the ends M' of the vectors $\overrightarrow{MM'} = h\boldsymbol{n}$ determine a new surface S'. Noting by $\boldsymbol{r}' = \overrightarrow{OM'}$ the position vector of the point $M' \in S'$, we can write the formula

$$\boldsymbol{r}' = \boldsymbol{r} + h\boldsymbol{n}, \tag{4.42}$$

that expresses the relation between the position vectors of the corresponding points M and M' belonging to the surfaces S and S'.

We assume that the two surfaces S and S' (Figure 4.24), located at a distance $h = MM'$, have distributed electric charges of opposite sign, such that the surface densities $-\sigma$ and σ', $\sigma, \sigma' > 0$, at the corresponding points M and M' satisfy the property

$$\sigma \mathrm{d}S = \sigma' \mathrm{d}S', \tag{4.43}$$

where $\mathrm{d}S, \mathrm{d}S'$ are the area elements at M and M', respectively.

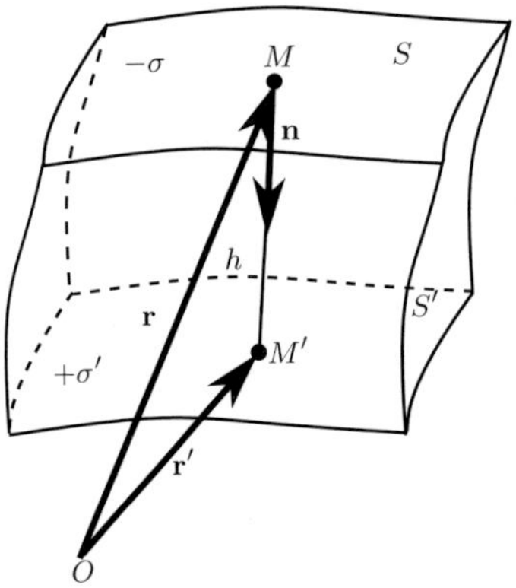

Figure 4.24

We note that the unit vector $\boldsymbol{n}$ is the common normal unit vector to the surfaces S and S' at the corresponding points M, M'. This unit vector has the sense from the negative electric charge to the positive electric one (Figure 4.24).

The relation (4.43) shows that the electric charges $\mathrm{d}q$ and $\mathrm{d}q'$ corresponding to the area elements $\mathrm{d}S$ and $\mathrm{d}S'$ are equal in absolute value, so that $\mathrm{d}q' = -\mathrm{d}q$.

As regards the surface densities $-\sigma$ and σ' at the points M and M', respectively, one has the relations $-\sigma = \mathrm{d}q/\mathrm{d}S$, $\sigma' = \mathrm{d}q'/\mathrm{d}S'$.

Definition 4.8 We shall call double electric layer the limit, in the sense of distributions, of the set of electric charges distributed on the surfaces S and S' when $h \to 0$, if the expression $\mu = h\sigma$ remains constant.

The magnitude $\mu = h\sigma > 0$ is called the moment of the electric double layer.

Proposition 4.6 The volume density $\rho \in \mathcal{D}'(\mathbb{R}^3)$ of an electric double layer has the expression

$$\rho(\boldsymbol{r}) = \rho(x, y, z) = -\frac{\partial}{\partial n}(\mu(\boldsymbol{r})\delta_S), \tag{4.44}$$

where μ is the moment of the double layer and δ_S is the Dirac delta distribution concentrated on the surface S.

Proof: The volume densities of the electric charges distributed on the surfaces S and S' are

$$(S): \rho_-(\boldsymbol{r}) = -\sigma(\boldsymbol{r})\delta_S; (S'): \rho_+(\boldsymbol{r}') = \sigma(\boldsymbol{r}')\delta_{S'} .$$

Consequently, the volume density of the double layer is

$$\rho(\boldsymbol{r}) = \lim_{h\to+0}[\rho_-(\boldsymbol{r}) + \rho_+(\boldsymbol{r}')] = \lim_{h\to+0}[\sigma'(\boldsymbol{r}')\delta_{S'} - \sigma(\boldsymbol{r})\delta_S] .$$

Taking into account the relation (4.43) and the definition of the Dirac delta distribution concentrated on a surface, $\forall\varphi \in \mathcal{D}(\mathbb{R}^3)$, we have

$$(\rho, \varphi) = \lim_{h\to+0}(\sigma'\delta_{S'} - \sigma\delta_S, \varphi) = \lim_{h\to+0}\left[\int_{S'}\sigma'\varphi\,\mathrm{d}S' - \int_S \sigma\varphi\,\mathrm{d}S\right]$$

$$= \lim_{h\to+0}\int_S \sigma(\boldsymbol{r})[\varphi(\boldsymbol{r}') - \varphi(\boldsymbol{r})]\mathrm{d}S .$$

Because the quantity $\mu(\boldsymbol{r}) = h\sigma(\boldsymbol{r})$ is constant for $h \to +0$, the previous relation can be written in the form

$$(\rho, \varphi) = \int_S \mu(\boldsymbol{r}) \lim_{h\to+0}\frac{\varphi(\boldsymbol{r}') - \varphi(\boldsymbol{r})}{h}\mathrm{d}S . \tag{4.45}$$

On the other hand, based on the relation (4.42), we have

$$\lim_{h\to+0}\frac{\varphi(\boldsymbol{r}')-\varphi(\boldsymbol{r})}{h}=\frac{\partial}{\partial n}\varphi(\boldsymbol{r})=\boldsymbol{n}\cdot\operatorname{grad}\varphi(\boldsymbol{r})\,,$$

where $\partial/\partial n$ is the normal derivative operator along $\boldsymbol{n}$ to the surface S (Figure 4.24).
Consequently, (4.45) becomes

$$(\rho,\varphi)=\int_S \mu(\boldsymbol{r})\frac{\partial}{\partial n}\varphi(\boldsymbol{r})\mathrm{d}S=\left(\delta_S,\mu\frac{\partial\varphi}{\partial n}\right)=\left(\mu(\boldsymbol{r})\delta_S,\frac{\partial\varphi}{\partial n}(\boldsymbol{r})\right),$$

namely

$$(\rho,\varphi)=\left(-\frac{\partial}{\partial n}\left[\mu(\boldsymbol{r})\delta_S\right],\varphi\right),\quad \forall\varphi\in\mathcal{D}(\mathbb{R}^3)\,,$$

which results in the formula (4.44). □

Using the formula (4.37), we shall call electric potential of double layer the distribution $V(\boldsymbol{r})\in\mathcal{D}'(\mathbb{R}^3)$ defined by the formula

$$V(\boldsymbol{r})=V(x,y,z)=-\frac{1}{r}*\frac{\partial}{\partial n}\left[\mu(\boldsymbol{r})\delta_S\right].$$

Taking into account (4.41), this formula can be written in the explicit form

$$V(x,y,z)=-\int_S \mu(\zeta,\eta,\tau)\frac{\partial}{\partial n(\zeta,\eta,\tau)}\frac{1}{R}\mathrm{d}S(\zeta,\eta,\tau)\,,$$

where $R=\sqrt{(x-\zeta)^2+(y-\eta)^2+(z-\tau)^2}$.

4.6
Operational Form of the Constitutive Law of One-Dimensional Viscoelastic Solids

The distribution theory is an effective tool in the study of the properties of elastic and viscoelastic solids, [20, 33–35].

A perfectly elastic solid is a deformable solid characterized by the fact that, under the action of external loads, it is deformed and it returns to its original shape once those loads are removed.

A deformable solid is distinguished from another by the constitutive law, that is, the relation that exists between stresses and strains.

Thus, the elastic linear solid is characterized by Hooke's law, which is expressed by linear relations independent of time between stresses and strains, which ensures a bijective correspondence between the state of strain and the state of stress.

In the case of a one-dimensional elastic solid, according to Hooke's law, between the stress σ and the strain ε we obtain the relation $\sigma=E\varepsilon$, where E is a constant characterizing the elastic solid, called longitudinal modulus of elasticity.

We note that this constant has the dimension of a stress, because the strain ε is a dimensionless quantity. Consequently, the dimensional equation for E is

$$[E] = [\sigma] = \frac{\mathrm{MLT}^{-2}}{\mathrm{L}^2} = \mathrm{ML}^{-1}\mathrm{T}^{-2} ,$$

representing the force per unit area.

The mechanical correspondent for the one-dimensional elastic solid is the spring represented in Figure 4.25.

The deformable solids which after unloading do not recover their initial shape are called nonelastic solids.

A model of nonelastic solid is the viscoelastic solid, which exhibits both viscous and elastic characteristics when undergoing deformation.

The viscous fluids which constitute a deformable medium are characterized by Newton's law

$$\sigma = \eta \frac{\mathrm{d}\varepsilon}{\mathrm{d}t} ,$$

where $\dot{\varepsilon} = \mathrm{d}\varepsilon/\mathrm{d}t$ is the deformation speed and η the viscosity coefficient whose dimensional equation is

$$[\eta] = \frac{[\sigma]}{[\dot{\varepsilon}]} = \mathrm{ML}^{-1}\mathrm{T}^{-1} .$$

The mechanical correspondent of the viscous fluid is the pneumatic damper represented in Figure 4.26, which consists of a perforated piston that can move without friction in a cylinder in which there is a viscous liquid.

Consequently, in mechanical terms, the viscoelastic solids can be represented by springs and dampers pneumatic systems. After their connecting mode in series and in parallel, different models of one-dimensional viscoelastic solids are obtained.

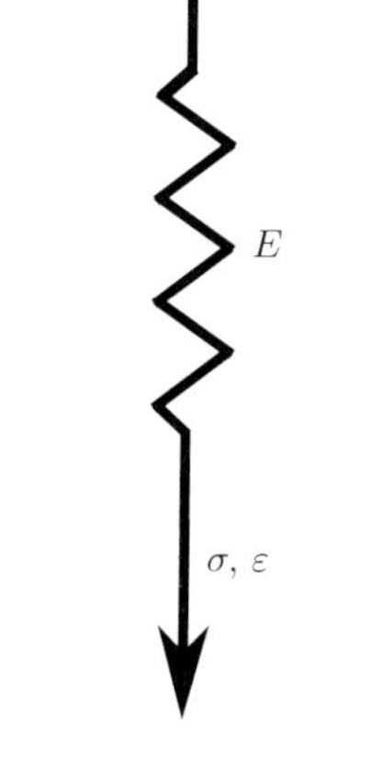

Figure 4.25

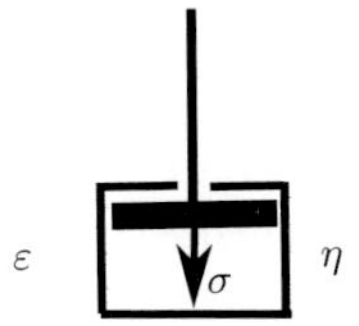

Figure 4.26

Thus, the Kelvin–Voigt model can be represented by a purely viscous damper and purely elastic spring connected in parallel as shown in the Figure 4.27.

The constitutive law for this model is

$$\sigma = E\varepsilon + \eta \frac{\partial \varepsilon}{\partial t} . \tag{4.46}$$

In general, for the one-dimensional viscoelastic solids, the correspondent of Hooke's law are Boltzmann's integral relations

$$\sigma(x,t) = \int_{-\infty}^{t} \frac{\partial \varepsilon(x,\tau)}{\partial \tau} \psi(t-\tau)\mathrm{d}\tau , \quad \varepsilon(x,t) = \int_{-\infty}^{t} \frac{\partial \sigma(x,\tau)}{\partial \tau} \varphi(t-\tau)\mathrm{d}\tau , \tag{4.47}$$

where φ and ψ are the creep and the relaxation functions, respectively.

With these functions depending only on the temporal variable $t \in \mathbb{R}$, the mechanical properties of the viscoelastic solid are expressed.

Using the convolution product, the formulas (4.47) can be written in the form

$$\sigma = H\psi * \frac{\partial \varepsilon}{\partial t} , \quad \varepsilon = H\varphi * \frac{\partial \sigma}{\partial t} , \tag{4.48}$$

where H is the Heaviside step function.

Indeed, we may write

$$H\psi * \frac{\partial \varepsilon}{\partial t} = \int_{-\infty}^{\infty} \frac{\partial \varepsilon(\tau)}{\partial \tau} H(t-\tau)\psi(t-\tau)\mathrm{d}\tau = \int_{-\infty}^{t} \frac{\partial \varepsilon(\tau)}{\partial \tau} \psi(t-\tau)\mathrm{d}\tau = \sigma(x,t) ,$$

because for $\tau > t$ we have $H(t-\tau) = 0$.

In order to write the equations in distributions, we assume that σ and ε are distributions with respect to the temporal variable $t \in \mathbb{R}$, with the supports in $\mathbb{R}_+ = [0, \infty)$, $x \in \mathbb{R}$ being considered a parameter.

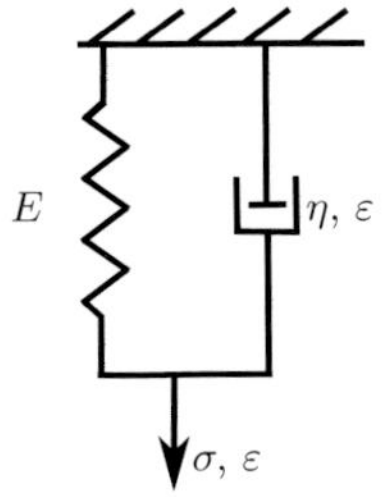

Figure 4.27

This means that $\sigma, \varepsilon \in \mathcal{D}'_+$ with respect to the variable $t \in \mathbb{R}$ and $x \in \mathbb{R}$ is considered a parameter, wherefrom it results $\sigma = 0$, $\varepsilon = 0$ for $t < 0$.

These relations show that the viscoelastic solid is in its natural state for $t < 0$.

These assumptions imply that φ and ψ are distributions from $\mathcal{D}'_+$, hence $\varphi, \psi \in \mathcal{D}'_+$, $\operatorname{supp}(\varphi), \operatorname{supp}(\psi) \subset [0, \infty)$.

Between the distributions φ and ψ, creep distribution and relaxation distribution, respectively, the following relation results:

$$\psi'(t) * \varphi'(t) = (\psi * \varphi)'' = \delta(t) , \tag{4.49}$$

where δ is the Dirac delta distribution.

Consequently, the formulas (4.48) become

$$\sigma = \psi * \frac{\partial \varepsilon}{\partial t} = \varepsilon * \frac{d\psi}{dt} , \tag{4.50}$$

$$\varepsilon = \varphi * \frac{\partial \sigma}{\partial t} = \sigma * \frac{d\varphi}{dt} , \tag{4.51}$$

where $\sigma(x, t), \varepsilon(x, t), \psi(t), \varphi(t) \in \mathcal{D}'_+ \subset \mathcal{D}'(\mathbb{R})$, the variable $x \in \mathbb{R}$ being considered a parameter.

Equations 4.50 and 4.51 represent the equivalent constitutive laws of the homogeneous, isotropic one-dimensional viscoelastic solid.

In particular, if the viscoelastic solid does not present the relaxation phenomenon, then $\psi(t) = EH(t)$, hence $\psi'(t) = E\delta(t)$ and from (4.50) we obtain

$$\sigma(x, t) = E\varepsilon(x, t) . \tag{4.52}$$

Obviously, this relation expresses Hooke's law for a one-dimensional elastic solid.

Similarly, if the viscoelastic solid does not present the creep phenomenon, then $\varphi(t) = H(t)/E$, hence $\varphi'(t) = \delta(t)/E$ and from (4.51) we obtain $\varepsilon = \sigma/E$, a relation that is equivalent to (4.52).

From the above considerations, it results that the elastic solid is a particular case of a viscoelastic solid, that is, in the case in which the latter one does not have the phenomenon of relaxation or creep.

Example 4.3 Determine the distributions of creep and relaxation $\varphi, \psi \in \mathcal{D}'_+$ when the constitutive law (4.46) is known for the Kelvin–Voigt model of viscoelastic solid.

From (4.46) and (4.50) we obtain $\varepsilon * \dot{\psi} = E\varepsilon + \eta\dot{\varepsilon}$. This equation can be written in the form

$$\varepsilon * \left(\dot{\psi} - E\delta(t) - \eta\dot{\delta}(t)\right) = 0 . \tag{4.53}$$

Because the convolution algebra $\mathcal{D}'_+$ has no divisors of zero, from (4.53) it results that

$$\dot{\psi} = E\delta(t) + \eta\dot{\delta}(t) , \tag{4.54}$$

namely $\psi(t) = EH(t) + \eta\delta(t) \in \mathcal{D}'_+$.

Using (4.49), we can determine the creep distribution $\varphi \in \mathcal{D}'_+$ as well.

Thus, taking into account (4.54), we obtain $\psi' * \varphi' = (E\delta + \eta\dot{\delta}) * \varphi' = E\varphi' + \eta\varphi'' = \delta$, namely $\mathrm{d}(E\varphi + \eta\varphi')/\mathrm{d}t = \mathrm{d}H/\mathrm{d}t$.

Because $\mathrm{supp}(\varphi) \subset [0, \infty)$, it results $\eta\varphi' + E\varphi = H(t)$. The solution of this linear equation in $[0, \infty)$ is

$$\varphi(t) = \frac{H(t)\,(1 - \exp(-Et/\eta))}{E}\,, \qquad t \in \mathbb{R}\,.$$

5
Applications of the Distribution Theory in Mechanics

5.1
Newtonian Model of Mechanics

The mathematical model of Newtonian mechanics, within which we operate, is based on introducing the concepts of space, time and mass, the corresponding quantities being independent from each other, [17]. Essentially, this mechanics is based on three laws, enunciated by Isaac Newton, that is,

1. First law: The velocity of a material point remains constant unless the material point is acted upon by an external force.
2. Second law: The acceleration $\tilde{d}^2 \boldsymbol{r}/dt^2$ of a body is collinear and in direct proportion to the given force $\boldsymbol{F}$ and in inverse proportion to the mass m, that is,

$$m \frac{\tilde{d}^2 \boldsymbol{r}}{dt^2} = \boldsymbol{F}\,, \quad t > 0\,, \tag{5.1}$$

where $\boldsymbol{F}$ is the given force (the vector sum of all the forces acting on a body), m is the mass, constant in time, and $\boldsymbol{r} = \boldsymbol{r}(t), t \geq 0$, is the position vector of that point with respect to a fixed reference point in an inertial reference frame (Figure 5.1); $\tilde{d}/dt$ is the derivative in the usual sense.
3. Third law: The mutual forces of action and reaction between two bodies are equal in magnitude, opposite and collinear.

In general, it is considered that the position vector $\boldsymbol{r} = \boldsymbol{r}(t), t \geq 0$, and gives the first and second-order derivatives (its components are functions of class C^2 for $t \geq 0$). It is also assumed that the force $\boldsymbol{F} = \boldsymbol{F}(\boldsymbol{r}, \tilde{d}\boldsymbol{r}/dt, t)$ is continuous with respect to the position vector, velocity and time.

To the equation of motion (5.1) are added the following initial conditions

$$\boldsymbol{r}_0 = \boldsymbol{r}(0) = \lim_{t \to +0} \boldsymbol{r}(t)\,, \quad \boldsymbol{v}_0 = \boldsymbol{v}(0) = \lim_{t \to +0} \frac{\tilde{d}\boldsymbol{r}(t)}{dt}\,. \tag{5.2}$$

We call the Cauchy problem corresponding to (5.1) the determination of the vector function $\boldsymbol{r} \in C^2([0, \infty))$, which for $t \geq 0$ verifies (5.1) and for $t = 0$ the initial conditions (5.2).

Distribution Theory, First Edition. Petre P. Teodorescu, Wilhelm W. Kecs, and Antonela Toma

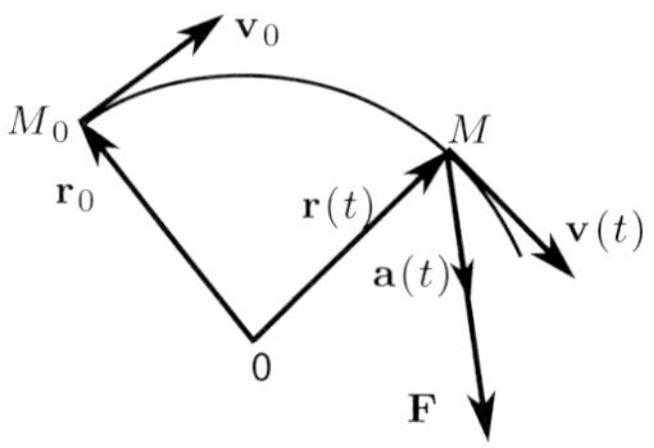

Figure 5.1

To use the distribution theory for solving mechanics problems, it is necessary to rewrite the equation of motion (5.1) in the distributions space.

To this end, we introduce vector distributions of function type $\boldsymbol{\rho}$ and $\mathcal{F}$, $t \in \mathbb{R}$, from $\mathcal{D}'_+$, by extension with null values for $t < 0$, both for the solution $\boldsymbol{r} = \boldsymbol{r}(t)$, $t \geq 0$, and for the force $\boldsymbol{F}$, by the relations

$$\boldsymbol{\rho}(t) = (H\boldsymbol{r})(t) = \begin{cases} 0\,, & t < 0 \\ \boldsymbol{r}(t)\,, & t \geq 0 \end{cases}, \quad \mathcal{F}(t) = (H\boldsymbol{F})(t) = \begin{cases} 0\,, & t < 0 \\ \boldsymbol{F}(t)\,, & t \geq 0 \end{cases},$$

where H is the Heaviside step function.

We note that $\boldsymbol{\rho}$ is a vector function of class $C^2(\mathbb{R})$ except at the origin where it has a discontinuity of first order.

The initial conditions (5.2) are the jumps of the function $\boldsymbol{\rho}$ and of the speed $\boldsymbol{v} = \tilde{\mathrm{d}}\boldsymbol{r}/\mathrm{d}t$ at the time $t = 0$.

Thus, we have

$$s_0(\boldsymbol{\rho}) = \boldsymbol{\rho}(+0) - \boldsymbol{\rho}(-0) = \boldsymbol{r}_0,\ s_0\left(\frac{\tilde{\mathrm{d}}\boldsymbol{\rho}}{\mathrm{d}t}\right) = \frac{\tilde{\mathrm{d}}\boldsymbol{\rho}}{\mathrm{d}t}(+0) - \frac{\tilde{\mathrm{d}}\boldsymbol{\rho}}{\mathrm{d}t}(-0) = \boldsymbol{v}_0\,.$$

Consequently, by differentiation, in the sense of distributions, we get

$$\begin{aligned} \frac{\mathrm{d}\boldsymbol{\rho}}{\mathrm{d}t} &= \frac{\tilde{\mathrm{d}}\boldsymbol{\rho}}{\mathrm{d}t} + s_0(\boldsymbol{\rho})\delta(t) = \frac{\tilde{\mathrm{d}}\boldsymbol{\rho}}{\mathrm{d}t} + \boldsymbol{r}_0\delta(t)\,, \\ \frac{\mathrm{d}^2\boldsymbol{\rho}}{\mathrm{d}t^2} &= \frac{\tilde{\mathrm{d}}^2\boldsymbol{\rho}}{\mathrm{d}t^2} + s_0\left(\frac{\tilde{\mathrm{d}}\boldsymbol{\rho}}{\mathrm{d}t}\right)\delta(t) + \boldsymbol{r}_0\delta'(t) = \frac{\tilde{\mathrm{d}}^2\boldsymbol{\rho}}{\mathrm{d}t^2} + \boldsymbol{v}_0\delta(t) + \boldsymbol{r}_0\delta'(t)\,. \end{aligned} \tag{5.3}$$

Because

$$\frac{\tilde{\mathrm{d}}^2\boldsymbol{\rho}}{\mathrm{d}t^2} = \begin{cases} 0\,, & t < 0 \\ \dfrac{\tilde{\mathrm{d}}^2\boldsymbol{r}}{\mathrm{d}t^2}\,, & t \geq 0 \end{cases} = H\frac{\tilde{\mathrm{d}}^2\boldsymbol{r}}{\mathrm{d}t^2}\,,$$

(5.1) is then equivalent to the equation

$$m\frac{\tilde{\mathrm{d}}^2\boldsymbol{\rho}}{\mathrm{d}t^2} = \mathcal{F}\,. \tag{5.4}$$

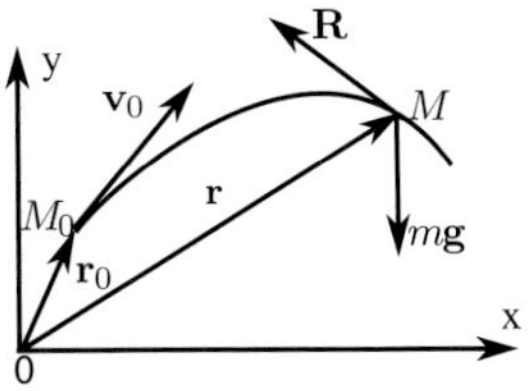

Figure 5.2

Taking into account (5.4) and (5.3), we obtain

$$m\frac{\mathrm{d}^2\boldsymbol{\rho}}{\mathrm{d}t^2} = \boldsymbol{\mathcal{F}} + m\boldsymbol{v}_0\delta(t) + m\boldsymbol{r}_0\delta'(t)\,, \quad \boldsymbol{\rho}, \boldsymbol{\mathcal{F}} \in \mathcal{D}'_+\,. \tag{5.5}$$

This equation written in the distributions space $\mathcal{D}'_+$ has the advantage to include the initial conditions of the problem and for $t > 0$ coincides with the original equation of motion (5.1).

Consequently, solving the Cauchy problem corresponding to (5.1) is equivalent to solving in $\mathcal{D}'_+$ of (5.5), and the solution obtained $\boldsymbol{\rho} = \boldsymbol{\rho}(t)$, $t \in \mathbb{R}$ is unique.

The effective solving of (5.5) depends on the concrete form of the force $\boldsymbol{F} = \boldsymbol{F}(\boldsymbol{r}, \boldsymbol{v}, t)$, $t \geq 0$. For this reason, to the solution determination we can apply the Laplace and Fourier transformations in distributions and properties of the convolution algebra $\mathcal{D}'_+$.

5.2 The Motion of a Heavy Material Point in Air

A material point M of mass m is released from the point M_0, having the radius vector $\boldsymbol{r}_0$, with the speed $\boldsymbol{v}_0$. We consider that the air resistance is a force the magnitude of which is proportional to the speed, of the form $\boldsymbol{R} = -km\boldsymbol{v}$, $k > 0$ (Figure 5.2).

We will determine the motion of the material point with respect to the reference system Oxy, where Ox and Oy are the horizontal and the vertical axes, respectively.

The equation of motion is

$$m\frac{\tilde{\mathrm{d}}^2\boldsymbol{r}}{\mathrm{d}t^2} = m\boldsymbol{g} - km\frac{\tilde{\mathrm{d}}\boldsymbol{r}}{\mathrm{d}t}\,, \quad t > 0\,,$$

namely

$$\frac{\tilde{\mathrm{d}}^2\boldsymbol{r}}{\mathrm{d}t^2} + k\frac{\tilde{\mathrm{d}}\boldsymbol{r}}{\mathrm{d}t} = \boldsymbol{g}\,, \quad t > 0\,, \tag{5.6}$$

where $\boldsymbol{g}(0, -g)$ is the gravitational acceleration.

The initial conditions are

$$\boldsymbol{r}_0 = \boldsymbol{r}(0) = \boldsymbol{r}|_{t=+0}\,, \quad \boldsymbol{v}_0 = \boldsymbol{v}(0) = \left.\frac{\tilde{\mathrm{d}}\boldsymbol{r}}{\mathrm{d}t}\right|_{t=+0}\,. \tag{5.7}$$

We introduce the radius vector $\boldsymbol{\rho}$ defined for $t \in \mathbb{R}$ by the relation

$$\boldsymbol{\rho}(t) = \begin{cases} 0\,, & t < 0\,, \\ \boldsymbol{r}\,, & t \geq 0\,. \end{cases}$$

Thus, the equation of motion (5.6) becomes

$$\frac{\tilde{\mathrm{d}}^2\boldsymbol{\rho}}{\mathrm{d}t^2} + k\frac{\tilde{\mathrm{d}}\boldsymbol{\rho}}{\mathrm{d}t} = H\boldsymbol{g}\,. \tag{5.8}$$

Differentiating in the sense of the distribution theory the radius vectors $\boldsymbol{\rho}$ we obtain

$$\frac{\mathrm{d}\boldsymbol{\rho}}{\mathrm{d}t} = \frac{\tilde{\mathrm{d}}\boldsymbol{\rho}}{\mathrm{d}t} + s_0(\boldsymbol{\rho})\delta(t) = \frac{\tilde{\mathrm{d}}\boldsymbol{\rho}}{\mathrm{d}t} + \boldsymbol{r}_0\delta(t)\,,$$

$$\frac{\mathrm{d}^2\boldsymbol{\rho}}{\mathrm{d}t^2} = \frac{\tilde{\mathrm{d}}^2\boldsymbol{\rho}}{\mathrm{d}t^2} + s_0\left(\frac{\tilde{\mathrm{d}}\boldsymbol{\rho}}{\mathrm{d}t}\right)\delta(t) + \boldsymbol{r}_0\delta'(t) = \frac{\tilde{\mathrm{d}}^2\boldsymbol{\rho}}{\mathrm{d}t^2} + \boldsymbol{v}_0\delta(t) + \boldsymbol{r}_0\delta'(t)\,. \tag{5.9}$$

Taking into account (5.9) the equation of motion (5.8) takes the form

$$\frac{\mathrm{d}^2\boldsymbol{\rho}}{\mathrm{d}t^2} + k\frac{\tilde{\mathrm{d}}\boldsymbol{\rho}}{\mathrm{d}t} = H\boldsymbol{g} + (\boldsymbol{v}_0 + k\boldsymbol{r}_0)\delta(t) + \boldsymbol{r}_0\delta'(t)\,. \tag{5.10}$$

This is the corresponding equation of motion (5.8) written in the space of distributions $\mathcal{D}'_+$.

Using the operator $P(\mathrm{d}/\mathrm{d}t) = \mathrm{d}^2/\mathrm{d}t^2 + k\mathrm{d}/\mathrm{d}t$, (5.10) can be written as a convolution equation in $\mathcal{D}'_+$

$$P\left(\frac{\mathrm{d}}{\mathrm{d}t}\right)\delta * \boldsymbol{\rho} = H\boldsymbol{g} + (\boldsymbol{v}_0 + k\boldsymbol{r}_0)\delta(t) + \boldsymbol{r}_0\delta'(t)\,. \tag{5.11}$$

Let $E \in \mathcal{D}'_+$ be the fundamental solution of the operator $P(\mathrm{d}/\mathrm{d}t)$, hence $P(\mathrm{d}/\mathrm{d}t)E = \delta$.

Then the unique solution of equation (5.11) and (5.10) will be

$$\boldsymbol{\rho} = E * (H\boldsymbol{g} + (\boldsymbol{v}_0 + k\boldsymbol{r}_0)\delta(t) + \boldsymbol{r}_0\delta'(t))\,,$$

namely

$$\boldsymbol{\rho} = (E * H)\boldsymbol{g} + E(\boldsymbol{v}_0 + k\boldsymbol{r}_0) + \boldsymbol{r}_0 E'\,. \tag{5.12}$$

The fundamental solution E has the expression

$$E = HY\,, \tag{5.13}$$

where Y is the solution of the homogeneous equation

$$P\left(\frac{\mathrm{d}}{\mathrm{d}t}\right)Y = Y'' + kY' = 0\,,$$

which satisfies the conditions $Y(0) = 0$, $Y'(0) = 1$.

We obtain

$$Y(t) = \frac{(1 - \exp(-kt)/k)}{k} \,, \qquad t \in \mathbb{R} \,.$$

Taking into account (5.13), we obtain for the fundamental solution E the expression $E(t) = k^{-1}H(t)(1 - \exp(-kt))$, $t \in \mathbb{R}$.

It follows $E'(t) = \mathrm{d}E/\mathrm{d}t = H(t)\exp(-kt)$ and thus the solution (5.12) becomes

$$\boldsymbol{\rho} = (E * H)\mathbf{g} + k^{-1}H(\boldsymbol{v}_0 + k\boldsymbol{r}_0) - k^{-1}H\boldsymbol{v}_0\exp(-kt) \,. \tag{5.14}$$

To calculate the function type distribution $E * H \in \mathcal{D}'_+$ we remark that

$$\begin{aligned}(E * H)(t) &= k^{-1}H(t)(1 - \exp(-kt)) * H(t) \\ &= k^{-1}\int_{\mathbb{R}} (1 - \exp(-k(t - \tau)))H(1 - \tau)H(\tau)\mathrm{d}\tau \\ &= k^{-1}\int_0^{\infty} (1 - \exp(-k(t - \tau)))H(t - \tau)\mathrm{d}\tau \,.\end{aligned}$$

Making the substitution $t - \tau = u$, we obtain

$$\begin{aligned}(E * H)(t) &= k^{-1}\int_{-\infty}^{t} (1 - \exp(-ku))H(u)\mathrm{d}u \\ &= \begin{cases} 0 \,, & t < 0 \,, \\ k^{-1}\displaystyle\int_0^t (1 - \exp(-ku))\mathrm{d}u \,, & t \geq 0 \,. \end{cases}\end{aligned}$$

Thus, we get

$$\begin{aligned}(E * H)(t) &= \begin{cases} 0 \,, & t < 0 \,, \\ k^{-2}(kt + \exp(-kt) - 1) \,, & t \geq 0 \,, \end{cases} \\ &= k^{-2}(kt + \exp(-kt) - 1)H(t) \,.\end{aligned}$$

With this, the expression (5.14) becomes

$$\boldsymbol{\rho}(t) = H(t)\left(k^{-2}\mathbf{g}\,(kt + \exp(-kt) - 1) + k^{-1}\boldsymbol{v}_0\,(1 - \exp(-kt)) + \boldsymbol{r}_0\right) \,, \qquad t \in \mathbb{R} \,,$$

which is the solution of (5.10) in $\mathcal{D}'_+$.

For $t > 0$ we obtain the solution of the Cauchy problem corresponding to (5.6) and the initial conditions (5.7)

$$\boldsymbol{r}(t) = k^{-2}\mathbf{g}(kt + \exp(-kt) - 1) + k^{-1}\boldsymbol{v}_0(1 - \exp(-kt)) + \boldsymbol{r}_0 \,, \quad t \geq 0 \,. \tag{5.15}$$

Particular case We consider the initial conditions $\boldsymbol{r}_0 = \boldsymbol{r}_0(0, h)$ and $\boldsymbol{v}_0 = \boldsymbol{v}_0(v_0, 0)$. This means that the material point is launched horizontally with initial speed v_0 from the height h. Because $\boldsymbol{r}(t) = \boldsymbol{r}(x(t), y(t))$, from (5.15) we shall obtain for the material point $M(x, y)$ the equations of motion

$$x = k^{-1} v_0 (1 - \mathrm{e}^{-kt}) \,, \quad y = -k^{-2} g(tk + \mathrm{e}^{-kt} - 1) \,, \quad t \geq 0 \,. \tag{5.16}$$

From the first equation we get $\exp(-kt) = (v_0 - kx)/v_0$, where $kt = \ln(v_0(v_0 - kx)^{-1})$.

By introducing these functions in the expression of the ordinate y given by (5.16), we obtain for the material point trajectory the equation

$$y = h - \frac{g}{k^2} \ln \frac{v_0}{v_0 - kx} + \frac{gx}{kv_0} \,, \quad x \in [0, \ell] \,,$$

where the condition $y(\ell) = 0$ is satisfied at ℓ.

5.3
Linear Oscillator

The oscillatory motion is often met in technical sciences, because the engine parts are made of materials that possess a certain degree of elasticity, and due to this elasticity, they have the ability to vibrate.

In practical cases, the vibrations of a system of material points take place. Very often, the essential characteristics of vibrations of a system of material points are identical with the vibration characteristics of a single material point.

5.3.1
The Cauchy Problem and the Phenomenon of Resonance

We consider a material point M which is moving on the Ox-axis, driven by an elastic force of attraction $F(x) = -kx$, $k > 0$ (Figure 5.3) and a disturbing force $Q(t) = mq(t)$, $t \geq 0$, where m is the mass point and k the elastic coefficient (which characterizes from the mechanical point of view, the elastic spring that develops, according to Hooke's law, the elastic attractive force F proportional to the elongation $\overline{OM} = x$).

According to the second law of mechanics, we obtain the equation of motion of the linear oscillator

$$\frac{\tilde{\mathrm{d}}^2}{\mathrm{d}t^2} x(t) + \omega^2 x(t) = q(t) \,, \quad \omega^2 = \frac{k}{m} > 0 \,, \quad t > 0 \,. \tag{5.17}$$

The number ω is called the pulsation of the linear oscillator.

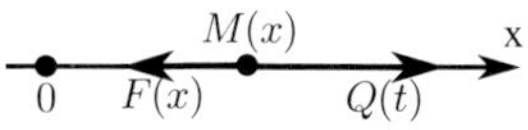

Figure 5.3

If the disturbing force Q is missing, hence $q(t) = 0$, then we obtain the equation of free (harmonic) oscillations

$$\frac{\tilde{d}^2}{dt^2} x(t) + \omega^2 x(t) = 0\,, \quad t > 0\,.$$

For this reason, (5.17) is called the equation of forced oscillations, the disturbing force per unit mass being q. The function q is assumed to be locally integrable on $[0, \infty)$.

For (5.17) we associate the initial conditions

$$x_0 = x(0) = \lim_{t \to +0} x(t), v_0 = v(0) = \lim_{t \to +0} \frac{\tilde{d}x(t)}{dt}\,. \tag{5.18}$$

The Cauchy problem for (5.17) consists in the determination of the function $x \in C^2([0, \infty))$, which for $t > 0$ verifies (5.17), and for $t = 0$ the initial conditions (5.18).

Making an extension with null values for $t < 0$, we introduce the functions

$$\overline{x}(t) = H(t)x(t)\,, \quad \overline{q}(t) = H(t)q(t)\,, \quad t \in \mathbb{R}\,.$$

Differentiating in the sense of distributions, we obtain

$$\begin{aligned}\frac{d}{dt}\overline{x}(t) &= \frac{\tilde{d}}{dt}\overline{x}(t) + x_0\delta(t)\,,\\ \frac{d^2}{dt^2}\overline{x}(t) &= \frac{\tilde{d}^2}{dt^2}\overline{x}(t) + v_0\delta(t) + x_0\delta'(t)\,.\end{aligned}$$

The equation of motion (5.17) takes in $\mathcal{D}'_+$ the form

$$\frac{d^2}{dt^2}\overline{x}(t) + \omega^2\overline{x}(t) = \overline{q}(t) + v_0\delta(t) + x_0\delta'(t)\,. \tag{5.19}$$

We determine the solution of this equation by applying the Laplace transformation in distributions.

Noting $L[\overline{x}](p) = \tilde{x}(p)$, $L[\overline{q}](p) = \tilde{q}(p)$, the operational equation corresponding to (5.19) is

$$(p^2 + \omega^2)\tilde{x}(p) = \tilde{q}(p) + v_0 + p\,x_0\,,$$

where

$$\tilde{x}(p) = x_0\frac{p}{p^2 + \omega^2} + \frac{v_0}{\omega}\frac{\omega}{p^2 + \omega^2} + \frac{\tilde{q}(p)}{\omega}\frac{\omega}{p^2 + \omega^2}\,. \tag{5.20}$$

Taking into account the relations

$$L[H(t)\cos\omega t](p) = \frac{p}{p^2 + \omega^2}\,, \quad L[H(t)\sin\omega t](p) = \frac{\omega}{p^2 + \omega^2}\,,$$

$$L[H(t)q(t) * H(t)\sin\omega t](p) = \tilde{q}(p)\frac{\omega}{p^2 + \omega^2}$$

and applying the inverse Laplace transform to (5.20), we obtain

$$\overline{x}(t) = \mathrm{L}^{-1}[\tilde{x}(p)] = \frac{H(t)}{\omega} q(t) * H(t) \sin \omega t + x_0 H(t) \cos \omega t + \frac{v_0}{\omega} H(t) \sin \omega t \,,$$

$t \in \mathbb{R}$.

For $t \geq 0$ we have the Cauchy problem solution corresponding to the linear oscillator

$$x(t) = \frac{1}{\omega} \int_0^t q(\tau) \sin \omega (t - \tau) \mathrm{d}\tau + x_0 \cos \omega t + \frac{v_0}{\omega} \sin \omega t \,. \tag{5.21}$$

From a practical point, of interest is the case in which the disturbing force $Q(t) = mq(t), t \geq 0$ is periodic and has the form $Q(t) = ma \sin \omega t$, $a > 0$.

In this case, the proper pulsation $\omega = \sqrt{km^{-1}}$ of the oscillator coincides with the pulsation of the disturbing force.

Because $q(t) = a \sin \omega t$, the differential equation (5.17) becomes

$$\frac{\tilde{\mathrm{d}}^2}{\mathrm{d}t^2} x(t) + \omega^2 x(t) = a \sin \omega t \,, \quad t > 0 \,,$$

and, according to the formula (5.21), the solution of this equation is

$$x(t) = x_0 \cos \omega t + \frac{v_0}{\omega} \sin \omega t + a \frac{\sin \omega t}{2\omega^2} - \frac{at}{2\omega} \cos \omega t \,, \quad t > 0 \,, \tag{5.22}$$

because

$$\int_0^t \sin \omega \tau \sin \omega (t - \tau) \mathrm{d}\tau = \frac{\sin \omega t}{2\omega} - \frac{t \cos \omega t}{2} \,.$$

We note that the first three terms of the solution (5.22) are bounded periodic functions. The fourth term which is due to the disturbing force is not a periodic function and it is not bounded either, because $\lim_{t \to \infty} |t \cos \omega t| = \infty$. This means that the elongation module x increases unboundedly and this represents the phenomenon of resonance.

Remark 5.1 In the case of an elastic repulsive force $F = kx, k > 0$ (Figure 5.4) the equation of motion becomes

$$\frac{\tilde{\mathrm{d}}^2}{\mathrm{d}t^2} x(t) - \omega^2 x(t) = q(t) \,, \quad t > 0 \,, \tag{5.23}$$

where $\omega^2 = km^{-1} > 0$.

Proceeding as in the case of an attractive force, we get

$$\tilde{x}(p) = x_0 \frac{p}{p^2 - \omega^2} + \frac{v_0}{\omega} \frac{\omega}{p^2 - \omega^2} + \frac{\tilde{q}(p)}{\omega} \frac{\omega}{p^2 - \omega^2} \,.$$

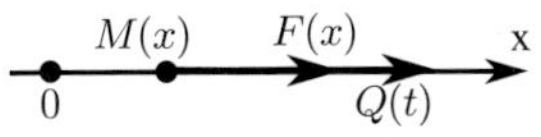

Figure 5.4

Based on the relations

$$L[H(t)\cosh\omega t](p) = p(p^2-\omega^2)^{-1}\,,\quad L[H(t)\sinh\omega t](p) = \omega(p^2-\omega^2)^{-1}$$

we obtain for the solution of the Cauchy problem the expression

$$x(t) = \frac{1}{\omega}\int_0^t q(\tau)\sinh\omega(t-\tau)\mathrm{d}\tau + x_0\cosh\omega t + \frac{v_0}{\omega}\sinh\omega t\,,\qquad t \geq 0\,.$$

5.4
Two-Point Problem

The Cauchy problem considered until now is a local problem, the initial conditions are placed at a single point in the interval where the equation is defined.

We refer to the two-point problem for the equation of motion (5.17) of the linear oscillator the determining of the function $x \in C^2([0,\infty))$, which satisfies (5.17) and at the times $t_1, t_2 \in (0,\infty)$ it takes the values

$$x(t_1) = x_1, x(t_2) = x_2\,. \tag{5.24}$$

We note that, in general, a two-point problem may not have a solution or may not have a unique solution.

Taking into account (5.21), we consider for the two-point problem a solution of the form

$$x(t) = \frac{1}{\omega}\int_{t_1}^{t_2} q(\tau)\sin\omega(t-\tau)\mathrm{d}\tau + A\sin\omega(t-t_1) + x_1\cos\omega(t-t_1)\,,\qquad t\in[0,\infty)\,, \tag{5.25}$$

where $y(t) = A\sin\omega(t-t_1) + x_1\cos\omega(t-t_1)$ represents the homogeneous solution of the equation

$$\frac{\tilde{\mathrm{d}}^2 y(t)}{\mathrm{d}t^2} + \omega^2 y(t) = 0\,,$$

and $Y(t) = (1/\omega)\int_{t_1}^{t} q(\tau)\sin\omega(t-\tau)\mathrm{d}\tau$ a particular solution of the equation of motion

$$\frac{\tilde{\mathrm{d}}^2}{\mathrm{d}t^2}Y(t) + \omega^2 Y(t) = q(t)\,,\qquad t \geq 0\,.$$

The verification is made directly using Leibniz's formula of differentiation of an integral depending on a parameter.

We observe that the solution (5.25) satisfies the first two-point condition $x(t_1) = x_1$. For the second condition, we get

$$x_2 = \frac{1}{\omega}\int_{t_1}^{t_2} q(\tau)\sin\omega(t_2-\tau)\mathrm{d}\tau + A\sin\omega(t_2-t_1) + x_1\cos\omega(t_2-t_1)\,. \quad (5.26)$$

Eliminating the unknown A between (5.25) and (5.26), we obtain the solution of the two-point problem under the form

$$x(t) = \frac{1}{\omega\sin\omega(t_2-t_1)}\left[\sin\omega(t_2-t_1)\int_{t_1}^{t} q(\tau)\sin\omega(t-\tau)\mathrm{d}\tau \right.$$
$$\left. -\sin\omega(t-t_1)\int_{t_1}^{t_2} q(\tau)\sin\omega(t_2-\tau)\mathrm{d}\tau + \omega(x_1\sin\omega(t_2-t) + x_2\sin\omega(t-t_1))\right]\,. \quad (5.27)$$

This result is valid if $t_2 \neq t_1 + n\pi/\omega$, $n \in \mathbb{N}$.

If we put the two-point conditions of the form (5.24) to (5.23), corresponding to a repulsive elastic force, and proceed similarly, then the problem solution will be that obtained from (5.27) if, instead of circular trigonometric functions sin and cos, we use the hyperbolic functions sinh and cosh, respectively.

5.5
Bending of the Straight Bars

Bodies which have one dimension (length) considerably greater than the other two (cross-section) will be called bars. The two dimensions (average) a and b of the cross-section will be of the same order of magnitude and satisfy the condition $a, b \ll \ell$, where ℓ is the length of the bar axis. According to the shape of the axis, the bars may be straight or curved.

We consider the problem of bending of a homogeneous and isotropic straight elastic bar with constant cross-section, under the assumption of Bernoulli–Euler model concerning plane cross-sections.

In the theory of strength of materials, a straight bar is reduced, from the deflection point of view, to its axis; the deflected axis of the bar is also called the elastic line of the bar (or the deflected curve).

Acknowledging that the straight bar is subjected to bending due to a normal load which can be represented by a distribution q and taking the bar axis as a coordinate axis Ox, we may write the differential equation of elastic line in the form

$$\frac{\mathrm{d}^4 v(x)}{\mathrm{d}x^4} = \frac{1}{EI}q(x)\,, \quad x \in \mathbb{R} \quad (5.28)$$

where the derivative is considered in the sense of distributions and $v, q \in \mathcal{D}'(\mathbb{R})$; it was noted with E the longitudinal modulus of elasticity and with I the moment of inertia of the cross-section with respect to the neutral axis (which passes through the bar axis and is normal to the plane of loading). The product EI represents the bending rigidity of the bar.

To determine the deflection $v \in \mathcal{D}'(\mathbb{R})$ of the bar, four conditions must be added to (5.28), according to the supporting of the bar.

In these conditions occur, also, the expressions of the shear force T and of the bending moment M by the formulas

$$T(x) = \frac{\tilde{\mathrm{d}} M(x)}{\mathrm{d}x} = -EI\frac{\tilde{\mathrm{d}}^3 v(x)}{\mathrm{d}x^3}\,, \quad M(x) = -EI\frac{\tilde{\mathrm{d}}^2 v(x)}{\mathrm{d}x^2}\,, \quad x \in \mathbb{R}\,,$$

where M is considered positive if it acts in the direct sense (Figure 5.5).

Because the fundamental solution in $\mathcal{D}'_+$ of the operator $\mathrm{d}^2/\mathrm{d}x^2$ is $E^*(x) = H(x)x^3/6$, the general solution of the elastic line equation (5.28) will be expressed in the form

$$v(x) = \frac{1}{EI}\left(H(x)\frac{x^3}{6} + A_1 + A_2 x + \frac{1}{2}A_3 x^2 + \frac{1}{6}A_4 x^3\right) * q(x)\,, \quad x \in \mathbb{R}\,, \tag{5.29}$$

where the constants A_1, A_2, A_3, A_4 are determined from the supporting conditions of the bar.

For example, let there be a straight bar OA, of length a (Figure 5.6), having a built-in end at the point O and the other end free (cantilever bar); we assume that a concentrated force P acts at the free end. The equivalent load will be $q(x) = P\delta(x-a)$.

Taking into account (5.29), the elastic line equation becomes

$$v(x) = \frac{P}{EI}\left(H(x-a)\frac{(x-a)^3}{6} + A_1 + A_2(x-a) + \frac{A_3}{2}(x-a)^2 + \frac{A_4}{6}(x-a)^3\right)\,,$$

where $x \in \mathbb{R}$.

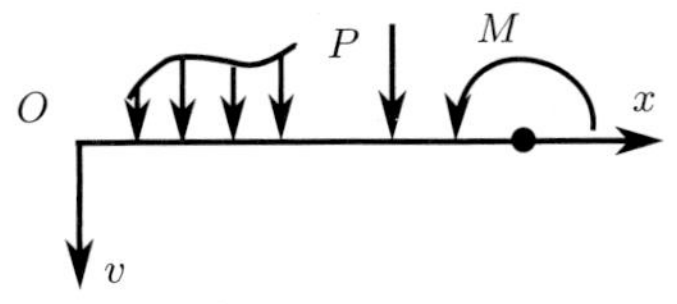

Figure 5.5

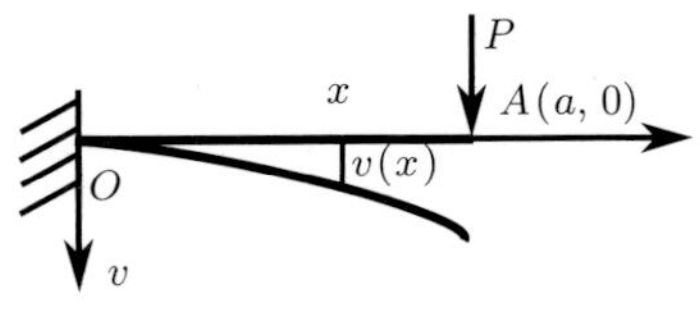

Figure 5.6

The constants A_1, A_2, A_3, A_4 are determined from the conditions at the built-in end

$$x = 0 : v(x) = 0 \,, \quad \frac{\tilde{\mathrm{d}} v(x)}{\mathrm{d}x} = 0 \,,$$

and from the conditions concerning the bending moment and the shearing force at the built-in cross-section

$$x = 0 : EI \frac{\tilde{\mathrm{d}}^2 v(x)}{\mathrm{d}x^2} = aP \,, \quad EI \frac{\tilde{\mathrm{d}}^3 v(x)}{\mathrm{d}x^3} = -P \,.$$

Thus, we obtain the constants $A_1 = a^3/3$, $A_2 = a^2/2$, $A_3 = 0$, $A_4 = -1$. Consequently, for $x \in [0, a]$ the equation of the elastic line is $v(x) = Px^2(3a - x)/6EI$.

6
Applications of the Distribution Theory to the Mechanics of the Linear Elastic Bodies

6.1
The Mathematical Model of the Linear Elastic Body

Under the action of external loads, the particles that make up a solid body change (in time) the position which they had before the action of these loads with respect to a considered fixed system of reference. If, by a translational and a rotational motion, we can make the particles of the body, subjected to the action of external loads, to get the position they had before the implementation of these loads, the motion is said to be of a rigid body. Otherwise, the body undergoes a deformation.

The totality of the deformations to which a particle of the body has been subjected constitutes the state of deformation at a point (center of mass of the particle). The totality of the states of deformation corresponding to all points (particles) of the solid body constitutes its state of deformation of the body.

Bodies which can be subjected only to rigid body displacements are said to be rigid solids; all the other solid bodies are termed deformable solids.

In the following, we shall make a simplifying hypothesis: we shall assume that we have to deal with small deformations and rotations only, which are negligible when referred to unity.

We shall use an orthogonal system of coordinate axes $Ox_1x_2x_3$. The state of deformation and the state of displacement are characterized by the symmetric tensors of the second order

$$T_\varepsilon = (\varepsilon_{ij}) \,, \quad \varepsilon_{ij} = \varepsilon_{ji} \,, \quad \varepsilon_{ij} = \varepsilon_{ij}(x_1, x_2, x_3) \,, \quad i, j = 1, 2, 3 \,,$$
$$T_\sigma = (\sigma_{ij}) \,, \quad \sigma_{ij} = \sigma_{ji} \,, \quad \sigma_{ij} = \sigma_{ij}(x_1, x_2, x_3) \,, \quad i, j = 1, 2, 3 \,.$$

We note that the components ε_{ii}, ε_{ij}, $i \neq j$, correspond to the linear strains and to the angular strains, respectively.

Similarly, the components σ_{ii} and σ_{ij}, $i \neq j$, are the normal stresses and the tangential stresses, respectively.

The state of displacement is given by the displacement vector $\boldsymbol{u}$, the components of which are $u_1 = u_1(x_1, x_2, x_3)$, $u_2 = u_2(x_1, x_2, x_3)$, $u_3 = u_3(x_1, x_2, x_3)$.

Distribution Theory, First Edition. Petre P. Teodorescu, Wilhelm W. Kecs, and Antonela Toma

In the linear case considered, between the tensor components T_ε and the vector components $\boldsymbol{u}$, Cauchy's relations

$$\varepsilon_{ij} = \frac{(u_{i,j} + u_{j,i})}{2}, \quad i, j = 1, 2, 3, \tag{6.1}$$

take place, where $(.)_{,j}$ subscript is a shorthand for $\partial(.)/\partial x_j$.

Equations 6.1, regarded as a system of equations in the unknown functions u_i, $i = 1, 2, 3$, will have to be compatible with one another. The conditions of compatibility of this system, which are at the same time conditions of continuity of the deformations, have been stated by B. de Saint-Venant in the form

$$\varepsilon_{ik,js} - \varepsilon_{jk,is} = \varepsilon_{is,jk} - \varepsilon_{js,ik}, \quad i, j, k, s = 1, 2, 3.$$

Specifying, we have the following equations of compatibility

$$\begin{aligned}
\varepsilon_{11,22} + \varepsilon_{22,11} &= 2\varepsilon_{12,12}, \\
\varepsilon_{11,33} + \varepsilon_{33,11} &= 2\varepsilon_{13,13}, \\
\varepsilon_{33,22} + \varepsilon_{22,33} &= 2\varepsilon_{23,23}, \\
\varepsilon_{11,23} &= (-\varepsilon_{23,1} + \varepsilon_{13,2} + \varepsilon_{12,3})_{,1}, \\
\varepsilon_{22,13} &= (-\varepsilon_{31,2} + \varepsilon_{21,3} + \varepsilon_{23,1})_{,2}, \\
\varepsilon_{33,12} &= (-\varepsilon_{12,3} + \varepsilon_{32,1} + \varepsilon_{31,2})_{,3}.
\end{aligned}$$

Hence, it follows that the displacements u_i, $i = 1, 2, 3$, must be functions of class C^3, while the deformations must be at least functions of class C^2.

In the dynamic case, these functions are also dependent on the time variable t, the displacements being thus of the form $u_i = u_i(x_1, x_2, x_3, t)$, $i = 1, 2, 3$.

The constitutive law of a solid emphasizes the physical properties of the material; from a mathematical viewpoint, such a law furnishes the relations existing between the strains and the stresses.

In the following we shall make some simplifying assumptions that determine the constitutive law we will use (corresponding to linear elastic bodies):

1. The solid body is isotropic, that is, it has the same mechanical and physical properties in any direction in the neighborhood of any of its points.
2. The solid body is homogeneous if it possesses the same mechanical and physical properties at every point. Based on these properties, the mechanical coefficients of the material which are involved in the constitutive law are constant with respect to the spatial variables.
3. The solid under investigation is perfectly elastic. Under the action of external loads, the body undergoes deformation; when the action of the loads ceases, the body returns to its initial position and shape (without the occurrence of any hysteresis effect); the deformation is reversible. Hence the distorted shape of a solid body will be affected only by the external loads acting upon it at that moment. Between stress and strain there will be a bijective connection; thus

the state of stress at a point of the solid body will depend only on the state of strain in the neighborhood of that point. In this way, we are in the field of the elasticity theory. We use a linear relation as constitutive law (Hooke's law). This hypothesis often corresponds satisfactorily to the physical phenomenon and leads, on the other hand, to important simplifications of the mathematical calculations.

Hooke's law, which will serve as the constitutive law, is written in the form

$$\sigma_{ij} = \lambda \varepsilon \delta_{ij} + 2\mu \varepsilon_{ij} , \quad i, j = 1, 2, 3 , \tag{6.2}$$

where λ and μ are Lamé's elastic constants,

$$\delta_{ij} = \begin{cases} 1 , & i = j , \\ 0 , & i \neq j , \end{cases}$$

and $\varepsilon = \varepsilon_{kk} = \varepsilon_{11} + \varepsilon_{22} + \varepsilon_{33}$.

By the contraction of the tensor σ_{ij} given by (6.2) we obtain

$$\sigma = (3\lambda + 2\mu)\varepsilon ,$$

where $\sigma = \sigma_{kk} = \sigma_{11} + \sigma_{22} + \sigma_{33}$.

Because the system of equations given by Hooke's law (6.2) acknowledges a unique solution with respect to the strain components ε_{ij}, we obtain

$$\varepsilon_{ij} = \frac{\sigma_{ij}}{2\mu} - \frac{\lambda \sigma \delta_{ij}}{2\mu(2\mu + 3\lambda)} , i, j = 1, 2, 3 ,$$

which involves the conditions $\mu \neq 0, 2\mu + 3\lambda \neq 0$.

The dependence among Lamé's elastic constants, longitudinal elasticity modulus E of Young and the Poisson ratio ν is given by

$$\nu = \frac{\lambda}{2(\lambda + \mu)} , \quad E = \frac{\mu(2\mu + 3\lambda)}{\lambda + \mu} = 2(1 + \nu)\mu . \tag{6.3}$$

Because $\nu \in (0, 1/2)$, the system of equations (6.3) is compatible determined with respect to Lamé's elastic constants, and we obtain

$$\lambda = \frac{\nu E}{(1 + \nu)(1 - 2\nu)} , \quad \mu = \frac{E}{2(1 + \nu)} .$$

We note that we have $\lambda > 0, \mu > 0$, and for the limit case $\nu = 1/2$ we get $E = 0$, which shows the incompressibility of the body. The elastic constants λ and μ are dependent on point, for the elastic body. In the case of the isotropic and homogeneous elastic body, λ and μ are constants from the point of view of the elastic properties. This remark is obviously true for the pair of elastic constants E and ν.

6.2
Equations of the Elasticity Theory

The elastic state of a body is completely determined by the tensors T_σ, T_ε and the displacement vector $\boldsymbol{u}$, quantities which determine at each point of the elastic body the states of stress, of deformation and of displacement. These three quantities T_σ, T_ε, $\boldsymbol{u}$ contain a total of 15 $(6 + 6 + 3)$ unknowns which depend on the spatial coordinates x_i and the temporal variable t. The 15 unknowns $\sigma_{ij}, \varepsilon_{ij}$ and u_i are connected by three groups of 15 independent equations, namely:

Equations of motion

$$\sigma_{ij,j} + X_i = \rho \ddot{u}_i \,, \quad (x_1, x_2, x_3) \in \Omega_0 \,, \quad t > t_0 \,, \tag{6.4}$$

where $X_i = X_i(x_1, x_2, x_3, t)$ are the components of the body force and $\ddot{u}_i = \partial^2 u_i / \partial t^2$;

Cauchy's equations (geometric equations)

$$\varepsilon_{ij} = \frac{(u_{i,j} + u_{j,i})}{2} \,, \quad (x_1, x_2, x_3) \in \overline{\Omega}_0 \,, \quad t \geq t_0 \,; \tag{6.5}$$

Hooke's constitutive law (physical equations)

$$\sigma_{ij} = \lambda \varepsilon \delta_{ij} + 2\mu \varepsilon_{ij} \,, \quad (x_1, x_2, x_3) \in \overline{\Omega}_0 \,, \quad t \geq t_0 \,, \quad i, j = 1, 2, 3 \,, \tag{6.6}$$

valid for isotropic elastic bodies.

Equations 6.4, 6.5, and 6.6 form the complete system of equations of elastodynamic, because they determine uniquely the quantities T_σ, T_ε, $\boldsymbol{u}$, hence the elastic state of the body, knowing the mass forces, the initial conditions and the boundary conditions.

The relation between the elastic body and the other bodies with which it is in contact occurs on a part or on the whole frontier S of the elastic body, connections which form the boundary conditions.

Because the body deformation occurs in time, the boundary conditions must be supplemented with initial conditions

$$u_i(x_1, x_2, x_3, t_0) = u_{i0}, \dot{u}(x_1, x_2, x_3, t_0) = \dot{u}_{i0} \,, \quad (x_1, x_2, x_3) \in \overline{\Omega}_0 = \Omega_0 \cup S \,, \tag{6.7}$$

where the frontier $S = \partial \Omega_0$ of the body is allowed to be piecewise smooth.

In relation to the boundary conditions, there are three basic types of boundary value problems of the theory of elasticity.

1) *The first fundamental problem (boundary conditions in displacements)* This problem consists of determining the solution $(\sigma_{ij}, \varepsilon_{ij}, u_i)$ of the system (6.4)–(6.6) which verifies the initial conditions (6.7) and the boundary conditions

$$u_i(x_1, x_2, x_3, t) = u_i^*(x_1, x_2, x_3, t) \,, \quad (x_1, x_2, x_3) \in S \,, \quad t \geq 0 \,, \tag{6.8}$$

where $u_i \in C^{3,2}(\Omega_0 \times [t_0, T]) \cap C^{1,1}(\overline{\Omega}_0 \times [t_0, T])$, $\sigma_{ij} \in C^{2,0}(\Omega_0 \times [t_0, T]) \cap C^{0,0}(\overline{\Omega}_0 \times [t_0, T])$.

The density X_i of the mass force per unit volume is assumed to be known.

If the elastic body is unbounded, the boundary conditions in displacement (6.8) is completed with regularity conditions at infinity, consisting of the assumption that σ_{ij} and u_i tend to zero for $|R| = \sqrt{x_i^2 + y_i^2 + z_i^2} \to \infty$. The behavior at infinity can be confirmed of the form $\lim_{|R|\to\infty}(\sigma_{ij}, u_i) = 0$.

2) *The second fundamental problem (boundary conditions in stresses)* In this case, the components $p_i(x_j, t)$ of the density $p(x_j, t)$ of the surface loads acting on the boundary S of the elastic body, and the components σ_{ij} of the stresses satisfy the relations

$$\sigma_{ij}(x_1, x_2, x_3, t)n_j(x_1, x_2, x_3) = p_i(x_1, x_2, x_3, t) , \quad (x_1, x_2, x_3) \in S , \quad t \geq t_0 , \tag{6.9}$$

where n_j are the director cosines of the external normal vector on the boundary S, considered fixed, that is, independent of time.

3) *Mixed fundamental problem* In this type of problem it is acknowledged that on one side S_u of the boundary S we specify displacements u_i^*, and on the remaining boundary S_σ, $S_u \cup S_\sigma = S$, there are specified components p_i of the density $\boldsymbol{p}(x_j, t)$ of the surface loads. Thus, on the boundary $S = S_u \cup S_\sigma$ we have the conditions

$$\begin{aligned} &u_i(x_1, x_2, x_3, t) = u_i^*(x_1, x_2, x_3, t) , \quad (x_1, x_2, x_3) \in S_u , \quad t \geq t_0 , \\ &\sigma_{ij}(x_1, x_2, x_3, t)n_j(x_1, x_2, x_3) = p_i(x_1, x_2, x_3, t) , \quad (x_1, x_2, x_3) \in S_\sigma . \end{aligned} \tag{6.10}$$

The mixed fundamental boundary value problem consists of determining the elastic state of the body, hence of the solution $(\sigma_{ij}, \varepsilon_{ij}, u_i)$ of the complete system of equations (6.4)–(6.6) which must satisfy the initial conditions (6.7) and the boundary conditions (6.10), knowing the density X_i of the mass force per unit volume.

Solving an elastodynamic problem of the three types previously formulated, from a mathematical point of view, consists of the integration of the system of the partial differential equations (6.4)–(6.6) with initial conditions and boundary conditions. The uniqueness of the solution of this problem was given by F. Neumann (1885), V.D. Kupradze, Gurtin and Sternberg [36], and Gurtin and Toupin [37].

From the theoretical and practical point of view, it is important to know whether the obtained solution of a fundamental problem of elastodynamics is unique.

The uniqueness of the solution of the mixed problem, in the assumption of its existence, involves the solution unique for the first and the second fundamental problem of elastodynamics.

In elastostatics the equations of motion (6.4) become

$$\sigma_{ij,j} + X_i = 0, (x_1, x_2, x_3) \in \Omega_0 , \quad i, j = 1, 2, 3 ,$$

that is, equations of equilibrium, where $X_i(x_1, x_2, x_3)$, $i = 1, 2, 3$, are the components of the mass force.

With no initial conditions, we have only boundary conditions, and the geometric (6.5) and physical (6.6) equations are preserved.

The first demonstration of the uniqueness of the solution of elastostatic fundamental problems was given by G. Kirchhoff (1858).

Thus, we can state that (6.4)–(6.6) form a complete system, because, using the initial and boundary conditions, the elastic state is uniquely determined, that is, $(\sigma_{ij}, \varepsilon_{ij}, u_i)$.

The existence of the fundamental solution of elastostatic problem was demonstrated in sufficiently general terms by A. Korn, I. Fredholm, G. Lauricela, L. Lichtenstein, D.I. Sherman, K. Fridrichs, S.L. Sobolev, V.D. Kupradze [38], S. Mihlin, G. Fichera [39] and others.

In the complete system of equations of elastodynamics (6.4)–(6.6) appear the unknowns $(\sigma_{ij}, \varepsilon_{ij}, u_i)$.

Choosing the main unknown components u_i of the displacement $\boldsymbol{u}$, the equations of motion can be expressed in terms of these unknowns and the equations thus obtained are called Lamé's equations.

On the assumption of the homogeneous and isotropic elastic body, Lamé's equations are

$$\mu u_{i,jj} + (\lambda + \mu) u_{j,jj} + X_i = \rho \ddot{u}_i$$

or, in vector form,

$$(\lambda + \mu)\nabla \boldsymbol{\varepsilon} + \mu \Delta \boldsymbol{u} + \boldsymbol{X} = \rho \ddot{\boldsymbol{u}} .$$

Solving the boundary value problem is done by adding the initial and the boundary conditions to Lamé's equations. In this way, we can solve, in displacements, the fundamental problems of elastodynamics. As regards the compatibility equations, they are no longer necessary; they are satisfied identically, because they were obtained using Cauchy's equations which contain the components u_i of the displacements.

The use of Lamé's equation is efficient to solve the first fundamental problem of elastodynamics.

6.3
Generalized Solution in $\mathcal{D}'(\mathbb{R}^2)$ Regarding the Plane Elastic Problems

Let $X(x, y)$, $Y(x, y)$ be the mass loads acting on the homogeneous and isotropic elastic plane. We shall denote by

$$T_\sigma = \begin{pmatrix} \sigma_{xx} & \sigma_{xy} \\ \sigma_{yx} & \sigma_{yy} \end{pmatrix}$$

the symmetric stress tensor, $\sigma_{xy} = \sigma_{yx}$. The components of this tensor must satisfy the equilibrium equations

$$\frac{\partial \sigma_{xx}}{\partial x} + \frac{\partial \sigma_{xy}}{\partial y} + X = 0\,, \quad \frac{\partial \sigma_{xy}}{\partial x} + \frac{\partial \sigma_{yy}}{\partial y} + Y = 0\,. \tag{6.11}$$

In the case of a plane state of strain, the equation of compatibility is

$$\Delta(\sigma_{xx} + \sigma_{yy}) = -\frac{1}{1-\nu}\left(\frac{\partial X}{\partial x} + \frac{\partial Y}{\partial y}\right)\,, \tag{6.12}$$

where $\Delta = \partial^2/\partial x^2 + \partial^2/\partial y^2$ and ν is the Poisson coefficient.

The problem, expressed in stresses, in the distributions space $\mathcal{D}'(\mathbb{R}^2)$ for the elastic plan consists in the determination of the elastic tensor $T_\sigma \in \left(\mathcal{D}'(\mathbb{R}^2)\right)^{2\times 2}$, which satisfies (6.11) and (6.12), considering that the mass loads X, Y are distributions from $\mathcal{D}'(\mathbb{R}^2)$.

We shall denote by

$$(Q) = \begin{pmatrix} X \\ Y \end{pmatrix} \in \left(\mathcal{D}'(\mathbb{R}^2)\right)^{2\times 1}$$

the matrix corresponding to the mass loads and by

$$(\sigma) = \begin{pmatrix} \sigma_{xx} \\ \sigma_{xy} \\ \sigma_{yy} \end{pmatrix} \in \left(\mathcal{D}'(\mathbb{R}^2)\right)^{2\times 1}$$

the matrix of the unknown stresses.

We consider the solution of the form

$$\sigma_{xx} = u_1 * X + u_2 * Y\,, \quad \sigma_{xy} = v_1 * X + v_2 * Y\,, \quad \sigma_{yy} = w_1 * X + w_2 * Y\,, \tag{6.13}$$

where the symbol $*$ represents the convolution product with respect to all variables (x, y) in $\mathcal{D}'(\mathbb{R}^2)$.

To the unknown distributions $u_i, v_i, w_i \in \mathcal{D}'(\mathbb{R}^2), i = 1, 2$, we associate the matrix

$$(u) = \begin{pmatrix} u_1 & u_2 \\ v_1 & v_2 \\ w_1 & w_2 \end{pmatrix} \in \left(\mathcal{D}'(\mathbb{R}^2)\right)^{3\times 2}\,.$$

Thus, the solution (6.13) in matrix writing becomes

$$(\sigma) = (u) * (Q)\,. \tag{6.14}$$

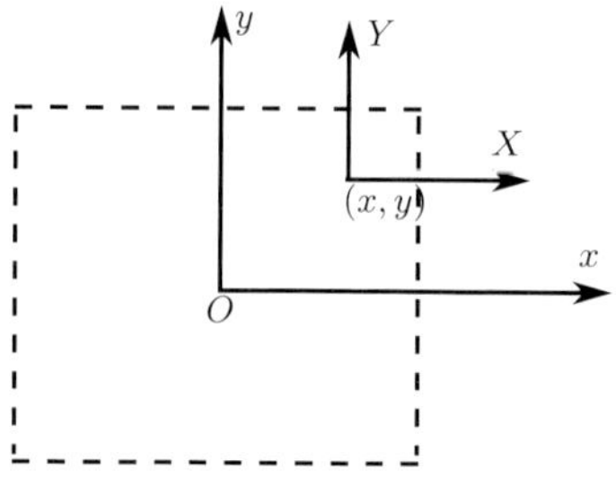

Figure 6.1

To determine the unknown distributions $u_i, v_i, w_i, i = 1, 2$, it is sufficient to consider the cases

$$1. \quad X = \delta(x, y), \quad Y = 0, \tag{6.15}$$

$$2. \quad X = 0, \quad Y = \delta(x, y), \tag{6.16}$$

where $\delta(x, y) \in \mathcal{D}'(\mathbb{R}^2)$ is the Dirac delta distribution concentrated at the origin of the frame Oxy (Figure 6.1).

Case 1 From (6.13) we obtain

$$\sigma_{xx} = u_1, \quad \sigma_{xy} = v_1, \quad \sigma_{yy} = w_1.$$

Equations 6.11 and 6.12 become

$$\frac{\partial u_1}{\partial x} + \frac{\partial v_1}{\partial y} + \delta(x, y) = 0, \quad \frac{\partial v_1}{\partial x} + \frac{\partial w_1}{\partial y} = 0, \tag{6.17}$$

$$\Delta(u_1 + w_1) = -\frac{1}{1-\nu}\frac{\partial \delta(x, y)}{\partial x}. \tag{6.18}$$

Applying the Fourier transform in $\mathcal{D}'(\mathbb{R}^2)$ to (6.17) and (6.18), we obtain

$$\alpha \hat{u}_1 + \beta \hat{v}_1 + \mathrm{i} = 0, \quad \alpha \hat{v}_1 + \beta \hat{w}_1 = 0, \tag{6.19}$$

$$\hat{u}_1 + \hat{v}_1 = -\frac{1}{1-\nu}\frac{\mathrm{i}\alpha}{\alpha^2 + \beta^2}, \tag{6.20}$$

because $\mathrm{F}[\delta(x, y)](\alpha, \beta) = 1$, $\mathrm{F}[u_k](\alpha, \beta) = \hat{u}_k$, $\mathrm{F}[v_k](\alpha, \beta) = \hat{v}_k$, $\mathrm{F}[w_k](\alpha, \beta) = \hat{w}_k$, $k = 1, 2$.

From (6.19) and (6.20), we obtain

$$\hat{u}_1 = -\frac{\mathrm{i}\alpha}{\alpha^2 + \beta^2} - \frac{\mathrm{i}}{1-\nu}\frac{\alpha\beta^2}{(\alpha^2 + \beta^2)^2}, \quad \hat{v}_1 = -\frac{\mathrm{i}\beta}{\alpha^2 + \beta^2} + \frac{\mathrm{i}}{1-\nu}\frac{\alpha^2\beta}{(\alpha^2 + \beta^2)^2},$$

$$\hat{w}_1 = \frac{\mathrm{i}\alpha}{\alpha^2 + \beta^2} - \frac{\mathrm{i}}{1-\nu}\frac{\alpha^3}{(\alpha^2 + \beta^2)^2}.$$

Case 2 Considering $X = 0$ and $Y = \delta(x, y)$, from (6.13) we obtain $\sigma_{xx} = u_2$, $\sigma_{xy} = v_2$, $\sigma_{yy} = w_2$.

Equations 6.11 and 6.12 become

$$\frac{\partial u_2}{\partial x} + \frac{\partial v_2}{\partial y} = 0\,, \quad \frac{\partial v_2}{\partial x} + \frac{\partial w_2}{\partial y} + \delta(x, y) = 0\,,$$
$$\Delta(u_2 + w_2) = -\frac{1}{1-\nu}\frac{\partial \delta(x, y)}{\partial x}\,. \tag{6.21}$$

Applying the Fourier transform in $\mathcal{D}'(\mathbb{R}^2)$, we get

$$\alpha\hat{u}_2 + \beta\hat{v}_2 = 0\,, \quad \alpha\hat{v}_2 + \beta\hat{w}_2 + \mathrm{i} = 0\,, \quad \hat{u}_2 + \hat{w}_2 = -\frac{1}{1-\nu}\frac{\mathrm{i}\beta}{\alpha^2+\beta^2}\,.$$

It follows that

$$\hat{u}_2 = \frac{\mathrm{i}\beta}{\alpha^2+\beta^2} - \frac{1}{1-\nu}\frac{\mathrm{i}\beta^3}{(\alpha^2+\beta^2)^2}\,, \quad \hat{v}_2 = -\frac{\mathrm{i}\alpha}{\alpha^2+\beta^2} + \frac{1}{1-\nu}\frac{\mathrm{i}\alpha\beta^2}{(\alpha^2+\beta^2)^2}\,,$$
$$\hat{w}_2 = -\frac{\mathrm{i}\beta}{\alpha^2+\beta^2} - \frac{1}{1-\nu}\frac{\mathrm{i}\alpha^2\beta}{(\alpha^2+\beta^2)^2}\,.$$

To obtain the expressions of the distributions u_k, v_k, w_k, $k = 1, 2$, we apply the inverse Fourier transform to the complex functions $\hat{u}_k, \hat{v}_k, \hat{w}_k$, $k = 1, 2$.

For this purpose, we use the formulas

$$\mathrm{F}^{-1}\left[\frac{\alpha}{\alpha^2+\beta^2}\right](x, y) = \frac{1}{2\pi\mathrm{i}}\frac{x}{x^2+y^2}\,,$$
$$\mathrm{F}^{-1}\left[\frac{\beta}{\alpha^2+\beta^2}\right](x, y) = \frac{1}{2\pi\mathrm{i}}\frac{y}{x^2+y^2}\,,$$
$$\mathrm{F}^{-1}\left[\frac{\alpha^3}{(\alpha^2+\beta^2)^2}\right](x, y) = \frac{1}{2\pi\mathrm{i}}\frac{x}{x^2+y^2} - \frac{1}{4\pi\mathrm{i}}\left(\frac{x}{x^2+y^2} - \frac{2xy^2}{(x^2+y^2)^2}\right)$$
$$= \frac{1}{4\pi\mathrm{i}}\frac{x(3y^2+x^2)}{(x^2+y^2)^2}\,,$$
$$\mathrm{F}^{-1}\left[\frac{\alpha^2\beta}{(\alpha^2+\beta^2)^2}\right](x, y) = \frac{1}{4\pi\mathrm{i}}\left(\frac{y}{x^2+y^2} - \frac{2x^2y}{(x^2+y^2)^2}\right) = \frac{1}{4\pi\mathrm{i}}\frac{y(y^2-x^2)}{(x^2+y^2)^2}\,,$$
$$\mathrm{F}^{-1}\left[\frac{\alpha\beta^2}{(\alpha^2+\beta^2)^2}\right](x, y) = \frac{1}{4\pi\mathrm{i}}\left(\frac{x}{x^2+y^2} - \frac{2y^2x}{(x^2+y^2)^2}\right) = \frac{1}{4\pi\mathrm{i}}\frac{x(x^2-y^2)}{(x^2+y^2)^2}\,,$$
$$\mathrm{F}^{-1}\left[\frac{\beta^3}{(\alpha^2+\beta^2)^2}\right](x, y) = \frac{1}{4\pi\mathrm{i}}\left(\frac{y}{x^2+y^2} + \frac{2x^2y}{(x^2+y^2)^2}\right) = \frac{1}{4\pi\mathrm{i}}\frac{y(3x^2+y^2)}{(x^2+y^2)^2}\,. \tag{6.22}$$

Consequently, we obtain

$$
\begin{aligned}
u_1(x,y) &= -\frac{1}{4\pi}\left[\frac{2x}{x^2+y^2} + \frac{1}{1-\nu}\frac{x(x^2-y^2)}{(x^2+y^2)^2}\right],\\
v_1(x,y) &= \frac{1}{4\pi}\left[\frac{-2y}{x^2+y^2} + \frac{1}{1-\nu}\frac{y(y^2-x^2)}{(x^2+y^2)^2}\right],\\
w_1(x,y) &= \frac{1}{4\pi}\left[\frac{2x}{x^2+y^2} - \frac{1}{1-\nu}\frac{x(3y^2+x^2)}{(x^2+y^2)^2}\right],\\
u_2(x,y) &= \frac{1}{4\pi}\left[\frac{2y}{x^2+y^2} - \frac{1}{1-\nu}\frac{y(3x^2+y^2)}{(x^2+y^2)^2}\right],\\
v_2(x,y) &= \frac{1}{4\pi}\left[\frac{-2x}{x^2+y^2} + \frac{1}{1-\nu}\frac{x(x^2-y^2)}{(x^2+y^2)^2}\right],\\
w_2(x,y) &= -\frac{1}{4\pi}\left[\frac{2y}{x^2+y^2} + \frac{1}{1-\nu}\frac{y(y^2-x^2)}{(x^2+y^2)^2}\right].
\end{aligned}
$$

All the elements u_k, v_k, w_k, $k = 1, 2$, of the matrix $(u) \in (\mathcal{D}'(\mathbb{R}^2))^{3\times 2}$ being determined, the generalized solution in stresses, in the distributions space $\mathcal{D}'(\mathbb{R}^2)$, is given by the system (6.13) or, in matrix writing, (6.14), valid for any load $X, Y \in \mathcal{D}'(\mathbb{R}^2)$.

The matrix $(u) \in (\mathcal{D}'(\mathbb{R}^2))^{3\times 2}$ can be called the fundamental matrix (fundamental solution) for the system of three equations (6.11) and (6.12). The components of this matrix satisfy (6.17), (6.18), and (6.21).

The generalization of the solution (6.13) is thus justified.

From (6.13) we obtain

$$
\begin{aligned}
\frac{\partial \sigma_{xx}}{\partial x} + \frac{\partial \sigma_{xy}}{\partial y} + X &= \left(\frac{\partial u_1}{\partial x} + \frac{\partial v_1}{\partial y}\right) * X + \left(\frac{\partial u_2}{\partial x} + \frac{\partial v_2}{\partial y}\right) * Y + X\,,\\
\frac{\partial \sigma_{xy}}{\partial x} + \frac{\partial \sigma_{yy}}{\partial y} + Y &= \left(\frac{\partial v_1}{\partial x} + \frac{\partial w_1}{\partial y}\right) * X + \left(\frac{\partial v_2}{\partial x} + \frac{\partial w_2}{\partial y}\right) * Y + Y\,,\\
\Delta\left(\sigma_{xx} + \sigma_{yy}\right) &= \Delta(u_1 + w_1) * X + \Delta(u_2 + w_2) * Y\,.
\end{aligned}
$$

Taking into account (6.17), (6.18), and (6.21), we have

$$
\begin{aligned}
\frac{\partial \sigma_{xx}}{\partial x} + \frac{\partial \sigma_{xy}}{\partial y} + X &= -\delta(x,y) * X + X = 0\,,\\
\frac{\partial \sigma_{xy}}{\partial x} + \frac{\partial \sigma_{yy}}{\partial y} + Y &= -\delta(x,y) * Y + Y = 0\,,\\
\Delta\left(\sigma_{xx} + \sigma_{yy}\right) &= -\frac{1}{1-\nu}\frac{\partial \delta(x,y)}{\partial x} * X - \frac{1}{1-\nu}\frac{\partial \delta(x,y)}{\partial y} * Y\\
&= -\frac{1}{1-\nu}\left(\frac{\partial X}{\partial x} + \frac{\partial Y}{\partial y}\right).
\end{aligned}
$$

Consequently, (6.11) and (6.12) are satisfied by the expressions (6.13), for any distribution X, Y with compact support, hence $X, Y \in \mathcal{E}'(\mathbb{R}^2)$.

Remark 6.1 The formulas (6.22) were determined using the relation $\Delta \ln(x^2 + y^2) = 4\pi\delta(x, y)$, where we obtain

$$F\left[\frac{\partial}{\partial x}\ln(x^2+y^2)\right](\alpha,\beta) = (-\mathrm{i}\alpha)F\left[\ln(x^2+y^2)\right](\alpha,\beta) = 4\pi\mathrm{i}\frac{\alpha}{\alpha^2+\beta^2} \, .$$

We also use the derivation formula, expressed in distributions, of the homogeneous functions of degree -1, where the origin is a singular point.

Example 6.1 At the point $A(x_0, y_0)$ of the elastic plane acts the concentrated force $\boldsymbol{F}(P_1, P_2)$; the equivalent load, represented in the distribution space $\mathcal{D}'(\mathbb{R}^2)$, is given by distribution $\boldsymbol{Q}(X, Y) = \boldsymbol{F}\delta(x - x_0, y - y_0)$, where $X(x, y) = P_1\delta(x - x_0, y - y_0)$, $Y(x, y) = P_2\delta(x - x_0, y - y_0)$.

Using the formulas (6.13), we obtain for the stresses the expressions

$$\begin{aligned}
\sigma_{xx} &= P_1 u_1(x - x_0, y - y_0) + P_2 u_2(x - x_0, y - y_0) \, , \\
\sigma_{xy} &= P_1 v_1(x - x_0, y - y_0) + P_2 v_2(x - x_0, y - y_0) \, , \\
\sigma_{yy} &= P_1 w_1(x - x_0, y - y_0) + P_2 w_2(x - x_0, y - y_0) \, .
\end{aligned}$$

Example 6.2 We acknowledge that the elastic plane is driven at the origin of the axes $x_0 = y_0 = 0$, by a rotational concentrated moment (center of rotation) of magnitude M (Figure 6.2); according to formula (4.18), the equivalent load in distributions is

$$\begin{aligned}
X(x, y) &= \frac{1}{2}M\frac{\partial\delta(x, y)}{\partial y} = \frac{1}{2}M\delta'_y(x, y) \, , \\
Y(x, y) &= -\frac{1}{2}M\frac{\partial\delta(x, y)}{\partial x} = -\frac{1}{2}M\delta'_y(x, y) \, .
\end{aligned}$$

For the stress state, we obtain

$$\sigma_{xx} = u_1 * \frac{1}{2}M\delta'_y(x, y) + u_2 * \left(-\frac{1}{2}M\delta'_y(x, y)\right) = \frac{M}{2}\left(\frac{\partial u_1}{\partial y} - \frac{\partial u_2}{\partial x}\right) ,$$

$$\sigma_{xy} = \frac{M}{2}\left(\frac{\partial v_1}{\partial y} - \frac{\partial v_2}{\partial x}\right) , \quad \sigma_{yy} = \frac{M}{2}\left(\frac{\partial w_1}{\partial y} - \frac{\partial w_2}{\partial x}\right) .$$

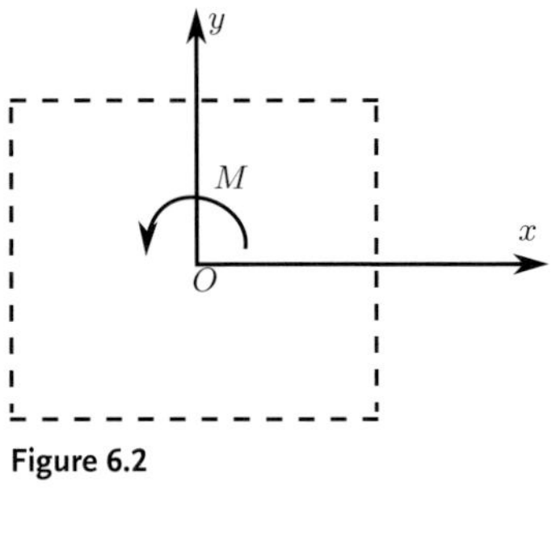

Figure 6.2

6.4
Generalized Solution in $\mathcal{D}'(\mathbb{R})$ of the Static Problem of the Elastic Half-Plane

We consider the homogeneous and isotropic elastic half-plane related to the reference system Oxy and we denote by $p(x), q(x), x \in \mathbb{R}$ the loads which act on the boundary of the elastic half-plane, along the Ox- and Oy-axis (Figure 6.3).

The loads p and q are expressed by distributions from $\mathcal{D}'(\mathbb{R})$, hence $p, q \in \mathcal{D}'(\mathbb{R})$. Due to the action of the loads p and q, at every point inside the elastic half-plane is created a state of stress characterized by the symmetrical stress tensor

$$T_\sigma = \begin{pmatrix} \sigma_{xx} & \sigma_{xy} \\ \sigma_{yx} & \sigma_{yy} \end{pmatrix}, \quad \sigma_{xy} = \sigma_{yx} .$$

The components of this tensor are considered distributions from $\mathcal{D}'(\mathbb{R})$ with respect to the variable $x \in \mathbb{R}$, depending on the parameter $y > 0$.

In the absence of mass forces, the components of the tensor T_σ at each point $(x, y) \in \mathbb{R} \times (0, \infty)$ satisfy the equilibrium equations

$$\frac{\partial \sigma_{xx}}{\partial x} + \frac{\partial \sigma_{xy}}{\partial y} = 0 , \quad \frac{\partial \sigma_{xy}}{\partial x} + \frac{\partial \sigma_{yy}}{\partial y} = 0 , \tag{6.23}$$

where the derivatives are considered in the sense of distributions.

In the case of a state of plane strain, the normal stresses σ_{xx}, σ_{yy} satisfy the equation of continuity (compatibility)

$$\Delta\left(\sigma_{xx} + \sigma_{yy}\right) = 0 , \quad \Delta = \frac{\partial^2}{\partial x^2} + \frac{\partial^2}{\partial y^2} . \tag{6.24}$$

The problem, expressed in stresses, in the distributions space $\mathcal{D}'(\mathbb{R})$, for the elastic half-plane, consists in the determination of the tensor $T_\sigma \in (\mathcal{D}'(\mathbb{R}))^{2\times 2}$, depending on the parameter $y > 0$, which for $y > 0$ satisfies (6.23) and (6.24), as well as the boundary conditions (conditions on the frontier Ox) and the regularity condition (at infinity)

$$\lim_{y \to +0} \sigma_{xy} = -p(x) , \quad \lim_{y \to +0} \sigma_{yy} = -q(x) , \tag{6.25}$$

$$\lim_{y \to \infty} \sigma_{xy} = \lim_{y \to \infty} \sigma_{xx} = \lim_{y \to \infty} \sigma_{yy} = 0 . \tag{6.26}$$

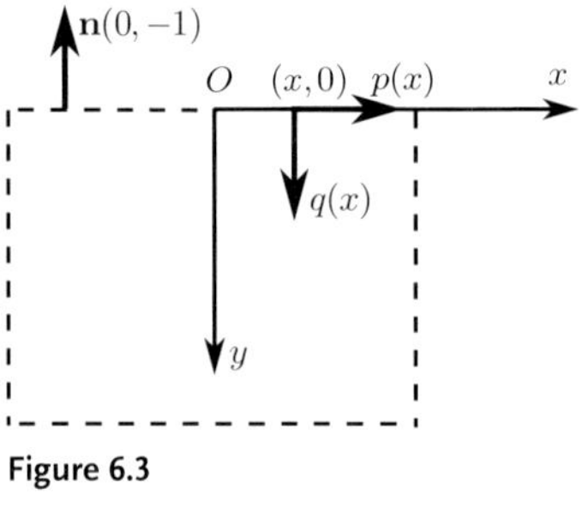

Figure 6.3

We introduce the matrices

$$(\sigma) = \begin{pmatrix} \sigma_{xx} \\ \sigma_{xy} \\ \sigma_{yy} \end{pmatrix} \in (\mathcal{D}'(\mathbb{R}))^{3\times 1}, \quad (u) = \begin{pmatrix} u_1 & u_2 \\ v_1 & v_2 \\ w_1 & w_2 \end{pmatrix} \in (\mathcal{D}'(\mathbb{R}))^{3\times 2},$$

$$(Q) = \begin{pmatrix} p(x) \\ q(x) \end{pmatrix} \in (\mathcal{D}'(\mathbb{R}))^{2\times 1},$$

where $u_k, v_k, w_k \in \mathcal{D}'(\mathbb{R})$, $k = 1, 2$, depend on the parameter $y > 0$.

With these matrices of distributions, the solution of the elastic half-plane is considered to be of the form

$$(\sigma) = (u) * (Q), \tag{6.27}$$

namely

$$\begin{aligned} \sigma_{xx}(x, y) &= u_1(x, y) * p(x) + u_2(x, y) * q(x), \\ \sigma_{xy}(x, y) &= v_1(x, y) * p(x) + v_2(x, y) * q(x), \\ \sigma_{yy}(x, y) &= w_1(x, y) * p(x) + w_2(x, y) * q(x), \end{aligned} \tag{6.28}$$

where the symbol $*$ represents the convolution product with respect to the variable $x \in \mathbb{R}$.

To determine the unknown distributions u_k, v_k, w_k, $k = 1, 2$, it is sufficient to consider the cases

1. $p(x) = \delta(x)$, $q(x) = 0$,
2. $p(x) = 0$, $q(x) = \delta(x)$.

Case 1 From (6.28) we obtain $\sigma_{xx} = u_1, \sigma_{xy} = v_1, \sigma_{yy} = w_1$. Applying the Fourier transform with respect to the variable $x \in \mathbb{R}$, to (6.23), (6.24), and the conditions (6.25) and (6.26), we obtain

$$(-i\alpha)\hat{u}_1(\alpha, y) + \frac{d}{dy}\hat{v}_1(\alpha, y) = 0, \quad (-i\alpha)\hat{v}_1(\alpha, y) + \frac{d}{dy}\hat{w}_1(\alpha, y) = 0, \tag{6.29}$$

$$(-i\alpha)^2(\hat{u}_1 + \hat{w}_1) + \frac{d^2}{dy^2}(\hat{u}_1 + \hat{w}_1) = 0, \tag{6.30}$$

$$\lim_{y\to+0} \hat{v}_1(\alpha, y) = -1, \quad \lim_{y\to+0} \hat{w}_1(\alpha, y) = 0, \tag{6.31}$$

$$\lim_{y\to\infty} \hat{v}_1(\alpha, y) = \lim_{y\to\infty} \hat{w}_1(\alpha, y) = \lim_{y\to\infty} \hat{u}_1(\alpha, y) = 0, \tag{6.32}$$

where

$$\begin{aligned} &\hat{u}_k(\alpha, y) = F[u_k(x, y)](\alpha), \quad \hat{v}_k(\alpha, y) = F[v_k(x, y)](\alpha), \\ &\hat{w}_k(\alpha, y) = F[w_k(x, y)](\alpha), F[\delta(x)] = 1. \end{aligned}$$

Taking into account (6.29), (6.30) becomes

$$\frac{d^4}{dy^4}\hat{w}_1 - 2\alpha^2\frac{d^2}{dy^2}\hat{w}_1 + \alpha^4\hat{w}_1 = 0. \tag{6.33}$$

To this equation we add four conditions. Thus, from (6.31) and (6.32) we get the conditions

$$\lim_{y\to+0} \hat{w}_1(\alpha, y) = 0\,, \qquad \lim_{y\to\infty} \hat{w}_1(\alpha, y) = 0\,. \tag{6.34}$$

Also, from (6.29) and (6.31) we obtain

$$\lim_{y\to+0} \frac{\mathrm{d}}{\mathrm{d}y}\hat{w}_1 = (\mathrm{i}\alpha) \lim_{y\to+0} \hat{v}_1 = -\mathrm{i}\alpha\,, \qquad \lim_{y\to\infty} \frac{\mathrm{d}}{\mathrm{d}y}\hat{w}_1 = (\mathrm{i}\alpha) \lim_{y\to\infty} \hat{v}_1 = 0\,. \tag{6.35}$$

Because the characteristic equation of (6.33) is $(r^2 - \alpha^2)^2 = 0$, for the general solution of (6.33) we obtain

$$\hat{w}_1(\alpha, y) = (A + By)\exp(-|\alpha|y) + (C + Dy)\exp(|\alpha|y)\,, \quad \alpha \in \mathbb{R}\,, \quad y > 0\,. \tag{6.36}$$

Based on the conditions (6.34), we get $A = C = D = 0$, and from (6.35) it results $B = -\mathrm{i}\alpha$.

Consequently, for $\hat{w}_1(\alpha, y)$ we have the expression

$$\hat{w}_1(\alpha, y) = (-\mathrm{i}\alpha)y \exp(-|\alpha|y)\,, \quad \alpha \in \mathbb{R}\,, \quad y > 0\,. \tag{6.37}$$

To calculate the distributions u_k, v_k, w_k, $k = 1, 2$, we use the Fourier transforms

$$F\left[\frac{1}{\pi}\frac{y}{x^2 + y^2}\right](\alpha) = \exp(-|\alpha|y)\,, \quad F\left[\frac{1}{\pi}\frac{x^2 - y^2}{(x^2 + y^2)^2}\right](\alpha) = |-\alpha|\exp(-|\alpha|y)\,,$$
$$F\left[-\frac{1}{\pi}\frac{2xy}{(x^2 + y^2)^2}\right](\alpha) = (-\mathrm{i}\alpha)\exp(-|\alpha|y)\,. \tag{6.38}$$

Applying the inverse Fourier transform to the relation (6.37), we get

$$w_1(x, y) = -\frac{2}{\pi}\frac{xy^2}{(x^2 + y^2)^2}\,, \quad x \in \mathbb{R}\,, \quad y > 0\,.$$

From (6.29) we obtain $\hat{v}_1 = (\mathrm{i}\alpha)^{-1}\mathrm{d}\hat{w}_1/\mathrm{d}y = -\exp(-|\alpha|y) + |\alpha|y\exp(-|\alpha|y)$, where it results,

$$\begin{aligned} v_1(x, y) &= -F^{-1}[\exp(-|\alpha|y)](x) + yF^{-1}[|\alpha|\exp(-|\alpha|y)](x) \\ &= -\frac{2}{\pi}\frac{x^2 y}{(x^2 + y^2)^2}\,, \quad x \in \mathbb{R}\,, \quad y > 0\,. \end{aligned}$$

Also, taking into account $\hat{u}_1 = -(-\mathrm{i}\alpha)^{-1}\mathrm{d}\hat{v}_1/\mathrm{d}y$ and applying the inverse Fourier transform, we obtain

$$u_1(x, y) = -\frac{\partial}{\partial y}\int_{-\infty}^{x} v_1(t, y)\mathrm{d}t = \frac{2}{\pi}\frac{\partial}{\partial y}\left[y\int_{-\infty}^{x}\frac{t^2\mathrm{d}t}{(t^2 + y^2)^2}\right]\,.$$

Because

$$\int_{-\infty}^{x} \frac{t^2 \mathrm{d}t}{(t^2+y^2)^2} = \frac{1}{2y} \arctan\frac{x}{y} + \frac{\pi}{4y} - \frac{x}{2(x^2+y^2)}$$

we get

$$u_1(x,y) = -\frac{2}{\pi}\frac{x^3}{(x^2+y^2)^2}\,, \qquad x \in \mathbb{R}\,, \quad y > 0\,.$$

Thus, for the distributions u_1, v_1 and w_1 there result the expressions

$$u_1(x,y) = -\frac{2}{\pi}\frac{x^3}{(x^2+y^2)^2}\,, \qquad v_1(x,y) = -\frac{2}{\pi}\frac{x^2 y}{(x^2+y^2)^2}\,,$$
$$w_1(x,y) = -\frac{2}{\pi}\frac{x y^2}{(x^2+y^2)^2}\,, \qquad x \in \mathbb{R}\,, \quad y > 0\,.$$

Case 2 From (6.28) we obtain $\sigma_{xx} = u_2, \sigma_{xy} = v_2, \sigma_{yy} = w_2$. Consequently, (6.23), (6.24) and the conditions (6.25), (6.26) become

$$\frac{\partial u_2}{\partial x} + \frac{\partial v_2}{\partial y} = 0\,, \qquad \frac{\partial v_2}{\partial x} + \frac{\partial w_2}{\partial y} = 0\,, \qquad \Delta(u_2 + w_2) = 0\,, \tag{6.39}$$

$$\lim_{y\to+0} v_2 = 0\,, \qquad \lim_{y\to+0} w_2 = -\delta(x)\,, \qquad \lim_{y\to\infty} u_2 = \lim_{y\to\infty} v_2 = \lim_{y\to\infty} w_2 = 0\,. \tag{6.40}$$

Applying the Fourier transform with respect to the variable $x \in \mathbb{R}$, to (6.39) and eliminating the Fourier transforms $\hat{u}_2$ and $\hat{v}_2$ among these equations, we obtain the differential equation

$$\frac{\mathrm{d}^4}{\mathrm{d}y^4}\hat{w}_2 - 2\alpha^2\frac{\mathrm{d}^2}{\mathrm{d}y^2}\hat{w}_2 + \alpha^4\hat{w}_2 = 0\,.$$

The general solution of this equation is

$$\hat{w}_2(\alpha,y) = (A + By)\exp(-|\alpha|y) + (C + Dy)\exp(|\alpha|y)\,, \quad \alpha \in \mathbb{R}\,, \quad y > 0\,. \tag{6.41}$$

Taking into account the condition (6.40), we obtain $A = -1$, $B = -|\alpha|$, $C = D = 0$.

As a consequence, the formula (6.41) becomes

$$\hat{w}_2(\alpha,y) = -(1 + |\alpha|y)\exp(-|\alpha|y)\,, \quad \alpha \in \mathbb{R}\,, \quad y > 0\,.$$

On the basis of the relations (6.38) we get

$$w_2(x,y) = -\frac{2}{\pi}\frac{y^3}{(x^2+y^2)^2}\,, \qquad x \in \mathbb{R}\,, \quad y > 0\,.$$

Taking into account $\hat{v}_2 = (\mathrm{i}\alpha)^{-1}\mathrm{d}\hat{w}_2/\mathrm{d}y$, $\hat{u}_2 = (\mathrm{i}\alpha)^{-1}\mathrm{d}\hat{v}_2/\mathrm{d}y$ and applying the inverse Fourier transform, we obtain

$$v_2(x, y) = -\frac{\partial}{\partial y}\int_{-\infty}^{x} w_2(t, y)\mathrm{d}t = \frac{2}{\pi}\frac{\partial}{\partial y}\left[y^3\int_{-\infty}^{x}\frac{\mathrm{d}t}{(t^2 + y^2)^2}\right] = -\frac{2}{\pi}\frac{xy^2}{(x^2 + y^2)^2},$$

$$u_2(x, y) = -\frac{\partial}{\partial y}\int_{-\infty}^{x} v_2(t, y)\mathrm{d}t = \frac{2}{\pi}\frac{\partial}{\partial y}\left[y^2\int_{-\infty}^{x}\frac{t\mathrm{d}t}{(t^2 + y^2)^2}\right] = -\frac{2}{\pi}\frac{x^2 y}{(x^2 + y^2)^2},$$

where $x \in \mathbb{R}, y > 0$.

The matrix $(u) \in (\mathcal{D}'(\mathbb{R}))^{3\times 2}$ being determined, the system (6.28) gives the solution, expressed in stresses, for the elastic half-plane for any load $p, q \in \mathcal{D}'(\mathbb{R})$ which acts on its boundary.

The matrix (u) can be called a fundamental solution for the elastic half-plane problem, expressed in stresses, namely of (6.23), (6.24) and of the conditions (6.25), (6.26).

For the existence of the convolution products, the distributions p, q must be with compact support, hence $p, q \in \mathcal{E}'(\mathbb{R})$.

The generality of the solution (6.28) is verified as in the case of the elastic plane.

The verification of the boundary conditions results in considering the family of functions $f_y(x) = (1/\pi)(y/(x^2 + y^2))$, $x \in \mathbb{R}$, $y > 0$ is a representative Dirac sequence, hence $\lim_{y\to+0}(1/\pi)(y/(x^2 + y^2)) = \delta(x)$.

The relation $x\delta(x) = 0$ is also used, which results in $x\delta'(x) = -\delta(x)$, where $\lim_{y\to+0}((-2/\pi)(xy/(x^2 + y^2)^2)) = \delta'(x)$.

Example 6.3 Let there be the elastic half-plane $y \geq 0$, under the action of the concentrated force $\boldsymbol{F}(P_1, P_2)$ at the origin. The equivalent load, expressed in distributions, of the concentrated force is $p(x) = P_1\delta(x), q(x) = P_2\delta(x)$.

Using the solution (6.28), we obtain for the state of stress the expressions

$$\sigma_{xx}(x, y) = P_1 u_1(x, y) + P_2 u_2(x, y) = -\frac{2}{\pi}\frac{x^2(P_1 x + P_2 y)}{(x^2 + y^2)^2},$$

$$\sigma_{xy}(x, y) = P_1 v_1(x, y) + P_2 v_2(x, y) = -\frac{2}{\pi}\frac{xy(P_1 x + P_2 y)}{(x^2 + y^2)^2},$$

$$\sigma_{yy}(x, y) = P_1 w_1(x, y) + P_2 w_2(x, y) = -\frac{2}{\pi}\frac{y^2(P_1 x + P_2 y)}{(x^2 + y^2)^2},$$

where $x \in \mathbb{R}, y > 0$.

6.5
Generalized Solution, in Displacements, for the Static Problem of the Elastic Space

Let $u_i(x) = u_i(x_1, x_2, x_3)$, $X_i(x)$, $i = 1, 2, 3$, $x \in \mathbb{R}^3$ be the components of the displacements and of the body forces, respectively, which are acting upon the homogeneous and isotropic elastic space.

The equilibrium equations, in displacements, of the elasticity theory are given by Lamé's equations

$$\mu \Delta u_i + (\lambda + \mu)\frac{\partial \varepsilon}{\partial x_i} = -X_i \,, \quad i = 1, 2, 3 \,, \tag{6.42}$$

where $\varepsilon = \sum_{i=1}^{3} \partial u_i / \partial x_i$ is the bulk strain, λ and μ are Lamé's elastic constants, Δ is the Laplace operator in $\mathbb{R}^3$.

The equilibrium equations (6.42) can be considered in the distribution space $\mathcal{D}'(\mathbb{R}^3)$, assuming that both the displacements u_i and the volume forces are expressed by distributions.

Thus, the solving in displacements of the problem of the elastic space consists of determining the distributions $u_i \in \mathcal{D}'(\mathbb{R}^3)$ which satisfy the equilibrium equations.

If the solutions u_i are a function type distribution, then they must verify the boundary conditions (of regularity at infinity)

$$\lim_{\|x\| \to \infty} u_i = 0 \,, \quad i = 1, 2, 3 \,. \tag{6.43}$$

In order to rewrite (6.42) in matrix formulation, we introduce the differential operators

$$\mathrm{D}_{ii} = (\lambda + \mu)\frac{\partial^2}{\partial x_i^2} + \mu \Delta \,, \quad \mathrm{D}_{ij} = (\lambda + \mu)\frac{\partial^2}{\partial x_i \partial x_j} \,, \quad i, j = 1, 2, 3 \,,$$

as well as the matrices

$$(u) = \begin{pmatrix} u_1 \\ u_2 \\ u_3 \end{pmatrix} , \quad (X) = \begin{pmatrix} -X_1 \\ -X_2 \\ -X_3 \end{pmatrix} , \quad (a) = (\mathrm{D}_{ij}\delta) \,, \quad \delta \in \mathcal{D}'(\mathbb{R}^3) \,.$$

Using these matrices, (6.42) will be written in the form a convolution matrix equation, namely

$$(a) * (u) = (X) \,. \tag{6.44}$$

Definition 6.1 We call the fundamental solution corresponding to (6.44), that is, of the matrix $(a) \in (\mathcal{E}'(\mathbb{R}^3))^{3\times 3}$, the distribution $(E) \in (\mathcal{D}'(\mathbb{R}^3))^{3\times 3}$ which satisfies the relations

$$(a) * (E) = (E) * (a) = (\delta) \,, \tag{6.45}$$

where (δ) is the unit element in the convolution algebra $(\mathcal{E}'(\mathbb{R}^3))^{3\times 3}$.

Denoting the components of the matrix (E) by E_{ij}, $i, j = 1, 2, 3$, then, for example, the components E_{i1} are the solutions (u) of (6.44), considering for the right side (X), the column matrix $(X)_1$, with the expression

$$(X)_1 = \begin{pmatrix} \delta \\ 0 \\ 0 \end{pmatrix} .$$

The fundamental solution (E) is determined considering successively for the right-hand side of (6.44) the matrices

$$(X)_2 = \begin{pmatrix} 0 \\ \delta \\ 0 \end{pmatrix}, (X)_3 = \begin{pmatrix} 0 \\ 0 \\ \delta \end{pmatrix}. \tag{6.46}$$

Once the fundamental solution (E) is determined, we obtain for the solution of (6.44), the expression

$$(u) = (E) * (X), \tag{6.47}$$

which will represent the generalized solution, in displacements, for the problem of the elastic space, in the distribution space $\mathcal{D}'(\mathbb{R}^3)$.

In order to determine the fundamental solution (E), we apply the Fourier transform, with respect to $(x_1, x_2, x_3) \in \mathbb{R}^3$, to (6.44), considering that the right-hand side of (6.44) is replaced by $(X)_1$.

We obtain thus the equations

$$(\lambda + \mu)\xi_i \left(\sum_{i=1}^{3} \xi_i \hat{E}_{i1} \right) + \mu \|\xi\|^2 \hat{E}_{i1} = Y_{i1}, \quad i = 1, 2, 3, \tag{6.48}$$

where $\hat{E}_{i1} = F[E_{i1}]$, $Y_{11} = -1$, $Y_{21} = 0$, $Y_{31} = 0$.

Multiplying (6.48) by ξ_i and summing up for all the is, we obtain

$$\sum_{i=1}^{3} \xi_i \hat{E}_{i1} = \frac{1}{(\lambda + 2\mu)\|\xi\|^2} \sum_{i=1}^{3} \xi_i Y_{i1} = \frac{-\xi_1}{(\lambda + 2\mu)\|\xi\|^2}.$$

By substituting this expression in (6.48) we get for $\hat{E}_{i1}$ the expression

$$\hat{E}_{i1} = \frac{Y_{i1}}{\mu\|\xi\|^2} + \frac{\xi_1 \xi_i}{\mu(\lambda + 2\mu)\|\xi\|^4}, \quad i = 1, 2, 3. \tag{6.49}$$

In order to apply the inverse Fourier transform, we notice that, because $\Delta r^{-1} = -4\pi\delta$, $r = \|x\|$, $x \in \mathbb{R}^3$, we have

$$F\left[\frac{1}{4\pi r}\right] = \frac{1}{\|\xi\|^2}, \quad \frac{\partial}{\partial \xi_j} \frac{1}{\|\xi\|^2} = \frac{i}{4\pi} F\left[\frac{x_j}{r}\right], \tag{6.50}$$

hence

$$F\left[\frac{x_j}{r}\right] = 8\pi i \frac{\xi_j}{\|\xi\|^4}. \tag{6.51}$$

Similarly, we have

$$F\left[\frac{\partial}{\partial x_j}\left(\frac{x_j}{r}\right)\right] = F\left[\frac{1}{r}\right] - F\left[\frac{x_j^2}{r^3}\right] = -i\xi_j F\left[\frac{x_j}{r}\right] = 8\pi i \frac{\xi_j^2}{\|\xi\|^4},$$
$$F\left[\frac{\partial}{\partial x_k}\left(\frac{x_j}{r}\right)\right] = -F\left[\frac{x_j x_k}{r^3}\right] = -i\xi_k F\left[\frac{x_j}{r}\right] = 8\pi \frac{\xi_j \xi_k}{\|\xi\|^4}, \quad j \neq k. \tag{6.52}$$

Applying the inverse Fourier transform to the expression (6.48), and taking into account (6.50)–(6.52), we obtain

$$E_{i1} = \frac{Y_{i1}}{4\pi\mu}\frac{1}{r} - \frac{1}{8\pi\mu(\lambda + 2\mu)}\frac{x_1 x_i}{r^3}\,, \quad i = 1, 2, 3\,, \tag{6.53}$$

that is,

$$E_{11} = -\frac{1}{4\pi\mu}\frac{1}{r} - \frac{1}{8\pi\mu(\lambda + 2\mu)}\frac{x_1^2}{r^3}\,,$$

$$E_{21} = -\frac{1}{8\pi\mu(\lambda + 2\mu)}\frac{x_1 x_2}{r^3}\,, \quad E_{31} = -\frac{1}{8\pi\mu(\lambda + 2\mu)}\frac{x_1 x_3}{r^3}\,.$$

In a similar manner, considering that the right-side of (6.44) is replaced by $(X)_2, (X)_3$, defined by the formulas (6.46), we obtain

$$E_{i2} = \frac{Y_{i2}}{4\pi\mu}\frac{1}{r} - \frac{1}{8\pi\mu(\lambda + 2\mu)}\frac{x_2 x_i}{r^3}\,, \quad E_{i3} = -\frac{Y_{i3}}{4\pi\mu}\frac{1}{r} - \frac{1}{8\pi\mu(\lambda + 2\mu)}\frac{x_3 x_i}{r^3}\,, \tag{6.54}$$

where

$$Y_{ij} = \begin{cases} 0\,, & i \neq j\,, \\ -1\,, & i = j\,. \end{cases}$$

The components of the fundamental solution (E) corresponding to the solution in displacements of the static problem of the elastic space are determined from the formulas (6.53) and (6.54).

These formulas can be written together in the form

$$E_{ij} = -\frac{\delta_{ij}}{4\pi\mu}\frac{1}{r} - \frac{1}{8\pi\mu(\lambda + 2\mu)}\frac{x_i x_j}{r^3}\,, \quad i, j = 1, 2, 3\,, \tag{6.55}$$

where δ_{ij} is Kronecker's delta.

We notice that both (a) and its inverse $(E) = (a)^{-1}$ are symmetric matrices; based on this property, we can easily verify that the relation (6.45) is satisfied, hence the matrix (E), whose components are given by (6.55), is indeed the fundamental solution of (6.44).

Also, we note that the functions E_{ij} are locally integrable, hence, the fundamental solution $(E) \in (\mathcal{D}'(\mathbb{R}^3))^{3\times 3}$ represents a function type matrix distribution, with the property that for $r \to +\infty$, $(E) \to (0)$.

If the elements of the force matrix (X) were compact-support distributions, then the expression (6.47) would exist; hence, the solution in displacements of the elastic space is given by (6.47) or, in the explicit form, by the formulas

$$u_i = -E_{i1} * X_1 - E_{i2} * X_2 - E_{i3} * X_3\,, \quad i = 1, 2, 3\,. \tag{6.56}$$

Example 6.4 If a force $\boldsymbol{F}(P_1, P_2, P_3)$ acts at the origin $x_0 = 0$ of the elastic space, then this action is equivalent to a body force with the components $X_i(x_1, x_2, x_3) = P_i\delta(x_1, x_2, x_3)$, $i = 1, 2, 3$.

Based on the formulas (6.56), we obtain for the displacement components the expressions

$$u_i(x) = -P_1 E_{i1}(x) - P_2 E_{i2}(x) - P_3 E_{i3}(x)\,, \quad i = 1, 2, 3\,, \tag{6.57}$$

which obviously satisfy the boundary conditions (6.43).

Specifying the formulas (6.57), we get

$$\begin{aligned}
u_1 &= P_1\left(\frac{1}{4\pi\mu}\frac{1}{r} + \frac{1}{8\pi\mu(\lambda+2\mu)}\frac{x_1^2}{r^3}\right) + P_2\frac{1}{8\pi\mu(\lambda+2\mu)}\frac{x_1x_2}{r^3}\\
&\quad + P_3\frac{1}{8\pi\mu(\lambda+2\mu)}\frac{x_1x_3}{r^3}\,,\\
u_2 &= P_1\frac{1}{8\pi\mu(\lambda+2\mu)}\frac{x_1x_2}{r^3} + P_2\left(\frac{1}{4\pi\mu}\frac{1}{r} + \frac{1}{8\pi\mu(\lambda+2\mu)}\frac{x_2^2}{r^3}\right)\\
&\quad + P_3\frac{1}{8\pi\mu(\lambda+2\mu)}\frac{x_2x_3}{r^3}\,,\\
u_3 &= P_1\frac{1}{8\pi\mu(\lambda+2\mu)}\frac{x_1x_3}{r^3} + P_2\frac{1}{8\pi\mu(\lambda+2\mu)}\frac{x_2x_3}{r^3}\\
&\quad + P_3\left(\frac{1}{4\pi\mu}\frac{1}{r} + \frac{1}{8\pi\mu(\lambda+2\mu)}\frac{x_3^2}{r^3}\right).
\end{aligned}$$

7
Applications of the Distribution Theory to Linear Viscoelastic Bodies

7.1
The Mathematical Model of a Linear Viscoelastic Solid

The viscoelastic solid combines the features of a perfectly elastic medium with the properties of a viscous fluid.

The equations of motion are the same regardless of the deformable solid nature. The constitutive law, that is, the relation between stress and strain, is what distinguishes the deformable solids from one another.

Thus, the linear elastic solid is characterized by Hooke's law expressed by linear relations, independent of time, between stress and strain, which ensures a one-to-one correspondence between the states of strain and stress.

For the linear elastic solid, the work done by the external forces towards its deformation, is entirely accumulated as strain energy; after the action of the external forces is removed, the accumulated strain work is restored to the elastic body in order to bring it to its original state, both in terms of shape and of stress and strain states.

Consequently, the linear elastic bodies are characterized in terms of energy by their ability to accumulate the entire mechanical energy and not to dissipate it. On the other hand, if upon a linear elastic body we act suddenly on its boundary with loads (sudden loads) and then we maintain the loads as constant, this instantly creates a state of stress and a state of strain that remain unchanged.

Thus, the state of strain, performed instantly, is independent of time, which is also a characteristic of the mechanical behavior of the linear elastic body upon a sudden application of forces and the keeping them constant.

Some bodies show that after a sudden application of some loads and keeping them constant for a long time, instantaneous deformations occur which do not remain constant but display a time-dependent limited/unlimited increase. This phenomenon is called creep.

Another important phenomenon is the phenomenon of relaxation, which represents the stress variation in the body with respect to time, by keeping the strain and the temperature constant.

Distribution Theory, First Edition. Petre P. Teodorescu, Wilhelm W. Kecs, and Antonela Toma

The viscoelastic solids are characterized by their ability to accumulate and diffuse mechanical energy, hence they simultaneously exhibit properties of an elastic solid and a viscous fluid.

These solids belong to the class of bodies with memory, because the current state of stress (hence at the time t) depends on the entire history of deformation, that is, on the strain states on the interval $(-\infty, t]$.

Further, we are situated within the framework of the linear theory of deformations to characterize the linear viscoelastic solid; for the deduction of the constitutive law we adopt the axiomatic method used by M. Gurtin, E. Sternberg [40] and I. Rabotnov [41].

Let $\sigma(\boldsymbol{r}, t)$ and $\varepsilon(\boldsymbol{r}, t)$, $(\boldsymbol{r}, t) \in \Omega \times \mathbb{R} \subset \mathbb{R}^3 \times \mathbb{R}$ be the tensor fields of the stress and strain, respectively, whose components we acknowledge to be functions of class $C^{0,0}(\Omega \times \mathbb{R})$. They satisfy the conditions $\sigma_{ij} = 0$, $\varepsilon_{ij} = 0$ for $t < 0$, $\forall \boldsymbol{r}(x_h) \in \Omega \subset \mathbb{R}^3$, which correspond to the natural state of the solid before the initial moment $t = 0$.

We acknowledge that the stress $\sigma_{ij}(\boldsymbol{r}, t)$ and the strain $\varepsilon_{kl}(\boldsymbol{r}, t)$ are related under the functional form

$$\sigma_{ij}(\boldsymbol{r}, t) = \mathrm{L}_{ij}(\varepsilon_{kl}) , \quad k, l = 1, 2, 3 ,$$

where L_{ij} are symmetric linear functionals, $\mathrm{L}_{ij} = \mathrm{L}_{ji}$, continuous and zero for $t < 0$.

For isotropic viscoelastic solids, the constitutive law in the distribution space $\mathcal{D}'_+$ has the expression

$$\sigma_{ij}(\boldsymbol{r}, t) = \lambda^*(\boldsymbol{r}, t)\delta_{ij} \underset{t}{*} \varepsilon(\boldsymbol{r}, t) + 2\mu^*(\boldsymbol{r}, t) \underset{t}{*} \varepsilon_{ij}(\boldsymbol{r}, t) , \tag{7.1}$$

where δ_{ij} is Kronecker's symbol, $\varepsilon = \varepsilon_{kk}$, and $\sigma_{ij}, \varepsilon_{ij}, \lambda^*, \mu^* \in \mathcal{D}'_+$, $\boldsymbol{r} \in \Omega$ being a parameter. The symbol $*_t$ represents the convolution product with respect to the time variable $t \in \mathbb{R}$.

We note that the constitutive law (7.1) of a viscoelastic isotropic solid is analogous as structure to Hooke's law

$$\sigma_{ij}(\boldsymbol{r}) = \lambda(\boldsymbol{r})\varepsilon\delta_{ij} + 2\mu(\boldsymbol{r})\varepsilon_{ij}$$

for isotropic elastic solids.

We call the relaxation distributions of a viscoelastic isotropic solid the distributions $\psi_1(\boldsymbol{r}, t)$, $\psi_2(\boldsymbol{r}, t) \in \mathcal{D}'_+$, depending on the parameter $\boldsymbol{r} \in \Omega$, defined by the relations

$$2\mu^*(\boldsymbol{r}, t) = \frac{\partial \psi_1(\boldsymbol{r}, t)}{\partial t} , \quad 3K^*(\boldsymbol{r}, t) = \frac{\partial \psi_2(\boldsymbol{r}, t)}{\partial t} .$$

The distribution $\mu^* \in \mathcal{D}'_+$ is called the relaxation shear modulus, and the distribution $K^* \in \mathcal{D}'_+$ is the relaxation bulk modulus and has the expression

$$K^*(\boldsymbol{r}, t) = \lambda^*(\boldsymbol{r}, t) + \frac{2}{3}\mu^*(\boldsymbol{r}, t) .$$

Using the distributions of relaxation $\psi_1, \psi_2 \in \mathcal{D}'_+$, the constitutive law (7.1) becomes

$$\sigma_{ij}(\mathbf{r}, t) = \frac{\psi_2 - \psi_1}{3}\delta_{ij} \underset{t}{*} \frac{\partial \varepsilon(\mathbf{r}, t)}{\partial t} + \psi_1 \underset{t}{*} \frac{\partial \varepsilon_{ij}(\mathbf{r}, t)}{\partial t}.$$

The creep distributions $\varphi_i(\mathbf{r}, t) \in \mathcal{D}'_+$, $i = 1, 2$ are related to the distributions of relaxation $\psi_i(\mathbf{r}, t) \in \mathcal{D}'_+$, $i = 1, 2$, by the relations

$$\varphi_i \underset{t}{*} \psi_i = tH(t) \Leftrightarrow \frac{\partial^2}{\partial t^2}\left(\varphi_i \underset{t}{*} \psi_i\right) = \delta(t)\,, \quad i = 1, 2\,,$$

where H is the Heaviside distribution.

In the case of a homogeneous and isotropic viscoelastic solid, both the distributions λ^*, μ^*, K^* and the creep and relaxation distributions $\varphi_i, \psi_i, i = 1, 2$, depend only on the temporal variable $t \in \mathbb{R}$.

Consequently, the constitutive law of viscoelastic homogeneous and isotropic solids has the form

$$\sigma_{ij}(\mathbf{r}, t) = \lambda^*(t)\delta_{ij} \underset{t}{*} \varepsilon(\mathbf{r}, t) + 2\mu^*(t) \underset{t}{*} \varepsilon_{ij}(\mathbf{r}, t)\,,$$

where

$$\varepsilon = \varepsilon_{kk}\,, \quad 2\mu^*(t) = \psi_1'(t)\,, \quad 3K^*(t) = \psi_2'(t)\,, \quad K^*(t) = \lambda^*(t) + \frac{2}{3}\mu^*(t)\,.$$

Hence $\sigma_{ij}, \varepsilon_{ij}, \varphi_i, \psi_i \in \mathcal{D}'_+$ with respect to the parameter $t \in \mathbb{R}$, $\mathbf{r} \in \Omega$ being a parameter as well.

7.2
Models of One-Dimensional Viscoelastic Solids

Let us consider a one-dimensional homogeneous and isotropic viscoelastic solid. In this case, the constitutive law can be written in the equivalent forms

$$\sigma(x, t) = \varepsilon(x, t) * \psi'(t) = \frac{\partial \varepsilon(x, t)}{\partial t} * \psi(t)\,, \tag{7.2}$$

$$\varepsilon(x, t) = \sigma(x, t) * \varphi'(t) = \frac{\partial \sigma(x, t)}{\partial t} * \varphi(t)\,. \tag{7.3}$$

The symbol $*$ represents the convolution product with respect to the variable $t \in \mathbb{R}$.

The dependence between the distributions of creep and of relaxation φ and ψ is given by the equivalent relations

$$\varphi * \psi = tH(t) \Leftrightarrow \varphi * \psi'' = \varphi'' * \psi = \varphi' * \psi' = \delta(t)\,.$$

Because $\varphi, \psi \in \mathcal{D}'_+$, we have $\text{supp}(\varphi), \text{supp}(\psi) \subset [0, \infty)$ and thus $\sigma = \varepsilon = 0$ for $t < 0$, $\forall x \in \mathbb{R}$. These relations express the fact that, at the initial moment $t = 0$, the viscoelastic solid is in the natural state.

If σ and ε are functions of class $C^1(\mathbb{R})$, and φ, ψ are continuous functions with respect to the variable $t \in \mathbb{R}$, then we have

$$\sigma(x,t) = \int_{-\infty}^{t} \frac{\partial \varepsilon(x,\tau)}{\partial \tau} \psi(t-\tau)\mathrm{d}\tau \, , \quad \varepsilon(x,t) = \int_{-\infty}^{t} \frac{\partial \sigma(x,\tau)}{\partial \tau} \varphi(t-\tau)\mathrm{d}\tau \, ,$$

which are Boltzmann's integral relations.

Consequently, (7.2) and (7.3) are a generalization of Boltzmann's relations in the distribution space $\mathcal{D}'_+$, corresponding to the viscoelastic solid in its natural state at time $t = 0$.

Let us suppose that the distribution of relaxation $\psi \in \mathcal{D}'_+$ is of the form $\psi(t) = EH(t), t \in \mathbb{R}$. This means that the viscoelastic solid does not display the relaxation phenomenon. In this case, from (7.2) we obtain

$$\sigma = \varepsilon * E\delta(t) = E\varepsilon \, ,$$

that is, Hooke's law.

We can say that the elastic solid is a special case of viscoelastic solid, namely in which the distributions of creep and relaxation are of the form $\psi = EH(t), \varphi = H(t)/E, t \in \mathbb{R}$.

In this case, the viscoelastic solid doesn't present the phenomenon of relaxation or of creep.

We note that the constitutive laws (7.2) and (7.3) can be represented in the differential form, thus

$$P(\mathrm{D})\sigma = Q(\mathrm{D})\varepsilon \, , \quad \mathrm{D} = \frac{\mathrm{d}}{\mathrm{d}t} \, ,$$

where P and Q are linear differential operators with constant coefficients.

The constitutive laws (7.2) and (7.3) can be extended even in the case in which the stress and the strain are distributions from $\mathcal{D}'(\mathbb{R}^2)$.

Thus, if $\sigma(x,t), \varepsilon(x,t) \in \mathcal{D}'(\mathbb{R}^2)$, $\mathrm{supp}(\sigma), \mathrm{supp}(\varepsilon) \subset \mathbb{R} \times [0,\infty)$, then (7.2) and (7.3) can be generalized in the form

$$\sigma(x,t) = \varepsilon(x,t) \underset{t}{*} \psi'(t) = \frac{\partial \varepsilon}{\partial t} \underset{t}{*} \psi(t) \, , \tag{7.4}$$

$$\varepsilon(x,t) = \sigma(x,t) \underset{t}{*} \varphi'(t) = \frac{\partial \sigma}{\partial t} \underset{t}{*} \varphi(t) \, , \tag{7.5}$$

where the symbol $*_t$ represents the partial convolution product with respect to the temporal variable $t \in \mathbb{R}$.

Because $\mathrm{supp}(\psi), \mathrm{supp}(\varphi) \subset [0,\infty)$ and $\mathrm{supp}(\sigma), \mathrm{supp}(\varepsilon) \subset \mathbb{R} \times [0,\infty)$ the convolution products exist.

We note that, both in $\mathcal{D}'_+$ and in $\mathcal{D}'(\mathbb{R}^2)$, the constitutive laws express the hypothesis that at the initial moment $t = 0$ the viscoelastic solid is in its natural state of rest, that is, $\sigma(x,t) = 0, \varepsilon(x,t) = 0$ for $t < 0$.

To show the equivalence of (7.4) and (7.5), we perform the convolution product between (7.4) and $\varphi'(t)$, obtaining

$$\sigma(x,t) \underset{t}{*} \varphi'(t) = \left[\varepsilon(x,t) \underset{t}{*} \psi'(t)\right] \underset{t}{*} \varphi'(t)$$
$$= \varepsilon(x,t) \underset{t}{*} (\psi' * \varphi') = \varepsilon(x,t) \underset{t}{*} \delta(t) = \varepsilon(x,t) ,$$

that is, the relation (7.5).

We have pointed out that the viscoelastic solids constitute media which possess properties from both elastic solids and properties of the viscous media: having the property to store the mechanical energy, a phenomenon characteristic of elastic solids, as well as the ability to diffuse mechanical energy, a characteristic property of the Newtonian viscous fluid.

The mechanical correspondent of the one-dimensional elastic solid is the elastic spring, which verifies Hooke's law

$$\sigma = E\varepsilon , \tag{7.6}$$

where E is the modulus of elasticity.

As regards the perfectly viscous liquids, they are characterized by Newton's law

$$\sigma = \eta\dot{\varepsilon} , \tag{7.7}$$

where $\dot{\varepsilon} = \partial\varepsilon/\partial t$ is the strain rate and η is the viscosity coefficient.

The mechanical correspondent of viscous fluids is the dashpot. Therefore, the elastic properties of the one-dimensional viscoelastic solid are described by springs and the viscous properties by dashpots. By connecting these mechanical elements in series and parallel, we obtain mechanical systems whose behavior depends on the choice of the parameters E and η.

We note that the elastic spring and the dashpot can be considered limiting cases (degeneracy) of the viscoelastic solid and we can determine the distributions of creep and of relaxation of these limits of the viscous solid.

Thus, in the case of an elastic spring, Hooke's law (7.6) can be written in the form

$$\sigma = E\delta(t) \underset{t}{*} \varepsilon = E\frac{\mathrm{d}H}{\mathrm{d}t} \underset{t}{*} \varepsilon = EH(t) \underset{t}{*} \frac{\partial\varepsilon}{\partial t} . \tag{7.8}$$

Because the convolution algebra $\mathcal{D}'_+$ has no divisors of zero, comparing (7.8) with (7.2) and (7.3) we obtain for the distribution of relaxation ψ the expression

$$\psi(t) = EH(t) .$$

Also, writing Hooke's law in the form

$$\varepsilon = \frac{1}{E}\sigma = \frac{1}{E}H(t) \underset{t}{*} \frac{\partial\sigma}{\partial t} ,$$

and comparing with (7.2) and (7.3) we obtain the distribution of creep

$$\varphi(t) = \frac{H(t)}{E} .$$

Proceeding similarly, in the case of the dashpot, we observe that Newton's law (7.7) in $\mathcal{D}'_+$ can be written in the form

$$\sigma = \eta \frac{\partial \varepsilon}{\partial t} = \eta \delta(t) \underset{t}{*} \frac{\partial \varepsilon}{\partial t} . \tag{7.9}$$

It follows that the distribution of relaxation

$$\psi = \eta \delta(t)$$

is a singular distribution and, from the mechanical point of view, it is equivalent to a sudden intensity η.

The distribution of creep is

$$\varphi = \frac{t H(t)}{\eta} .$$

The Kelvin–Voigt Model This model is obtained by a viscous damper and an elastic spring, connected in parallel as shown in the Figure 7.1. The parameter E/η is called the rate of relaxation.

We note that these two components of the Kelvin–Voigt model, under the action of stress σ corresponds the same strain ε which represents the total deformation of the model. Corresponding to the strain ε, the stresses from the spring and the dashpot are $\sigma_1 = E\varepsilon, \sigma_2 = \eta\dot{\varepsilon}$.

Because the stress σ is balanced by the two parallel stresses σ_1 and σ_2, we have $\sigma = \sigma_1 + \sigma_2$, wherefrom the constitutive law of Kelvin–Voigt model reads as

$$\sigma = E\varepsilon + \eta \frac{\partial \varepsilon}{\partial t} . \tag{7.10}$$

The obtained stress–strain relation is considered in the distribution space $\mathcal{D}'_+$, because $\sigma, \varepsilon \in \mathcal{D}'_+$, which means that for $t < 0$ the viscoelastic model is in its natural state.

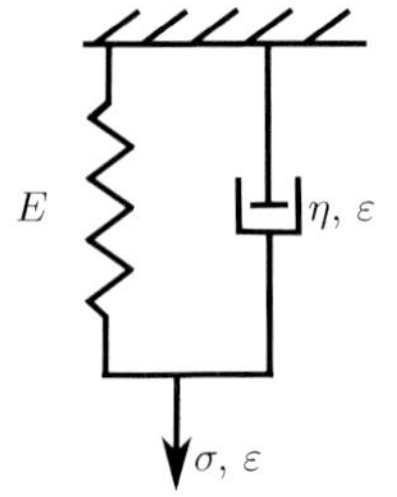

Figure 7.1

We note that the constitutive law (7.10) is written in the differential form, that is, $P(\mathrm{D})\sigma = Q(\mathrm{D})\varepsilon$, where the differential operators have the expressions

$$P(\mathrm{D}) = 1\,, \quad Q(\mathrm{D}) = \eta\,\mathrm{D} + E\,; \quad \mathrm{D} = \frac{\mathrm{d}}{\mathrm{d}t}\,.$$

To determine the distribution of creep and relaxation $\varphi, \psi \in \mathcal{D}'_+$, we apply the Laplace transform in distributions to (7.10) and (7.4).

We obtain

$$\tilde{\sigma}(x,p) = E\tilde{\varepsilon}(x,p) + \eta p \tilde{\varepsilon}(x,p) = p\tilde{\varepsilon}(x,p)\tilde{\psi}(p)\,, \tag{7.11}$$

where $\tilde{\sigma} = \mathrm{L}[\sigma](p), \tilde{\varepsilon} = \mathrm{L}[\varepsilon](p), \tilde{\psi} = \mathrm{L}[\psi](p)$.

From (7.11) it follows $\tilde{\psi}(p) = E/p + \eta$.

Applying the inverse Laplace transform, we have for the distribution of relaxation the expression

$$\psi(t) = EH(t) + \eta\delta(t)\,.$$

We use the relation $\varphi * \psi'' = \delta$ to determine the distribution of creep $\varphi \in \mathcal{D}'_+$. By applying the Laplace transform we get $\tilde{\varphi}p^2\tilde{\psi} = 1$, wherefrom it follows

$$\tilde{\varphi} = \frac{1}{p(E+\eta p)} = \frac{1}{E}\left(\frac{1}{p} - \frac{1}{p + E/\eta}\right)\,.$$

Consequently, by applying the inverse Laplace transform we obtain

$$\varphi(t) = \frac{1}{E}H(t)\left[1 - \exp\left(-\frac{Et}{\eta}\right)\right]\,, \quad t \in \mathbb{R}\,.$$

We note that the creep distribution φ is a function type distribution from $\mathcal{D}'_+$, while the distribution of relaxation ψ is a singular distribution.

The Maxwell Model This model is obtained by a series connection of an elastic spring with a dashpot (Figure 7.2).

We observe that the stress σ acting on this model will be equal to the stress of the component elements.

As regards the deformations produced in the spring and the dashpot, namely ε_1 and ε_2, will be different. The total strain ε will be equal to the strain sum of those two elements, that is,

$$\varepsilon = \varepsilon_1 + \varepsilon_2\,. \tag{7.12}$$

For the stress σ of the two elements, spring and dashpot, we can write

$$\sigma = E\varepsilon_1 + \eta\frac{\partial\varepsilon_2}{\partial t}\,. \tag{7.13}$$

Differentiating the relation (7.12) and taking into account (7.13), we obtain for the Maxwell model the following constitutive law

$$\frac{\partial\varepsilon}{\partial t} = \frac{1}{E}\frac{\partial\sigma}{\partial t} + \frac{1}{\eta}\sigma\,. \tag{7.14}$$

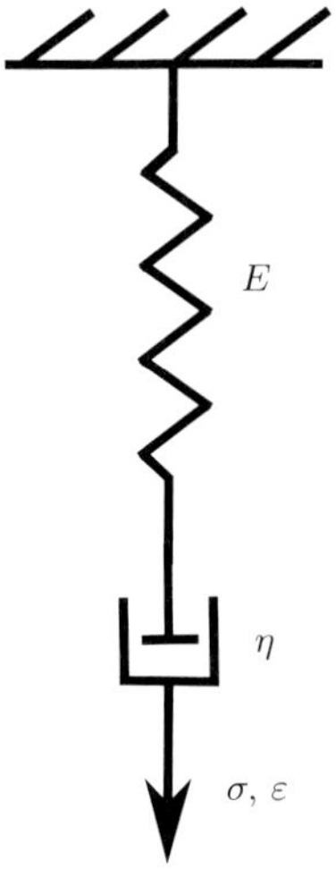

Figure 7.2

This constitutive law is of the differential type $P(\mathrm{D})\sigma = Q(\mathrm{D})\varepsilon$, where

$$P(\mathrm{D}) = \frac{1}{E}\,\mathrm{D} + \frac{1}{\eta}\,, \qquad Q(\mathrm{D}) = \mathrm{D}\;.$$

Using (7.14) as well as the relation $\partial\varepsilon/\partial t = \sigma * \varphi''$, we obtain

$$\sigma * \left(\varphi'' - \frac{1}{E}\delta' - \frac{1}{\eta}\delta\right) = 0\,,$$

wherefrom it follows

$$\varphi'' = \frac{1}{E}\delta' + \frac{1}{\eta}\delta\;.$$

Consequently, we have

$$\varphi(t) = \frac{1}{E}H(t) + \frac{1}{\eta}t\,H(t)\,, \qquad t \in \mathbb{R}\;.$$

Based on the relation $\psi * \varphi'' = \delta$, we get

$$\frac{1}{E}\psi' + \frac{1}{\eta}\psi = \delta\;.$$

Using the Laplace transform, we obtain for the distribution of relaxation the expression

$$\psi(t) = E\,H(t)\exp\left(-\frac{Et}{\eta}\right)\,, \qquad t \in \mathbb{R}\;.$$

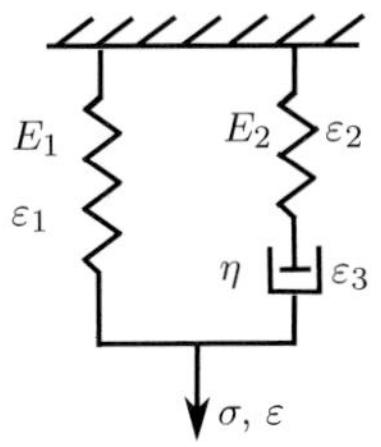

Figure 7.3

Zener Model The Zener model shown in Figure 7.3 is obtained by connecting in parallel a spring with a Maxwell model. Let σ_1 and σ_2 be the stresses in the two branches and $\varepsilon_1, \varepsilon_2, \varepsilon_3$ the strains.

Denoting by σ and ε the stress and the strain of the Zener model, respectively, we have the relations

$$\sigma = \sigma_1 + \sigma_2, \sigma_1 = E_1 \varepsilon_1 \,, \quad \sigma_2 = E_2 \varepsilon_2 = \eta \dot{\varepsilon}_3 \,, \quad \varepsilon = \varepsilon_1 + \varepsilon_2 + \varepsilon_3 \,,$$

wherefrom it follows

$$\dot{\sigma}_2 = E_2(\dot{\varepsilon} - \dot{\varepsilon}_3) = E_2 \left(\dot{\varepsilon} - \frac{\sigma_2}{\eta} \right) , \quad \sigma_2 = \sigma - E_1 \varepsilon$$

hence

$$\dot{\sigma} + E_2 \frac{\sigma}{\eta} = E_1 E_2 \frac{\varepsilon}{\eta} + (E_1 + E_2) \dot{\varepsilon} \,.$$

This relation is the constitutive law of the Zener model, which has the differential form $P(\mathrm{D})\sigma = Q(\mathrm{D})\varepsilon$, where the differential operators have the expressions

$$P(\mathrm{D}) = \mathrm{D} + \frac{E_2}{\eta} , \quad Q(\mathrm{D}) = (E_1 + E_2)\,\mathrm{D} + \frac{E_1 E_2}{\eta} \,.$$

For the relaxation and the creep distributions we obtain the expressions

$$\psi(t) = H(t) \left(E_1 + E_2 \exp\left(-\frac{E_2 t}{\eta} \right) \right) ,$$

$$\varphi(t) = \frac{H(t)}{E_1} \left(1 - \frac{E_1}{E_1 + E_2} \exp\left(-\frac{E_1 E_2 t}{\eta (E_1 + E_2)} \right) \right) .$$

The distributions of creep and of relaxation are function type distributions.

7.3 Viscoelasticity Theory Equations: Correspondence Principle

7.3.1 Viscoelasticity Theory Equations

Because the viscoelastic solid differs from the elastic solid only by the constitutive law, it follows that the fundamental problems formulated for the elastic solid will be the same for the viscoelastic solid.

Taking into account that the constitutive equation of the viscoelastic solid is written using the convolution product in the distribution space $\mathcal{D}'_+$ with respect to time, $\boldsymbol{r} \in \Omega$ being a parameter, it is necessary that all mechanical and geometrical quantities, as well as the equations associated to the study of the viscoelastic solid motion should be rewritten in the distribution space $\mathcal{D}'_+$.

For this purpose, we note

$$\begin{aligned} \overline{\sigma}_{ij}(\boldsymbol{r},t) &= H(t)\sigma_{ij}(\boldsymbol{r},t)\,, \quad \overline{X}_i(\boldsymbol{r},t) = H(t)X_i(\boldsymbol{r},t)\,, \\ \overline{u}_i(\boldsymbol{r},t) &= H(t)u_i(\boldsymbol{r},t)\,, \quad \overline{\varepsilon}_{ij}(\boldsymbol{r},t) = H(t)\varepsilon_{ij}(\boldsymbol{r},t)\,. \end{aligned} \tag{7.15}$$

These quantities are function type distributions from $\mathcal{D}'_+$ with respect to $t \in \mathbb{R}$, depending on the parameter $\boldsymbol{r} \in \Omega$ and with discontinuities of the first kind at the origin $t = 0$.

We note that the derivatives in the usual sense of a certain quantity with respect to the space variable will coincide with the derivatives considered in the sense of distributions.

As regards the derivatives with respect to $t \in \mathbb{R}$, it is necessary to take into account the dependence between the derivative in the ordinary sense and the derivative in the distribution theory meaning.

The equations of motion in the distribution space $\mathcal{D}'_+$ are

$$\overline{\sigma}_{ij}(\boldsymbol{r},t) + \overline{X}_i(\boldsymbol{r},t) = \rho\ddot{\overline{u}}_i - \rho\left(\dot{u}_{i0}\delta(t) + u_{i0}\delta'(t)\right)\,, \tag{7.16}$$

where u_{i0} and $\dot{u}_{i0}$ are the initial conditions

$$u_{i0} = u_i(\boldsymbol{r},t)|_{t=+0}\,, \quad \dot{u}_{i0} = \left.\frac{\tilde{\partial} u_i(\boldsymbol{r},t)}{\partial t}\right|_{t=+0}\,, \quad \boldsymbol{r} \in \Omega\,.$$

In order to have the complete system of equations of a homogeneous and isotropic viscoelastic solid, to the equations of motion (7.16) we must add

1. Constitutive law (physical equations)

$$\overline{\sigma}_{ij,j}(\boldsymbol{r},t) = \lambda^*(t)\delta_{ij} \underset{t}{*} \overline{\varepsilon}(\boldsymbol{r},t) + 2\mu^*(t) \underset{t}{*} \overline{\varepsilon}_{ij}(\boldsymbol{r},t)\,,$$

   where

$$2\mu^*(t) = \frac{\mathrm{d}\psi_1(t)}{\mathrm{d}t}\,, \quad 3K^*(t) = \frac{\mathrm{d}\psi_2(t)}{\mathrm{d}t}\,,$$
$$K^* = \lambda^* + \frac{2}{3}\mu^*\,, \quad \mu^*, \lambda^*, K^*, \psi_1, \psi_2 \in D'_+\,;$$

2. Cauchy equations (geometrical equations)

$$\overline{\varepsilon}_{ij} = \frac{(\overline{u}_{i,j} + \overline{u}_{j,i})}{2}\,;$$

3. Equations of compatibility

$$\overline{\varepsilon}_{ik,kl} + \overline{\varepsilon}_{lk,ij} = \overline{\varepsilon}_{jk,il} + \overline{\varepsilon}_{il,kj}\,.$$

The initial conditions are included in the equations of motion (7.16), such that for the formulation of a viscoelasticity problem we must give only the boundary conditions.

Using distributions of creep and of relaxation, the equations of motion in displacements are written in the form

$$\psi_1'(t) \underset{t}{*} \overline{u}_{i,jj} + \frac{\psi_1' + 2\psi_2'}{3} \underset{t}{*} \overline{u}_{j,ji} + 2\overline{X}_i = 2\rho\ddot{\overline{u}}_i - 2\rho\left[\dot{u}_{i0}\delta(t) + u_{i0}\delta'(t)\right] .$$

7.3.2
Correspondence Principle

Among the methods of solving physical-mathematical problems (in particular of the linear viscoelasticity problems) the integral transformation method, in particular Fourier and Laplace transforms, play an important role. The elastic solid is different from the viscoelastic solid only by the constitutive law.

Between the constitutive laws of the two models of solids there is a strong dependence, namely the Laplace (or Fourier) image, in distribution, of the viscoelastic solid constitutive law coincides as mathematical structure with Hooke's law corresponding to the elastic solid, the complex variable p of the Laplace transform having the role of a parameter like $\boldsymbol{r} \in \Omega \subset \mathbb{R}^3$.

These observations led T. Alfrey [42] and E.H. Lee [43] to formulate a method for solving the viscoelasticity problems called the correspondence principle.

This principle, which is a method of solving viscoelasticity problems, was generalized by W.T. Read [44]. H.S. Tsien [45], and J. Mandel contributed significantly to the enhancement of this method.

To view what the constitutes the correspondence principle, as well as its applications, we write the system of equations common to the elastic solid and to the viscoelastic homogeneous and isotropic solid.

Thus, the equations of motion, Cauchy's equations and compatibility equations, are

$$\overline{\sigma}_{ij,j}(\boldsymbol{r},t) + \overline{X}_i(\boldsymbol{r},t) = \rho\ddot{\overline{u}}_i - \rho\left(\dot{u}_{i0}\delta(t) + u_{i0}\delta'(t)\right) , \tag{7.17}$$

$$\overline{\varepsilon}_{ij} = \frac{(\overline{u}_{i,j} + \overline{u}_{j,i})}{2} , \tag{7.18}$$

$$\overline{\varepsilon}_{ik,kl} + \overline{\varepsilon}_{lk,ij} = \overline{\varepsilon}_{jk,il} + \overline{\varepsilon}_{il,kj} , \tag{7.19}$$

where $\overline{\sigma}_{ij}, \overline{\varepsilon}_{ij}, \overline{u}_i, \overline{X}_i$ are considered distributions from $\mathcal{D}'_+$, and $\boldsymbol{r} \in \Omega \subset \mathbb{R}^3$ is a parameter.

The initial conditions of an elastic or viscoelastic problem, namely u_{i0} and $\dot{u}_{i0}$, are included in the equations of motion (7.17).

To these common equations are added the constitutive laws of the two bodies, elastic and viscoelastic, that is,

$$\overline{\sigma}_{ij}(\boldsymbol{r},t) = \lambda\delta_{ij}\overline{\varepsilon}(\boldsymbol{r},t) + 2\mu(t)\overline{\varepsilon}_{ij}(\boldsymbol{r},t) \tag{7.20}$$

$$\overline{\sigma}_{ij}(\boldsymbol{r},t) = \lambda^*(t)\delta_{ij} \underset{t}{*} \overline{\varepsilon}(\boldsymbol{r},t) + 2\mu^*(t) \underset{t}{*} \overline{\varepsilon}_{ij}(\boldsymbol{r},t) . \tag{7.21}$$

Applying the Laplace transformation with respect to the variable $t \in \mathbb{R}$ to the two constitutive laws, we obtain

$$\tilde{\overline{\sigma}}_{ij}(\boldsymbol{r},p) = \lambda\delta_{ij}\tilde{\overline{\varepsilon}}(\boldsymbol{r},p) + 2\mu\tilde{\overline{\varepsilon}}_{ij}(\boldsymbol{r},p) , \tag{7.22}$$

$$\tilde{\overline{\sigma}}_{ij}(\boldsymbol{r},p) = \tilde{\lambda}^*(p)\tilde{\overline{\varepsilon}}(\boldsymbol{r},p) + 2\tilde{\mu}^*(p)\tilde{\overline{\varepsilon}}_{ij}(\boldsymbol{r},p) , \tag{7.23}$$

where, for example, $\tilde{\overline{\sigma}}_{ij}(\boldsymbol{r},p) = \mathrm{L}[\overline{\sigma}_{ij}(\boldsymbol{r},t)](p)$ is the Laplace transform and p is the complex variable having the role of a parameter.

Comparing (7.22) and (7.23), it results that they are identical as structure, which determines the bijections

$$\lambda \leftrightarrow \tilde{\lambda}^*(p) , \quad \mu \leftrightarrow \tilde{\mu}^*(p)$$

wherefrom it follows the bijection

$$K = \lambda + \frac{2}{3}\mu \leftrightarrow \tilde{K}^*(p) = \tilde{\lambda}^*(p) + \frac{2}{3}\tilde{\mu}^*(p) .$$

The remaining equations and boundary conditions, by Laplace transformations, are common to both elastic and viscoelastic media.

Particularly, for the one-dimensional elastic and viscoelastic solids the constitutive laws are

$$\sigma(x,t) = E\varepsilon(x,t) , \quad \sigma(x,t) = \varepsilon(x,t) * \psi'(t) .$$

Applying the Laplace transform with respect to the variable $t \in \mathbb{R}$, we obtain

$$\tilde{\sigma}(x,p) = E\tilde{\varepsilon}(x,p) , \quad \tilde{\sigma}(x,p) = \tilde{\varepsilon}(x,p)p\tilde{\psi}(p) .$$

It follows the bijections

$$E \leftrightarrow p\tilde{\psi}(p) \quad \text{and} \quad \frac{1}{E} \leftrightarrow p\tilde{\varphi}(p) ,$$

where $\varphi(t) \in \mathcal{D}'_+$ represents the distribution of creep.

The above considerations allow us to state the principle of correspondence in the following form:

To solve a problem of viscoelasticity, the corresponding problem from elastodynamics is solved and we consider the Laplace image of the solution thus obtained; then the elastic constants K and μ are replaced by $\tilde{K}^*(p), \tilde{\mu}^*(p)$ representing the Laplace images of $K^*(t)$ and $\mu^*(t)$, where $3K^*(t) = \mathrm{d}\psi_2(t)/\mathrm{d}t$, $2\mu^*(t) = \mathrm{d}\psi_1(t)/\mathrm{d}t$, $K^*(t) = \lambda^*(t) + (2/3)\mu^*(t)$.

The Laplace image of the solution of the viscoelasticity problem is thus obtained. By applying the inverse Laplace transform, the problem solution is obtained.

We note that, for the application of this principle, it is necessary that the boundary of the viscoelastic solid should be fixed, hence, time independent, and the Laplace image of the considered quantities should exist. If elastic constants other than K and μ appear in the problem solution from elastodynamics, then these, upon the application of the correspondence principle, are replaced with the Laplace images of the corresponding quantities from viscoelasticity.

8
Applications of the Distribution Theory in Electrical Engineering

8.1
Study of the RLC Circuit: Cauchy Problem

We consider a conductor of section A, through which passes an electric current of charge $q = q(t)$, $t \geq 0$, where $t = 0$ is the initial moment from which the electric current starts flowing through the conductor.

For $t < 0$, no electricity passes through the conductor. The electric current intensity is given by

$$i(t) = \frac{\tilde{\mathrm{d}}}{\mathrm{d}t} q(t) , \quad t \geq 0 , \tag{8.1}$$

where the derivative is considered in the usual sense.

We denote

$$q(0) = q_0 , \left. \frac{\tilde{\mathrm{d}}}{\mathrm{d}t} q(t) \right|_{t=0} = i(0) = i_0 . \tag{8.2}$$

We introduce the functions

$$\overline{q}(t) = H(t)q(t) , \overline{i}(t) = H(t)i(t) , \quad t \in \mathbb{R} , \tag{8.3}$$

where H is the Heaviside distribution.

For $t < 0$ we have $\overline{q}(t) = 0, \overline{i}(t) = 0$, which means that no electricity passes through the conductor.

The relation (8.1) can be written now in the form

$$\overline{i}(t) = \frac{\tilde{\mathrm{d}}}{\mathrm{d}t} \overline{q}(t) , \quad t \in \mathbb{R} . \tag{8.4}$$

Distribution Theory, First Edition. Petre P. Teodorescu, Wilhelm W. Kecs, and Antonela Toma

Assuming that $q \in C^2([0,\infty))$ and differentiating in the distribution theory sense, it results

$$\frac{\mathrm{d}}{\mathrm{d}t}\overline{q}(t) = \frac{\tilde{\mathrm{d}}}{\mathrm{d}t}\overline{q}(t) + q_0\delta(t) ,$$

$$\frac{\mathrm{d}^2}{\mathrm{d}t^2}\overline{q}(t) = \frac{\tilde{\mathrm{d}}^2}{\mathrm{d}t^2}\overline{q}(t) + \left.\frac{\tilde{\mathrm{d}}}{\mathrm{d}t}\overline{q}(t)\right|_{t=0}\delta(t) + q_0\delta'(t)$$

$$= \frac{\tilde{\mathrm{d}}^2}{\mathrm{d}t^2}\overline{q}(t) + i_0\delta(t) + q_0\delta'(t) . \tag{8.5}$$

An RLC circuit is an electrical circuit consisting of a resistor, an inductor, and a capacitor, connected in series (Figure 8.1) or in parallel. The corresponding parameters are the resistance R, the inductance L and the capacity C.

In the following we study a series RLC circuit (Figure 8.1).

We denote by $q(t), t \geq 0$ the variable positive load on the capacitor plate and by $E(t), t \geq 0, E \in C^0([0,\infty))$ the electromotive tension with which the circuit is fed.

In accordance with (8.3), we introduce the function

$$\overline{E}(t) = H(t)E(t) , \quad t \in \mathbb{R} . \tag{8.6}$$

After closing the circuit, a current of variable intensity $i(t), t \geq 0$ appears.

Having a closed circuit, according to Kirchhoff's law, the electromotive tension which supplies the circuit is equivalent to the sum of the potential drops.

Consequently, the following equation corresponds to the RLC circuit:

$$E(t) = L\frac{\tilde{\mathrm{d}}i(t)}{\mathrm{d}t} + Ri(t) + \frac{q(t)}{C} = L\frac{\tilde{\mathrm{d}}i(t)}{\mathrm{d}t} + Ri(t) + \frac{1}{C}\int i(\tau)\mathrm{d}\tau , \quad t > 0 , \tag{8.7}$$

because $i(t) = \tilde{\mathrm{d}}q(t)/\mathrm{d}t$.

Equation 8.7 is an integrodifferential equation with respect to the unknown $i(t), t \geq 0$.

This equation can be written as a second-order differential equation with respect to the charge $q(t), t \geq 0$,

$$L\frac{\tilde{\mathrm{d}}^2}{\mathrm{d}t^2}q(t) + R\frac{\tilde{\mathrm{d}}}{\mathrm{d}t}q(t) + \frac{1}{C}q(t) = E(t) , \quad t > 0 . \tag{8.8}$$

Multiplying this equation with $H(t)$ we get

$$L\frac{\tilde{\mathrm{d}}^2}{\mathrm{d}t^2}\overline{q}(t) + R\frac{\tilde{\mathrm{d}}}{\mathrm{d}t}\overline{q}(t) + \frac{1}{C}\overline{q}(t) = \overline{E}(t) , \quad t \in \mathbb{R} . \tag{8.9}$$

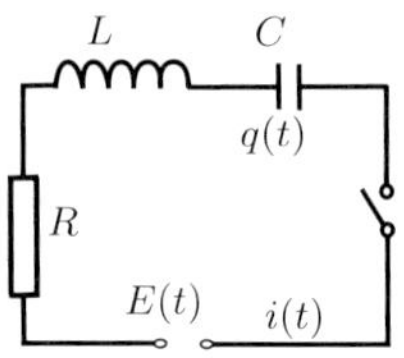

Figure 8.1

To rewrite (8.9) in the distributions space $\mathcal{D}'_+$ we use the functions (8.3), (8.6), the initial conditions (8.2) and the formulas (8.5).

Thus [16, 20], the differential equation in the distributions space $\mathcal{D}'_+$ of the RLC circuit is

$$L\frac{d^2}{dt^2}\overline{q}(t) + R\frac{d}{dt}\overline{q}(t) + \frac{1}{C}\overline{q}(t) = \overline{E}(t) + (Li_0 + Rq_0)\delta(t) + Lq_0\delta'(t) . \quad (8.10)$$

We observe that the Cauchy problem for the RLC circuit consists of determining the function $q(t)$, $t \geq 0$, which verifies (8.8) and the initial conditions (8.2).

In the distributions space $\mathcal{D}'_+$, the Cauchy problem consists of determining the function type distribution $\overline{q}(t) \in \mathcal{D}'_+$, which verifies the differential equation (8.10), including the initial conditions (8.2).

We introduce the differential operator

$$\mathrm{D} = L\frac{d^2}{dt^2} + R\frac{d}{dt} + \frac{1}{C} , \quad (8.11)$$

and the distribution

$$F(t) = \overline{E}(t) + (Li_0 + Rq_0)\delta(t) + Lq_0\delta'(t) \in \mathcal{D}'_+ .$$

Equation 8.10 can be written in $\mathcal{D}'_+$ in the form of a convolution equation

$$\mathrm{D}\,\delta(t) * \overline{q}(t) = F(t) . \quad (8.12)$$

Let $\mathcal{E}(t) \in \mathcal{D}'_+$ be the fundamental solution of the operator (8.11). We have

$$\mathcal{E}(t) = (\mathrm{D}\,\delta(t))^{-1} , \quad \mathcal{E}^{-1}(t) = \mathrm{D}\,\delta(t) \Leftrightarrow \mathrm{D}\,\mathcal{E}(t) = \delta(t) .$$

The solution of (8.12) of the RLC circuit in the $\mathcal{D}'_+$ space is unique and reads

$$\overline{q}(t) = \mathcal{E}(t) * F(t) .$$

Specifying, we obtain

$$\overline{q}(t) = \mathcal{E}(t) * \overline{E}(t) + (Li_0 + Rq_0)\mathcal{E}(t) + Lq_0\mathcal{E}'(t) . \quad (8.13)$$

Particularly, if the initial conditions (8.2) are zero, we obtain

$$\overline{q}(t) = \mathcal{E}(t) * \overline{E}(t) . \quad (8.14)$$

In the following, we determine the concrete form of the fundamental solution $\mathcal{E}(t) \in \mathcal{D}'_+$.

The fundamental solution $\mathcal{E}(t)$ satisfies the equation

$$\left(L\delta''(t) + R\delta'(t) + \frac{1}{C}\delta(t)\right) * \mathcal{E}(t) = \delta(t) ;$$

applying the Laplace transform, we obtain

$$\mathrm{L}[\mathcal{E}(t)](p) = \frac{1}{Lp^2 + Rp + 1/C} = \frac{1}{L(p - p_1)(p - p_2)} \tag{8.15}$$

where $p_1 = -\alpha + \beta$, $p_2 = -\alpha - \beta$ and $\alpha = R/2L, \beta = \sqrt{R^2 - 4L/C}/2L$.
Depending on the roots p_1, p_2, we distinguish three cases, that is,

1. $R^2 > 4L/C$,
2. $R^2 < 4L/C$,
3. $R^2 = 4L/C$.

Case 1 We have $\beta^2 > 0$ and $\beta > 0$. The Laplace transform (8.15) will be written in the form

$$\mathrm{L}[\mathcal{E}(t)](p) = \frac{1}{2L\beta}\left(\frac{1}{p - p_1} - \frac{1}{p - p_2}\right) ,$$

wherefrom it follows

$$E(t) = \frac{H(t)}{2L\beta}(\exp(p_1 t) - \exp(p_2 t)) = \frac{1}{L\beta} H(t) \sinh \beta t , \quad t \in \mathbb{R} . \tag{8.16}$$

Substituting in (8.13), we obtain the Cauchy problem solution corresponding to the RLC circuit in the form

$$\begin{aligned}\overline{q}(t) &= \frac{1}{L\beta} H(t) \exp(-\alpha t) \sinh(\beta t) * \overline{E}(t) \\ &+ \frac{1}{2L\beta} H(t) \exp(-\alpha t) \left[(2Li_0 + Rq_0) \sinh(\beta t) + 2L\beta q_0 \cosh(\beta t)\right] .\end{aligned}$$

If $\overline{E}(t)$, $t \in \mathbb{R}$ is a locally integrable function, then for $t \geq 0$ we obtain

$$\begin{aligned}q(t) &= \frac{1}{L\beta} \int_0^t \exp(-\alpha\tau) E(t - \tau) \sinh(\beta\tau) \mathrm{d}\tau \\ &+ \frac{1}{2L\beta} \exp(-\alpha t) \left[(2Li_0 + Rq_0) \sinh(\beta t) + 2L\beta q_0 \cosh(\beta t)\right] ,\end{aligned} \tag{8.17}$$

where $\sinh(\beta t) = (\exp(\beta t) - \exp(-\beta t))/2$.

Case 2 Because $R^2 < 4L/C$ we have $\beta = \mathrm{i}\omega$, $\omega = \sqrt{4L/C - R^2}/(2L)$. The fundamental solution is

$$\mathcal{E}(t) = \frac{1}{L\omega} H(t) \exp(-\alpha t) \sin(\omega t) , \quad t \in \mathbb{R} .$$

In this case, the Cauchy problem solution will be

$$\begin{aligned}\overline{q}(t) &= \frac{1}{L\omega} H(t) \exp(-\alpha t) \sin(\omega t) * \overline{E}(t) \\ &+ \frac{1}{2L\omega} H(t) \exp(-\alpha t) \left[(2Li_0 + Rq_0) \sin(\omega t) + 2L\omega q_0 \cos(\omega t)\right] .\end{aligned}$$

If $\overline{E}$ is a locally integrable function, we obtain

$$q(t) = \frac{1}{L\omega} \int_0^t \exp(-\alpha\tau) E(t-\tau) \sin(\omega\tau) \mathrm{d}\tau$$
$$+ \frac{1}{2L\omega} \exp(-\alpha t) \left[(2Li_0 + Rq_0)\sin(\omega t) + 2L\omega q_0 \cos(\omega t)\right] , \quad t \geq 0 .$$

This case corresponds to an oscillating circuit.

Case 3 Because $R^2 = 4L/C$, we have $\beta = 0$ and thus $\mathrm{L}[\mathcal{E}(t)](p) = L^{-1}(p+\alpha)^{-2}$; the fundamental solution is $\mathcal{E}(t) = tH(t)\exp(-\alpha t)/L$, $t \in \mathbb{R}$.

The solution of the Cauchy problem will be written in the form

$$\overline{q}(t) = \frac{1}{L} tH(t)\exp(-\alpha t) * \overline{E}(t)$$
$$+ \frac{1}{2L} H(t)\exp(-\alpha t)\left[(2Li_0 + Rq_0)t + 2Lq_0\right] , \quad t \in \mathbb{R} .$$

If $\overline{E}(t)$, $t \in \mathbb{R}$ is a locally integrable function, we obtain

$$q(t) = \frac{1}{L} \int_0^t \tau \exp(-\alpha\tau) E(t-\tau) \mathrm{d}\tau$$
$$+ \frac{1}{2L} \exp(-\alpha t)\left[(2Li_0 + Rq_0)t + 2Lq_0\right] , \quad t \geq 0 .$$

Example 8.1 Let us consider the particular case when the initial conditions are zero, $q_0 = i_0 = 0$, and the emf $E(t)$ is constant for $t > 0$, hence $\overline{E}(t) = E_0 H(t)$, $t \in \mathbb{R}$.

We consider the case when $R^2 > 4L/C$. Using the formula (8.17) we obtain

$$q(t) = \frac{E_0}{L\beta} \int_0^t \exp(-\alpha\tau) \sinh(\beta\tau) \mathrm{d}\tau$$
$$= \frac{E_0}{2L\beta} \left(\frac{\exp(\tau(\beta-\alpha))}{\beta-\alpha} + \frac{\exp(-\tau(\alpha+\beta))}{\beta+\alpha} \right)\Bigg|_0^t$$
$$= \frac{E_0}{L\beta} \frac{1}{\beta^2-\alpha^2} \left[\exp(-\alpha t)(\alpha \sinh(\beta t) + \beta \cosh(\beta t)) - \beta\right] , \quad t \geq 0 .$$

Because $\beta^2 - \alpha^2 = -1/CL$, we get

$$q(t) = -\frac{CE_0}{\beta}[\exp(-\alpha t)(\alpha \sinh(\beta t) + \beta \cosh(\beta t)) - \beta] , \quad t \geq 0 .$$

Taking into account (8.1) for the intensity $i(t)$, $t \geq 0$, we obtain

$$i(t) = \frac{E_0}{L\beta} \exp(-\alpha t) \sinh(\beta t) , \quad t \geq 0 .$$

8.2
Coupled Oscillating Circuit: Cauchy Problem

Let us consider two identical RLC circuits (Figure 8.2), inductively coupled. Such a circuit is called a coupled oscillating circuit. We denote by M the mutual inductance of two circuits and with R, L and C the resistance, the inductance and the capacitance, respectively.

Let $E(t)$ be an emf (electromotive force) acting on the circuit at the initial moment $t = 0$, $i_1(t), i_2(t)$ the currents in the two circuits and $q_1(t), q_2(t)$ the electric charges on the capacitor plates.

Applying Kirchhoff's law to the two circuits, we get the system of equations

$$\begin{aligned} &Ri_1(t) + L\frac{\tilde{d}i_1(t)}{dt} + M\frac{\tilde{d}i_2(t)}{dt} + \frac{1}{C}q_1(t) = E(t)\,, \\ &Ri_2(t) + L\frac{\tilde{d}i_2(t)}{dt} + M\frac{\tilde{d}i_2(t)}{dt} + \frac{1}{C}q_2(t) = 0\,. \end{aligned} \tag{8.18}$$

We obtained thus a system of two differential equations in the unknowns i_1, i_2, q_1, q_2. Taking into account the relations

$$i_1(t) = \frac{\tilde{d}q_1(t)}{dt}\,, \quad i_2(t) = \frac{\tilde{d}q_2(t)}{dt}$$

and multiplying the system (8.18) with the Heaviside function H, we may write

$$\begin{aligned} &L\frac{\tilde{d}^2}{dt^2}\overline{q}_1(t) + M\frac{\tilde{d}^2}{dt^2}\overline{q}_2(t) + R\frac{\tilde{d}}{dt}\overline{q}_1(t) + \frac{1}{C}\overline{q}_1(t) = \overline{E}(t)\,, \\ &M\frac{\tilde{d}^2}{dt^2}\overline{q}_1(t) + L\frac{\tilde{d}^2}{dt^2}\overline{q}_2(t) + R\frac{\tilde{d}}{dt}\overline{q}_2(t) + \frac{1}{C}\overline{q}_2(t) = 0\,, \end{aligned} \tag{8.19}$$

where

$$\overline{q}_1(t) = H(t)q_1(t)\,, \quad \overline{q}_2(t) = H(t)q_2(t)\,. \tag{8.20}$$

We acknowledge that the initial conditions are zero

$$q_1(0) = i_1(0) = 0\,, \quad q_2(0) = i_2(0) = 0\,.$$

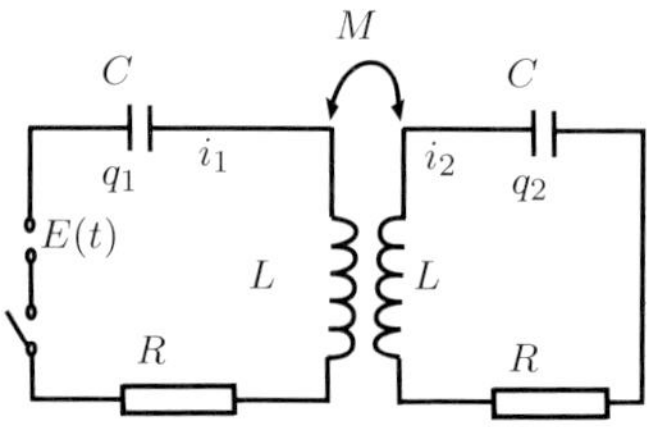

Figure 8.2

Because $q_j \in C^2([0,\infty))$, $j = 1, 2$, it follows

$$\frac{\mathrm{d}}{\mathrm{d}t}\overline{q}_j(t) = \frac{\tilde{\mathrm{d}}}{\mathrm{d}t}\overline{q}_j(t)\,, \quad \frac{\mathrm{d}^2}{\mathrm{d}t^2}\overline{q}_j(t) = \frac{\tilde{\mathrm{d}}^2}{\mathrm{d}t^2}\overline{q}_j(t)\,, \quad j = 1, 2\,.$$

Consequently, (8.19) in the distributions space $\mathcal{D}'_+$ become

$$\begin{aligned} L\frac{\mathrm{d}^2}{\mathrm{d}t^2}\overline{q}_1(t) + M\frac{\mathrm{d}^2}{\mathrm{d}t^2}\overline{q}_2(t) + R\frac{\mathrm{d}}{\mathrm{d}t}\overline{q}_1(t) + \frac{1}{C}\overline{q}_1(t) &= \overline{E}(t)\,, \\ M\frac{\mathrm{d}^2}{\mathrm{d}t^2}\overline{q}_1(t) + L\frac{\mathrm{d}^2}{\mathrm{d}t^2}\overline{q}_2(t) + R\frac{\mathrm{d}}{\mathrm{d}t}\overline{q}_2(t) + \frac{1}{C}\overline{q}_2(t) &= 0\,. \end{aligned} \tag{8.21}$$

If the initial conditions are not zero on the right-hand side of (8.21), then expressions that depend on the initial conditions are added.

The Cauchy problem for a coupled oscillating circuit is reduced to determining the distributions $\overline{q}_1, \overline{q}_2 \in \mathcal{D}'_+$ which verify the system (8.21).

The intensities $\overline{i}_1(t) = H(t)i_1(t)$, $\overline{i}_2(t) = H(t)i_2(t)$ are determined from the relations

$$\overline{i}_1(t) = \frac{\mathrm{d}}{\mathrm{d}t}\overline{q}_1(t)\,, \quad \overline{i}_2(t) = \frac{\mathrm{d}}{\mathrm{d}t}\overline{q}_2(t)\,. \tag{8.22}$$

Applying the Laplace transform in distributions to the system (8.21), we obtain

$$\begin{aligned} &\left(Lp^2 + Rp + \frac{1}{C}\right)\tilde{\overline{q}}_1(p) + Mp^2\tilde{\overline{q}}_2(p) = \tilde{\overline{E}}(p)\,, \\ &Mp^2\tilde{\overline{q}}_1(p) + \left(Lp^2 + Rp + \frac{1}{C}\right)\tilde{\overline{q}}_2(p) = 0\,, \end{aligned} \tag{8.23}$$

where

$$\tilde{\overline{q}}_1(p) = \mathrm{L}[\overline{q}_1(t)](p)\,, \quad \tilde{\overline{q}}_2(p) = \mathrm{L}[\overline{q}_2(t)](p)\,, \quad \tilde{\overline{E}}(p) = \mathrm{L}[\overline{E}(t)](p)\,.$$

From (8.23) we find

$$\begin{aligned} \tilde{\overline{q}}_1(p) &= -\frac{Lp^2 + Rp + C^{-1}}{M^2p^4 - (Lp^2 + Rp + C^{-1})^2}\tilde{\overline{E}}(p)\,, \\ \tilde{\overline{q}}_2(p) &= \frac{Mp^2}{M^2p^4 - (Lp^2 + Rp + C^{-1})^2}\tilde{\overline{E}}(p)\,. \end{aligned} \tag{8.24}$$

Assuming that $R^2 < 4(L - M)C^{-1}$ and $L > M$, the roots of the denominators of the expressions (8.24) will be

$$p_{1,2} = -\alpha_1 \pm \mathrm{i}\omega_1\,, \quad p_{3,4} = -\alpha_2 \pm \mathrm{i}\omega_2\,,$$

where

$$\alpha_{1,2} = \frac{R}{2(L \pm M)}\,, \quad \omega_{1,2} = \frac{\sqrt{4C^{-1}(L \pm M) - R^2}}{2(L \pm M)}\,.$$

In these conditions we can write

$$-\frac{Lp^2+Rp+C^{-1}}{M^2p^4-(Lp^2+Rp+C^{-1})^2} = \frac{1}{2}\left[\frac{1}{(L+M)p^2+Rp+C^{-1}}+\frac{1}{(L-M)p^2+Rp+C^{-1}}\right],$$

$$\frac{Mp^2}{M^2p^4-(Lp^2+Rp+C^{-1})^2} = \frac{1}{2}\left[\frac{1}{(L+M)p^2+Rp+C^{-1}}-\frac{1}{(L-M)p^2+Rp+C^{-1}}\right].$$

Applying the inverse Laplace transform to the expressions (8.24), we obtain for the unknown distributions $\overline{q}_1, \overline{q}_2$ the expressions

$$\overline{q}_1(t) = \overline{E}(t) * \frac{H(t)}{2}\left[\frac{1}{(L+M)\omega_1}\exp(-\alpha_1 t)\sin\omega_1 t + \frac{1}{(L-M)\omega_2}\exp(-\alpha_2 t)\sin\omega_2 t\right],$$

$$\overline{q}_2(t) = \overline{E}(t) * \frac{H(t)}{2}\left[\frac{1}{(L+M)\omega_1}\exp(-\alpha_1 t)\sin\omega_1 t - \frac{1}{(L-M)\omega_2}\exp(-\alpha_2 t)\sin\omega_2 t\right].$$

For the intensities $\overline{i}_1(t) = H(t)i_1(t), \overline{i}_2(t) = H(t)i_2(t)$ we get

$$\overline{i}_1(t) = \frac{d\overline{q}_1}{dt} = \frac{d\overline{E}}{dt} * \frac{H(t)}{2}\left[\frac{1}{(L+M)\omega_1}\exp(-\alpha_1 t)\sin\omega_1 t + \frac{1}{(L-M)\omega_2}\exp(-\alpha_2 t)\sin\omega_2 t\right],$$

$$\overline{i}_2(t) = \frac{d\overline{q}_2}{dt} = \frac{d\overline{E}}{dt} * \frac{H(t)}{2}\left[\frac{1}{(L+M)\omega_1}\exp(-\alpha_1 t)\sin\omega_1 t - \frac{1}{(L-M)\omega_2}\exp(-\alpha_2 t)\sin\omega_2 t\right].$$

Particularly, if at the initial moment an electromotive tension $\overline{E}(t) = E_0 H(t)$ is applied, which remains constant for $t > 0$, then, since $d\overline{E}/dt = E_0\delta(t)$, we obtain

$$\overline{i}_1(t) = E_0\frac{H(t)}{2}\left[\frac{1}{(L+M)\omega_1}\exp(-\alpha_1 t)\sin\omega_1 t + \frac{1}{(L-M)\omega_2}\exp(-\alpha_2 t)\sin\omega_2 t\right],$$

$$\overline{i}_2(t) = E_0\frac{H(t)}{2}\left[\frac{1}{(L+M)\omega_1}\exp(-\alpha_1 t)\sin\omega_1 t - \frac{1}{(L-M)\omega_2}\exp(-\alpha_2 t)\sin\omega_2 t\right].$$

For $t > 0$, one has

$$i_1(t) = \frac{E_0}{2}\left[\frac{1}{(L+M)\omega_1}\exp(-\alpha_1 t)\sin\omega_1 t + \frac{1}{(L-M)\omega_2}\exp(-\alpha_2 t)\sin\omega_2 t\right],$$
$$i_2(t) = \frac{E_0}{2}\left[\frac{1}{(L+M)\omega_1}\exp(-\alpha_1 t)\sin\omega_1 t - \frac{1}{(L-M)\omega_2}\exp(-\alpha_2 t)\sin\omega_2 t\right].$$

Therefore

$$\begin{aligned} q_1(t) = -\frac{1}{2}C E_0 \Bigg[&\frac{1}{\omega_1}\exp(-\omega_1 t)\,(\omega_1\cos\omega_1 t + \alpha_1\sin\omega_1 t) \\ &+\frac{1}{\omega_2}\exp(-\omega_2 t)\,(\omega_2\cos\omega_2 t + \alpha_2\sin\omega_2 t) - 2\Bigg], \quad t > 0, \\ q_2(t) = -\frac{1}{2}C E_0 \Bigg[&\frac{1}{\omega_1}\exp(-\omega_1 t)\,(\omega_1\cos\omega_1 t + \alpha_1\sin\omega_1 t) \\ &-\frac{1}{\omega_2}\exp(-\omega_2 t)\,(\omega_2\cos\omega_2 t + \alpha_2\sin\omega_2 t)\Bigg], \quad t > 0. \end{aligned}$$

In the case of a tension pulse at the initial moment, expressed in the form $\overline{E}(t) = E_0\delta(t)$, we obtain a transient phenomenon, for which

$$q_1(t) = \frac{E_0}{2}\left[\frac{1}{(L+M)\omega_1}\exp(-\alpha_1 t)\sin\omega_1 t + \frac{1}{(L-M)\omega_2}\exp(-\alpha_2 t)\sin\omega_2 t\right], \quad t > 0,$$
$$q_2(t) = \frac{E_0}{2}\left[\frac{1}{(L+M)\omega_1}\exp(-\alpha_1 t)\sin\omega_1 t - \frac{1}{(L-M)\omega_2}\exp(-\alpha_2 t)\sin\omega_2 t\right], \quad t > 0,$$

and

$$\begin{aligned} i_1(t) = \frac{1}{2}E_0 \Bigg[&\frac{1}{(L+M)\omega_1}\exp(-\alpha_1 t)\,(\omega_1\cos\omega_1 t - \alpha_1\sin\omega_1 t) \\ &+\frac{1}{(L-M)\omega_2}\exp(-\alpha_2 t)\,(\omega_2\cos\omega_2 t - \alpha_2\sin\omega_2 t)\Bigg], \quad t > 0, \\ i_2(t) = \frac{1}{2}E_0 \Bigg[&\frac{1}{(L+M)\omega_1}\exp(-\alpha_1 t)\,(\omega_1\cos\omega_1 t - \alpha_1\sin\omega_1 t) \\ &-\frac{1}{(L-M)\omega_2}\exp(-\omega_2 t)\,(\omega_2\cos\omega_2 t - \alpha_2\sin\omega_2 t)\Bigg], \quad t > 0. \end{aligned}$$

At the limit, we obtain

$$\lim_{t\to\infty} q_1(t) = \lim_{t\to\infty} q_2(t) = \lim_{t\to\infty} i_1(t) = \lim_{t\to\infty} i_2(t) = 0.$$

8.3
Admittance and Impedance of the RLC Circuit

If $\mathcal{E}(t) \in \mathcal{D}'_+$ is the fundamental solution of the operator $\mathrm{D} = L(\mathrm{d}^2/\mathrm{d}t^2) + R(\mathrm{d}/\mathrm{d}t) + C^{-1}$, hence $\mathrm{D}\,\mathcal{E}(t) = \delta(t) \Leftrightarrow \mathcal{E}(t) = (\mathrm{D}\,\delta(t))^{-1}$, then, according to the formula (8.14), we can determine the charge $\overline{q}$ depending on the emf $\overline{E} \in \mathcal{D}'_+$

$$\overline{q}(t) = \mathcal{E}(t) * \overline{E}(t) = (\mathrm{D}\,\delta(t))^{-1} * \overline{E}(t) . \tag{8.25}$$

For the intensity $\overline{i} \in \mathcal{D}'_+$ of the RLC circuit, $i_0 = q_0 = 0$, we obtain

$$\overline{i}(t) = \frac{\mathrm{d}}{\mathrm{d}t}\overline{q}(t) = \frac{\mathrm{d}\mathcal{E}(t)}{\mathrm{d}t} * \overline{E}(t) . \tag{8.26}$$

By definition, the distribution $\mathcal{E}(t) = (\mathrm{D}\,\delta(t))^{-1}$ is called the admittance of the circuit relative to the charge and is denoted by

$$A_q(t) = \mathcal{E}(t) = (\mathrm{D}\,\delta(t))^{-1} \in \mathcal{D}'_+ . \tag{8.27}$$

The distribution denoted by

$$A_i(t) = \frac{\mathrm{d}\mathcal{E}(t)}{\mathrm{d}t} \in \mathcal{D}'_+ \tag{8.28}$$

will be called the admittance of the RLC circuit with respect to the intensity.

From (8.25), (8.26), (8.27), and (8.28) result the relations

$$\overline{q}(t) = A_q(t) * \overline{E} , \quad \overline{i}(t) = A_i(t) * \overline{E} , \tag{8.29}$$

$$A_i = \frac{\mathrm{d}}{\mathrm{d}t} A_q . \tag{8.30}$$

The inverse in $\mathcal{D}'_+$ of the admittances A_q and A_i, denoted by $Z_q, Z_i \in \mathcal{D}'_+$, respectively, are called the impedances of the circuit with respect to charge and intensity, respectively.

We have

$$Z_q = A_q^{-1} = \mathcal{E}^{-1} \tag{8.31}$$

$$Z_i = A_i^{-1} = (\mathcal{E}')^{-1} . \tag{8.32}$$

It follows the relations

$$Z_q * A_q = \delta(t) , \quad Z_i * A_i = \delta(t) .$$

On the other hand, from (8.29) we obtain

$$\overline{E} = A_q^{-1} * \overline{q} = Z_q * \overline{q} , \quad \overline{E} = A_i^{-1} * \overline{i} = Z_i * \overline{i} .$$

Example 8.2 Let the impedances and admittances of the RLC circuit be determined in the case $R^2 > 4LC^{-1}$.

According to the formula (8.16), the fundamental solution of the operator $\mathrm{D} = L(\mathrm{d}^2/\mathrm{d}t^2) + R(\mathrm{d}/\mathrm{d}t) + C^{-1}$ is

$$\mathcal{E}(t) = \frac{1}{L\beta} H(t) \sinh \beta t \,, \quad t \in \mathbb{R} \,.$$

Consequently, for the admittances A_q and A_i of the circuit we have

$$A_q = \mathcal{E}(t) = \frac{1}{L\beta} H(t) \sinh \beta t \,, \quad t \in \mathbb{R} \,,$$

$$A_i = \frac{\mathrm{d}\mathcal{E}(t)}{\mathrm{d}t} = \frac{1}{L} H(t) \cosh \beta t \,, \quad t \in \mathbb{R} \,.$$

As regards the impedances Z_q and Z_i, we have

$$Z_q = A_q^{-1} = L\delta''(t) - \beta^2 L\delta(t) \,,$$

$$Z_i = A_i^{-1} = L\delta'(t) - \beta^2 LH(t) \,.$$

8.4 Quadrupoles

A quadrupole is an electric circuit having two input (1 and 1′) and two output terminals (2 and 2′). The block diagram of a quadrupole is presented in Figure 8.3, where $i_1(t)$, $E_1(t)$, $i_2(t)$, $E_2(t)$ are currents and voltages at the input and output terminals, respectively.

There are two types of quadrupoles – active and passive – according to whether they contain a source of energy or not.

In the following, we consider only passive quadrupoles and we assume that their parameters, consisting of resistors, inductances and capacities are constant.

A quadrupole is said to be linear if the dependence between its input and output can be expressed by linear functions.

For instance, if the input variable $\varphi_1(t)$ and the output variable $\varphi_2(t)$ are related by the relation

$$\varphi_2(t) = \int_{-\infty}^{\infty} K_1(t, \tau)\varphi_1(\tau)\mathrm{d}\tau \,, \tag{8.33}$$

then the corresponding quadrupole is linear.

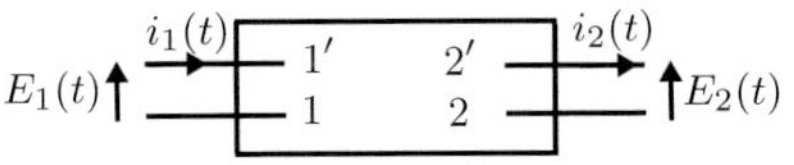

Figure 8.3

Obviously, the relation (8.33) defines an integral operator where $K_1(t,\tau)$ is the kernel of the integral transformation.

If the kernel $K_1(t,\tau)$ is of the form $K_1(t,\tau) = K(t-\tau)$, then the interdependence between input and output can be expressed by a convolution, namely

$$\varphi_2(t) = K(t) * \varphi_1(t) . \tag{8.34}$$

For the sake of generality, (8.34) will be considered in the distributions space $\mathcal{D}'_+$.

The quadrupoles for which their inputs and outputs satisfy (8.34) are called linear and stationary.

The distribution $K(t)$ is called a weighting distribution and its expression depends on the nature of the corresponding variables. For $\varphi_1(t) = \delta(t)$, we obtain $\varphi_2(t) = K(t)$.

Hence, the weighting distribution is the response of the quadrupole to the unit impulse δ.

The Laplace transform of the weighting distribution $K(t)$, if it exists, is called the transfer function of the quadrupole.

Let us consider now two stationary quadrupoles with the weighting distributions $K_1(t)$ and $K_2(t)$. By connecting them in series (Figure 8.4), we also get a quadrupole.

Denoting the output to the first quadrupole by $\varphi(t)$, we have

$$\varphi = K_1 * \varphi_1 . \tag{8.35}$$

Because this variable represents the input to the second quadrupole, we can write

$$\varphi_2 = K_2 * \varphi \tag{8.36}$$

under the assumption that the distributions $\varphi_1, \varphi_2, K_1, K_2$ belong to the space $\mathcal{D}'_+$.

From the relations (8.35) and (8.36), we obtain

$$\varphi_2 = (K_1 * K_2) * \varphi_1 .$$

Consequently, by series connection of two linear stationary quadrupoles, the weighting distribution has the expression

$$K = K_1 * K_2 . \tag{8.37}$$

Example 8.3 Consider a stationary quadrupole Figure 8.5 obtained by the series connection of two stationary quadrupoles illustrated in Figure 8.6.

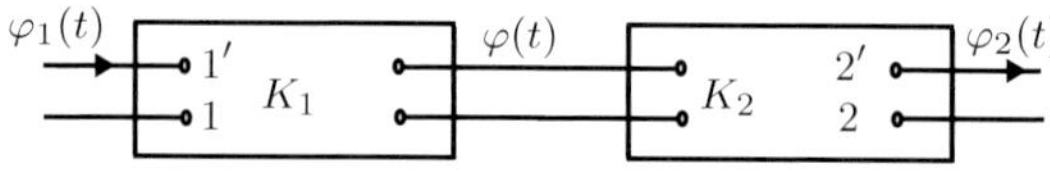

Figure 8.4

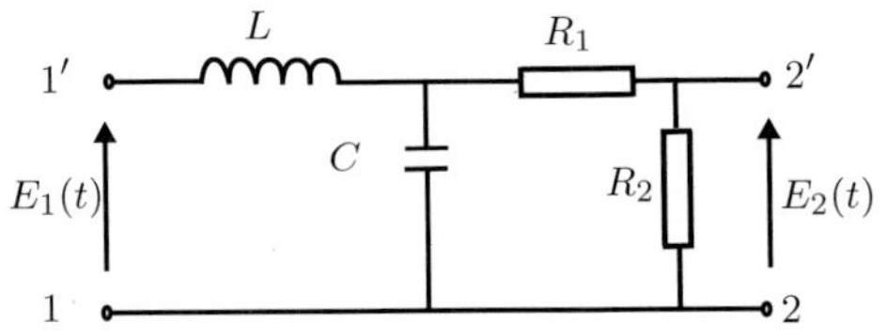

Figure 8.5

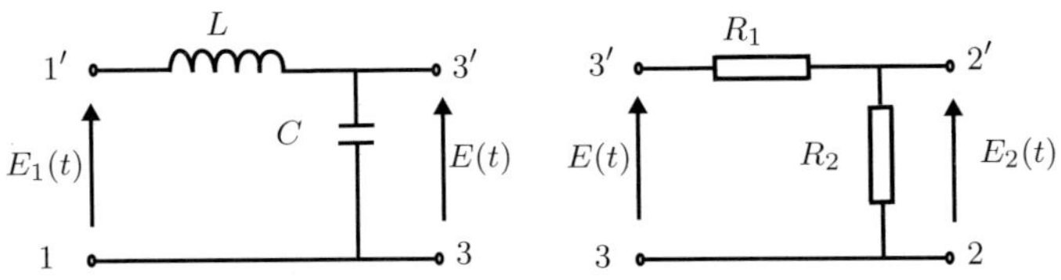

Figure 8.6

For the first quadrupole shown in Figure 8.6, we can write the equations

$$E_1 = L\frac{\mathrm{d}}{\mathrm{d}t}i(t) + \frac{1}{C}q(t)\,, \quad E(t) = \frac{q(t)}{C}\,, \quad i = \frac{\mathrm{d}q}{\mathrm{d}t}\,.$$

Eliminating the electric charge q, we obtain

$$LC\frac{\mathrm{d}^2}{\mathrm{d}t^2}E(t) + E(t) = E_1\,. \tag{8.38}$$

The weighting distribution $K_1(t)$ of this quadrupole is the fundamental solution of (8.38), because it is obtained considering $E_1(t) = \delta(t)$.

Consequently, applying the Laplace transform, we obtain

$$\mathrm{L}[K_1(t)] = \frac{1}{LCp^2 + 1}\,,$$

where

$$K_1(t) = \frac{1}{\sqrt{CL}}H(t)\sin\frac{t}{\sqrt{CL}}\,.$$

The voltage $E(t)$, representing the output of the first quadrupole, will have the expression

$$E(t) = K_1(t) * E_1(t)\,.$$

Similarly, for the second quadrupole shown in Figure 8.6, we obtain the equations

$$E(t) = (R_1 + R_2)i_2(t)\,, \quad E_2(t) = R_2 i(t)\,,$$

where

$$E_2(t) = \frac{R_2}{R_1 + R_2}E(t) = \frac{R_2}{R_1 + R_2}\delta(t) * E(t)\,.$$

Thus, the distribution K_2 of the second quadrupole is

$$K_2(t) = \frac{R_2}{R_1 + R_2}\delta(t) .$$

According to (8.37), for the weighting distribution of the quadrupole from Figure 8.5, we shall obtain the expression

$$K(t) = K_1(t) * K_2(t) = \frac{R_2}{R_1 + R_2}\frac{H(t)}{\sqrt{CL}}\sin\frac{t}{\sqrt{CL}} .$$

Thus, the output voltage E_2 is

$$E_2(t) = K(t) * E_1(t) = \frac{R_2}{R_1 + R_2}\frac{H(t)}{\sqrt{CL}}\sin\frac{t}{\sqrt{CL}} * E_1(t) .$$

Example 8.4 RC high frequency filter. Weighting distribution The quadrupole represented in Figure 8.7 is called a high frequency RC filter.

The corresponding differential equations are

$$E_1(t) = R\frac{\mathrm{d}}{\mathrm{d}t}q(t) + \frac{1}{C}q(t) , \quad E_2 = R\frac{\mathrm{d}}{\mathrm{d}t}q(t) .$$

Eliminating the electric charge q, we obtain

$$RC\frac{\mathrm{d}}{\mathrm{d}t}E_2(t) + E_2(t) = RC\frac{\mathrm{d}}{\mathrm{d}t}E_1(t) . \tag{8.39}$$

Between the input, the emf E_1 and the output of the emf E_2, we have the relation

$$E_2(t) = K(t) * E_1(t) , \tag{8.40}$$

where $K \in \mathcal{D}'_+$ is the weighting distribution of the circuit.

For $E_1(t) = \delta(t)$ we obtain the weighting distribution of K; from (8.39) it follows

$$RCK'(t) + K(t) = RC\delta'(t) .$$

Applying the Laplace transform, we obtain

$$\tilde{K}(p) = 1 - \frac{1}{RC}\frac{1}{p + R^{-1}C^{-1}} ,$$

where $\tilde{K}(p) = \mathrm{L}[K(t)](p)$.

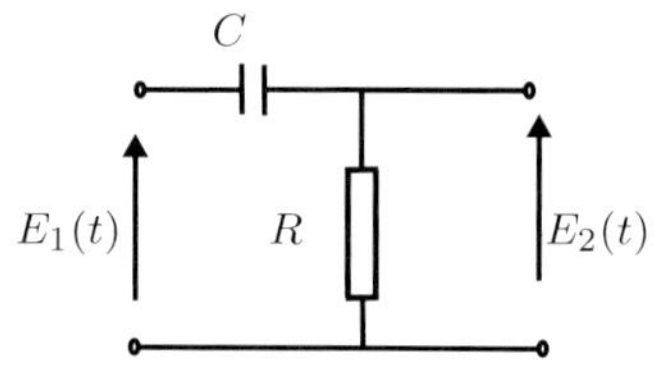

Figure 8.7

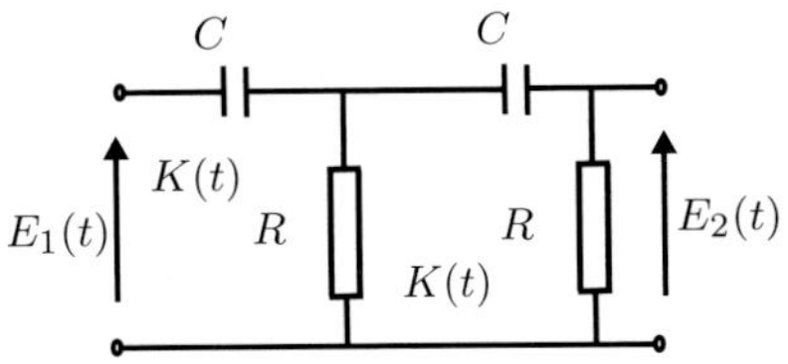

Figure 8.8

By applying the inverse Laplace transform, we obtain for the weighting distribution the expression

$$K(t) = \delta(t) - \frac{1}{RC} H(t) \exp\left(-t R^{-1} C^{-1}\right) .$$

Taking into account (8.40), the output of the emf has the expression

$$E_2(t) = E_1(t) - H(t) R^{-1} C^{-1} \exp\left(-t R^{-1} C^{-1}\right) * E_1(t) .$$

Particularly, if $E_1(t) = E_0 H(t)$, then we obtain

$$E_2(t) = E_0 H(t) \exp\left(-t R^{-1} C^{-1}\right) = \begin{cases} 0 , & t < 0 , \\ E_0 \exp\left(-t R^{-1} C^{-1}\right) , & t \geq 0 . \end{cases}$$

If we connect two such filters in series (Figure 8.8), the weighting distribution $K_1 \in \mathcal{D}'_+$ will have the expression

$$\begin{aligned} K_1(t) &= K(t) * K(t) = (K(t))^{*2} \\ &= \delta(t) - \frac{2}{RC} H(t) \exp\left(-t R^{-1} C^{-1}\right) + \frac{1}{R^2 C^2} t H(t) \exp\left(-t R^{-1} C^{-1}\right) , \end{aligned}$$

because $(H(t) \exp(\alpha t))^{*2} = t H(t) \exp(\alpha t)$.

The output of the emf E_2 is given by

$$E_2(t) = K_1(t) * E_1(t) = E_1(t) - R^{-1} C^{-1} \exp(-t R^{-1} C^{-1}) \left(2 H(t) - \frac{t H(t)}{RC}\right) * E_1 .$$

9
Applications of the Distribution Theory in the Study of Elastic Bars

9.1
Longitudinal Vibrations of Elastic Bars

9.1.1
Equations of Motion Expressed in Displacements and in Stresses: Formulation of the Problems with Boundary–Initial Conditions

We consider a bar of length ℓ for which the bar axis is considered to be the Ox-axis with one end at the origin and the other end at the point A (Figure 9.1).

A first approximation of modeling the longitudinal vibrations phenomenon of bars is the Bernoulli–Euler model. It is accepted that the bar has strains only along its axis, and the parallel transverse planes remain parallel during the motion.

Therefore, the equation of motion of the Bernoulli–Euler type bars is

$$E\frac{\tilde{\partial}^2}{\partial x^2}u(x,t) - \rho\frac{\tilde{\partial}^2}{\partial t^2}u(x,t) + X(x,t) = 0\,, \tag{9.1}$$

where $u \in C^{2,2}([0,\ell] \times \mathbb{R}_+)$ is the displacement, $\tilde{\partial}/\partial x, \tilde{\partial}/\partial t$ are the derivatives in the classical sense, E is the modulus of elasticity, ρ is the density and $X \in C^{0,0}([0,\ell] \times \mathbb{R}_+)$ represents the body force.

A.E.H. Love [46], taking into consideration the influence of transverse contraction, sets for longitudinal vibrations the equation

$$E\frac{\tilde{\partial}^2}{\partial x^2}u(x,t) + \nu^2 r_0^2\rho\frac{\tilde{\partial}^4}{\partial x^2\partial t^2}u(x,t) - \rho\frac{\tilde{\partial}^2}{\partial t^2}u(x,t) + X(x,t) = 0\,, \tag{9.2}$$

where the term $\nu^2 r_0^2 \rho\tilde{\partial}^4 u(x,t)/\partial x^2\partial t^2$ is due to the influence of the transverse contraction, r_0 is the radius of gyration, $S r_0^2 = \iint_\Omega (y^2 + z^2)\mathrm{d}y\mathrm{d}z$, S is the cross-section area, and Ω is the domain corresponding to the cross-section.

Figure 9.1

Distribution Theory, First Edition. Petre P. Teodorescu, Wilhelm W. Kecs, and Antonela Toma

Taking into consideration the influence of the shear stresses σ_{xy}, σ_{xz}, [47], one obtains the generalized equation of the longitudinal vibrations

$$E\frac{\tilde{\partial}^2}{\partial x^2}u(x,t) - \frac{E\nu^2 r_0^2}{2(1+\nu)}\frac{\tilde{\partial}^4}{\partial x^4}u(x,t) + \nu^2 r_0^2 \rho \frac{\tilde{\partial}^4}{\partial x^2 \partial t^2}u(x,t)$$
$$-\rho\frac{\tilde{\partial}^2}{\partial t^2}u(x,t) + X(x,t) = 0\,, \tag{9.3}$$

where the term $-(E\nu^2 r_0^2)/(2(1+\nu))(\tilde{\partial}^4/\partial x^4)u(x,t)$ incorporates the influence of the shear stresses σ_{xy}, σ_{xz}.

Next we consider only Bernoulli–Euler type bars. The Bernoulli–Euler equation (9.1) is written with respect to the displacement $u(x,t)$, $x \in [0,\ell]$, $t \geq 0$.

Thus, we formulate the problem of determining the displacement u to verify the (9.1), the initial conditions

$$u(x,0) = u_0(x)\,, \quad \dot{u}(x,0) = \frac{\tilde{\partial}}{\partial t}u(x,t)\bigg|_{t=0} = u_1(x)\,, \quad u_0, u_1 \in C^0([0,\ell])\,, \tag{9.4}$$

and the boundary conditions

$$u(0,t) = f(t)\,, \quad u(\ell,t) = g(t)\,, \quad f, g \in C^0(\mathbb{R}_+)\,. \tag{9.5}$$

The longitudinal vibrations equation (9.1) can also be written with respect to the stresses $\sigma(x,t)$, $x \in [0,\ell]$, $t \geq 0$.

To this end, differentiating (9.1) with respect to x and using the relation $\sigma(x,t) = E\tilde{\partial}u(x,t)/\partial x$, we obtain

$$\frac{\tilde{\partial}^2}{\partial x^2}\sigma(x,t) - \frac{\rho}{E}\frac{\tilde{\partial}^2}{\partial t^2}\sigma(x,t) + \frac{\tilde{\partial}}{\partial t}X(x,t) = 0\,. \tag{9.6}$$

This equation is the longitudinal vibrations equation expressed in stresses.

To this equation we must add the initial conditions

$$\sigma(x,0) = \sigma_0(x)\,, \quad \frac{\tilde{\partial}}{\partial t}\sigma(x,t)\bigg|_{t=0} = \dot{\sigma}(x,0) = \sigma_1(x)\,, \quad \sigma_0, \sigma_1 \in C^0([0,\ell])\,, \tag{9.7}$$

and the boundary conditions

$$\sigma(0,t) = f_1(t)\,, \quad \sigma(\ell,t) = f_2(t)\,, \quad f_1, f_2 \in C^0(\mathbb{R}_+)\,. \tag{9.8}$$

In this way, the vibration problem is to determine the stresses which verify (9.6) as well as the conditions (9.7) and (9.8).

9.1.2
Forced Longitudinal Vibrations of a Bar with Boundary Conditions Expressed in Displacements

We consider a bar uniformly acted upon by a load distributed along its length of the form

$$X(x,t) = X_0 H(t)\,, \quad x \in [0,\ell]\,, \quad t \geq 0\,, \quad X_0 = \text{const}\,. \tag{9.9}$$

Let us determine the displacement $u(x,t) \in C^{2,2}([0,\ell] \times \mathbb{R}_+)$, which satisfies (9.1), the initial conditions

$$u(x,0) = u_0\,, \quad \dot{u}(x,0) = u_1, u_0, u_1 = \text{const}\,. \tag{9.10}$$

and the boundary conditions

$$u(0,t) = f_1(t)\,, \quad u(\ell,t) = f_2(t)\,, \quad f_1, f_2 \in C^0(\mathbb{R}_+)\,. \tag{9.11}$$

The equation of longitudinal vibrations (9.1), expressed in displacements, can be written in the form

$$\frac{\tilde{\partial}^2}{\partial t^2} u(x,t) = c^2 \frac{\tilde{\partial}^2}{\partial x^2} u(x,t) + \frac{X_0 H(t)}{\rho}\,, \quad x \in [0,\ell]\,, \quad t \geq 0\,, \tag{9.12}$$

where $c^2 = E/\rho$.

We rewrite this equation in the distributions space $\mathcal{D}'_+$. We denote by $\overline{u}_x(t) = u(x,t)H(t)$ a function type distribution with respect to the variable $t \in \mathbb{R}$ from the space $\mathcal{D}'_+$, x being the parameter. We denote by $\partial/\partial t, \partial/\partial x$ the derivatives in the sense of distributions.

Because

$$\frac{\tilde{\partial}\overline{u}_x(t)}{\partial t} = H(t)\frac{\tilde{\partial} u(x,t)}{\partial t}\,, \quad \frac{\tilde{\partial}^2\overline{u}_x(t)}{\partial t^2} = H(t)\frac{\tilde{\partial}^2 u(x,t)}{\partial t^2}\,, \quad \frac{\tilde{\partial}}{\partial x} = \frac{\partial}{\partial x}\,,$$

(9.12) becomes

$$\frac{\tilde{\partial}^2\overline{u}_x(t)}{\partial t^2} = c^2 \frac{\tilde{\partial}^2\overline{u}_x(t)}{\partial x^2} + \frac{X_0 H(t)}{\rho}\,.$$

Differentiating in the sense of distributions, we obtain

$$\begin{aligned}
\frac{\partial\overline{u}_x(t)}{\partial t} &= \frac{\tilde{\partial}\overline{u}_x(t)}{\partial t} + \overline{u}_x(+0)\delta(t) = \frac{\tilde{\partial}\overline{u}_x(t)}{\partial t} + u_0\delta(t)\,,\\
\frac{\partial^2\overline{u}_x(t)}{\partial t^2} &= \frac{\tilde{\partial}^2\overline{u}_x(t)}{\partial t^2} + \left.\frac{\tilde{\partial}\overline{u}_x(t)}{\partial t}\right|_{t=+0} \delta(t) + u_0\delta'(t)\\
&= \frac{\tilde{\partial}^2\overline{u}_x(t)}{\partial t^2} + u_1\delta(t) + u_0\delta'(t)\,,
\end{aligned}$$

where $\overline{u}_x(+0) = u_0$ is the jump of $\overline{u}_x(t)$ at the origin.

We note that, to calculate derivatives in the sense of distributions, we used the initial conditions (9.10).

Thus, the longitudinal vibrations equation (9.12) in the distributions space is

$$\frac{\partial^2 \overline{u}_x(t)}{\partial t^2} = c^2 \frac{\partial^2 \overline{u}_x(t)}{\partial x^2} + \frac{X_0 H(t)}{\rho} + u_1 \delta(t) + u_0 \delta'(t)\,, \quad x \in (0, \ell)\,. \quad (9.13)$$

In this case, the problem with initial and boundary conditions consists in the determination of the distribution $\overline{u}_x(t) \in \mathcal{D}'_+(\mathbb{R})$, x being the parameter, which satisfies (9.13) and boundary conditions in displacements

$$\overline{u}_x(t)|_{x=0} = f_1(t)\,, \quad \overline{u}_x(t)|_{x=\ell} = f_2(t)\,, \quad f_1, f_2 \in \mathcal{D}'_+(\mathbb{R})\,. \quad (9.14)$$

We assume that $\overline{u}_x(t)$, $f_1(t)$ and $f_2(t)$ have Laplace images in distributions. Applying the Laplace transform in distributions to (9.13) and to boundary conditions (9.14), we obtain for the Laplace image of the displacement the expression

$$\begin{aligned}\tilde{\overline{u}}_x(p) = &\frac{\sinh((\ell - x)p/c)}{\sinh(\ell p/c)} \left[\tilde{f}_1(p) - \frac{u_0}{p} - \frac{u_1}{p^2} - \frac{X_0}{\rho p^3} \right] \\ &+ \frac{\sinh(xp/c)}{\sinh(\ell p/c)} \left[\tilde{f}_2(p) - \frac{u_0}{p} - \frac{u_1}{p^2} - \frac{X_0}{3p^3} \right] + \frac{u_0}{p} + \frac{u_1}{p^2} + \frac{X_0}{\rho p^3}\,. \end{aligned} \quad (9.15)$$

Applying the inverse Laplace transform to the relation (9.15), we determine the displacement in the form

$$\begin{aligned}\overline{u}_x(t) = &h_{c,\ell}(\ell - x, t) \underset{t}{*} \left[f_1(t) - u_0 H(t) - u_1 t H(t) - \frac{X_0\, t^2 H(t)}{2\rho} \right] + u_1 t H(t) \\ &+ h_{c,\ell}(x, t) \underset{t}{*} \left[f_2(t) - u_0 H(t) - u_1 t H(t) - \frac{X_0 t^2 H(t)}{2\rho} \right] + u_0 H(t) + \frac{X_0 t^2 H(t)}{2\rho}\,, \end{aligned} \quad (9.16)$$

where $*_t$ is the convolution product with respect to t in the distribution space, and

$$h_{c,\ell}(x, t) = \mathrm{L}^{-1} \left[\frac{\sinh(xp/c)}{\sinh(\ell p/c)} \right] = \frac{2cH(t)}{\ell} \sum_{k=1}^{\infty} (-1)^{k-1} \sin \frac{k\pi x}{\ell} \sin \frac{k\pi ct}{\ell} \quad (9.17)$$

is a distribution from $\mathcal{D}'_+(\mathbb{R})$ with respect to t.

We can give another expression for the displacement using the following result

$$\begin{aligned}f(x, t) &= \mathrm{L}^{-1} \left[\frac{\sinh(xp/c)}{p^2 \sinh(\ell p/c)} \right] \\ &= H(t) \frac{tx}{\ell} + \frac{2\ell}{c\pi^2} H(t) \sum_{k=1}^{\infty} \frac{(-1)^k}{k^2} \sin \frac{k\pi x}{\ell} \sin \frac{kc\pi t}{\ell}\,. \end{aligned} \quad (9.18)$$

Thus, we have

$$\overline{u}_x(t) = f(\ell - x, t) \underset{t}{*} f_1''(t) + f(x, t) \underset{t}{*} f_2''(t)$$
$$- u_0 \left[\frac{\partial f(\ell - x, t)}{\partial t} + \frac{\partial f(x, t)}{\partial t} - H(t) \right] - u_1 \left[f(\ell - x, t) + f(x, t) - t H(t) \right]$$
$$- \frac{X_0}{\rho} \left[f(\ell - x, t) \underset{t}{*} H(t) + f(x, t) \underset{t}{*} H(t) - \frac{t^2 H(t)}{2} \right], \tag{9.19}$$

where $f(x, t)$ is a function type distribution from $\mathcal{D}'_+(\mathbb{R})$ given by (9.18).

Developing the convolution product from (9.19), we obtain

$$\overline{u}_x(t) = f(\ell - x, t) \underset{t}{*} f_1''(t) + f(x, t) \underset{t}{*} f_2''(t)$$
$$- u_0 \frac{4H(t)}{\pi} \sum_{m=1}^{\infty} \frac{(-1)^m}{2m - 1} \cos \frac{(2m - 1)c\pi t}{\ell} \cos \frac{(2m - 1)\pi(x - \ell/2)}{\ell}$$
$$- u_1 \frac{4\ell H(t)}{c\pi^2} \sum_{m=1}^{\infty} \frac{(-1)^m}{(2m - 1)^2} \sin \frac{(2m - 1)c\pi t}{\ell} \cos \frac{(2m - 1)\pi(x - \ell/2)}{\ell}$$
$$- \frac{X_0}{\rho} \frac{4\ell^2 H(t)}{c^2\pi^3} \sum_{m=1}^{\infty} \frac{(-1)^m}{(2m - 1)^3} \cos \frac{(2m - 1)\pi(x - \ell/2)}{\ell} \left(1 - \cos \frac{(2m - 1)c\pi t}{\ell} \right). \tag{9.20}$$

Example 9.1 We consider that $u_0 = u_1 = X_0 = 0$ and that the boundary conditions are

$$\overline{u}_x(t)|_{x=0} = f_1(t) = a H(t) \sin \omega t \,, \overline{u}_x(t)|_{x=\ell} = f_2(t) = 0 \,, \quad \omega \in \mathbb{R}_+ \,.$$

In these conditions, from (9.20) we have

$$\overline{u}_x(t) = f(\ell - x, t) \underset{t}{*} f_1''(t) = \frac{\partial^2 f(\ell - x, t)}{\partial t^2} \underset{t}{*} f_1(t) \,. \tag{9.21}$$

From (9.21) we get

$$\frac{\partial^2 f(\ell - x, t)}{\partial t^2} = \frac{2cH(t)}{\ell} \sum_{k=1}^{\infty} (-1)^{k+1} \sin \frac{k\pi(\ell - x)}{\ell} \sin \frac{kc\pi t}{\ell} \,. \tag{9.22}$$

Substituting (9.22) in (9.21), we obtain

$$\overline{u}_x(t) = \frac{2caH(t)}{\ell} \sum_{k=1}^{\infty} (-1)^{k+1} \sin \frac{k\pi(\ell - x)}{\ell} \int_0^t \sin \frac{kc\pi(t - \tau)}{\ell} \sin(\omega\tau) d\tau \,, \tag{9.23}$$

namely

$$\overline{u}_x(t) \\ = 2caH(t)\sum_{k=1}^{\infty}\frac{(-1)^{k+1}}{\ell^2\omega^2 - k^2c^2\pi^2}\sin\frac{k\pi(\ell-x)}{\ell}\left(\ell\omega\sin\frac{k\pi ct}{\ell} - k\pi c\sin(\omega t)\right). \tag{9.24}$$

Expanding the function $\sin(\omega x/c)$ in Fourier series of sines on the interval $(0, \ell)$, we get

$$\sin\frac{\omega x}{c} = 2c^2\pi\sin\frac{\omega\ell}{c}\sum_{k=1}^{\infty}\frac{(-1)^k k}{\omega^2\ell^2 - k^2c^2\pi^2}\sin\frac{k\pi x}{\ell}.$$

Thus, (9.24) becomes

$$\overline{u}_x(t) = aH(t)\frac{\sin(\omega(\ell-x)/c) - \sin(\omega t)}{\sin(\omega\ell/c)} \\ + H(t)\sum_{k=1}^{\infty}\frac{2\ell\omega ca}{\ell^2\omega^2 - k^2c^2\pi^2}\sin\frac{k\pi x}{\ell}\sin\frac{k\pi ct}{\ell}, \quad x \in [0, \ell]. \tag{9.25}$$

We note that the expression (9.25) is identical to the solution given by I. Sneddon [48, see p. 253, problems 3–72].

Example 9.2 Let us determine the displacements in case of forced longitudinal vibrations of a bar of length ℓ with embedded ends, acted upon by a uniformly distributed load along its length, of the form $X(x, t) = X_0H(t)$, $X_0 = \text{const}$, knowing that at the initial moment the bar is at rest.

We determine the displacement $u(x, t) \in C^{2,2}([0, \ell] \times \mathbb{R}_+)$ which satisfies (9.13) in distributions, the initial conditions $u(x, 0) = 0$, $\dot{u}(x, 0) = 0$ and the boundary conditions $u(0, t) = 0$, $u(\ell, t) = 0$.

Using the solution in distributions (9.20), we obtain

$$u(x, t) = \frac{4X_0\ell^2}{\rho c^2\pi^3}\sum_{n=1,3,5,\ldots}\frac{1}{n^3}\left(1 - \cos\frac{cn\pi t}{\ell}\right)\sin\frac{n\pi x}{\ell}.$$

9.1.3 Forced Vibrations of a Bar with Boundary Conditions Expressed in Stresses

Let us determine the displacement $\overline{u}_x(t) \in \mathcal{D}'_+$ represented by a function type distribution with respect to the variable $t \in \mathbb{R}$, x being a parameter, which satisfies the longitudinal vibrations equation (9.13) in the distributions space and the boundary conditions expressed in stresses

$$\sigma_x(t)|_{x=0} = -f_1(t), \quad \sigma_x(t)|_{x=\ell} = -f_2(t), \quad f_1, f_2 \in \mathcal{D}'_+(\mathbb{R}), \tag{9.26}$$

where $\sigma_x(t) = E(\tilde{\partial}\overline{u}_x(t)/\partial x) \in \mathcal{D}'_+(\mathbb{R})$.

We assume that $\overline{u}_x(t)$, $f_1(t)$, $f_2(t)$ have Laplace images in distributions. Applying the Laplace transform to (9.13) and to the boundary conditions (9.26), we obtain

$$\frac{\mathrm{d}^2\tilde{\overline{u}}_x(p)}{\mathrm{d}x^2} - \frac{p^2}{c^2}\tilde{\overline{u}}_x(p) = -\frac{p\,u_0 + u_1 + \rho^{-1}p^{-1}X_0}{c^2}\,, \tag{9.27}$$

$$E\left.\frac{\mathrm{d}\tilde{\overline{u}}_x(p)}{\mathrm{d}x}\right|_{x=0} = -\tilde{f}_1(p)\,,\quad E\left.\frac{\mathrm{d}\tilde{\overline{u}}_x(p)}{\mathrm{d}x}\right|_{x=\ell} = -\tilde{f}_2(p)\,, \tag{9.28}$$

where $\tilde{\overline{u}}_x(p) = \mathrm{L}[\overline{u}_x(t)](p)$, $\tilde{f}_i(p) = \mathrm{L}[f_i(t)](p)$, $i = 1, 2$.

The solution of (9.27) with the conditions (9.28) is

$$\tilde{\overline{u}}_x(p) = \frac{c}{E}\left[p\,\tilde{f}_1(p)\frac{\cosh\left((\ell - x)cp^{-1}\right)}{p^2\sinh(\ell p c^{-1})} - p\,\tilde{f}_2(p)\frac{\cosh(x p c^{-1})}{p^2\sinh(\ell p c^{-1})}\right] + \frac{u_0}{p} + \frac{u_1}{p^2} + \frac{X_0}{\rho p^3}\,. \tag{9.29}$$

Applying the inverse Laplace transform and taking into account that

$$g(x,t) = \mathrm{L}^{-1}\left[\frac{\cosh(x p c^{-1})}{p^2\sinh(\ell p c^{-1})}\right] = \frac{cH(t)}{2\ell}\left[\frac{x^2}{c^2} + t^2 - \frac{\ell^2}{3c^2}\right] + \frac{2\ell H(t)}{c\pi^2}\sum_{k=1}^{\infty}\frac{(-1)^{k-1}}{k^2}\cos\frac{k\pi x}{\ell}\sin\frac{k\pi c t}{\ell}\,, \tag{9.30}$$

we get

$$\overline{u}_x(t) = \mathrm{L}^{-1}[\tilde{\overline{u}}_x(p)](t) = \frac{c}{E}\left[f_1'(t) \underset{t}{*} g(\ell - x, t) - f_2'(t) \underset{t}{*} g(x,t)\right] + u_0 H(t) + u_1 t H(t) + \frac{X_0 t^2 H(t)}{2\rho}\,,\quad x \in [0,\ell]\,. \tag{9.31}$$

This last relation can be written in the form

$$\overline{u}_x(t) = u_x^c(t) + u_x^r(t)\,,\quad x \in [0,\ell]\,, \tag{9.32}$$

where

$$u_x^c(t) = \frac{c}{E}\frac{\partial}{\partial t}\left[f_1(t) \underset{t}{*} g(\ell - x, t) - f_2(t) \underset{t}{*} g(x,t)\right]\,, \tag{9.33}$$

$$u_x^r(t) = H(t)\left[\frac{X_0 t^2}{2\rho} + u_1 t + u_0\right]\,. \tag{9.34}$$

We note that the term $u_x^r(t)$ is a function type distribution and that it represents the rigid displacement corresponding to a uniform varied motion of the bar; for $X_0 > 0$ we have a uniformly accelerated motion and for $X_0 < 0$ we have a uniformly slow motion. Because the density of the mass force X_0 has the dimension $[X_0] = \mathrm{MLT}^{-2}/\mathrm{L}^3$, we have $[X_0/\rho] = \mathrm{LT}^{-2}$, hence X_0/ρ has the dimension of an acceleration.

Taking into account the space law in uniformly varied motion, $s = at^2/2 + v_0 t + s_0$, and comparing with $u_x^r(t)$, it follows that $u_x^r(t)$ gives a rigid translational motion along the bar axis.

In conclusion, we can say that the solution (9.32) is the sum between the proper elastic displacement $u_x^e(t)$ of the bar and the rigid body displacement $u_x^r(t)$.

Example 9.3 In particular, if $u_0 = u_1 = X_0 = 0$ and the boundary conditions are $\sigma_x(t)|_{x=0} = -BH(t)$, $\sigma_x(t)|_{x=\ell} = 0$, $B \in \mathbb{R}_+$, then, taking into account (9.31) we obtain for the displacement the expression

$$\overline{u}_x(t) = \frac{BcH(t)}{E}\left[\frac{c}{2\ell}\left((\ell - x)^2 + t^2 - \frac{\ell^2}{3c^2}\right) - \frac{2\ell}{c\pi^2}\sum_{k=1}^{\infty}\frac{\ell}{k^2}\cos\frac{k\pi x}{\ell}\sin\frac{k\pi ct}{\ell}\right].$$

9.2 Transverse Vibrations of Elastic Bars

9.2.1 Differential Equation in Distributions of Transverse Vibrations of Elastic Bars

We consider a homogeneous and isotropic straight elastic bar with constant cross-section. We acknowledge the Bernoulli–Euler hypothesis of the plane sections and we use the linear theory of deformation. Under these assumptions, the straight bar is reduced to a medium fiber, called the elastic line of the bar, which practically coincides with the symmetry axis of it, representing the locus of the centers of gravity of the cross-sections. The bar axis is considered as the Ox-axis (Figure 9.2).

The way in which the bar is fixed involves the introduction of appropriate constraint forces, which are assumed to act normally to the bar axis and in the plane of symmetry Oxv of the bar.

We denote by $T(x,t)$, $M(x,t)$, $(x,t) \in [a,b] \times \mathbb{R}_+$, the shear force and the bending moment at the point $x \in [a,b]$ of the bar, respectively.

We note that the shear force $T(x,t)$ is the resultant of the given and the constraint forces on the segment $(x,b]$. The forces are considered positive if they act in the positive direction of Ov-axis.

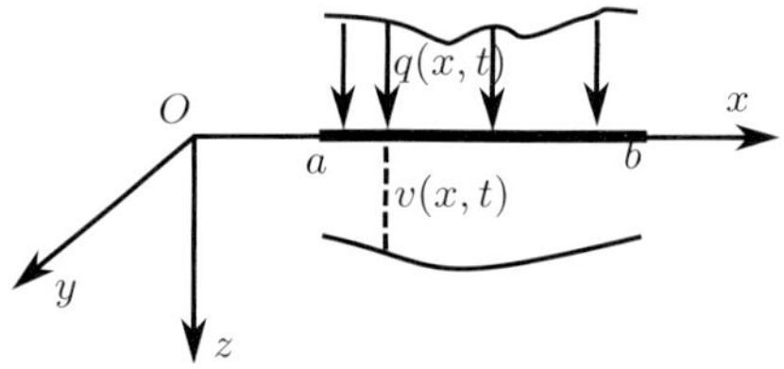

Figure 9.2

The bending moment $M(x,t)$ at a point $x \in [a,b]$ is the algebraic sum of moments of the given and constraint forces and of the concentrated moments on the portion $(x,b]$ with respect to the point $x \in [a,b]$.

The moments are considered positive if they have a direct rotation direction (counterclockwise).

For the deduction of the transverse vibrations equation of the elastic bars Hamilton's principle is applied [34].

This provides the equation

$$EI\frac{\tilde{\partial}^4}{\partial x^4}v(x,t) + \rho\frac{\tilde{\partial}^2}{\partial t^2}v(x,t) = q(x,t)\,, \tag{9.35}$$

where EI is the bending stiffness of the bar, $I = I_y$ is the inertia moment of the cross-section with respect to the neutral axis of it, ρ is the mass per unit length of the bar and $q(x,t)$ is the intensity of the distributed forces perpendicular to the bar.

Following the action of the load $q(x,t) \in C^0([a,b] \times \mathbb{R}_+)$, every point of the bar axis will deflect perpendicular by Ox with $v(x,t) \in C^{4,2}([a,b] \times \mathbb{R}_+)$.

We mention that (9.35) is obtained by eliminating the shear force $T(x,t)$ and the bending moment $M(x,t)$ from the equations

$$\begin{aligned}
&\frac{\tilde{\partial}}{\partial x}T(x,t) + q(x,t) - \rho\frac{\tilde{\partial}^2}{\partial t^2}v(x,t) = 0\,,\\
&T(x,t) = \frac{\tilde{\partial}}{\partial x}M(x,t)\,, \quad M(x,t) = -EI\frac{\tilde{\partial}^2}{\partial x^2}v(x,t)\,.
\end{aligned} \tag{9.36}$$

Equations 9.36 represent the complete system of equations of transverse vibrations of elastic bars with constant cross-section. These equations establish relations between the fundamental quantities of the vibrations of elastic bars, namely:

- the deflection $v(x,t)$;
- the shear force $T(x,t)$;
- the bending moment $M(x,t)$;
- the intensity $q(x,t)$ of the forces perpendicular distributed on the bar axis.

The transverse vibration problem of the elastic bars is to determine the deflection v which satisfies (9.35), the initial conditions

$$v(x,0) = v_0(x) \in C^0([a,b])\,, \quad \left.\frac{\tilde{\partial}}{\partial t}v(x,t)\right|_{t=0} = v_1(x) \in C^0([a,b]) \tag{9.37}$$

and the boundary conditions.

The boundary conditions are derived from the method of attachment of the bar at its ends. Thus, for example, at the end $x = d$ ($d = a$ or $d = b$) we can have:

1. Built-in support
   - the deflection is zero $v(d,t) = 0$,
   - the rotation is null

$$\left.\frac{\tilde{\partial} v(x,t)}{\partial x}\right|_{x=d} = 0 \; ;$$

2. Simple support
   - the deflection is zero $v(d,t) = 0$,
   - the bending moment is zero

$$\left.\frac{\tilde{\partial}^2 v(x,t)}{\partial x^2}\right|_{x=d} = 0 \; ;$$

3. Free end
   - the shear force is zero

$$\left.\frac{\tilde{\partial}^3 v(x,t)}{\partial x^3}\right|_{x=d} = 0 \; ,$$

   - the bending moment is zero

$$\left.\frac{\tilde{\partial}^2 v(x,t)}{\partial x^2}\right|_{x=d} = 0 \; ;$$

4. Elastic fixing
   - the deflection is zero $v(d,t) = 0$,
   - the bending moment is proportional with the rotation of the cross-section,

$$EI \left.\frac{\tilde{\partial}^2 v(x,t)}{\partial x^2}\right|_{x=d} = k \left.\frac{\tilde{\partial} v(x,t)}{\partial x}\right|_{x=d} \; ;$$

5. Simple elastic support
   - the deflection is proportional to the reaction from the support,

$$EI \left.\frac{\tilde{\partial}^3 v(x,t)}{\partial x^3}\right|_{x=d} = k v(d,t),$$

   - the bending moment is zero,

$$\left.\frac{\tilde{\partial}^2 v(x,t)}{\partial x^2}\right|_{x=d} = 0 \; ,$$

where I is the moment of inertia of the cross-section with respect to the neutral Oy-axis.

To rewrite (9.35) of the transverse vibrations of elastic bars in the distribution space $\mathcal{D}'_+$, we introduce the function type distributions with respect to $t \in \mathbb{R}$ from $\mathcal{D}'_+$, x being the parameter,

$$\bar{v}_x(t) = H(t)v(x,t) = \begin{cases} v(x,t)\,, & (x,t) \in [a,b] \times \mathbb{R}_+\,, \\ 0\,, & \text{otherwise}\,, \end{cases}$$

$$\overline{q}_x(t) = H(t)q(x,t) = \begin{cases} q(x,t)\,, & (x,t) \in [a,b] \times \mathbb{R}_+\,, \\ 0\,, & \text{otherwise}\,. \end{cases} \tag{9.38}$$

By differentiating, in the sense of distributions, we have

$$\frac{\partial^4}{\partial x^4}\bar{v}_x(t) = \frac{\tilde{\partial}^4}{\partial x^4}\bar{v}_x(t)\,,$$

$$\frac{\partial}{\partial t}\bar{v}_x(t) = \frac{\tilde{\partial}}{\partial t}\bar{v}_x(t) + \bar{v}_x(0)\delta(t) = \frac{\tilde{\partial}}{\partial t}\bar{v}_x(t) + v_0(x)\delta(t)\,,$$

$$\frac{\partial^2}{\partial t^2}\bar{v}_x(t) = \frac{\tilde{\partial}^2}{\partial t^2}\bar{v}_x(t) + \left.\frac{\tilde{\partial}}{\partial t}\bar{v}_x(t)\right|_{t=0}\delta(t) + v_0(x)\delta'(t)$$

$$= \frac{\tilde{\partial}^2}{\partial t^2}\bar{v}_x(t) + v_1(x)\delta(t) + v_0(x)\delta'(t)\,, \tag{9.39}$$

where $v_0, v_1 \in C^0([a,b])$ are the initial values given by (9.37).

As a result, the transverse vibrations equation (9.35) in the distributions space $\mathcal{D}'_+$ is

$$\frac{\partial^2}{\partial t^2}\bar{v}_x(t) + c^2\frac{\partial^4}{\partial x^4}\bar{v}_x(t) = \frac{\overline{q}_x(t)}{\rho} + v_1(x)\delta(t) + v_0(x)\delta'(t)\,, \tag{9.40}$$

where $c^2 = EI/\rho$, $x \in [a,b]$ being a parameter.

9.2.2 Free Vibrations of an Infinite Bar

In this case we use (9.40), considering $x \in \mathbb{R}$ and $\overline{q}_x(t) = 0$.

We take the initial conditions (9.37) of the form

$$v(x,0) = v_0(x),\ \left.\frac{\tilde{\partial}}{\partial t}v(x,t)\right|_{t=0} = v_1(x) = c\frac{\tilde{d}^2}{dx^2}h(x)\,. \tag{9.41}$$

To obtain the solution of (9.40), we apply the Laplace transformation with respect to the time variable $t \in \mathbb{R}$ and the Fourier transformation with respect to $x \in \mathbb{R}$.

We obtain

$$p^2\hat{\bar{v}}(\alpha,p) + c^2(-\mathrm{i}\alpha)^4\hat{\bar{v}}(\alpha,p) = \hat{v}_1(\alpha) + p\hat{v}_0(\alpha) = -c\alpha^2\hat{h}(\alpha) + p\hat{v}_0(\alpha)\,, \tag{9.42}$$

where

$$\hat{\bar{v}}(\alpha,p) = \mathrm{L}[\mathrm{F}[\bar{v}_x(t)]](\alpha,p)\,, \quad \hat{v}_1(\alpha) = \mathrm{F}[v_1(x)](\alpha) = -c\alpha^2\hat{h}(\alpha)\,,$$
$$\hat{h}(\alpha) = \mathrm{F}[h(x)](\alpha)\,, \quad \hat{v}_0(\alpha) = \mathrm{F}[v_0(x)](\alpha)\,.$$

From (9.42) it follows

$$\hat{\bar{v}}(\alpha,p) = \hat{v}_0(\alpha)\frac{p}{p^2+c^2\alpha^4} - \hat{h}(\alpha)\frac{c\alpha^2}{p^2+c^2\alpha^4}\,.$$

Applying the inverse Laplace transform and taking into account the formulas

$$\mathrm{L}[H(t)\cos\omega t](p) = \frac{p}{p^2+\omega^2}\,, \quad \mathrm{L}[H(t)\sin\omega t](p) = \frac{\omega}{p^2+\omega^2}\,,$$

we can write

$$\begin{aligned}\mathrm{F}[\bar{v}_x(t)](\alpha) &= \mathrm{F}[v_0(x)](\alpha)H(t)\cos(\alpha^2 ct)\\ &\quad - \mathrm{F}[h(x)](\alpha)H(t)\sin(\alpha^2 ct)\,.\end{aligned} \tag{9.43}$$

For the application of the inverse Fourier transform we use the formulas

$$\mathrm{F}\left[\cos\frac{x^2}{4ct} + \sin\frac{x^2}{4ct}\right](\alpha) = 2\sqrt{2\pi ct}\cos(\alpha^2 ct)\,,$$
$$\mathrm{F}\left[\cos\frac{x^2}{4ct} - \sin\frac{x^2}{4ct}\right](\alpha) = 2\sqrt{2\pi ct}\sin(\alpha^2 ct)\,,$$
$$\mathrm{F}[f(x) * g(x)](\alpha) = \mathrm{F}[f](\alpha)\mathrm{F}[g](\alpha)\,.$$

From (9.43) we obtain the solution of the problem in the form

$$\bar{v}(x,t) = \frac{H(t)}{2\sqrt{2\pi ct}}\left[v_0(x) * \left(\cos\frac{x^2}{4ct} + \sin\frac{x^2}{4ct}\right) - h(x) * \left(\cos\frac{x^2}{4ct} - \sin\frac{x^2}{4ct}\right)\right].$$

For $t > 0$ one has

$$v(x,t) = \frac{1}{2\sqrt{2\pi ct}}\left[v_0(x) * \left(\cos\frac{x^2}{4ct} + \sin\frac{x^2}{4ct}\right) - h(x) * \left(\cos\frac{x^2}{4ct} - \sin\frac{x^2}{4ct}\right)\right].$$

9.2.3 Forced Transverse Vibrations of the Bars

Let OA be an elastic bar of length ℓ, with simply supported ends (Figure 9.3).

We acknowledge that the bar performs forced vibration under the action of the load

$$\overline{q}_x(t) = H(t)f(t)g(x) = \begin{cases} 0\,, & t<0\,, \quad x\in[0,\ell]\,,\\ f(t)g(x)\,, & (x,t)\in[0,\ell]\times\mathbb{R}_+\,.\end{cases} \tag{9.44}$$

The initial conditions are given by (9.37)

$$v(x,0) = v_0(x) \in C^0([0,\ell]) , \quad \left. \frac{\tilde{\partial}}{\partial t} v(x,t) \right|_{t=0} = v_1(x) \in C^0([0,\ell]) . \quad (9.45)$$

Because the bar is simply supported at the ends, the boundary conditions are

$$v(x,t)|_{x=0} = 0 , \quad \left. \frac{\tilde{\partial}^2}{\partial x^2} v(x,t) \right|_{x=0} = 0 ,$$
$$v(x,t)|_{x=\ell} = 0 , \quad \left. \frac{\tilde{\partial}^2}{\partial x^2} v(x,t) \right|_{x=\ell} = 0 . \quad (9.46)$$

The equation of transverse vibrations in $\mathcal{D}'_+$, is given by (9.40) where $t \in \mathbb{R}$ and $x \in [0,\ell]$ is a parameter.

In this way the problem of transverse vibrations of the bar is reduced to determining the deflection $\bar{v}_x(t) = H(t)v(x,t)$, $(x,t) \in [0,\ell] \times \mathbb{R}$, which satisfies (9.40) and the boundary conditions (9.46); the initial conditions (9.45) are incorporated in (9.40).

We have

$$\frac{\partial^2}{\partial t^2} \bar{v}_x(t) + c^2 \frac{\partial^4}{\partial x^4} \bar{v}_x(t) = \frac{\overline{f}(t)}{\rho} g(x) + v_1(x)\delta(t) + v_0(x)\delta'(t) , \quad (9.47)$$

where $c^2 = EI/\rho$, $\overline{f}(t) = H(t) f(t)$, $(x,t) \in [0,\ell] \times \mathbb{R}$.

The requested solution $\bar{v}_x(t)$ is of the form

$$\bar{v}_x(t) = H(t) \sum_{n=1}^{\infty} A_n(t) \sin \frac{n\pi x}{\ell} . \quad (9.48)$$

We note that for any $t \in \mathbb{R}$, the deflection $\bar{v}_x$, satisfies the boundary conditions (9.46) and therefore we have to determine the functions $A_n(t)$, $t \geq 0$, $n \in \mathbb{N}$, so that $\bar{v}_x$ should satisfy (9.47).

Differentiating in the sense of distributions we obtain

$$\frac{\partial}{\partial t} \bar{v}_x(t) = \delta(t) \sum_{n=1}^{\infty} A_n(0) \sin \frac{n\pi x}{\ell} + H(t) \sum_{n=1}^{\infty} A''_n(t) \sin \frac{n\pi x}{\ell} , \quad (9.49)$$

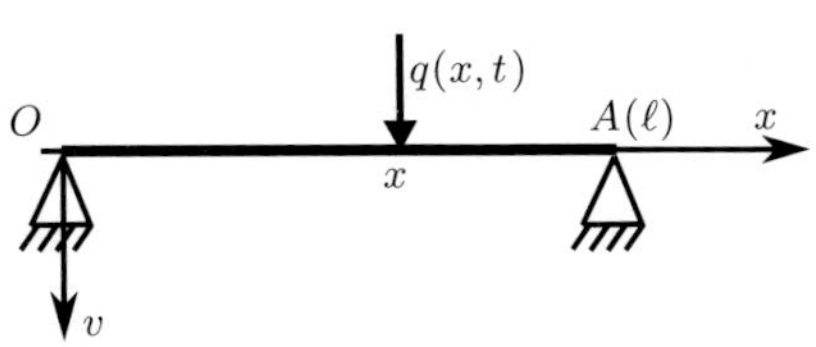

Figure 9.3

$$\frac{\partial^2}{\partial t^2}\bar{v}_x(t) = \delta'(t)\sum_{n=1}^{\infty} A_n(0)\sin\frac{n\pi x}{\ell}$$
$$+\,\delta(t)\sum_{n=1}^{\infty} A'_n(0)\sin\frac{n\pi x}{\ell} + H(t)\sum_{n=1}^{\infty} A''_n(t)\sin\frac{n\pi x}{\ell}\,, \tag{9.50}$$
$$\frac{\partial^4}{\partial x^4}\bar{v}_x(t) = H(t)\sum_{n=1}^{\infty}\omega_n^2 A_n(t)\sin\frac{n\pi x}{\ell}\,,\qquad \omega_n = \frac{\pi^2 n^2}{\ell^2}\,. \tag{9.51}$$

Substituting the expressions (9.48), (9.50), and (9.51) in (9.47), we obtain, by identification, the equations

$$H(t)\sum_{n=1}^{\infty}\left(A''_n(t) + \omega_n^2 A_n(t)\right)\sin\frac{n\pi x}{\ell} = H(t)\frac{f(t)}{\rho}g(x)\,, \tag{9.52}$$
$$\sum_{n=1}^{\infty} A_n(0)\sin\frac{n\pi x}{\ell} = v_0(x)\,, \tag{9.53}$$
$$\sum_{n=1}^{\infty} A'_n(0)\sin\frac{n\pi x}{\ell} = v_1(x)\,. \tag{9.54}$$

Developing the functions g, v_0, v_1 after Fourier sine series on $[0, \ell]$, we get

$$g(x) = \sum_{n=1}^{\infty} B_n\sin\frac{n\pi x}{\ell}\,,\qquad v_0(x) = \sum_{n=1}^{\infty} a_n\sin\frac{n\pi x}{\ell}\,,$$
$$v_1(x) = \sum_{n=1}^{\infty} b_n\sin\frac{n\pi x}{\ell}\,, \tag{9.55}$$

where

$$B_n = \frac{2}{\ell}\int_0^{\ell} g(x)\sin\frac{n\pi x}{\ell}\mathrm{d}x\,,$$
$$a_n = \frac{2}{\ell}\int_0^{\ell} v_0(x)\sin\frac{n\pi x}{\ell}\mathrm{d}x\,,\quad b_n = \frac{2}{\ell}\int_0^{\ell} v_1(x)\sin\frac{n\pi x}{\ell}\mathrm{d}x\,. \tag{9.56}$$

Taking into account of (9.52), (9.53), and (9.54), we obtain

$$A''_n(t) + \omega_n^2 A_n(t) = \frac{f(t)}{\rho}B_n\,,\quad t \ge 0\,, \tag{9.57}$$

$$A_n(0) = a_n\,,\quad A'_n(0) = b_n\,. \tag{9.58}$$

Equation 9.57 with the initial conditions (9.58) is a Cauchy problem whose solution is unique and we have

$$A_n(t) = a_n\cos(\omega_n t) + \frac{b_n}{\omega_n}\sin(\omega_n t) + \frac{B_n}{\rho\omega_n}\int_0^t f(\tau)\sin\omega_n(t-\tau)\mathrm{d}\tau\,,\quad t \ge 0\,. \tag{9.59}$$

Substituting this function into (9.48), we obtain the deflection of the transverse forced vibrations of the elastic bar as a series converging in $\mathcal{D}'_+$

$$v(x,t) = \sum_{n=1}^{\infty} \left[a_n \cos(\omega_n t) + \frac{b_n}{\omega_n} \sin(\omega_n t) \right] \sin \frac{n\pi x}{\ell} + \frac{B_n}{\rho\omega_n} \int_0^t f(\tau) \sin \omega_n (t-\tau) d\tau \,. \tag{9.60}$$

Example 9.4 We consider the force $\overline{q}_x(t)$ of the form $\overline{q}_x(t) = H(t)P\delta(x-c)$ which is a force concentrated at the point $c \in (0,\ell)$, of constant intensity P that acts for $t \geq 0$.

Taking into account the restriction of the development in Fourier series after sines of the distribution $\delta(x-c)$ on the interval $x \in (0,\ell)$, $c \in (0,\ell)$, we have

$$\delta(x-c) = \frac{2}{\ell} \sum_{n=1}^{\infty} \sin \frac{n\pi c}{\ell} \sin \frac{n\pi x}{\ell} \,. \tag{9.61}$$

Comparing this development in Fourier series with (9.55), we obtain

$$B_n = \frac{2}{\ell} \sin \frac{n\pi c}{\ell} \,.$$

Consequently, the displacement expression (9.60) takes in this case the form

$$v(x,t) = \sum_{n=1}^{\infty} \left(a_n \cos(\omega_n t) + \frac{b_n}{\omega_n} \sin(\omega_n t) \right) \sin \frac{n\pi x}{\ell} + \frac{2P}{\ell\rho\omega_n^2} \sin \frac{n\pi c}{\ell} (1 - \cos \omega_n t) \,, \quad (x,t) \in [0,\ell] \times \mathbb{R}_+ \,. \tag{9.62}$$

Remark 9.1 Instead of the force $\overline{q}_x(t)$, we can consider the load $\overline{q}_\varepsilon(x,t)$ depending on the parameter $\varepsilon > 0$, with the expression $\overline{q}_\varepsilon(x,t) = H(t)Pg_\varepsilon(x)$, where

$$g_\varepsilon(x) = \begin{cases} \varepsilon^{-1} \,, & x \in [c, c+\varepsilon] \,, \\ 0 \,, & \text{otherwise} \,. \end{cases}$$

Then, in the sense of distributions we have $\lim_{\varepsilon \to +0} q_\varepsilon(x) = \delta(x-c)$. Therefore, if instead of $g(x)$ we consider $q_\varepsilon(x)$, then the deflection $v(x,t) = v_\varepsilon(x,t)$ will depend on the parameter $\varepsilon > 0$. Finally, for $\varepsilon \to +0$ we obtain the result given by (9.62).

9.2.4 Bending of Elastic Bars on Elastic Foundation

The bending of elastic bars can be considered as a special case of transverse vibrations of elastic bars when the loads which act on the bar do not depend on a temporal variable. Consequently, the quantities, the deflection v, the intensity q of the distributed loads, the shear force T and the bending moment M will depend only on the spatial variable $x \in [a, b]$.

If the bar lies on an elastic foundation, then, adopting Winkler's hypothesis, the reaction of the elastic medium $q_e(x)$, $x \in [a, b]$, is directly proportional to the common deflection of the bar and of the elastic foundation, that is,

$$q_e(x) = -kv(x) , \quad x \in [a, b] . \tag{9.63}$$

The coefficient $k = E_0 k_0 > 0$ is called the rigidity coefficient of the elastic foundation, where E_0 is the modulus of elasticity of the foundation.

We note that the load q_e has the dimension of a force per length and therefore it must be regarded as an additional load with respect to the given ones, acting normally to the elastic line of the bar.

Taking into account (9.63) the complete system of equations of elastic bars on an elastic foundation is

$$\frac{\tilde{\mathrm{d}}}{\mathrm{d}x} T(x) + q(x) - kv(x) = 0 , \quad T(x) = \frac{\tilde{\mathrm{d}}}{\mathrm{d}x} M(x) ,$$
$$M(x) = -EI \frac{\tilde{\mathrm{d}}^2}{\mathrm{d}x^2} v(x) . \tag{9.64}$$

We acknowledge that the bending of the bar occurs due to the action of distributed loads q, as well as to the concentrated forces P_i and to the concentrated moments m_i applied at the points $c_1 = a, c_2, c_3, \ldots, c_{n-1}, c_n = b$.

We note that the quantities P_i and m_i acting at the point $x = c_i$ of the bar (Figure 9.4), can lead to constraint forces and moments with respect to the fixing of the bar.

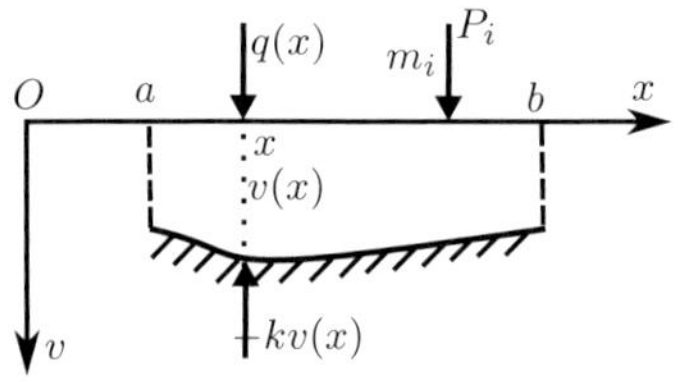

Figure 9.4

To rewrite the system of (9.64) in the distributions space $\mathcal{D}'_+$ we define the following function type distributions:

$$\bar{v}(x) = v(x)\chi(x) = \begin{cases} 0\,, & x \notin [a,b]\,, \\ v(x)\,, & x \in [a,b]\,, \end{cases}$$

$$\overline{q}(x) = q(x)\chi(x) = \begin{cases} 0\,, & x \notin [a,b]\,, \\ q(x)\,, & x \in [a,b]\,, \end{cases}$$

$$\overline{T}(x) = T(x)\chi(x) = \begin{cases} 0\,, & x \notin [a,b]\,, \\ T(x)\,, & x \in [a,b]\,, \end{cases}$$

$$\overline{M}(x) = M(x)\chi(x) = \begin{cases} 0\,, & x \notin [a,b]\,, \\ M(x)\,, & x \in [a,b]\,, \end{cases} \tag{9.65}$$

where

$$\chi(x) = \begin{cases} 1\,, & x \notin [a,b] \\ 0\,, & x \in [a,b] \end{cases}$$

is the characteristic function corresponding to the interval $[a, b]$.

Therefore the system of (9.64), [28], becomes

$$\frac{\tilde{\mathrm{d}}}{\mathrm{d}x}\overline{T}(x)+\overline{q}(x)-k\bar{v}(x) = 0\,, \quad \overline{T}(x) = \frac{\tilde{\mathrm{d}}}{\mathrm{d}x}\overline{M}(x)\,, \quad \overline{M}(x) = -EI\frac{\tilde{\mathrm{d}}^2}{\mathrm{d}x^2}\bar{v}(x)\,. \tag{9.66}$$

We observe that the point of action of a concentrated force is a point of discontinuity of the first order for the shear force $\overline{T}$ and for the derivative of the bending moment $\overline{M}$.

Also, the point of action of a concentrated moment is a point of discontinuity of first order for the bending moment $\overline{M}$ and an ordinary point for the shear force.

Consequently, the jumps of the bending moment and of the shear force at a point c_i, $i = \overline{i, n}$, have the expressions

$$S_{c_i}[\overline{T}(x)] = \overline{T}(c_i + 0) - \overline{T}(c_i - 0) = -P_i\,,$$
$$S_{c_i}[\overline{M}(x)] = \overline{M}(c_i + 0) - \overline{M}(c_i - 0) = -m_i\,.$$

Differentiating in the sense of distributions we obtain

$$\frac{\mathrm{d}}{\mathrm{d}x}\overline{T}(x) = \frac{\tilde{\mathrm{d}}}{\mathrm{d}x}\overline{T}(x) - \sum_{i=1}^{n} P_i\delta(x - c_i)\,,$$

$$\frac{\mathrm{d}}{\mathrm{d}x}\overline{M}(x) = \frac{\tilde{\mathrm{d}}}{\mathrm{d}x}\overline{M}(x) - \sum_{i=1}^{n} m_i\delta(x - c_i)\,,$$

$$\frac{\mathrm{d}}{\mathrm{d}x}\bar{v}(x) = \frac{\tilde{\mathrm{d}}}{\mathrm{d}x}\bar{v}(x) + S_a[\bar{v}]\delta(x - a) + S_b[\bar{v}]\delta(x - b)\,,$$

$$\frac{\mathrm{d}^2}{\mathrm{d}x^2}\bar{v}(x) = \frac{\tilde{\mathrm{d}}^2}{\mathrm{d}x^2}\bar{v}(x) + S_a[\bar{v}]\delta'(x - a) + S_b[\bar{v}]\delta'(x - b) + S_a\left[\frac{\tilde{\mathrm{d}}}{\mathrm{d}x}\bar{v}(x)\right]\delta(x - a) + S_b\left[\frac{\tilde{\mathrm{d}}}{\mathrm{d}x}\bar{v}(x)\right]\delta(x - b)\,. \quad (9.67)$$

Replacing in (9.66) the derivatives in the ordinary sense by the derivatives in the sense of distributions given by the relations (9.67), we obtain

$$\frac{\mathrm{d}}{\mathrm{d}x}\overline{T}(x) + \overline{q}(x) - k\bar{v}(x) = -\sum_{i=1}^{n} P_i\delta(x - c_i)\,, \quad (9.68)$$

$$\frac{\mathrm{d}}{\mathrm{d}x}\overline{M}(x) - \overline{T}(x) = -\sum_{i=1}^{n} m_i\delta(x - c_i)\,, \quad (9.69)$$

$$EI\frac{\mathrm{d}^2}{\mathrm{d}x^2}\bar{v}(x) + \overline{M}(x) = EI\left(S_a[\bar{v}]\delta'(x - a) + S_b[\bar{v}]\delta'(x - b) + S_a\left[\frac{\tilde{\mathrm{d}}}{\mathrm{d}x}\bar{v}(x)\right]\delta(x - a) + S_b\left[\frac{\tilde{\mathrm{d}}}{\mathrm{d}x}\bar{v}(x)\right]\delta(x - b)\right)\,. \quad (9.70)$$

Eliminating between (9.68)–(9.70) the shear force $\overline{T}$ and the bending moment $\overline{M}$, we obtain

$$EI\frac{\mathrm{d}^4}{\mathrm{d}x^4}\bar{v}(x) + k\bar{v}(x) = \overline{q}_1(x) + EI\left(S_a[\bar{v}]\delta'''(x - a) + S_b[\bar{v}]\delta'''(x - b) + S_a\left[\frac{\tilde{\mathrm{d}}}{\mathrm{d}x}\bar{v}(x)\right]\delta''(x - a) + S_b\left[\frac{\tilde{\mathrm{d}}}{\mathrm{d}x}\bar{v}(x)\right]\delta''(x - b)\right)\,, \quad x \in \mathbb{R} \quad (9.71)$$

where $\overline{q}_1(x)$ has the expression

$$\overline{q}_1(x) = \overline{q}(x) + \sum_{i=1}^{n} P_i\delta(x - c_i) + \sum_{i=1}^{n} m_i\delta'(x - c_i)\,. \quad (9.72)$$

Equation 9.71 is the differential equation, in deflections, in the distributions space $\mathcal{D}'_+$ of the bending of the elastic bar on a Winkler-type elastic foundation.

We introduce the matrices $(A) \in (\mathcal{D}'_+)^{3\times 3}$, $(X) \in (\mathcal{D}'_+)^{3\times 1}$, $(B) \in (\mathcal{D}'_+)^{3\times 1}$ with the expressions

$$(A) = \begin{pmatrix} EI\delta''(x) & 0 & \delta(x) \\ -k\delta(x) & \delta'(x) & 0 \\ 0 & -\delta(x) & \delta'(x) \end{pmatrix},$$

$$(X) = \begin{pmatrix} \bar{v}(x) \\ \overline{T}(x) \\ \overline{M}(x) \end{pmatrix}, \quad (B) = \begin{pmatrix} B_1(x) \\ B_2(x) \\ B_3(x) \end{pmatrix}. \tag{9.73}$$

With the help of the convolution product $*$, (9.68)–(9.70) can be written in a matrix form as

$$(A) * (X) = (B), \tag{9.74}$$

where $B_1, B_2, B_3 \in \mathcal{D}'_+$ have the expressions

$$B_1(x) = EI\left(S_a[\bar{v}]\delta'(x-a) + S_b[\bar{v}]\delta'(x-b) \right.$$
$$\left. + S_a\left[\frac{\tilde{\mathrm{d}}}{\mathrm{d}x}\bar{v}(x)\right]\delta(x-a) + S_b\left[\frac{\tilde{\mathrm{d}}}{\mathrm{d}x}\bar{v}(x)\right]\delta(x-b)\right),$$

$$B_2(x) = -\bar{q}(x) - \sum_{i=1}^{n} P_i\delta(x-c_i), \quad B_3(x) = -\sum_{i=1}^{n} m_i\delta(x-c_i). \tag{9.75}$$

As regards the determinant $\Delta = \det(A)$ associated to the matrix $(A) \in (\mathcal{D}'_+)^{3\times 3}$, it is determined considering the products involved in its calculation as convolution products.

We have

$$\Delta = \det(A) = EI\delta^{(4)}(x) + k\delta(x) \in \mathcal{D}'_+. \tag{9.76}$$

Denoting

$$\omega = \sqrt[4]{\frac{k}{4EI}}, \tag{9.77}$$

we obtain

$$\Delta = EI(\delta^{(4)}(x) + 4\omega^4\delta(x)) \in \mathcal{D}'_+. \tag{9.78}$$

The inverse $\Delta^{-1} \in \mathcal{D}'_+$ has the expression

$$\Delta^{-1} = \frac{H(x)}{4EI\omega^3}[\cosh\omega x \sin\omega x - \sinh\omega x \cos\omega x] \in \mathcal{D}'_+. \tag{9.79}$$

Considering Example 3.5 results in the solution of (9.74), which is unique and has the expression

$$(X) = (E) * (B) = (A)^{-1} * (B) . \tag{9.80}$$

Specifying, we obtain

$$\begin{aligned} \bar{v} &= E_{11} * B_1 + E_{12} * B_2 + E_{13} * B_3 , \\ \overline{T} &= E_{21} * B_1 + E_{22} * B_2 + E_{23} * B_3 , \\ \overline{M} &= E_{31} * B_1 + E_{32} * B_2 + E_{33} * B_3 , \end{aligned} \tag{9.81}$$

where $E_{ij} = \Delta^{-1} * \alpha_{ji}, i, j = 1, 2, 3$ are the fundamental components of the matrix (E) and α_{ij} is the algebraic complement of the element A_{ij} from $\det(A)$.

We introduce the real functions $u, u_1, u_2, u_3 \in C^\infty(\mathbb{R})$ having the expressions

$$\begin{aligned} u(x) &= \cosh \omega x \sin \omega x - \sinh \omega x \cos \omega x , \\ u_1(x) &= u'(x) = 2\omega \sinh \omega x \sin \omega x , \\ u_2(x) &= u''(x) = 2\omega^2 (\cosh \omega x \sin \omega x + \sinh \omega x \cos \omega x) , \\ u_3(x) &= u'''(x) = 4\omega^3 \cosh \omega x \cos \omega x , \end{aligned} \tag{9.82}$$

where

$$\begin{aligned} u^{(4k)}(x) &= (-4\omega^4)^k u(x) , \quad u^{(4k+1)}(x) = (-4\omega^4)^k u_1(x) , \\ u^{(4k+2)}(x) &= (-4\omega^4)^k u_2(x) , \quad u^{(4k+3)}(x) = (-4\omega^4)^k u_3(x) , \quad k \in \mathbb{N} . \end{aligned} \tag{9.83}$$

Because every natural number $n \geq 4$ can be written as $n = 4k + p$, $p = 0, 1, 2, 3$, $k \in \mathbb{N}$, we can state the following.

Proposition 9.1 Any derivative of order $n \geq 4$ of the function $u \in C^\infty(\mathbb{R})$ is a multiple of one of the functions $u, u_1 = u', u_2 = u'', u_3 = u'''$, that is,

$$u^{(n)}(x) = \begin{cases} (-4\omega^4)^k u(x) , & n = 4k , \\ (-4\omega^4)^k u_1(x) , & n = 4k + 1 , \\ (-4\omega^4)^k u_2(x) , & n = 4k + 2 , \\ (-4\omega^4)^k u_3(x) , & n = 4k + 3 , \end{cases} \quad k \in \mathbb{N} . \tag{9.84}$$

On the basis of the relations $H'(x) = \delta(x)$, $g(x)\delta(x) = g(0)\delta(x)$ and using the functions (9.82), the element $\Delta^{-1} \in \mathcal{D}'_+$ given by (9.79) verifies the relations

$$\Delta^{-1} = \frac{H(x)}{4EI\omega^3}u(x)\,,$$

$$(\Delta^{-1})' = \frac{H(x)}{2EI\omega^2}(\sinh\omega x \sin\omega x) = \frac{H(x)}{4EI\omega^3}u_1(x)\,,$$

$$(\Delta^{-1})'' = \frac{H(x)}{2EI}(\cosh\omega x \sin\omega x + \sinh\omega x \cos\omega x) = \frac{H(x)}{4EI\omega^3}u_2(x)\,,$$

$$(\Delta^{-1})''' = \frac{H(x)}{EI}(\cosh\omega x \cos\omega x) = \frac{H(x)}{4EI\omega^3}u_3(x)\,,$$

$$(\Delta^{-1})^{(\mathrm{IV})} = \frac{\delta(x)}{EI} - \frac{\omega}{EI}H(x)u(x) = \frac{\delta(x)}{EI} - 4\omega^4\Delta^{-1}\,. \tag{9.85}$$

This last formula allows the direct verification of the relation $\Delta * \Delta^{-1} = \delta(x)$. Indeed, we have

$$\begin{aligned}\Delta * \Delta^{-1} &= (EI\delta^{(4)}(x) + k\delta(x)) * \Delta^{-1} = EI(\Delta^{-1})^{(4)} + k\Delta^{-1}\\ &= EI\left(\frac{\delta(x)}{EI} - 4\omega^4\Delta^{-1}\right) + k\Delta^{-1} = \delta(x) - 4EI\omega^4\Delta^{-1} + k\Delta^{-1}\\ &= \delta(x) - 4EI\frac{k\Delta^{-1}}{4EI} + k\Delta^{-1} = \delta(x)\,.\end{aligned}$$

Proposition 9.2 The fundamental solution $(E) = (A)^{-1} \in (\mathcal{D}'_+)^{3\times 3}$ of the elastic bar on an elastic foundation, that is, of the matrix equation (9.74) or of the system of (9.68)–(9.70), has the expression

$$(E) = (A)^{-1} = \begin{pmatrix} \dfrac{H(x)u_2(x)}{4EI\omega^3} & -\dfrac{H(x)u(x)}{4EI\omega^3} & -\dfrac{H(x)u_1(x)}{4EI\omega^3} \\ \dfrac{kH(x)u_1(x)}{4EI\omega^3} & \dfrac{H(x)u_3(x)}{4\omega^3} & -\dfrac{kH(x)u(x)}{4EI\omega^3} \\ \dfrac{kH(x)u(x)}{4EI\omega^3} & \dfrac{H(x)u_2(x)}{4\omega^3} & \dfrac{H(x)u_3(x)}{4\omega^3} \end{pmatrix}. \tag{9.86}$$

Indeed, taking into account (9.73) and (9.85), we obtain

$$E_{11} = \Delta^{-1} * \alpha_{11} = \Delta^{-1} * \delta''(x) = (\Delta^{-1})'' = \frac{H(x)u_2(x)}{4EI\omega^3},$$

$$E_{12} = \Delta^{-1} * \alpha_{21} = \Delta^{-1} * (-\delta(x)) = -\Delta^{-1} = \frac{-H(x)u(x)}{4EI\omega^3},$$

$$E_{13} = \Delta^{-1} * \alpha_{31} = \Delta^{-1} * (-\delta'(x)) = -(\Delta^{-1})' = \frac{-H(x)u_1(x)}{4EI\omega^3},$$

$$E_{21} = \Delta^{-1} * \alpha_{12} = \Delta^{-1} * (k\delta'(x)) = k(\Delta^{-1})' = \frac{kH(x)u_1(x)}{4EI\omega^3},$$

$$E_{22} = \Delta^{-1} * \alpha_{22} = \Delta^{-1} * (EI\delta'''(x)) = EI(\Delta^{-1})''' = \frac{H(x)u_3(x)}{4\omega^3},$$

$$E_{23} = \Delta^{-1} * \alpha_{32} = \Delta^{-1} * (-k\delta(x)) = -k\Delta^{-1} = \frac{-kH(x)u(x)}{4EI\omega^3},$$

$$E_{31} = \Delta^{-1} * \alpha_{13} = \Delta^{-1} * (k\delta(x)) = k\Delta^{-1} = \frac{kH(x)u(x)}{4EI\omega^3},$$

$$E_{32} = \Delta^{-1} * \alpha_{23} = \Delta^{-1} * (EI\delta''(x)) = EI(\Delta^{-1})'' = \frac{H(x)u_2(x)}{4\omega^3},$$

$$E_{33} = \Delta^{-1} * \alpha_{33} = \Delta^{-1} * (EI\delta'''(x)) = EI(\Delta^{-1})''' = \frac{H(x)u_3(x)}{4\omega^3}. \quad (9.87)$$

On the basis of the expressions (9.75) and of the formulas (9.81) for the deflection $\bar{v}$, we get

$$\bar{v}(x) = EI\left[S_a[\bar{v}]E'_{11}(x-a) + S_b[\bar{v}]E'_{11}(x-b) + S_a\left[\frac{\tilde{\mathrm{d}}}{\mathrm{d}x}\bar{v}\right]E_{11}(x-a)\right.$$
$$\left. + S_b\left[\frac{\tilde{\mathrm{d}}}{\mathrm{d}x}\bar{v}\right]E_{11}(x-b)\right] - E_{12} * \overline{q}(x) - \sum_{i=1}^{n} P_i E_{12}(x-c_i) - \sum_{i=1}^{n} m_i E_{13}(x-c_i).$$

On the other hand, using the property $h(x) * \delta^{(p)}(x-\alpha) = h^{(p)}(x-\alpha)$, $p = 1, 2, \ldots$, we have

$$E'_{11}(x) = \frac{1}{4EI\omega^3}(u_2(0)\delta(x) + H(x)u_3(x)) = \frac{H(x)u_3(x)}{4EI\omega^3}.$$

Thus, the deflection $\bar{v}$ becomes

$$\begin{aligned}\bar{v}(x) &= \frac{1}{4\omega^3}S_a[\bar{v}]H(x-a)u_3(x-a) + \frac{1}{4\omega^3}S_b[\bar{v}]H(x-b)u_3(x-b) \\ &+ \frac{1}{4\omega^3}S_a\left[\frac{\tilde{\mathrm{d}}}{\mathrm{d}x}\bar{v}\right]H(x-a)u_2(x-a) + \frac{1}{4\omega^3}S_b\left[\frac{\tilde{\mathrm{d}}}{\mathrm{d}x}\bar{v}\right]H(x-b)u_2(x-b) \\ &+ \frac{1}{4EI\omega^3}\left[H(x)u(x) * \overline{q}(x) + \sum_{i=1}^{n} P_i H(x-c_i)u(x-c_i)\right. \\ &\left. + \sum_{i=1}^{n} m_i H(x-c_i)u_1(x-c_i)\right] . \end{aligned} \tag{9.88}$$

It follows that the displacement support $\bar{v}$, $\mathrm{supp}(\bar{v})$, is not the interval $[a, b]$.
We note that $\bar{v}(x) = 0$ for $x < a = c_1$, because $H(x - c_i) = 0$, $i = \overline{1, n}$.
To have $\mathrm{supp}(\bar{v}) = [a, b]$, we require that $\bar{v} = 0$ for $x > b$.
We show that this condition is always possible, which allows us to obtain four relations, used to determine certain unknowns, such as constraint forces and moments.

Taking into account the formula

$$H(x)f(x) * \overline{q}(x) = \begin{cases} 0 , & x < a , \\ \displaystyle\int_a^x q(t)f(x-t)\mathrm{d}t , & x \in [a, b] , \\ \displaystyle\int_a^b q(t)f(x-t)\mathrm{d}t , & x > b , \end{cases} \tag{9.89}$$

the deflection $\bar{v}(x)$ for $x > b$ has the expression

$$\begin{aligned}\bar{v}(x) &= \frac{1}{4EI\omega^3}\left[\int_a^b q(t)u(x-t)\mathrm{d}t + \sum_{i=1}^{n} P_i u(x-c_i) + \sum_{i=1}^{n} m_i u_1(x-c_i)\right] \\ &+ \frac{1}{4\omega^3}\left[S_a[\bar{v}]u_3(x-a) + S_b[\bar{v}]u_3(x-b)\right. \\ &\left. + S_a\left[\frac{\tilde{\mathrm{d}}}{\mathrm{d}x}\bar{v}\right]u_2(x-a) + S_b\left[\frac{\tilde{\mathrm{d}}}{\mathrm{d}x}\bar{v}\right]u_2(x-b)\right] , \end{aligned} \tag{9.90}$$

because $H(x - c_i) = 1$, $i = \overline{1, n}$ for $x > b$.

We note that $\bar{v}(x)$ can be developed for $x > b$ as a Taylor series in powers of $x - b$ and we have

$$\begin{aligned}\bar{v}(x) &= \bar{v}(b+0) + \frac{1}{1!}\frac{\tilde{\mathrm{d}}}{\mathrm{d}x}\bar{v}(b+0)(x-b) + \frac{1}{2!}\frac{\tilde{\mathrm{d}}^2}{\mathrm{d}x^2}\bar{v}(b+0)(x-b)^2 \\ &+ \frac{1}{3!}\frac{\tilde{\mathrm{d}}^3}{\mathrm{d}x^3}\bar{v}(b+0)(x-b)^3 + \frac{1}{4!}\frac{\tilde{\mathrm{d}}^4}{\mathrm{d}x^4}\bar{v}(b+0)(x-b)^4 \ldots \end{aligned} \tag{9.91}$$

In order to have $\bar{v} = 0$ for $x > b$, it is necessary and sufficient that the following conditions should be fulfilled

$$\bar{v}(b+0) = 0\,, \quad \frac{\tilde{\mathrm{d}}}{\mathrm{d}x}\bar{v}(b+0) = 0\,, \quad \frac{\tilde{\mathrm{d}}^2}{\mathrm{d}x^2}\bar{v}(b+0) = 0\,, \quad \frac{\tilde{\mathrm{d}}^3}{\mathrm{d}x^3}\bar{v}(b+0) = 0\,. \tag{9.92}$$

Indeed, using the formula (9.91) and taking into account the Proposition 9.1, it follows that the derivatives $\tilde{\mathrm{d}}^4\bar{v}(x)/\mathrm{d}x^n$, $n \geq 4$ are expressed only using the functions u, $u_1 = u'$, $u_2 = u''$, $u_3 = u'''$.

From this, it follows that the conditions (9.92) imply the relations

$$\frac{\tilde{\mathrm{d}}^n}{\mathrm{d}x^n}\bar{v}(b+0) = 0\,, \quad n \geq 4\,. \tag{9.93}$$

Next, we explicit the conditions (9.92).

1. Because $u(0) = 0$, $u_1(0) = 0$, $u_2(0) = 0$, $u_3(0) = 4\omega^3$ and taking into account (9.84), the condition $\bar{v}(b+0) = 0$ leads to

$$\int_a^b q(t)u(b-t)\mathrm{d}t + \sum_{i=1}^n P_i u(b-c_i) + \sum_{i=1}^n m_i u_1(b-c_i)$$
$$+ EI\left(S_a[\bar{v}]u_3(b-a) + 4\omega^3 S_b[\bar{v}] + S_a\left[\frac{\tilde{\mathrm{d}}}{\mathrm{d}x}\bar{v}\right]u_2(b-a)\right) = 0\,.$$

2. Similarly, because $(\tilde{\mathrm{d}}/\mathrm{d}x)\bar{v}(b+0) = 0$, from (9.90) and using the formulas (9.84) we obtain

$$\int_a^b q(t)u_1(b-t)\mathrm{d}t + \sum_{i=1}^n P_i u_1(b-c_i) + \sum_{i=1}^n m_i u_2(b-c_i)$$
$$+ EI\left(-4\omega^4 S_a[\bar{v}]u(b-a) + S_a\left[\frac{\tilde{\mathrm{d}}}{\mathrm{d}x}\bar{v}\right]u_3(b-a) + 4\omega^3 S_b\left[\frac{\tilde{\mathrm{d}}}{\mathrm{d}x}\bar{v}\right]\right) = 0\,.$$

3. From (9.90), the condition $(\tilde{\mathrm{d}}^2/\mathrm{d}x^2)\bar{v}(b+0) = 0$ implies the relation

$$\int_a^b q(t)u_2(b-t)\mathrm{d}t + \sum_{i=1}^n P_i u_2(b-c_i) + \sum_{i=1}^n m_i u_3(b-c_i)$$
$$- k\left(S_a[\bar{v}]u_1(b-a) + S_a\left[\frac{\tilde{\mathrm{d}}}{\mathrm{d}x}\bar{v}\right]u(b-a)\right) = 0\,,$$

because $4\omega^4 EI = k$.

4. Proceeding analogously, from (9.90), the condition $(\tilde{d}^3/dx^3)\bar{v}(b+0) = 0$ gives

$$\int_a^b q(t)u_3(b-t)dt + \sum_{i=1}^n P_i u_3(b-c_i) - 4\omega^4 \sum_{i=1}^n m_i u(b-c_i)$$
$$- k\left(S_a[\bar{v}]u_2(b-a) + S_a\left[\frac{\tilde{d}}{dx}\bar{v}\right]u_1(b-a)\right) = 0\,.$$

In conclusion, we can state that the relations 1–4 express the necessary and sufficient conditions that the displacement support $\bar{v}(x), x \in \mathbb{R}$ given by (9.88) should be the interval $[a, b]$.

Proposition 9.3 The distribution $\bar{v} \in \mathcal{D}'_+$ with $\operatorname{supp}(\bar{v}) = [a, b]$, having the expression

$$\bar{v}(x) = \begin{cases} 0\,, & x < a\,,\\ w(x)\,, & x \in [a, b]\,,\\ 0\,, & x > b\,, \end{cases}$$

where

$$w(x) = \frac{1}{4EI\omega^3}\int_a^x q(t)u(x-t)dt + \frac{1}{4EI\omega^3}\sum_{i=1}^n P_i H(x-c_i)u(x-c_i)$$
$$+ \frac{1}{4EI\omega^3}\sum_{i=1}^n m_i H(x-c_i)u_1(x-c_i)$$
$$+ \frac{1}{4\omega^3}\left(S_a[\bar{v}]u_3(x-a) + S_a\left[\frac{\tilde{d}}{dx}\bar{v}\right]u_2(x-a)\right)\,,$$

is the deflection of the bending problem of the elastic bar on elastic foundation, written in matrix form $(A) * (X) = (B)$, given by (9.74), if the conditions 1–4 are fulfilled.

Using the formulas (9.81), (9.87) and proceeding as in the case of the deflection $\bar{v}$, we determine the expression of the shear force $\overline{T}$ and of the bending moment $\overline{M}$.

We can state:

Proposition 9.4 The distribution $\overline{T} \in \mathcal{D}'_+$ with $\operatorname{supp}(\overline{T}) = [a, b]$ has the expression

$$\overline{T}(x) = \begin{cases} 0\,, & x < a\,,\\ w_1(x)\,, & x \in [a, b]\,,\\ 0\,, & x > b\,, \end{cases}$$

where

$$w_1(x) = \frac{-1}{4\omega^3}\int_a^x q(t)u_3(x-t)\mathrm{d}t - \frac{1}{4\omega^3}\sum_{i=1}^{n} P_i H(x-c_i)u_3(x-c_i)$$
$$+ \frac{k}{4EI\omega^3}\sum_{i=1}^{n} m_i H(x-c_i)u(x-c_i)$$
$$+ \frac{k}{4\omega^3}\left(S_a[\bar{v}]u_2(x-a) + S_a\left[\frac{\tilde{\mathrm{d}}}{\mathrm{d}x}\bar{v}\right]u_1(x-a)\right) .$$

Proposition 9.5 The distribution $\overline{M} \in \mathcal{D}'_+$ with $\mathrm{supp}(\overline{M}) = [a, b]$ has the expression

$$\overline{M}(x) = \begin{cases} 0\,, & x < a\,, \\ w_2(x)\,, & x \in [a, b]\,, \\ 0\,, & x > b\,, \end{cases}$$

where

$$w_2(x) = \frac{-1}{4\omega^3}\int_a^x q(t)u_2(x-t)\mathrm{d}t - \frac{1}{4\omega^3}\sum_{i=1}^{n} P_i H(x-c_i)u_2(x-c_i)$$
$$- \frac{1}{4\omega^3}\sum_{i=1}^{n} m_i H(x-c_i)u_3(x-c_i)$$
$$+ \frac{k}{4\omega^3}\left(S_a[\bar{v}]u_1(x-a) + S_a\left[\frac{\tilde{\mathrm{d}}}{\mathrm{d}x}\bar{v}\right]u(x-a)\right) .$$

We emphasize that the conditions 1–4 established so that the displacement $\bar{v}$ should be with $\mathrm{supp}(\bar{v}) = [a, b]$, are also necessary and sufficient to have $\mathrm{supp}(\overline{T}) = [a, b]$, as well as $\mathrm{supp}(\overline{M}) = [a, b]$.

This may be justified by using the properties of the distributions support.

Let there be the distributions $f, g \in \mathcal{D}'(\mathbb{R})$ and $\alpha, \beta \in \mathbb{R}$. Then, we have

1. $\mathrm{supp}(\alpha f + \beta g) \subseteq \mathrm{supp}(f) \cup \mathrm{supp}(g)$;
2. if $\mathrm{supp}(f) \subseteq [a, b]$, then $\mathrm{supp}(f^{(k)}) \subseteq [a, b]\,, \quad k \in \mathbb{N}$.

We show that $\mathrm{supp}(\bar{v}) \subseteq [a, b]$ implies $\mathrm{supp}(\overline{T}) \subseteq [a, b]$ and $\mathrm{supp}(\overline{M}) \subseteq [a, b]$. Consequently, the conditions 1–4 obtained from the condition $\mathrm{supp}(\bar{v}) \subseteq [a, b]$ will be necessary and sufficient relations to have $\mathrm{supp}(\overline{T}) \subseteq [a, b]$ and $\mathrm{supp}(\overline{M}) \subseteq [a, b]$.

Indeed, from (9.70) we obtain

$$\mathrm{supp}(\overline{M}) = \mathrm{supp}\left(B_1 - EI\frac{\mathrm{d}^2}{\mathrm{d}x^2}\bar{v}\right) \subseteq \mathrm{supp}(B_1) \cup \mathrm{supp}\left(\frac{\mathrm{d}^2}{\mathrm{d}x^2}\bar{v}\right) .$$

Because $\operatorname{supp}(B_1) = \{a, b\}$ and $\operatorname{supp}((\mathrm{d}^2/\mathrm{d}x^2)\bar{v}) \subseteq [a, b]$, it follows that $\operatorname{supp}(\overline{M}) \subseteq \{a, b\} \cup [a, b] = [a, b]$, hence $\operatorname{supp}(\overline{M}) \subseteq [a, b]$.

Also, from (9.69) we have

$$\operatorname{supp}(\overline{T}) = \operatorname{supp}\left(\frac{\mathrm{d}}{\mathrm{d}x}\overline{M} - B_3\right) \subseteq \operatorname{supp}\left(\frac{\mathrm{d}}{\mathrm{d}x}\overline{M}\right) \cup \operatorname{supp}(B_3)\,.$$

Because $\operatorname{supp}((\mathrm{d}/\mathrm{d}x)\overline{M}) \subseteq \operatorname{supp}(\overline{M}) \subseteq [a, b]$ and $\operatorname{supp}(B_3) = \{a, c_2, \ldots, c_{n-1}, b\} \subseteq [a, b]$, we get $\operatorname{supp}(\overline{T}) \subseteq [a, b] \cup [a, b] = [a, b]$, that is, $\operatorname{supp}(\overline{T}) \subseteq [a, b]$.

Example 9.5 Let OA (Figure 9.5) be an elastic bar of length ℓ and a constant cross-section, the supports at the points O and A, lying on an elastic foundation.

We acknowledge that uniformly distributed loads of intensity q act on the bar, as well as a concentrated load of intensity P, applied at the point $c \in (0, \ell)$.

We shall apply Proposition 9.3 to determine the deflection $\bar{v}(x)$ at a point $x \in \mathbb{R}$. Due to the supports of the bar and because $c_1 = a = 0$, $c_n = b = \ell$, the boundary conditions are

$$\begin{aligned}
&\bar{v}(0+0) = 0\,, \quad \bar{v}(0-0) = 0\,, \quad \bar{v}(\ell+0) = 0\,, \quad \bar{v}(\ell-0) = 0\,,\\
&\frac{\tilde{\mathrm{d}}^2}{\mathrm{d}x^2}\bar{v}(0+0) = 0\,, \quad \frac{\tilde{\mathrm{d}}^2}{\mathrm{d}x^2}\bar{v}(0-0) = 0\,, \quad \frac{\tilde{\mathrm{d}}^2}{\mathrm{d}x^2}\bar{v}(\ell+0) = 0\,,\\
&\frac{\tilde{\mathrm{d}}^2}{\mathrm{d}x^2}\bar{v}(\ell-0) = 0\,.
\end{aligned} \tag{9.94}$$

The boundary conditions (9.94) can be written in the form

$$\begin{aligned}
&S_0[\bar{v}] = \bar{v}(0+0) - \bar{v}(0-0) = 0\,, \quad S_\ell[\bar{v}] = \bar{v}(\ell+0) - \bar{v}(\ell-0) = 0\,,\\
&S_0\left[\frac{\tilde{\mathrm{d}}^2}{\mathrm{d}x^2}\bar{v}\right] = 0\,,\ S_\ell\left[\frac{\tilde{\mathrm{d}}^2}{\mathrm{d}x^2}\bar{v}\right] = 0\,.
\end{aligned} \tag{9.95}$$

Because we have

$$\begin{aligned}
S_0\left[\frac{\tilde{\mathrm{d}}}{\mathrm{d}x}\bar{v}\right] &= \frac{\tilde{\mathrm{d}}}{\mathrm{d}x}\bar{v}(0+0) - \frac{\tilde{\mathrm{d}}}{\mathrm{d}x}\bar{v}(0-0) = \frac{\tilde{\mathrm{d}}}{\mathrm{d}x}\bar{v}(0+0)\,,\\
S_\ell\left[\frac{\tilde{\mathrm{d}}}{\mathrm{d}x}\bar{v}\right] &= \frac{\tilde{\mathrm{d}}}{\mathrm{d}x}\bar{v}(\ell+0) - \frac{\tilde{\mathrm{d}}}{\mathrm{d}x}\bar{v}(\ell-0) = -\frac{\tilde{\mathrm{d}}}{\mathrm{d}x}\bar{v}(\ell-0)\,,
\end{aligned} \tag{9.96}$$

due to the Proposition 9.3 we obtain for the deflection $\bar{v}$ the expression

$$\bar{v}(x) = \begin{cases} 0\,, & x \notin [0, \ell)\,,\\ v_1\,, & x \in [0, \ell]\,, \end{cases} \tag{9.97}$$

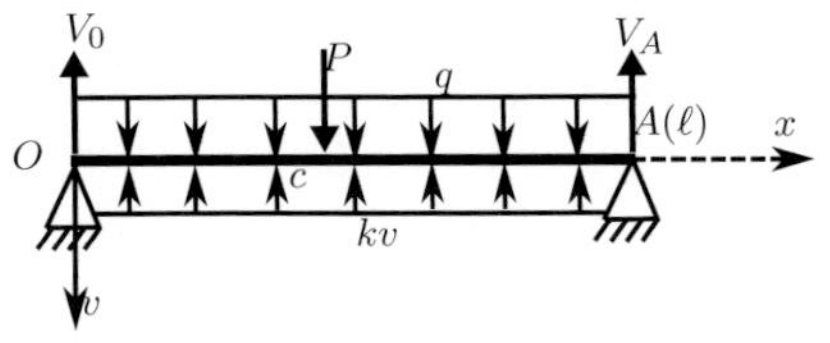

Figure 9.5

where

$$v_1 = \frac{q}{4EI\omega^3}\int_0^x u(x-t)\mathrm{d}t - \frac{V_0}{4EI\omega^3}H(x)u(x) + \frac{P}{4EI\omega^3}H(x-c)u(x-c)$$
$$- \frac{V_A}{4EI\omega^3}H(x-\ell)u(x-\ell) + \frac{1}{4\omega^3}\frac{\mathrm{d}\tilde{\bar{v}}(0+0)}{\mathrm{d}x}H(x)u_2(x)\,.$$

Taking into account (9.96) and the formula $\int_0^x f(x-t)\mathrm{d}t = \int_0^x f(t)\mathrm{d}t$, we obtain

$$\bar{v}(x) = \begin{cases} 0\,, & x \notin [0,\ell)\,, \\ v_{11}\,, & x \in [0,c)\,, \\ v_{12}\,, & x \in [c,\ell]\,, \end{cases} \tag{9.98}$$

where

$$v_{11} = \frac{q}{4EI\omega^3}\int_0^x u(t)\mathrm{d}t - \frac{V_0 u(x)}{4EI\omega^3} + \frac{1}{4\omega^3}\frac{\mathrm{d}\tilde{\bar{v}}(0+0)}{\mathrm{d}x}u_2(x)\,,$$

$$v_{12} = \frac{q}{4EI\omega^3}\int_0^x u(t)\mathrm{d}t - \frac{V_0 u(x)}{4EI\omega^3} + \frac{Pu(x-c)}{4EI\omega^3} + \frac{1}{4\omega^3}\frac{\mathrm{d}\tilde{\bar{v}}(0+0)}{\mathrm{d}x}u_2(x)\,.$$

We observe that in the expression of the deflection $\bar{v}$ appear only two unknowns, namely: the reaction V_0 at O and the rotation of bar to the right at the point O, $\mathrm{d}\tilde{\bar{v}}(0+0)/\mathrm{d}x$.

These unknowns, as well as the unknowns $V_A, \mathrm{d}\tilde{\bar{v}}(\ell-0)/\mathrm{d}x$ representing the reaction at the point A and the rotation of the bar to the left at the point A, respectively, will be determined from the conditions 1–4.

Using the formula $\int_0^x f(x-t)\mathrm{d}t = \int_0^x f(t)\mathrm{d}t$ and the relation $u(0)=0$, $u_1(0) = 0$, $u_2(0) = 0$, $u_3(0) = 4\omega^3$, the conditions 1–4 become

$$q\int_0^\ell u(t)\mathrm{d}t - V_0 u(\ell) + Pu(\ell-c) + EI\frac{\mathrm{d}\tilde{\bar{v}}(0+0)}{\mathrm{d}x}u_2(\ell) = 0\,, \tag{9.99}$$

$$q\int_0^\ell u_1(t)\mathrm{d}t - V_0 u_1(\ell) + Pu_1(\ell-c)$$
$$+ EI\left[\frac{\mathrm{d}\tilde{\bar{v}}(0+0)}{\mathrm{d}x}u_3(\ell) - 4\omega^3\frac{\mathrm{d}\tilde{\bar{v}}(\ell-0)}{\mathrm{d}x}\right] = 0\,, \tag{9.100}$$

$$q\int_0^\ell u_2(t)\mathrm{d}t - V_0 u_2(\ell) + Pu_2(\ell-c) - k\frac{\mathrm{d}\tilde{\bar{v}}(0+0)}{\mathrm{d}x}u(\ell) = 0\,, \tag{9.101}$$

$$q\int_0^\ell u_3(t)\mathrm{d}t - V_0 u_3(\ell) + Pu_3(\ell-c) - 4\omega^3 V_A - k\frac{\mathrm{d}\tilde{\bar{v}}(0+0)}{\mathrm{d}x}u_1(\ell) = 0\,. \tag{9.102}$$

We denote

$$u_1^*(x) = \frac{u_1(x)}{2\omega} = \sinh \omega x \sin \omega x \ ,$$
$$u_2^*(x) = \frac{u_2(x)}{2\omega^2} = \cosh \omega x \sin \omega x + \sinh \omega x \cos \omega x \ ,$$
$$u_3^*(x) = \frac{u_3(x)}{4\omega^3} = \cosh \omega x \cos \omega x \ , \tag{9.103}$$

and we can write the relations

$$I = \int_0^x u(t)\mathrm{d}t = \frac{1}{\omega}[1 - \cos \omega x \cosh \omega x] = \frac{1}{\omega}\left[1 - u_3^*(x)\right] \ ,$$
$$I_1 = \int_0^\ell u_1^*(t)\mathrm{d}t = \frac{u(\ell)}{2\omega}, \ I_3 = \int_0^\ell u_3^*(t)\mathrm{d}t = \frac{u_2^*(\ell)}{2\omega} \ ,$$
$$I_2 = \int_0^\ell u_2^*(t)\mathrm{d}t = \omega \sinh \omega\ell \sin \omega\ell = \omega u_1^*(\ell) \ . \tag{9.104}$$

From (9.99) and (9.101), we obtain for the unknowns V_0 and $\mathrm{d}\tilde{\bar{v}}(0+0)/\mathrm{d}x$ the expressions

$$\left[u^2(\ell) + u_2^{*2}(\ell)\right] V_0$$
$$= P\left[u(\ell)u(\ell - c) + u_2^*(\ell)u_2^*(\ell - c)\right] + q\left[u(\ell)\int_0^\ell u(t)\mathrm{d}t + u_2^*(\ell)\int_0^\ell u_2^*(t)\mathrm{d}t\right] \ ,$$
$$2EI\omega^2\left[u^2(\ell) + u_2^{*2}(\ell)\right]\frac{\mathrm{d}\tilde{\bar{v}}(0+0)}{\mathrm{d}x}$$
$$= P\left[u(\ell)u_2^*(\ell - c) - u_2^*(\ell)u(\ell - c)\right] + q\left[u(\ell)\int_0^\ell u_2^*(t)\mathrm{d}t - u_2^*(\ell)\int_0^\ell u(t)\mathrm{d}t\right] \ . \tag{9.105}$$

From (9.100) and (9.102) we obtain for the unknowns V_A and $\mathrm{d}\tilde{\bar{v}}(\ell-0)/\mathrm{d}x$ the expressions

$$V_A = q\int_0^\ell u_3^*(t)\mathrm{d}t - u_3^*(\ell)V_0 + Pu_3^*(\ell - c) - 2EI\omega^2 u_1^*(\ell)\frac{\mathrm{d}\tilde{\bar{v}}(0+0)}{\mathrm{d}x} \tag{9.106}$$

$$2EI\omega^2\frac{\mathrm{d}\tilde{\bar{v}}(\ell-0)}{\mathrm{d}x}$$
$$= q\int_0^\ell u_1^*(t)\mathrm{d}t - u_1^*(\ell)V_0 + Pu_1^*(\ell - c) + 2EI\omega^2 u_3^*(\ell)\frac{\mathrm{d}\tilde{\bar{v}}(0+0)}{\mathrm{d}x} \ . \tag{9.107}$$

Thus, the four unknowns V_0, V_A, $\mathrm{d}\tilde{\bar{v}}(0+0)/\mathrm{d}x$, $\mathrm{d}\tilde{\bar{v}}(\ell-0)/\mathrm{d}x$ are determined.
On the basis of the relations (9.104), the formula (9.98) of the deflection becomes

$$\tilde{v}(x) = \begin{cases} 0\,, & x \notin [0,\ell)\,, \\ v_{11}\,, & x \in [0,c)\,, \\ v_{12}\,, & x \in [c,\ell]\,, \end{cases} \tag{9.108}$$

where

$$\begin{aligned} v_{11} &= \frac{q}{4EI\omega^4}(1-u_3^*(x)) - \frac{u(x)V_0}{4EI\omega^3} + \frac{1}{4\omega^3}\frac{\mathrm{d}\tilde{\bar{v}}(0+0)}{\mathrm{d}x}u_2(x)\,, \\ v_{12} &= \frac{q}{4EI\omega^4}(1-u_3^*(x)) - \frac{u(x)V_0}{4EI\omega^3} + \frac{Pu(x-c)}{4EI\omega^3} + \frac{1}{4\omega^3}\frac{\mathrm{d}\tilde{\bar{v}}(0+0)}{\mathrm{d}x}u_2(x)\,, \end{aligned} \tag{9.109}$$

and the relations (9.105), (9.106), and (9.107) can be written in the form

$$\begin{aligned} \left[u^2(\ell) + u_2^{*2}(\ell)\right]V_0 &= P\left[u(\ell)u(\ell-c) + u_2^*(\ell)u_2^*(\ell-c)\right] \\ &\quad + q\left[\frac{u(\ell)}{\omega}(1-u_3^*(\ell)) + \omega u_1^*(\ell)u_2^*(\ell)\right]\,, \end{aligned} \tag{9.110}$$

$$\begin{aligned} &2EI\omega^2\left[u^2(\ell) + u_2^{*2}(\ell)\right]\frac{\mathrm{d}\tilde{\bar{v}}(0+0)}{\mathrm{d}x} \\ &= P\left[u(\ell)u_2^*(\ell-c) - u_2^*(\ell)u(\ell-c)\right] + q\left[\omega u(\ell)u_1^*(\ell) - \frac{u_2^*(\ell)}{\omega}(1-u_3^*(\ell))\right]\,, \end{aligned} \tag{9.111}$$

$$V_A = \frac{q}{2\omega}u_2^*(\ell) - u_3^*(\ell)V_0 + Pu_3^*(\ell-c) - 2\omega^2 EIu_1^*(\ell)\frac{\mathrm{d}\tilde{\bar{v}}(0+0)}{\mathrm{d}x}\,, \tag{9.112}$$

$$\begin{aligned} &2\omega^2 EI\frac{\mathrm{d}\tilde{\bar{v}}(\ell-0)}{\mathrm{d}x} \\ &= \frac{q}{2\omega}u(\ell) - u_1^*(\ell)V_0 + Pu_1^*(\ell-c) + 2\omega^2 EIu_3^*(\ell)\frac{\mathrm{d}\tilde{\bar{v}}(0+0)}{\mathrm{d}x}\,. \end{aligned} \tag{9.113}$$

Substituting the values V_0 and $\mathrm{d}\tilde{\bar{v}}(0+0)/\mathrm{d}x$ given by (9.110), (9.111) in (9.108) we obtain the expression of the deflection $\bar{v}(x)$, $x \in [0,\ell]$ for the elastic bar on an elastic foundation.

Consequently, the reactions V_0 and V_A at the points O and A, as well as the deflection of the bar on an elastic foundation have been determined.

9.3
Torsional Vibration of the Elastic Bars

9.3.1
Differential Equation of Torsional Vibrations of the Elastic Bars with Circular Cross-Section

The longitudinal and transverse vibrations of the elastic and viscoelastic bars are generally accompanied by torsional vibrations. Although these three types of mechanical phenomena are different from each other, one may shown that the torsional vibrations equation has the same structure as the longitudinal vibrations equation.

This allows us to establish the correspondence between mechanical and geometrical quantities involved in the study of the longitudinal vibrations and of the torsional vibrations of the elastic and viscoelastic bars.

This similarity between the mechanical and geometrical quantities takes place as regards to the initial and boundary conditions. On the basis of this analogy, the boundary value problems of the torsional vibrations are reduced to problems regarding the longitudinal vibrations of the bars.

Consequently, any solution of a longitudinal vibrations problem allows us to write the solution corresponding to the torsional vibrations of the elastic or the viscoelastic bar.

We consider thus a cylindrical elastic bar, homogeneous and isotropic, of circular cross-section and length ℓ. The bar is considered reported to the reference system from (Figure 9.6), where the Ox-axis is the symmetry axis of the cylindrical bar.

We denote by Ω the corresponding lower base of the cylinder, which is located in the Oyz-plane and which is supposed to be fixed. A couple (Figure 9.6) of moment M, oriented along the positive direction of the Ox-axis, is acting on the top base. This couple leads to the twist of the bar in the direct sense. Because the cylindrical bar has a circular cross-section, we acknowledge that, during the deformation by the torsion, due to the couple of moment M, all the cross-sections of the cylinder remain in their plane.

We will also acknowledge that the generator AB (Figure 9.6) remains after deformation a straight line, namely AB'. Therefore, if we denote by $\gamma = \alpha\ell$ the angle $\measuredangle BO''B'$ (Figure 9.6) with which has been rotated the top base of the cylinder, $x = \ell$, compared to the fixed bottom base, $x = 0$, then the angle of rotation of the cross-section at the distance x from the fixed base will be

$$\theta = \alpha x\,, \quad \theta = \measuredangle PO'P' \quad \text{(Figure 9.7)}\,.$$

Due to the torsion assumptions made, it follows that the torsional angle of a cross-section of the cylinder will be proportional with the distance between the cross-section and the fixed base.

We note that the angle $\alpha = \gamma/\ell$ is the angle of twist (torsion) per unit length of the bar, called the specific angle of torsion.

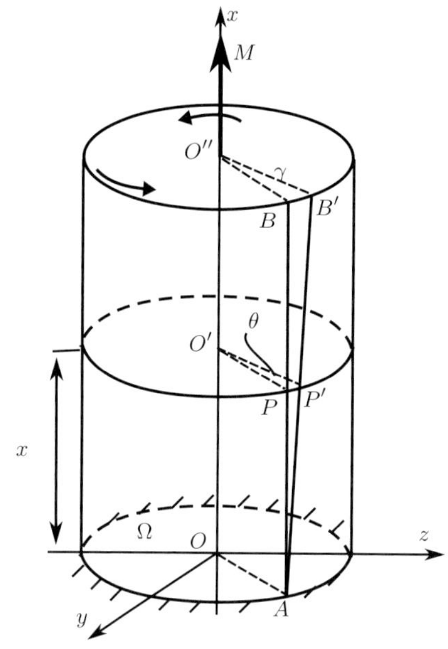

Figure 9.6

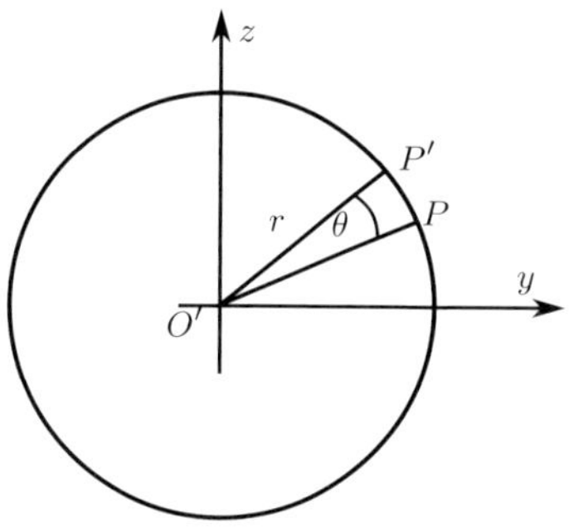

Figure 9.7

Because the couple of moment M acts on the top base of the cylinder, its magnitude is

$$M = \mu\alpha\frac{\pi R^4}{2}\,, \tag{9.114}$$

where R is the radius of the circle, and μ is Lamé's elastic constant.

The relation (9.114) can be written as

$$M = D\alpha\,, \tag{9.115}$$

where $D = \mu I_0$ is called the torsion bar stiffness, and $I_0 = \pi R^4/2$ is the moment of inertia of the cross-section with respect to the Ox-axis of torsion.

In the following, we neglect the mass loads and we acknowledge that continuously distributed couples act on the lateral surface of the cylinder, whose moments are oriented along the Ox-axis in the positive sense.

We denote by $m_r(x,t)$ the density per unit length of the distributed couples which act on the boundary of the cross-section and which are at the distance x at the time t (Figure 9.8).

The function $m_r(x,t)$, $(x,t) \in \mathbb{R} \times \mathbb{R}_+$, is considered continuous on its domain of definition. The distributed couples with the density $m_r(x,t)$ determine the torsional vibrations of the cylindrical bar.

Consequently, each point of the bar will perform oscillations on arcs of circles in the plane of the cross-section in which the point is considered. Hence, during the vibration, each cross-section of abscissa x rotates towards its position at the initial moment $t = 0$ with the angle $\theta = \theta(x,t)$, $(x,t) \in \mathbb{R} \times \mathbb{R}_+$.

Now let us consider two cross-sections of abscises x and $x + \Delta x$, respectively. Then, the cross-section of abscissa $x + \Delta x$ is rotated towards the cross-section of abscissa x with the angle $\Delta\theta = \theta(x + \Delta x, t) - \theta(x,t)$. It follows that the specific angle of rotation at the point x at time t has the expression

$$\alpha(x,t) = \lim_{\Delta x \to 0} \frac{\theta(x + \Delta x, t) - \theta(x,t)}{\Delta x} = \frac{\tilde{\partial}\theta(x,t)}{\partial x} . \qquad (9.116)$$

Hamilton's principle is applied to establish the differential equation of torsional vibrations of the elastic bar, according to which the Hamiltonian action is null, $\delta h = 0$.

Consequently, we obtain the equation

$$\frac{\tilde{\partial}^2}{\partial t^2}\theta(x,t) = \frac{\mu}{\rho}\frac{\tilde{\partial}^2}{\partial x^2}\theta(x,t) + \frac{m_r(x,t)}{\rho I_0} , \quad x \in \mathbb{R} , \quad t > 0 . \qquad (9.117)$$

The obtained equation is the torsional vibrations equation of the elastic bars of circular cross-section, on which distributed couples act with the intensity $m_r(x,t)$. The unknown function $\theta \in C^2(\mathbb{R} \times \mathbb{R}_+)$ is the angle of rotation of the cross-section of abscissa x.

Knowing the angle θ, we can determine the torsional moment $M_r(x,t)$ at any cross-section. Let us consider the cross-section of abscissa x of the elastic bar and let

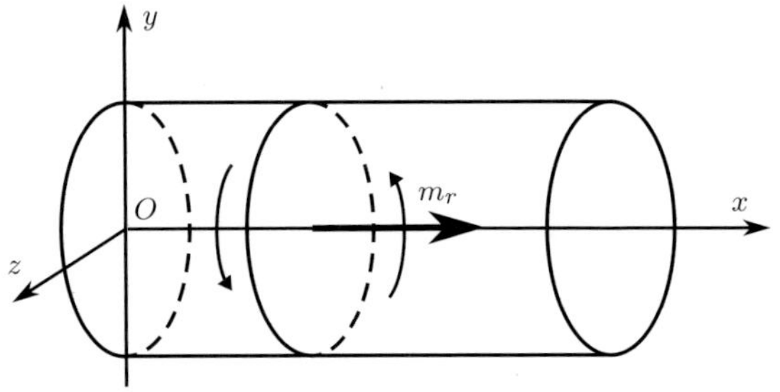

Figure 9.8

us remove the right side of the sectioned cylinder. Then, the mechanical action of the removed part over the remaining part is manifested by a couple of size $M_r(x,t)$, oriented along the Ox-axis (Figure 9.9).

For the moment $M_r(x,t)$ we have the expression

$$M_r(x,t) = \mu \frac{\tilde{\partial}\theta(x,t)}{\partial x} I_0 \,, \tag{9.118}$$

where I_0 is the moment of inertia of the cross-section with respect to the Ox-axis.

Because (9.117) of the torsional vibration is a second-order differential equation, it must be supplemented with initial and boundary conditions. Thus, the initial conditions are

$$\theta(x,t)|_{t=+0} = \theta_0(x) \,, \qquad \left.\frac{\tilde{\partial}\theta(x,t)}{\partial t}\right|_{t=+0} = \theta_1(x) \,,$$

where $\theta_0, \theta_1 \in C^0([0,\ell])$.

By these conditions, the rotation θ and, rotation speed $\partial\theta/\partial t$ at the initial moment $t = 0$ are specified for each section of the bar. Thus, if the bar is deformed first by applying a torsional couple M_0 and its action ceases at the time $t = 0$ then the torsional vibration begins at $t = 0$, and the function $\theta_0(x)$ expresses the deformed shape of the bar produced by M_0, and also $\theta_1(x) \equiv 0$ shows that the vibration started from the rest state, hence the speed at each point of the bar is zero.

The boundary conditions specify the constraints at the bar ends. We consider the following boundary conditions:

1. Free bar at both ends

$$M_r(x,t)|_{x=0} = \mu I_0 \left.\frac{\tilde{\partial}\theta}{\partial x}\right|_{x=0} = 0 \,, \quad M_r(x,t)|_{x=\ell} = \mu I_0 \left.\frac{\tilde{\partial}\theta}{\partial x}\right|_{x=\ell} = 0 \,,$$

   hence

$$\left.\frac{\tilde{\partial}\theta}{\partial x}\right|_{x=0} = 0 \,, \quad \left.\frac{\tilde{\partial}\theta}{\partial x}\right|_{x=\ell} = 0 \,, \quad t \in \mathbb{R}_+ \,;$$

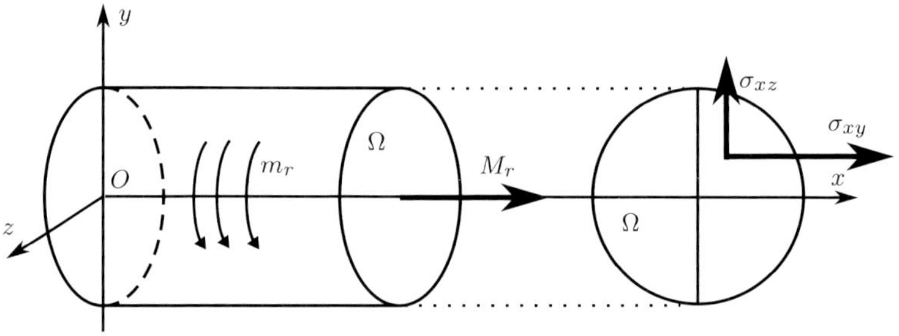

Figure 9.9

2. Embedded bar at both ends

$$\theta(x,t)|_{x=0} = 0\,, \quad \theta(x,t)|_{x=\ell} = 0\,, \quad t \in \mathbb{R}_+\,,$$

3. Bar embedded at one end and free at the other end

$$\theta(x,t)|_{x=0} = 0\,, \quad M_r(x,t)|_{x=\ell} = \mu\, I_0 \left.\frac{\tilde{\partial}\theta}{\partial x}\right|_{x=\ell} = 0\,, \quad t \in \mathbb{R}_+\,.$$

Hence,

$$\theta(x,t)|_{x=0} = 0\,, \left.\frac{\tilde{\partial}\theta}{\partial x}\right|_{x=\ell} = 0\,, \quad t \in \mathbb{R}_+\,.$$

Lamé's coefficient μ has the expression $\mu = E/(2(1+\nu)$ and is called the transverse modulus of elasticity. The number ν is called the Poisson ratio or the transverse contraction coefficient.

9.3.2
The Analogy between Longitudinal Vibrations and Torsional Vibrations of Elastic Bars

The longitudinal vibrations equation of elastic bars is

$$\rho \frac{\tilde{\partial}^2}{\partial t^2} u(x,t) = E \frac{\tilde{\partial}^2}{\partial x^2} u(x,t) + X(x,t)\,, \tag{9.119}$$

where ρ is the specific mass, E the longitudinal modulus, $u(x,t) \in C^{2,2}([0,\ell] \times \mathbb{R}_+)$ the bar displacement and $X(x,t) \in C^{0,0}([0,\ell] \times \mathbb{R}_+)$ the density of the mass forces.

Comparing (9.119) and (9.117), we see that they have the same structure mathematically, which allows us to establish a correspondence between the mechanical and geometrical quantities involved in the study of longitudinal and torsional vibrations of elastic bars of circular cross-section.

To this end, we write the two equations in the form

$$\frac{\tilde{\partial}^2}{\partial t^2}\theta(x,t) = \frac{\mu}{\rho}\frac{\tilde{\partial}^2}{\partial x^2}\theta(x,t) + \frac{m_r(x,t)}{\rho I_0}\,, \tag{9.120}$$

$$\frac{\tilde{\partial}^2}{\partial t^2}u(x,t) = \frac{E}{\rho}\frac{\tilde{\partial}^2}{\partial x^2}u(x,t) + \frac{X(x,t)}{\rho}\,. \tag{9.121}$$

Comparing the above equations, we can write the following correspondences

$$u \leftrightarrow \theta\,, \quad E \leftrightarrow \mu\,, \quad X \leftrightarrow \frac{m_r}{I_0}\,,$$

$$\sigma \leftrightarrow \frac{M_r}{I_0}\,, \quad \varepsilon = \frac{\tilde{\partial} u}{\partial x} \leftrightarrow \alpha = \frac{\tilde{\partial}\theta}{\partial x}\,. \tag{9.122}$$

The last two correspondences have been established taking into account the one-dimensional Hooke's law $\sigma = E\varepsilon$, as well as the formulas (9.118) and (9.116), where $\alpha(x, t)$ is the specific angle of rotation.

The correspondences established are valid with regard to the initial conditions and boundary conditions.

Thus, we have the correspondences between the initial conditions

$$\theta(x, t)|_{t=+0} = \theta_0(x) \leftrightarrow u(x, t)|_{t=+0} = u_0(x) ,$$
$$\frac{\tilde{\partial}}{\partial t}\theta(x, t)\Big|_{t=+0} = \theta_1(x) \leftrightarrow \frac{\tilde{\partial}}{\partial t}u(x, t)\Big|_{t=+0} = u_1(x) . \tag{9.123}$$

If both bars (of the same length) are embedded at both ends, then the correspondences follow:

$$\theta(x, t)|_{x=0} = 0 \leftrightarrow u(x, t)|_{x=0} = 0, \theta(x, t)|_{x=\ell} = 0 \leftrightarrow u(x, t)|_{x=\ell} = 0 . \tag{9.124}$$

9.3.3 Torsional Vibration of Free Bars Embedded at One End and Free at the Other End

We acknowledge the following initial conditions for the bar

$$\theta(x, t)|_{t=+0} = \theta_0(x) \in C^0([0, \ell]) , \quad \frac{\partial\theta(x, t)}{\partial t}\Big|_{t=+0} = 0 . \tag{9.125}$$

This means that the bar is twisted and the torsion vibrations start from the rest state, because the initial velocity is zero for each section of the bar.

As regards the boundary conditions, they are of the form

$$\theta(x, t)|_{x=0} = 0 , \quad \frac{\tilde{\partial}\theta(x, t)}{\partial x}\Big|_{x=\ell} = 0 , \tag{9.126}$$

that is, the bar is embedded at one end and free at the end $x = \ell$.

Thus, the free torsional vibrations are due to initial condition of deformation of the bar.

The solution will be sought in the series form

$$\theta(x, t) = \sum_{n=1}^{\infty} T_n(t) \sin\frac{(2n-1)\pi x}{2\ell} , \quad (x, t) \in [0, \ell] \times \mathbb{R}_+ . \tag{9.127}$$

We note that the sequence of functions $(\sin((2n-1)\pi x/2\ell))_{n\geq 1}$, $x \in [0, \ell]$, is an orthogonal series so that

$$\int_0^\ell \sin\frac{(2n-1)\pi x}{2\ell} \sin\frac{(2m-1)\pi x}{2\ell} \mathrm{d}x = \begin{cases} 0 , & m \neq n , \\ \dfrac{\ell}{2} , & m = n , \end{cases} \quad m, n \in \mathbb{N} . \tag{9.128}$$

It follows that any function $f(x)$, $x \in [0, \ell]$, which satisfies Dirichlet's conditions can be represented as a series

$$f(x) = \sum_{n=1}^{\infty} a_n \sin \frac{(2n-1)\pi x}{2\ell} \,, \qquad x \in [0, \ell] \,, \tag{9.129}$$

where

$$a_n = \frac{2}{\ell} \int_0^{\ell} f(x) \sin \frac{(2n-1)\pi x}{2\ell} \mathrm{d}x \,. \tag{9.130}$$

A direct verification shows that the adopted solution (9.127) satisfies the boundary conditions (9.126).

Because the distributed torsional moments do not act along the bar, $m_r = 0$, the equation of motion (9.117) becomes

$$\frac{\partial^2}{\partial t^2} \theta(x,t) = c^{*2} \frac{\partial^2}{\partial x^2} \theta(x,t) \,, \quad (x,t) \in [0,\ell] \times \mathbb{R}_+ \,, \tag{9.131}$$

where $c^* = \sqrt{\mu/\rho}$.

From this equation, taking into account (9.127), we obtain

$$T_n''(t) + \alpha_n^2 T_n(t) = 0 \,, \tag{9.132}$$

where $\alpha_n = (2n-1)\pi c^*/2\ell$.

On the other hand, from the initial conditions (9.125), it follows

$$\sum_{n=1}^{\infty} T_n(0) \sin \frac{(2n-1)\pi x}{2\ell} = \theta_0(x) \,, \quad x \in [0, \ell] \,, \tag{9.133}$$

$$T_n'(0) = 0 \,. \tag{9.134}$$

Expanding the function $\theta_0(x)$, $x \in [0, \ell]$, by orthogonal sequences $(\sin((2n-1)\pi x/2\ell))_{n \geq 1}$, we obtain

$$\theta_0(x) = \sum_{n=1}^{\infty} b_n \sin \beta_n x, \, x \in [0, \ell] \,, \tag{9.135}$$

where

$$b_n = \frac{2}{\ell} \int_0^{\ell} \theta_0(x) \sin(\beta_n x) \mathrm{d}x \,, \quad \beta_n = \frac{2n-1}{2\ell} \pi \,. \tag{9.136}$$

By identifying the expansions in series (9.133) and (9.135), we find

$$T_n(0) = b_n \,. \tag{9.137}$$

Equation 9.132, together with the conditions (9.134) and (9.137), constitute a Cauchy problem, whose solution is

$$T_n(t) = b_n \cos \alpha_n t \,, \quad t \in \mathbb{R} \,.$$

Substituting this expression in (9.127), we obtain the solution of the free torsional vibrations of the elastic bar

$$\theta(x, t) = \sum_{n=1}^{\infty} b_n \cos \alpha_n t \sin \beta_n x \,, \quad (x, t) \in [0, \ell] \times \mathbb{R} \,. \tag{9.138}$$

Example 9.6 Let us assume that the initial condition $\theta_0(x)$ is obtained by applying a torsional moment M_0 at the free end of the bar $x = \ell$. Then, on the basis of the torsional moment M_r, we have

$$M_0 = M_r(x)|_{x=\ell} = \mu I_0 \varphi_0 \,.$$

Consequently, the rotation angle $\theta_0(x)$ of the section of abscissa $x \in [0, \ell]$ has the expression

$$\theta_0(x) = \varphi_0 \frac{x}{\ell} \,, \quad x \in [0, \ell]$$

where $\varphi_0 = M_0/\mu I_0$ is the angle of rotation corresponding to the free end $x = \ell$.

Taking into account (9.136), we find

$$b_n = \frac{2}{\ell^2} \varphi_0 \frac{(-1)^{n-1}}{\beta_n^2} \,.$$

Thus, the solution of the problem, according to the formula (9.138), is

$$\theta(x, t) = \frac{2\varphi_0}{\ell^2} \sum_{n=1}^{\infty} \frac{(-1)^{n+1}}{\beta_n^2} \cos(\alpha_n t) \sin(\beta_n t) \,, \quad (x, t) \in [0, \ell] \times \mathbb{R} \,.$$

9.3.4
Forced Torsional Vibrations of a Bar Embedded at the Ends

We consider an elastic bar of circular cross-section and length ℓ, embedded at both ends (Figure 9.10).

We analyze the forced vibrations due to certain distributed moments along the bar, of magnitude $m_r(x, t)$, $(x, t) \in [0, \ell] \times \mathbb{R}_+$, per unit length of the bar.

Due to the assumptions made, the initial conditions and the boundary conditions are

$$\theta(x, t)|_{t=+0} = 0 \,, \qquad \left.\frac{\partial \theta(x, t)}{\partial t}\right|_{t=+0} = 0 \,, \tag{9.139}$$

$$\theta(x, t)|_{x=0} = 0 \,, \quad \theta(x, t)|_{x=\ell} = 0 \,. \tag{9.140}$$

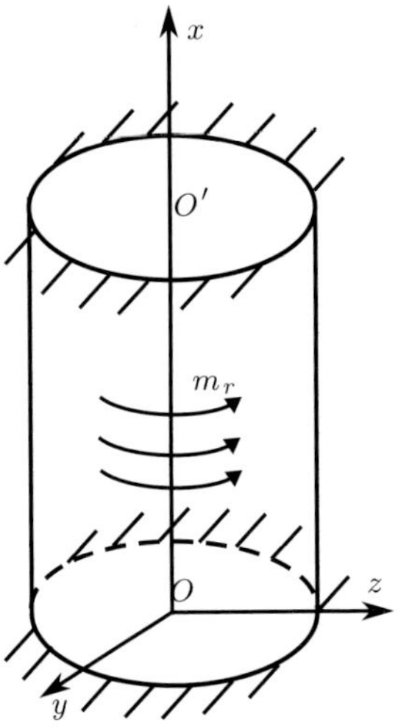

Figure 9.10

The differential equation of torsional vibrations is

$$\frac{\partial^2}{\partial t^2}\theta(x,t) = c^{*2}\frac{\partial^2}{\partial x^2}\theta(x,t) + \frac{m_r(x,t)}{\rho I_0}\,, \quad c^* = \sqrt{\frac{\mu}{\rho}}\,, \quad x \in \mathbb{R}\,, \quad t > 0\,. \tag{9.141}$$

The solution is sought in the form

$$\theta(x,t) = \sum_{n=1}^{\infty} T_n(t)\sin\frac{n\pi x}{\ell}\,, \tag{9.142}$$

with $(x,t) \in [0,\ell] \times \mathbb{R}_+$.

We acknowledge, for the generality of the results, that the series (9.142) of locally integrable functions converges in the sense of distributions of $\mathcal{D}'(\mathbb{R})$ with respect to the variable $x \in \mathbb{R}$, for all $t \in \mathbb{R}$ fixed.

Because the boundary conditions (9.140) are automatically checked, the determination of the functions T_n requires that the series (9.142) should satisfy the initial conditions (9.139), as well as the equation of motion (9.141).

From (9.142) we get

$$\frac{\partial\theta}{\partial t} = \sum_{n=1}^{\infty} T_n'(t)\sin\frac{n\pi x}{\ell}\,, \quad \frac{\partial^2\theta}{\partial t^2} = \sum_{n=1}^{\infty} T_n''(t)\sin\frac{n\pi x}{\ell}\,, \tag{9.143}$$

$$\frac{\partial^2\theta}{\partial x^2} = -\sum_{n=1}^{\infty}\frac{n^2\pi^2}{\ell^2}T_n(t)\sin\frac{n\pi x}{\ell}\,. \tag{9.144}$$

Using the initial conditions (9.139) we obtain

$$T_n(0) = 0\,, \quad T_n'(0) = 0\,. \tag{9.145}$$

Assuming that the function $m_r(x,t)$ is locally integrable on $[0,\ell] \times \mathbb{R}_+$, then we can write the Fourier series expansion

$$\frac{m_r(x,t)}{\rho I_0} = \sum_{n=1}^{\infty} a_n(t) \sin \frac{n\pi x}{\ell} , \tag{9.146}$$

where

$$a_n(t) = \frac{2}{\ell} \int_0^\ell \frac{m_r(x,t)}{\rho I_0} \sin \frac{n\pi x}{\ell} \mathrm{d}x . \tag{9.147}$$

We note that the equality (9.146) always occurs in the sense of convergence in the distributions space $\mathcal{D}'(\mathbb{R})$ for any t fixed.

Substituting the expressions (9.143), (9.144), and (9.146) in the equation of motion (9.141), we obtain the equation

$$T_n''(t) + \omega_n^2 T_n(t) = a_n(t) , \tag{9.148}$$

where $\omega_n = n\pi c^*/\ell$.

Adding to (9.148) the conditions (9.145), we obtain a Cauchy problem whose solution is

$$T_n(t) = \frac{1}{\omega_n} \int_0^\ell a_n(\tau) \sin \omega_n (t-\tau) \mathrm{d}\tau . \tag{9.149}$$

Thus, the solution (9.142) of the forced vibrations is determined and the existence and uniqueness of the solution is ensured in the distributions space $\mathcal{D}'(\mathbb{R})$ with respect to $x \in \mathbb{R}$, for any t fixed.

Example 9.7 In practice it is of interest to consider distributed moments which vary harmonically in time, hence

$$m_r(x,t) = m_0(x) \sin \omega^* t , \quad (x,t) \in [0,\ell] \times \mathbb{R}_+ . \tag{9.150}$$

Consequently, the formulas (9.147) and (9.149) become

$$a_n(t) = \frac{2}{\ell \rho I_0} \sin(\omega^* t) \int_0^\ell m_0(x) \sin \frac{n\pi x}{\ell} \mathrm{d}x , \tag{9.151}$$

$$T_n(t) = \frac{2}{\pi \rho I_0 c^* n} \int_0^\ell m_0(x) \sin \frac{n\pi x}{\ell} \mathrm{d}x \int_0^t \sin(\omega^* \tau) \sin(\omega_n(t-\tau))\, \mathrm{d}\tau , \tag{9.152}$$

where

$$\int_0^t \sin \omega^* \tau \sin \omega_n (t-\tau) \mathrm{d}\tau = \frac{1}{{\omega^*}^2 - \omega_n^2} \left(\omega^* \sin(\omega_n t) - \omega_n \sin(\omega^* t) \right) . \tag{9.153}$$

If the value ω^* of the distributed torsional moment $m_r(x,t)$ is equal to one of the eigenvalues ω_n, then $T_n \to \infty$, which means that the resonance phenomenon occurs.

Example 9.8 Another important case from a practical and theoretical point of view is that in which the torsional moment $m_r(x,t)$ is variable in time, but concentrated at a certain section $x = \xi \in (0,\ell)$ of the bar.

In the distributions space $\mathcal{D}'(\mathbb{R})$ with respect to the variable $x \in \mathbb{R}$, the torsional moment has the representation

$$m_r(x,t) = m(t)\delta(x-\xi)\,, \quad \xi \in (0,\ell)\,, \tag{9.154}$$

where $m(t) \in C^0(\mathbb{R})$.

Using a Fourier sine series expansion of the distribution $\delta(x-\xi)$, we have

$$\delta(x-\xi) = \frac{2}{\ell}\sum_{n=1}^{\infty} \sin\frac{n\pi\xi}{\ell}\sin\frac{n\pi x}{\ell}\,, \quad \xi \in (0,\ell)\,. \tag{9.155}$$

Therefore, taking into account (9.154), we obtain

$$\frac{m_r(x,t)}{\rho I_0} = \sum_{n=1}^{\infty} a_n(t)\sin\frac{n\pi x}{\ell}\,, \tag{9.156}$$

where

$$a_n(t) = \frac{m(t)}{\rho I_0}\frac{2}{\ell}\sin\frac{n\pi\xi}{\ell}\,. \tag{9.157}$$

Thus, we can apply the formula (9.149), which gives

$$T_n(t) = \frac{2}{\omega_n \ell \rho I_0}\sin\frac{n\pi\xi}{\ell}\int_0^t m(\tau)\sin\omega_n(t-\tau)\mathrm{d}\tau\,. \tag{9.158}$$

The solution is obtained from (9.142), namely

$$\theta(x,t) = \frac{2}{\rho I_0 \pi c^*}\sum_{n=1}^{\infty}\frac{1}{n}\sin\frac{n\pi\xi}{\ell}\sin\frac{n\pi x}{\ell}\int_0^t m(\tau)\sin\frac{n\pi c^*(t-\tau)}{\ell}\mathrm{d}\tau\,, \tag{9.159}$$

where $c^* = \sqrt{\mu/\rho}$.

We note that the series (9.159) is convergent not only in the distributions space $\mathcal{D}'(\mathbb{R})$ for any t fixed, but it is uniformly convergent on the basis of Weierstrass's criterion.

10
Applications of the Distribution Theory in the Study of Viscoelastic Bars

10.1
The Equations of the Longitudinal Vibrations of the Viscoelastic Bars in the Distributions Space $\mathcal{D}'_+$

In the linear deformations theory, the equations of motion corresponding to the deformable solid are independent from its mechanical properties. Consequently, both for the perfectly elastic body and for the viscoelastic one-dimensional body the equation of motion has the same expression.

We shall deduce the equation of longitudinal vibrations of viscoelastic bars in the distributions space $\mathcal{D}'_+$ starting from the longitudinal vibrations equation for a Bernoulli–Euler type rod in the space $\mathcal{D}'_+$.

Based on Hooke's law $\sigma = E\varepsilon = E\tilde{\partial} u/\partial x$, the longitudinal vibrations equation for a bar of Bernoulli–Euler type can be written in the form, [49],

$$\frac{\tilde{\partial}\sigma(x,t)}{\partial x} - \rho\frac{\tilde{\partial}^2 u(x,t)}{\partial t^2} + X(x,t) = 0 , \tag{10.1}$$

where $\sigma(x,t) \in C^{2,0}(\mathbb{R} \times \mathbb{R}_+)$ is the stress, $u(x,t) \in C^{3,2}(\mathbb{R} \times \mathbb{R}_+)$ is the displacement, $X(x,t) \in C^{0,0}(\mathbb{R} \times \mathbb{R}_+)$ is the mass force density per unit volume and $\rho(x,t) \in C^{0,0}(\mathbb{R} \times \mathbb{R}_+)$ is the mass density.

In the equation of motion (10.1), $\tilde{\partial}/\partial t$ and $\tilde{\partial}/\partial x$ are the derivatives in the ordinary sense.

Unlike them, the derivatives in the sense of distributions will be noted $\partial/\partial t$, $\partial/\partial x$.

The strain ε is given by

$$\varepsilon(x,t) = \frac{\tilde{\partial} u(x,t)}{\partial x} , \quad \varepsilon(x,t) \in C^{2,2}(\mathbb{R} \times \mathbb{R}_+) . \tag{10.2}$$

We acknowledge that the bar is infinite with constant cross-section, homogeneous, isotropic and $X(x,t) \in C^{1,0}(\mathbb{R} \times \mathbb{R}_+)$.

To rewrite the equation in the distributions space $\mathcal{D}'_+$, considering the variable x as a parameter, we define the function type distributions from $\mathcal{D}'_+$ by

$$\begin{aligned} \overline{\sigma}_x(t) &= H(t)\sigma(x,t) , \quad \overline{u}_x(t) = H(t)u(x,t) , \\ \overline{X}_x(t) &= H(t)X(x,t) , \quad \overline{\varepsilon}_x(t) = H(t)\varepsilon(x,t) . \end{aligned} \tag{10.3}$$

Distribution Theory, First Edition. Petre P. Teodorescu, Wilhelm W. Kecs, and Antonela Toma

These quantities are function type distributions from $\mathcal{D}'_+$ with respect to $t \in \mathbb{R}$, depending on the parameter $x \in I \subset \mathbb{R}$ and having discontinuities of the first kind at the origin $t = 0$.

We note that the derivatives in the usual sense of certain quantities with respect to the variable $x \in I \subset \mathbb{R}$ will coincide with the derivatives considered in the sense of distributions.

As regards the derivative in the distributions sense with respect to the temporal variable $t \in \mathbb{R}$, we can write

$$\frac{\partial \overline{u}_x(t)}{\partial t} = \frac{\tilde{\partial} \overline{u}_x(t)}{\partial t} + u_0(x)\delta(t) ,$$
$$\frac{\partial^2 \overline{u}_x(t)}{\partial t^2} = \frac{\tilde{\partial}^2 \overline{u}_x(t)}{\partial t^2} + u_0(x)\delta'(t) + u_1(x)\delta(t) , \tag{10.4}$$

where $u_0(x)$ and $u_1(x)$ are the initial conditions

$$u_0(x) = u(x,t)|_{t=+0} , \quad u_1(x) = \left. \frac{\tilde{\partial} u(x,t)}{\partial t} \right|_{t=+0} . \tag{10.5}$$

On the basis of the formulas (10.4), (10.1) in the distributions space $\mathcal{D}'_+$ takes the form

$$\frac{\partial \overline{\sigma}_x(t)}{\partial x} - \rho \frac{\partial^2 \overline{u}_x(t)}{\partial t^2} + \rho \left(u_1(x)\delta(t) + u_0(x)\delta'(t) \right) + \overline{X}_x(t) = 0 . \tag{10.6}$$

To this equation, we must add the constitutive equation of the one-dimensional viscoelastic solid

$$\overline{\sigma}_x(t) = \psi(t) * \frac{\partial \overline{\varepsilon}_x(t)}{\partial t} = \psi'(t) * \overline{\varepsilon}_x(t) , \tag{10.7}$$

equivalent to the equation

$$\overline{\varepsilon}_x(t) = \varphi(t) * \frac{\partial \overline{\sigma}_x(t)}{\partial t} = \varphi'(t) * \overline{\sigma}_x(t) , \tag{10.8}$$

where $\psi, \varphi \in \mathcal{D}'_+$ are the relaxation distribution and the creep distribution of the viscoelastic bar, respectively.

Taking into account the relations $\overline{\varepsilon}_x(t) = \partial \overline{u}_x(t)/\partial x$ and (10.7), (10.6) becomes

$$\psi'(t) * \frac{\partial^2 \overline{u}_x(t)}{\partial x^2} - \rho \frac{\partial^2 \overline{u}_x(t)}{\partial t^2} + \rho(u_1(x)\delta(t) + u_0(x)\delta'(t)) + \overline{X}_x(t) = 0 . \tag{10.9}$$

Equation 10.9 is the equation of longitudinal vibrations in displacements of the viscoelastic bars in the space $\mathcal{D}'_+$.

To obtain the equation of longitudinal vibrations in stresses, we differentiate (10.8) with respect to the parameter $x \in I \subset \mathbb{R}$; taking into account (10.2), it follows

$$\psi'(t) * \frac{\partial^2 \overline{\varepsilon}_x(t)}{\partial x^2} - \rho \frac{\partial^2 \overline{\varepsilon}_x(t)}{\partial t^2} + \rho(u_1'(x)\delta(t) + u_0'(x)\delta'(t)) + \frac{\partial \overline{X}_x(t)}{\partial x} = 0 , \tag{10.10}$$

where we have acknowledged that $u_0(x), u_1(x) \in C^1(I)$.

Using the formula (10.7) and the relation $\varphi'(t) * \psi'(t) = \delta(t)$, (10.10) becomes

$$(\psi'(t) * \varphi'(t)) * \frac{\partial^2 \overline{\sigma}_x(t)}{\partial x^2} - \rho\varphi'(t) * \frac{\partial^2 \overline{\sigma}_x(t)}{\partial t^2} + \rho\left(u_1'(x)\delta(t) + u_0'(x)\delta'(t)\right) + \frac{\partial \overline{X}_x(t)}{\partial x} = 0\,,$$

hence

$$\frac{\partial^2 \overline{\sigma}_x(t)}{\partial x^2} - \rho\varphi'(t) * \frac{\partial^2 \overline{\sigma}_x(t)}{\partial t^2} + \rho\left(u_1'(x)\delta(t) + u_0'(x)\delta'(t)\right) + \frac{\partial \overline{X}_x(t)}{\partial x} = 0\,, \tag{10.11}$$

which is the equation in stresses of the longitudinal vibrations of the homogeneous and isotropic viscoelastic bars in the distributions space $\mathcal{D}'_+$.

10.2 Longitudinal Vibrations of Maxwell Type Viscoelastic Bars: Solution in the Distributions Space $\mathcal{D}'_+$

The distributions of creep and relaxation for the Maxwell model of the viscoelastic bar have the expressions, [22, 47, 49]

$$\varphi(t) = \frac{1}{E}H(t) + \frac{1}{\eta}tH(t)\,, \quad \psi(t) = EH(t)\exp\left(-\frac{E}{\eta}t\right)\,,$$

where E is the modulus of elasticity and η is the dynamic coefficient of viscoelasticity.

Because $\mathrm{d}H(t)/\mathrm{d}t = \delta(t)$, $t\delta(t) = 0$, $\exp(-Et/\eta)\delta(t) = \delta(t)$, we have

$$\varphi'(t) = \frac{1}{E}\delta(t) + \frac{1}{\eta}H(t)\,, \quad \psi'(t) = E\delta(t) - \frac{E^2}{\eta}H(t)\exp\left(-\frac{E}{\eta}t\right)\,.$$

One can easily check the validity of the relation $\varphi'' * \psi = \varphi * \psi'' = \varphi' \times \psi' = \delta(t)$. Substituting the expression of $\varphi'(t)$ in (10.11), we obtain the equation

$$\frac{\partial^2 \overline{\sigma}_x(t)}{\partial x^2} - \frac{1}{c^2}\frac{\partial^2 \overline{\sigma}_x(t)}{\partial t^2} - \frac{1}{\lambda^2}\frac{\partial \overline{\sigma}_x(t)}{\partial t} + \frac{\partial \overline{X}_x(t)}{\partial x} + \rho\left(u_1'(x)\delta(t) + u_0'(x)\delta'(t)\right) = 0\,, \tag{10.12}$$

where $c^2 = E/\rho$, $\lambda^2 = \eta/\rho$.

Equation 10.12 is the equation in stresses of the longitudinal vibrations for Maxwell type viscoelastic bar in the distributions space $\mathcal{D}'_+$.

We study the longitudinal vibrations of the bar of Maxwell type, assuming that the stresses at the ends of the bar is zero.

We shall determine the stress $\overline{\sigma}_x(t) \in \mathcal{D}'_+$, $x \in [0, \ell]$, which satisfies the longitudinal vibrations equation (10.12) and the boundary conditions

$$\overline{\sigma}_x(t)|_{x=0} = 0\,, \overline{\sigma}_x(t)|_{x=\ell} = 0\,. \tag{10.13}$$

The initial conditions

$$\overline{u}_x(t)|_{t=+0} = u_0(x)\,, \qquad \left.\frac{\tilde{\partial}\overline{u}_x(t)}{\partial t}\right|_{t=+0} = u_1(x)\,, \quad u_0(x), u_1(x) \in C^1[0,\ell]\,,$$

as well as the mass forces $\overline{X}_x(t) \in \mathcal{D}'_+(\mathbb{R})$ are contained in (10.12).

We shall search the solution in the form of series of distributions

$$\overline{\sigma}_x(t) = H(t)\sum_{n=1}^{\infty} T_n(t)\sin\frac{n\pi x}{\ell}\,, \quad T_n(t) \in C^2(\mathbb{R})\,. \tag{10.14}$$

The distribution (10.14) satisfies the boundary conditions (10.13).

We determine $T_n(t)$ such that (10.14) does satisfy (10.12). From (10.14) we obtain

$$\frac{\partial^2\overline{\sigma}_x(t)}{\partial x^2} = -H(t)\sum_{n=1}^{\infty}\frac{n^2\pi^2}{\ell^2}T_n(t)\sin\frac{n\pi x}{\ell}\,, \tag{10.15}$$

$$\frac{\partial\overline{\sigma}_x(t)}{\partial t} = \delta(t)\sum_{n=1}^{\infty}T_n(0)\sin\frac{n\pi x}{\ell} + H(t)\sum_{n=1}^{\infty}T'_n(t)\sin\frac{n\pi x}{\ell}\,, \tag{10.16}$$

$$\frac{\partial^2\overline{\sigma}_x(t)}{\partial t^2} = \delta'(t)\sum_{n=1}^{\infty}T_n(0)\sin\frac{n\pi x}{\ell} + \delta(t)\sum_{n=1}^{\infty}T'_n(0)\sin\frac{n\pi x}{\ell} + H(t)\sum_{n=1}^{\infty}T''_n(t)\sin\frac{n\pi x}{\ell}\,. \tag{10.17}$$

We suppose that the density of the mass forces $\overline{X}_x(t) \in \mathcal{D}'_+(\mathbb{R})$ is of the form $\overline{X}_x(t) = H(t)X(x,t)$, where the function $X(x,t) \in C^{1,0}([0,\ell]\times\mathbb{R}_+)$ can be developed in a Fourier sine series on the interval $[0,\ell]$.

Consequently, we can write

$$\frac{\partial\overline{X}_x(t)}{\partial x} = H(t)\sum_{n=1}^{\infty}a_n(t)\sin\frac{n\pi x}{\ell}\,, \tag{10.18}$$

where

$$a_n(x) = \frac{2}{\ell}\int_0^{\ell}\frac{\partial X(x,t)}{\partial x}\sin\frac{n\pi x}{\ell}\mathrm{d}x = -\frac{2nx}{\ell}\int_0^{\ell}X(x,t)\cos\frac{n\pi x}{\ell}\mathrm{d}x\,. \tag{10.19}$$

We also acknowledge that the functions $u'_0(x)$, $u'_1(x) \in C^0([0,\ell])$ can be developed in Fourier sine series on the interval $[0,\ell]$. Thus, we can write

$$u'_0(x) = \sum_{n=1}^{\infty}b_n\sin\frac{n\pi x}{\ell}\,, \quad u'_1(x) = \sum_{n=1}^{\infty}c_n\sin\frac{n\pi x}{\ell}\,, \tag{10.20}$$

where

$$b_n = \frac{2}{\ell}\int_0^{\ell}u'_0(x)\sin\frac{n\pi x}{\ell}\mathrm{d}x\,, \quad c_n = \frac{2}{\ell}\int_0^{\ell}u'_1(x)\sin\frac{n\pi x}{\ell}\mathrm{d}x\,. \tag{10.21}$$

Substituting the expressions (10.15)–(10.18) and (10.20) in (10.12) and identifying the coefficients, we obtain the following relations

$$\frac{1}{c^2}T_n''(t) + \frac{1}{\lambda^2}T_n'(t) + \frac{n^2\pi^2}{\ell^2}T_n(t) = a_n(t)\,, \quad t > 0\,, \tag{10.22}$$

$$\frac{1}{c^2\rho}T_n'(0) + \frac{1}{\lambda^2\rho}T_n(0) = c_n\,, \tag{10.23}$$

$$T_n(0) = c^2\rho b_n\,. \tag{10.24}$$

Thus, to determine the function $T_n(t)$ we have to solve a Cauchy problem. The characteristic equation of the differential equation (10.22) is

$$\frac{r^2}{c^2} + \frac{r}{\lambda^2} + \frac{n^2\pi^2}{\ell^2} = 0\,, \tag{10.25}$$

and the discriminant of this equation has the expression

$$\Delta = \frac{1}{\lambda^4} - \frac{4}{c^2}\frac{n^2\pi^2}{\ell^2} = \left(\frac{1}{\lambda^2} - \frac{2n\pi}{\ell}\right)\left(\frac{1}{\lambda^2} + \frac{2n\pi}{\ell}\right)\,. \tag{10.26}$$

We have the following cases:

1. $\Delta \geq 0$ if $n \leq \ell\sqrt{\rho E}/2\pi\eta$;
2. $\Delta < 0$ if $n > \ell\sqrt{\rho E}/2\pi\eta$.

Let $n_0 = [\ell\sqrt{\rho E}/2\pi\eta]$ be the integer part of the number $\ell\sqrt{\rho E}/2\pi\eta$; then, in solving the characteristic equation (10.25), we distinguish two cases, as we have $n \geq n_0$ or $n < n_0$. According to these cases, we get different solutions for the functions.

With this, we can say that the solution of the Maxwell type bar with free ends exists, is unique and has the expression

$$\overline{\sigma}_x(t) = \sum_{n=1}^{n_0} T_n(t)\sin\frac{n\pi x}{\ell} + \sum_{n=n_0+1}^{\infty} T_n(t)\sin\frac{n\pi x}{\ell}\,, \quad t > 0\,, \quad x \in [0, \ell]\,.$$

The functions $T_n(t)$ are determined by solving the Cauchy problem (10.22)–(10.24), which can be done effectively by using a Laplace transform or by using Lagrange's method of variation of constants.

10.3 Steady-State Longitudinal Vibrations for the Maxwell Bar

An important class of problems concerning the longitudinal vibrations of elastic and viscoelastic bars are the so-called problems without initial conditions.

If the boundary conditions, the displacements or the stresses at the ends of the bar act for a sufficiently long time, then, due to the damping inherent in any real physical system, the influence of the initial conditions diminishes in time.

We shall call such problems without initial conditions steady-state boundary value problems.

We study the longitudinal vibrations for a Maxwell type bar of length ℓ considering the bar axis as the Ox-axis, where O is at the left end.

We acknowledge that the mass forces and the initial conditions are zero. This means that, at the initial moment, the bar is at rest. As regards the boundary conditions, we assume that the two ends of the bar are subject to time-varying stresses, that is,

$$\overline{\sigma}_x(t)|_{x=0} = f_1(t) , \quad \overline{\sigma}_x(t)|_{x=\ell} = f_2(t) , \quad f_1(t), f_2(t) \in \mathcal{D}'_+(\mathbb{R}) . \quad (10.27)$$

With these assumptions, the longitudinal vibrations equation (10.12) for Maxwell type bar takes the form

$$\frac{\partial^2 \overline{\sigma}_x(t)}{\partial x^2} - \frac{1}{c^2}\frac{\partial^2 \overline{\sigma}_x(t)}{\partial t^2} - \frac{1}{\lambda^2}\frac{\partial \overline{\sigma}_x(t)}{\partial t} = 0 , \quad c^2 = \frac{E}{\rho} , \quad \lambda^2 = \frac{\eta}{\rho} . \quad (10.28)$$

Equation 10.28 with the boundary conditions in stresses (10.27) is a steady-state boundary value problem.

We apply the Laplace transform in distributions as a solving method. Thus, assuming that $\overline{\sigma}_x(t), f_1(t), f_2(t) \in \mathcal{D}'_+(\mathbb{R})$ have Laplace images and denoting

$$\mathrm{L}[\overline{\sigma}_x(t)](p) = \tilde{\sigma}_x(p) , \quad \mathrm{L}[f_1(t)](p) = \tilde{f}_1(p) , \quad \mathrm{L}[f_2(t)](p) = \tilde{f}_2(p) , \quad (10.29)$$

(10.28) and (10.27) lead to

$$\frac{\mathrm{d}^2 \tilde{\sigma}_x(p)}{\mathrm{d}x^2} - \left(\frac{p^2}{c^2} + \frac{p}{\lambda^2}\right)\tilde{\sigma}_x(p) = 0 , \quad (10.30)$$

$$\tilde{\sigma}_x(p)|_{x=0} = \tilde{f}_1(p) , \quad \tilde{\sigma}_x(p)|_{x=\ell} = \tilde{f}_2(p) . \quad (10.31)$$

The general solution of (10.30) is

$$\tilde{\sigma}_x(p) = A(p)\exp(-x\omega) + B(p)\exp(x\omega) , \quad (10.32)$$

where ω has the expression

$$\omega(p) = \sqrt{\frac{p^2}{c^2} + \frac{p}{\lambda^2}} . \quad (10.33)$$

Taking into account the conditions (10.31), we obtain for the constants $A(p)$ and $B(p)$ the system of equations

$$\begin{aligned} A(p) + B(p) &= \tilde{f}_1(p) , \\ A(p)\exp(-\omega\ell) + B(p)\exp(\omega\ell) &= \tilde{f}_2(p) . \end{aligned} \quad (10.34)$$

The solution of this system is

$$A(p) = \frac{\tilde{f}_1(p)}{2} \frac{\exp(\omega \ell)}{\sinh(\omega \ell)} - \frac{\tilde{f}_2(p)}{2} \frac{1}{\sinh(\omega \ell)} ,$$

$$B(p) = -\frac{\tilde{f}_1(p)}{2} \frac{\exp(-\omega \ell)}{\sinh(\omega \ell)} + \frac{\tilde{f}_2(p)}{2} \frac{1}{\sinh(\omega \ell)} . \tag{10.35}$$

Substituting (10.35) in (10.32), we get the Laplace image of the steady-state problem for a Maxwell type viscoelastic bar, that is,

$$\tilde{\sigma}_x(p) = \tilde{f}_1(p) \frac{\sinh(\omega(\ell - x))}{\sinh(\omega \ell)} + \tilde{f}_2(p) \frac{\sinh(\omega x)}{\sinh(\omega \ell)} , \quad x \in [0, \ell] . \tag{10.36}$$

By applying the inverse Laplace transform, we obtain the searched solution

$$\tilde{\sigma}_x(t) = \mathrm{L}^{-1}[\tilde{\sigma}_x(p)] = f_1(t) \underset{t}{*} \mathrm{L}^{-1}\left[\frac{\sinh(\omega(\ell - x))}{\sinh(\omega \ell)}\right] + f_2(t) \underset{t}{*} \mathrm{L}^{-1}\left[\frac{\sinh(\omega x)}{\sinh(\omega \ell)}\right] , \tag{10.37}$$

where the symbol $*_t$ corresponds to the convolution product with respect to the variable $t \in \mathbb{R}$.

10.4 Quasi-Static Problems of Viscoelastic Bars

The equation of transverse vibrations of elastic bars with constant cross-section in the distributions space $\mathcal{D}'_+$ is

$$EI \frac{\partial^4 \overline{v}(x,t)}{\partial x^4} + \rho \frac{\partial^2 \overline{v}(x,t)}{\partial t^2} = \overline{q}(x,t) + \rho \left(v_1(x)\delta(t) + v_0(x)\delta'(t)\right) , \tag{10.38}$$

where

$$\overline{v}(x,t) = H(t)v(x,t) , \quad \overline{q}(x,t) = H(t)q(x,t) ,$$

$$v_0(x) = v(x,0) , \quad v_1(x) = \left.\frac{\tilde{\partial} v(x,t)}{\partial t}\right|_{t=0} .$$

The distributions $\overline{v}$ and $\overline{q}$ are distributions from $\mathcal{D}'_+$ with respect to $t \in \mathbb{R}$ and $x \in [0, \ell]$ is a parameter.

If the load $\overline{q}(x,t)$ varies slowly over time, then we can neglect the influence of the inertial force $\rho\partial^2\overline{v}/\partial t^2$ and of the initial conditions $v_0(x)$ and $v_1(x)$ in the equation of transverse vibrations (10.38).

In this case, [22, 49], we say that we have a quasi-static problem of bending of the elastic bar.

Consequently, the differential equation (10.38) becomes

$$E I \frac{\partial^4 \overline{v}(x,t)}{\partial x^4} = \overline{q}(x,t)\,, \quad x \in [0,\ell]\,, \quad t \in \mathbb{R}\,. \tag{10.39}$$

Thus, for example, if the load $\overline{q}(x,t)$ has the expression

$$\overline{q}(x,t) = q_1(x) H(t)\,, \quad x \in [0,\ell]\,, \quad t \in \mathbb{R}\,,$$

then the deflection is $\overline{v}(x,t) = v_s(x)H(t)$, where $v_s(x)$ is the static displacement of the elastic bar in a bending caused by the action of the load $q_1(x)$, $x \in [0,\ell]$.

It can also be considered as a quasi-static problem even if the load $\overline{q}(x,t)$ has the expression

$$\overline{q}(x,t) = \begin{cases} P f(t)\delta(x - v_0 t)\,, & 0 \le v_0 t \le \ell\,, \\ 0\,, & t \notin \left[0, \dfrac{\ell}{v_0}\right]\,, \end{cases}$$

where v_0 is the speed of a concentrated load along the bar with the magnitude $P f(t)$. The speed v_0 of the displacement of a moving load $\overline{q}(x,t)$ must be considered small, in order to approximate the vibrations equation (10.38) by (10.39).

We mention that we must add four boundary conditions independent of time depending on the constraints of the bar to the quasi-static equation (10.39).

10.4.1 Bending of the Viscoelastic Bars

To deduce the bending equation of the viscoelastic bar of length ℓ, we apply the correspondence principle developed by T. Alfrey and E.H. Lee. According to this principle we shall consider the quasi-static equation (10.39) of an elastic bar of length ℓ, with the same boundary conditions independent of time as the viscoelastic bar.

Applying the Laplace transform with respect to the variable $t \in \mathbb{R}$ to (10.39), we obtain

$$E I \frac{\mathrm{d}^4 \tilde{v}(x,p)}{\mathrm{d}x^4} = \tilde{q}(x,p)\,, \tag{10.40}$$

where $\tilde{v}(x,p) = \mathrm{L}[\overline{v}(x,t)](p)$, $\tilde{q}(x,p) = \mathrm{L}[\overline{q}(x,t)](p)$.

Because $p\tilde{\psi}(p)$, $\tilde{\psi}(p) = \mathrm{L}[\psi(t)](p)$ corresponds to the modulus of elasticity E in viscoelasticity, where $\psi(t) \in \mathcal{D}'_+(\mathbb{R})$ represents the distribution of relaxation, (10.40) becomes

$$p\tilde{\psi}(p) I \frac{\mathrm{d}^4 \tilde{w}(x,p)}{\mathrm{d}x^4} = \tilde{q}(x,p)\,. \tag{10.41}$$

The quantity $\overline{w}(x,t)$ is the deflection of the viscoelastic bar and $\overline{v}(x,t)$ is the deflection of the elastic bar.

Taking into account that between the creep distribution $\varphi(t) \in \mathcal{D}'_+(\mathbb{R})$ and the relaxation distribution $\psi(t) \in \mathcal{D}'_+(\mathbb{R})$ occurs the relation

$$\varphi'(t) * \psi'(t) = \delta(t) , \tag{10.42}$$

by applying the Laplace transform we obtain

$$\tilde{\varphi}(p)\tilde{\psi}(p) = p^{-2} . \tag{10.43}$$

Therefore, (10.41) can be written in the form

$$\frac{I}{p\tilde{\varphi}(p)} \frac{\mathrm{d}^4 \tilde{w}(x, p)}{\mathrm{d}x^4} = \tilde{q}(x, p) . \tag{10.44}$$

From (10.40) and (10.44) there results the relation between the Laplace images of the deflections of the viscoelastic bar and the deflections of the corresponding elastic bar

$$\frac{\mathrm{d}^4 \tilde{w}(x, p)}{\mathrm{d}x^4} = Ep\tilde{\varphi}(p)\frac{\mathrm{d}^4 \tilde{v}(x, p)}{\mathrm{d}x^4} . \tag{10.45}$$

Because both the viscoelastic and the elastic bars have the same boundary conditions, independent of time and null initial conditions, from (10.45) it follows

$$\tilde{w}(x, p) = p\, E\tilde{\varphi}(p)\tilde{v}(x, p) . \tag{10.46}$$

Applying the inverse Laplace transform, we obtain

$$\overline{w}(x, t) = E\varphi(t) \underset{t}{*} \frac{\partial \overline{v}(x, t)}{\partial t} = E\varphi'(t) \underset{t}{*} \overline{v}(x, t) , \tag{10.47}$$

which shows the relation between the deflections of the viscoelastic bar and of the elastic bar subjected to the same load and the same boundary conditions.

Particularly, if the load $\overline{q}(x, t)$ has the expression $\overline{q}(x, t) = q_1(x)H(t)$, then $\overline{v}(x, t) = v_s(x)H(t)$. Therefore, in this case the formula (10.47) gives

$$\begin{aligned} \overline{w}(x, t) &= E\varphi'(t) \underset{t}{*} v_s(x)H(t) = Ev_s(x)\varphi(t) * \frac{\mathrm{d}H(t)}{\mathrm{d}t} \\ &= Ev_s(x)\varphi(t) * \delta(t) = Ev_s(x)\varphi(t) , \end{aligned} \tag{10.48}$$

that is,

$$\overline{w}(x, t) = E\varphi(t)v_s(x) , \quad x \in [0, \ell] , \quad t \in \mathbb{R} . \tag{10.49}$$

We can say that the deflection in the case of the bending of a viscoelastic bar is in direct proportion to the corresponding deflection of the elastic bar, the proportionality factor being the distribution of creep $\varphi(t)$.

10.4.2
Bending of a Viscoelastic Bar of Kelvin–Voigt Type

The distribution of creep $\varphi(t) \in \mathcal{D}'_+(\mathbb{R})$ for this type of bar is

$$\varphi(t) = \frac{H(t)}{E}\left(1 - \exp\left(-\frac{E}{\eta}t\right)\right) , \quad t \in \mathbb{R} . \tag{10.50}$$

Taking into account (10.49), the deflection $\overline{w}(x,t)$ to bending of the Kelvin–Voigt type bar under the action of the load $\overline{q}(x,t) = q_1(x)H(t)$ has the expression

$$\overline{w}(x,t) = H(t)\left(1 - \exp\left(-\frac{E}{\eta}t\right)\right) v_s(x) , \quad t \in \mathbb{R} , \quad x \in [0,\ell] , \tag{10.51}$$

where $v_s(x)$ is the deflection of the elastic bar due to the load $q_1(x)$.

For $t \to \infty$ we get $\overline{w} = v_s$, so that the deflection of the Kelvin–Voigt bar increases and tends to the static deflection v_s. At the initial moment $t = 0$, $\overline{w}(x,t) = 0$.

10.4.3
Bending of a Viscoelastic Bar of Maxwell Type

In this case the distribution of creep $\varphi(t) \in \mathcal{D}'_+(\mathbb{R})$ is

$$\varphi(t) = H(t)\left(\frac{1}{E} + \frac{t}{\eta}\right) , \quad t \in \mathbb{R} . \tag{10.52}$$

Substituting in (10.49), we get the deflection $\overline{w}(x,t)$ to bending of the Maxwell bar

$$\overline{w}(x,t) = H(t)\left(1 + \frac{E}{\eta}t\right) v_s(x) , \quad t \in \mathbb{R} , \quad x \in [0,\ell] . \tag{10.53}$$

Hence, it follows

$$\overline{w}(x,t)|_{t=+\infty} = +\infty , \quad \overline{w}(x,t)|_{t=+0} v_s(x) ,$$

that is, at the initial moment the deflection of the Maxwell bar coincides with the corresponding deflection of the elastic bar; then the deflection of the Maxwell bar increases unboundedly.

From both practical and theoretical points of view the viscoelastic bar bending under the action of a fixed concentrated force or of a mobile concentrated force, moving uniformly along the bar is important. To apply the formulas (10.56) and (10.49), where one needs to know the deflection $\overline{w}(x,t)$ of the elastic bar subjected to the action of a concentrated force.

We consider an elastic bar simply supported at the ends and subjected to the action of concentrated forces of magnitude P (Figure 10.1).

The concentrated force is applied at the point $c \in (0,\ell)$ and, from the point of view of the distribution theory, is a distributed load of intensity

$$q(x) = P\delta(x - c) , \quad x \in \mathbb{R} . \tag{10.54}$$

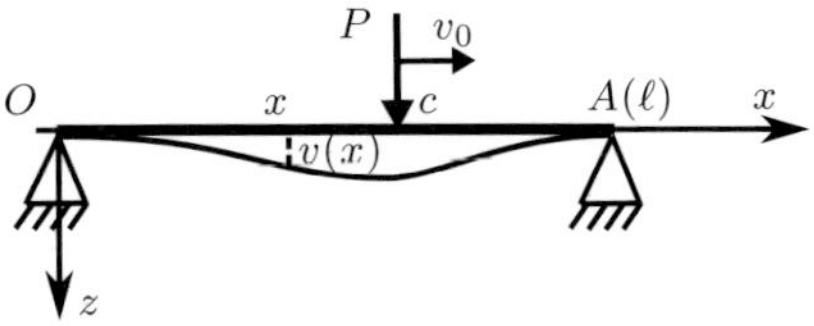

Figure 10.1

Because the load $q(t)$ is a distribution from the space $\mathcal{D}'(\mathbb{R})$, we write the equation of bending of the elastic bar in distributions, namely

$$EI\frac{\mathrm{d}^4 v(x)}{\mathrm{d}x^4} = q(x)\,, \quad x \in \mathbb{R}\,, \tag{10.55}$$

where $v(x) \in C^4((0,\ell))$ is considered a function type distribution from $\mathcal{D}'(\mathbb{R})$.

The boundary conditions are

$$v(0) = v(\ell) = 0\,, \quad v''(0) = v''(\ell) = 0\,. \tag{10.56}$$

Taking into account (10.54), (10.55) becomes

$$\frac{\mathrm{d}^4 v(x)}{\mathrm{d}x^4} = \frac{P}{EI}\delta(x-c)\,. \tag{10.57}$$

Because the fundamental solution of the operator $\mathrm{d}^4/\mathrm{d}x^4$ in $\mathcal{D}'_+(\mathbb{R})$ is $E(x) = H(x)x^3/6$, the general solution of (10.57) becomes

$$v(x) = \frac{H(x)P}{6EI}x^3 * \delta(x-c) + \alpha\frac{x^3}{6} + \beta\frac{x^2}{2} + \gamma x + \xi\,, \quad x \in \mathbb{R}\,, \tag{10.58}$$

where $\alpha, \beta, \gamma, \xi$ are constants which are determined from the fixing conditions (10.56).

Specifying the solution (10.58), we obtain

$$v(x) = \frac{6}{6EI}H(x-c)(x-c)^3 + \alpha\frac{x^3}{6} + \beta\frac{x^2}{2} + \gamma x + \xi\,, \quad x \in \mathbb{R}\,, \tag{10.59}$$

namely

$$v(x) = \begin{cases} \alpha\dfrac{x^3}{6} + \beta\dfrac{x^2}{2} + \gamma x + \xi\,, & x \le c\,, \\ \dfrac{P}{EI}\dfrac{(x-c)^3}{6} + \alpha\dfrac{x^3}{6} + \beta\dfrac{x^2}{2} + \gamma x + \xi\,, & c < x\,. \end{cases} \tag{10.60}$$

From the expression (10.60) it follows that $v \in C^2(\mathbb{R})$, which implies the equality of the first, second and third-order derivatives, in the sense of distributions and in the ordinary sense.

Hence, we can write

$$\frac{\mathrm{d}v}{\mathrm{d}x} = \frac{\tilde{\mathrm{d}}v}{\mathrm{d}x} = \begin{cases} \alpha\dfrac{x^2}{2} + \beta x + \gamma\,, & x < c\,, \\ \dfrac{P}{EI}\dfrac{(x-c)^2}{2} + \alpha\dfrac{x^2}{2} + \beta x + \gamma\,, & x \ge c\,, \end{cases} \tag{10.61}$$

$$\frac{d^2 v}{dx^2} = \frac{\tilde{d}^2 v}{dx^2} = \begin{cases} \alpha x + \beta\,, & x < c\,, \\ \dfrac{P}{EI}(x - c) + \alpha x + \beta\,, & x \geq c\,. \end{cases} \tag{10.62}$$

The boundary conditions (10.56) lead us to the following values for the constants $\alpha, \beta, \gamma, \xi$

$$\alpha = -\frac{Pb}{\ell EI}\,, \quad b = \ell - c\,, \quad \beta = 0\,, \quad \gamma = \frac{Pb(\ell^2 - b^2)}{6\ell EI}\,, \quad \xi = 0\,. \tag{10.63}$$

Substituting these values in (10.60), we obtain the solution

$$v(x) = \begin{cases} \dfrac{-Pb}{6\ell EI}x^3 + \dfrac{Pb}{6\ell EI}(\ell^2 - b^2)x\,, & x \leq c\,, \\ \dfrac{P}{6EI}(x - c)^3 - \dfrac{Pb}{6\ell EI}x^3 + \dfrac{Pb}{6\ell EI}(\ell^2 - b^2)x\,, & c < x\,, \end{cases} \tag{10.64}$$

which obviously is a function type distribution from $\mathcal{D}'(\mathbb{R})$.

Remark 10.1 If the concentrated force magnitude is variable in time, hence $P = P_1(t)H(t)$, and is moving with the speed v_0 from the end O of the bar, then the law of displacement of the point c of application of the force has the expression $c = v_0 t$. The deflection $\overline{w}(x, t)$ is determined by using the formula (10.64), so that we have

$$\overline{w}(x,t) = \begin{cases} \dfrac{-P_1(t)}{6\ell EI}(bx^3 - b(\ell^2 - b^2))\,, & x \leq v_0 t\,, \quad t \geq 0\,, \\ \dfrac{P_1(t)}{6EI}\left((x - v_0 t)^3 - \dfrac{b}{\ell}x^3 + \dfrac{b}{\ell}(\ell^2 - b^2)x\right)\,, & x > v_0 t\,, \quad t \geq 0\,, \\ 0\,, & x \in \mathbb{R}\,, \quad t \geq 0\,. \end{cases}$$

10.5
Torsional Vibrations Equation of Viscoelastic Bars in the Distributions Space $\mathcal{D}'_+$

To solve the boundary value problems of viscoelasticity, the correspondence principle stated by T. Alfrey and E.H. Lee is applied. For this purpose, we rewrite in the distributions space $\mathcal{D}'_+$, the torsional vibrations equation of the elastic bars

$$\frac{\tilde{\partial}^2 \theta(x,t)}{\partial t^2} = \frac{\mu}{\rho}\frac{\tilde{\partial}^2 \theta(x,t)}{\partial x^2} + \frac{m_r(x,t)}{\rho I_0}\,, \quad x \in \mathbb{R}\,, \quad t > 0\,. \tag{10.65}$$

We introduce the function type distributions from $\mathcal{D}'_+$ with respect to $t \in \mathbb{R}$, x being the parameter,

$$\overline{\theta}(x,t) = H(t)\theta(x,t)\,, \quad \overline{m}_r(x,t) = H(t)m_r(x,t)\,,$$

$$\overline{\alpha}(x,t) = H(t)\alpha(x,t)\,, \quad \alpha(x,t) = \frac{\tilde{\partial}\theta(x,t)}{\partial x}\,, \quad x \in [0,\ell]\,. \tag{10.66}$$

Denoting by $\partial/\partial t, \partial/\partial x$ the derivatives in the sense of distributions, we have

$$\frac{\partial \overline{\theta}(x,t)}{\partial t} = \frac{\tilde{\partial} \overline{\theta}(x,t)}{\partial t} + \theta_0(x)\delta(t) ,$$
$$\frac{\partial^2 \overline{\theta}(x,t)}{\partial t^2} = \frac{\tilde{\partial}^2 \overline{\theta}(x,t)}{\partial t^2} + \theta_0(x)\delta'(t) + \theta_1(x)\delta(t) , \tag{10.67}$$

where $\theta_0, \theta_1 \in C^1([0,\ell])$ are the initial conditions

$$\theta(x,t)|_{t=+0} = \theta_0(x) , \quad \left.\frac{\tilde{\partial}\theta(x,t)}{\partial t}\right|_{t=+0} = \theta_1(x) . \tag{10.68}$$

Multiplying (10.65) by $H(t)$ and taking into account (10.67), we get

$$\frac{\partial^2 \overline{\theta}(x,t)}{\partial t^2} = \frac{\mu}{\rho}\frac{\partial^2 \overline{\theta}(x,t)}{\partial x^2} + \frac{\overline{m}_r(x,t)}{\rho I_0} + \theta_1(x)\delta(t) + \theta_0(x)\delta'(t) ,$$
$$t \in \mathbb{R} , \quad x \in [0,\ell] . \tag{10.69}$$

The obtained equation is the torsional vibrations equation of the elastic bars written in the distributions space $\mathcal{D}'_+$ with respect to the temporal variable $t \in \mathbb{R}$, x being a parameter.

Applying the Laplace transform to (10.69), with respect to $t \in \mathbb{R}$, in the distributions space, we obtain

$$p^2\hat{\overline{\theta}}(x,p) = \frac{\mu}{\rho}\frac{\partial^2 \hat{\overline{\theta}}(x,p)}{\partial x^2} + \frac{\hat{\overline{m}}_r(x,p)}{\rho I_0} + \theta_1(x) + p\,\theta_0(x) , \tag{10.70}$$

where $\hat{\overline{\theta}}(x,p) = \mathrm{L}[\overline{\theta}(x,t)](p), \hat{\overline{m}}_r(x,p) = \mathrm{L}[\overline{m}_r(x,t)](p)$.

In accordance with the correspondence principle, we must replace the elastic constant μ by the corresponding quantity from one-dimensional viscoelasticity in (10.70).

As shown in Section 7.3.2, the complex function $p\hat{\psi}(p)$ from viscoelasticity corresponds to the constant E from elasticity.

We mention that $\hat{\psi}(p) = \mathrm{L}[\psi(t)](p)$, ψ being the distribution of relaxation of the viscoelastic bar.

Because between E and μ one has the relation $\mu = E/2(1+\nu)$, it follows that to the quantity μ corresponds, in viscoelasticity, the complex function $p\hat{\psi}(p)/2(1+\nu)$.

Consequently, substituting in (10.70) the quantity corresponding to μ and applying the inverse Laplace transform, we get the equation

$$\frac{\partial^2 \overline{\theta}(x,t)}{\partial t^2} = \frac{\psi'(t)}{2\rho(1+\nu)} \underset{t}{*} \frac{\partial^2 \overline{\theta}(x,t)}{\partial x^2} + \frac{\overline{m}_r(x,t)}{\rho I_0} + \theta_1(x)\delta(t) + \theta_0(x)\delta'(t) , \tag{10.71}$$

where the symbol $*_t$ corresponds to the convolution product, in the distributions sense, with respect to the variable $t \in \mathbb{R}$, x being a parameter.

The obtained equation (10.71) is the equation of the torsional vibrations of viscoelastic bars with circular cross-section. In this equation, the model of viscoelastic bar is indicated by the distribution of relaxation $\psi(t) \in \mathcal{D}'_+$.

We note that the formula of the torsional moment for the elastic bar

$$M_r(x,t) = \mu \frac{\tilde{\partial}\theta(x,t)}{\partial x} I_0 \tag{10.72}$$

corresponds for the viscoelastic bar as well.

Noting $\overline{M}_r(x,t) = H(t) M_r(x,t)$, (10.72) becomes

$$\overline{M}_r(x,t) = \mu I_0 \frac{\partial \overline{\theta}(x,t)}{\partial x} . \tag{10.73}$$

Using the correspondence principle, we find

$$\overline{M}_r(x,t) = \frac{\psi'(t)}{2(1+\nu)} \underset{t}{*} \frac{\partial \overline{\theta}(x,t)}{\partial x} I_0 , \tag{10.74}$$

where $\overline{M}_r(x,t) \in \mathcal{D}'_+$, x being a parameter.

The formula (10.74) allows us to determine the torsional distribution $\overline{M}_r(x,t)$ at the section x of the viscoelastic bar.

Obviously, the relation (10.74) is a generalization of formula (10.72) or (10.73) for the elastic bar.

Equation 10.71 of the torsional vibrations, written with respect to the distribution of relaxation $\psi(t) \in \mathcal{D}'_+$, can be expressed using the distribution of creep $\varphi(t) \in \mathcal{D}'_+$.

Thus, (10.71) is equivalent to the equation

$$\varphi'''(t) * \overline{\theta}(x,t) = \frac{1}{2\rho(1+\nu)} \frac{\partial^2 \overline{\theta}(x,t)}{\partial x^2} + \frac{1}{\rho I_0} \overline{m}_r(x,t) * \varphi'(t) + \theta_1(x)\varphi'(t) + \theta_0(x)\varphi''(t) . \tag{10.75}$$

Indeed, performing the convolution product of (10.71) with $\varphi'(t)$ and taking into account that $\varphi'(t) * \psi'(t) = \delta(t)$, we get

$$\varphi'(t) * \frac{\partial^2 \overline{\theta}(x,t)}{\partial t^2} = \frac{\psi'(t) * \varphi'(t)}{2\rho(1+\nu)} * \frac{\partial^2 \overline{\theta}(x,t)}{\partial x^2} + \frac{\overline{m}_r(x,t) * \varphi'(t)}{\rho I_0} + \theta_1(x)\varphi'(t) + \theta_0(x)\varphi''(t) ,$$

namely

$$\varphi'''(t) * \overline{\theta}(x,t) = \frac{1}{2\rho(1+\nu)} \frac{\partial^2 \overline{\theta}(x,t)}{\partial x^2} + \frac{1}{\rho I_0} \overline{m}_r(x,t) * \varphi'(t) + \theta_1(x)\varphi'(t) + \theta_0(x)\varphi''(t) .$$

10.6 Transverse Vibrations of Viscoelastic Bars on Viscoelastic Foundation

10.6.1 The Generalized Equation in Distributions Space $\mathcal{D}'(\mathbb{R}^2)$ of the Transverse Vibrations of Viscoelastic Bars on Viscoelastic Foundation

An elastic or viscoelastic bar resting on an elastic or viscoelastic foundation acted upon by external forces is deformed and, due to the boundary environment, is subjected to distributed constraints throughout its length. Consequently, the foundation may take both tensile and compression stresses.

We note that there are no friction forces between the foundation and the bar. In the case of the elastic bar placed on an elastic foundation, the Winkler hypothesis is adopted, according to which the reaction $\overline{q}_e(x,t)$, $(x,t) \in \mathbb{R}^2$ of the elastic environment is directly proportional to the common displacement of the bar and the elastic foundation, that is,

$$\overline{q}_e(x,t) = -E_0 k_0 \overline{v}(x,t)\,, \quad (x,t) \in \mathbb{R}^2\,. \tag{10.76}$$

The coefficient $E_0 k_0 > 0$ is called the stiffness coefficient of the elastic foundation, E_0 being the modulus of elasticity of the foundation.

Taking into account (9.40) for transverse vibrations of the infinite bars on elastic foundation, in the distributions space $\mathcal{D}'(\mathbb{R}^2)$, we obtain

$$EI\frac{\partial^4 \overline{v}(x,t)}{\partial x^4} + \rho\frac{\partial^2 \overline{v}(x,t)}{\partial t^2} + k_0 E_0 \overline{v}(x,t) = f(x,t)\,, \quad (x,t) \in \mathbb{R}^2 \tag{10.77}$$

where

$$f(x,t) = \overline{q}(x,t) + \rho v_1(x)\delta(t) + \rho v_0(x)\delta'(t)\,. \tag{10.78}$$

We mention that the displacement $\overline{v}$ and the density $\overline{q}$ of the loads normally distributed along the bar axis have the expressions

$$\overline{v}(x,t) = H(t)v(x,t) = \begin{cases} v(x,t)\,, & (x,t) \in \mathbb{R} \times \mathbb{R}_+\,, \\ 0\,, & \text{otherwise}\,, \end{cases}$$

$$\overline{q}(x,t) = H(t)q(x,t) = \begin{cases} q(x,t)\,, & (x,t) \in \mathbb{R} \times \mathbb{R}_+\,, \\ 0\,, & \text{otherwise}\,. \end{cases} \tag{10.79}$$

As regards the boundary conditions

$$v(x,0) = v_0(x)\,, \quad \left.\frac{\partial v(x,t)}{\partial t}\right|_{t=0} = v_1(x)\,, \quad x \in \mathbb{R}\,, \tag{10.80}$$

they are contained in the equation of motion (10.77).

To deduce the transverse vibrations equation of the viscoelastic bars on a viscoelastic foundation we shall apply the correspondence principle stated by Alfrey and Lee.

Thus, we apply the Laplace transform to (10.77) with respect to $t \in \mathbb{R}$; then the elastic constants E and E_0 are replaced by the corresponding constants for the viscoelastic bar and the viscoelastic foundation.

By applying the inverse Laplace transform L^{-1}, we obtain the equation of transverse vibrations, in $\mathcal{D}'(\mathbb{R}^2)$, of the viscoelastic infinite bars on a viscoelastic foundation.

Thus, from (10.77), by applying the operator L and substituting the elastic constants E and E_0, respectively, with the complex functions $p\tilde{\psi}(p)$, $p\tilde{\psi}_f(p)$, we obtain the equation

$$I p\tilde{\psi}(p)\frac{\mathrm{d}^4\tilde{\overline{v}}(x,p)}{\mathrm{d}x^4} + \rho p^2\tilde{\overline{v}}(x,p) + k_0 p\tilde{\psi}_f(p)\tilde{\overline{v}}(x,p) = \mathrm{L}[f(x,t)](x,p)\,, \tag{10.81}$$

where

$$\tilde{\overline{v}}(x,p) = \mathrm{L}[\overline{v}(x,p)](x,p)\,, \quad \tilde{\psi}(p) = \mathrm{L}[\psi(t)](p)\,, \quad \tilde{\psi}_f(p) = \mathrm{L}[\psi_f(t)](p) \tag{10.82}$$

and where $\psi(t), \psi_f(t) \in \mathcal{D}'_+$ are the relaxation distributions of the viscoelastic bar and of the viscoelastic foundation, respectively.

Taking into account the Laplace transform property regarding the partial convolution product, namely

$$\mathrm{L}[F(x,t) \underset{t}{*} g(t)](x,p) = \mathrm{L}[F(x,t)](x,p)\mathrm{L}[g(t)](p)\,,$$

(from (10.81)) by applying the inverse Laplace transform, we obtain the equation

$$\frac{\partial^4\overline{v}(x,t)}{\partial x^4} \underset{t}{*} I\psi'(t) + \rho\frac{\partial^2\overline{v}(x,t)}{\partial t^2}\overline{v}(x,t) + \overline{v}(x,t) \underset{t}{*} k_0\psi'_f(t) = f(x,t)\,, \quad (x,t) \in \mathbb{R}^2\,, \tag{10.83}$$

where the distribution $f(x,t) \in \mathcal{D}'(\mathbb{R}^2)$ is given by the formula (10.78)

Consequently, this equation is the generalized equation, in the distributions space $\mathcal{D}'(\mathbb{R}^2)$, of the transverse vibrations of viscoelastic infinite bars on a viscoelastic foundation.

From (10.83) it results that the resistance of the viscoelastic foundation to the vibrations of the viscoelastic bar is expressed by the term

$$\overline{q}_f(x,t) = -\overline{v}(x,t) \underset{t}{*} k_0\psi'_f(t) \in \mathcal{D}'(\mathbb{R}^2)\,, \tag{10.84}$$

which is a generalization of Winkler's hypothesis for the case of an elastic foundation.

Because the elastic solid is a limit case of the viscoelastic solid, namely when no relaxation phenomenon occurs, then, considering both the bar and the foundation to be elastic, we have

$$\psi(t) = EH(t)\,, \quad \psi_f(t) = E_0 H(t)\,, \quad t \in \mathbb{R}\,.$$

Consequently, it follows

$$\psi'(t) = E\delta(t)\,, \quad \psi'_f(t) = E_0\delta(t)$$

and thus (10.83) coincides with (10.77) which describes the transverse vibrations of elastic bars on elastic foundation.

The foundation viscoelastic resistance (10.84), becomes, in this limit case

$$\overline{q}_f(x,t) = -\overline{v}(x,t) \underset{t}{*} k_0 \psi'_f(t) = -\overline{v}(x,t) \underset{t}{*} k_0 E_0 \delta(t) = -k_0 E_0 \overline{v}(x,t)$$

namely, the foundation reaction is in direct proportion to the displacement of the bar, which corresponds to Winkler's hypothesis for elastic foundations.

In conclusion, (10.83) can be applied both for elastic bars and for viscoelastic ones, either on an elastic foundation or on a viscoelastic foundation.

We note that (10.83) can be written in a more general form in the distributions space $\mathcal{D}'(\mathbb{R}^2)$.

Thus, we consider that the density of the external loads $q(x,t)$ is a distribution from $\mathcal{D}'(\mathbb{R}^2)$ and may represent the distributed or concentrated loads (forces and moments).

The initial conditions $v_0(x)$ and $v_1(x)$ given by (10.80) can be considered as distributions from $\mathcal{D}'(\mathbb{R})$.

Consequently, the general form in $\mathcal{D}'(\mathbb{R}^2)$ of (10.83) of the transverse vibrations of viscoelastic bars on a viscoelastic foundation becomes

$$\frac{\partial^4 \overline{v}(x,t)}{\partial x^4} \underset{t}{*} I\psi(t) + \rho \frac{\partial^2 \overline{v}(x,t)}{\partial t^2} + \overline{v}(x,t) \underset{t}{*} k_0 \psi'_f(f) = F(x,t)\,, \quad (x,t) \in \mathbb{R}^2\,, \tag{10.85}$$

where $F(x,t) \in \mathcal{D}'(\mathbb{R}^2)$ has the expression

$$F(x,t) = \overline{q}(x,t) + \rho v_1(x) \times \delta(t) + \rho v_0(x) \times \delta'(t)\,. \tag{10.86}$$

The symbol $\times$ corresponds to the direct product of two distributions.

Because (10.85) includes the initial conditions $v_0(x), v_1(x) \in \mathcal{D}'(\mathbb{R})$, the solution $\overline{v}(x,t) \in \mathcal{D}'(\mathbb{R}^2)$ of (10.85) will be called the generalized Cauchy problem solution for transverse vibrations of viscoelastic bars on viscoelastic foundation.

10.6.2
Generalized Cauchy Problem Solution for Transverse Vibrations of the Elastic Bars on Kelvin–Voigt Type Viscoelastic Foundation

We consider an infinite elastic bar that rests on a viscoelastic Kelvin–Voigt type foundation.

Therefore, the relaxation distributions corresponding to the bar and the foundation have the expressions

$$\psi(t) = EH(t) , \quad \psi_f(t) = E_0 H(t) + \eta\delta(t) .$$

Consequently, (10.85) of transverse vibrations on a viscoelastic foundation becomes

$$EI\frac{\partial^4\overline{v}(x,t)}{\partial x^4} + \rho\frac{\partial^2\overline{v}(x,t)}{\partial t^2} + \overline{v}(x,t) \underset{t}{*} k_0(E_0\delta(t) + \eta\delta'(t)) = F(x,t) ,$$

namely

$$EI\frac{\partial^4\overline{v}(x,t)}{\partial x^4} + \rho\frac{\partial^2\overline{v}(x,t)}{\partial t^2} + k_0\eta\frac{\partial\overline{v}(x,t)}{\partial t} + k_0 E_0\overline{v}(x,t) = F(x,t) , \quad (x,t) \in \mathbb{R}^2 , \tag{10.87}$$

where

$$F(x,t) = \overline{q}(x,t) + \rho v_1(x) \times \delta(t) + \rho v_0(x) \times \delta'(t) . \tag{10.88}$$

The operator, [14], $P(\partial/\partial x, \partial/\partial t) : \mathcal{D}'(\mathbb{R}^2) \to \mathcal{D}'(\mathbb{R}^2)$ given by

$$P\left(\frac{\partial}{\partial x}, \frac{\partial}{\partial t}\right) = EI\frac{\partial^4}{\partial x^4} + \rho\frac{\partial^2}{\partial t^2} + k_0\eta\frac{\partial}{\partial t} + k_0 E_0 \tag{10.89}$$

describes the transverse vibrations of elastic bars on a viscoelastic Kelvin–Voigt type foundation.

With this, (10.87) becomes

$$P\left(\frac{\partial}{\partial x}, \frac{\partial}{\partial t}\right)\overline{v}(x,t) = F(x,t) , \quad (x,t) \in \mathbb{R}^2 . \tag{10.90}$$

We introduce the notations

$$EI = a \in \mathbb{R}_+ , \quad k_0\eta = b \in \mathbb{R}_+ , \quad k_0 E_0 = c \in \mathbb{R}_+ ,$$

where all real constants are positive.

The operator (10.89) becomes

$$P\left(\frac{\partial}{\partial x}, \frac{\partial}{\partial t}\right) = a\frac{\partial^4}{\partial x^4} + \rho\frac{\partial^2}{\partial t^2} + b\frac{\partial}{\partial t} + c . \tag{10.91}$$

Let $E(x,t) \in \mathcal{D}'(\mathbb{R}^2)$ be the fundamental solution of the operator (10.91), hence

$$P\left(\frac{\partial}{\partial x}, \frac{\partial}{\partial t}\right) E(x,t) = \delta(x,t) . \tag{10.92}$$

Then, for the Cauchy problem solution corresponding to (10.90) we obtain

$$\overline{v}(x,t) = E(x,t) * F(x,t) , \quad (x,t) \in \mathbb{R}^2 , \tag{10.93}$$

where the symbol $*$ denotes the convolution product with respect to the assembly of variables (x, t).

Consequently, the distribution $\overline{v}(x, t) \in \mathcal{D}'(\mathbb{R}^2)$ given by (10.93) is the generalized Cauchy problem solution for the transverse vibrations of infinite elastic bars on a Kelvin–Voigt type viscoelastic foundation.

To determine the fundamental solution $E(x, t) \in \mathcal{D}'(\mathbb{R}^2)$, we shall apply the Fourier transform F_x, with respect to the variable $x \in \mathbb{R}$, to (10.92).

Hence, we obtain the equation

$$\rho \frac{\mathrm{d}^2 \hat{E}(\alpha, t)}{\mathrm{d}t^2} + b \frac{\mathrm{d}\hat{E}(\alpha, t)}{\mathrm{d}t} + (a\alpha^4 + c)\hat{E}(\alpha, t) = \delta(t) , \tag{10.94}$$

where $\hat{E}(\alpha, t) = \mathrm{F}_x \left[E(x, t)\right](\alpha, t)$.

From the expression of (10.94) it follows that the distribution $\hat{E}(\alpha, t) \in \mathcal{D}'(\mathbb{R})$ is the fundamental solution of this equation.

Consequently, $\hat{E}(\alpha, t)$ has the expression

$$\hat{E}(\alpha, t) = H(t) Y(\alpha, t) , \tag{10.95}$$

where $Y(\alpha, t) \in C^\infty(\mathbb{R})$ satisfies the equation

$$\rho \frac{\mathrm{d}^2 Y}{\mathrm{d}t^2} + b \frac{\mathrm{d}Y}{\mathrm{d}t} + (a\alpha^4 + c) Y = 0 , \tag{10.96}$$

as well as the initial conditions

$$Y(\alpha, t)|_{t=0} = 0 , \qquad \left.\frac{\mathrm{d}Y(\alpha, t)}{\mathrm{d}t}\right|_{t=0} = \frac{1}{\rho} . \tag{10.97}$$

Denoting

$$\omega(\alpha) = \frac{1}{\rho}\sqrt{\rho(a\alpha^4 + c) - \frac{b^2}{4}} , \qquad \alpha \in \mathbb{R} , \tag{10.98}$$

the solution of (10.96) with the conditions (10.97) is

$$Y(\alpha, t) = \exp\left(-\frac{b}{2\rho}t\right) \frac{\sin t\omega(\alpha)}{\rho\omega(\alpha)} , \qquad (\alpha, t) \in \mathbb{R}^2 . \tag{10.99}$$

Taking into account (10.95), we obtain

$$\hat{E}(\alpha, t) = \mathrm{F}_x[E(x, t)](\alpha, t) = H(t) \exp\left(-\frac{b}{2\rho}t\right) \frac{\sin t\omega(\alpha)}{\rho\omega(\alpha)} , \qquad (\alpha, t) \in \mathbb{R}^2 . \tag{10.100}$$

Expanding in Taylor series the sine function, we obtain

$$\hat{E}(\alpha, t) = \frac{H(t)}{\rho} \exp\left(-\frac{b}{2\rho}t\right) \sum_{m=0}^{\infty} (-1)^m \frac{t^{2m+1}}{(2m+1)!} \omega^{2m}(\alpha) . \tag{10.101}$$

Applying the inverse Fourier transform F_x^{-1} to the relation (10.101), for the fundamental solution $E(x,t) \in \mathcal{D}'(\mathbb{R}^2)$ of the operator $P(\partial/\partial x, \partial/\partial t)$ defined by (10.91), we obtain the expression

$$\begin{aligned} E(x,t) &= \mathrm{F}_x^{-1}[\hat{E}(\alpha,t)](x,t) \\ &= \sum_{m=0}^{\infty} \frac{H(t)}{\rho} \exp\left(-\frac{b}{2\rho}t\right) \frac{(-1)^m t^{2m+1}}{(2m+1)!} \times \mathrm{F}_x^{-1}[\omega^{2m}(\alpha)](x) \,, \end{aligned} \tag{10.102}$$

where

$$\begin{aligned} \mathrm{F}_x^{-1}[\omega^{2m}(\alpha)](x) &= \underbrace{\mathrm{F}_x^{-1}[\omega^2(\alpha)](x) * \mathrm{F}_x^{-1}[\omega^2(\alpha)](x) * \cdots * \mathrm{F}_x^{-1}[\omega^2(\alpha)](x)}_{m\text{times}} \\ &= \left(\mathrm{F}_x^{-1}[\omega^2(\alpha)](x)\right)^{*m} \end{aligned} \tag{10.103}$$

and where $\times$ corresponds to the direct product symbol of two distributions.

Taking into account the formulas

$$\mathrm{F}_x[\delta(x)] = 1 \,, \quad \mathrm{F}_x[\delta^{(k)}(x)](\alpha) = (-\mathrm{i}\alpha)^k \,, \quad \omega^2(\alpha) = \left(\frac{c}{\rho} - \frac{b^2}{4\rho^2}\right) + \frac{a}{\rho}\alpha^4 \,,$$

we obtain

$$\mathrm{F}_x^{-1}[\omega^2(\alpha)](x) = \left(\frac{c}{\rho} - \frac{b^2}{4\rho^2}\right)\delta(x) + \frac{a}{\rho}\delta^{(4)}(x) \,.$$

Denoting

$$A = \frac{c}{\rho} - \frac{b^2}{4\rho^2} \,, \qquad B = \frac{a}{\rho} \,, \tag{10.104}$$

the expression (10.103) becomes

$$\begin{aligned} \mathrm{F}_x^{-1}[\omega^{2m}(\alpha)](x) &= (A\delta(x) + B\delta^{(4)}(x))^{*m} \\ &= \sum_{\alpha_1+\alpha_2=m} \frac{m!}{\alpha_1!\alpha_2!} A^{\alpha_1} B^{\alpha_2} \delta^{(4\alpha_2)}(x) \,, \end{aligned} \tag{10.105}$$

where we took into account Newton's binomial development $(x_1 + x_2)^m = \sum_{k+n=m} (m!/(k!n!)) x_1^k x_2^n$ and that $\delta(x) \in \mathcal{D}'_+$ is the unit element in the convolution algebra $\mathcal{D}'_+$.

Consequently, for the fundamental solution $E(x,t) \in \mathcal{D}'(\mathbb{R}^2)$ given by (10.102) we obtain the expression

$$\begin{aligned} &E(x,t) \\ &= \sum_{m=0}^{\infty} \sum_{\alpha_1+\alpha_2=m} \frac{(-1)^m}{(2m+1)!} \frac{m!}{\alpha_1!\alpha_2!} \left(\frac{c}{\rho} - \frac{b^2}{4\rho^2}\right)^{\alpha_1} \left(\frac{a}{\rho}\right)^{\alpha_2} G(t) \times \delta^{(4\alpha_2)}(x) \,, \end{aligned}$$

where

$$G(t) = H(t)t^{2m+1}\exp\left(-\frac{b}{2\rho}t\right) . \tag{10.106}$$

Taking into account (10.93) and (10.88), the solution of the generalized Cauchy problem for transverse vibrations of elastic bars on Kelvin–Voigt type viscoelastic foundation can be written in the form

$$\begin{aligned}\overline{v}(x,t) &= E(x,t) * F(x,t) = E(x,t) * \overline{q}(x,t)\\ &+ \rho\frac{\partial E(x,t)}{\partial t} \underset{x}{*} v_0(x) + \rho E(x,t) \underset{x}{*} v_1(x) ,\end{aligned} \tag{10.107}$$

where we used the formula

$$(f(x) * F(x)) \times (g(t) * G(t)) = (f(x) \times g(t)) \underset{(x,t)}{*} (F(x) \times G(t)) ,$$

symbols $*_{(x,t)}$ and $*_x$ denote the convolution product, with respect to the variables ensemble (x, t) and with respect to the variable x, respectively.

To perform the convolution product $E(x,t) * \overline{q}(x,t)$ from (10.107) we use the formula

$$\left(A(t) \times \delta^{(p)}(x)\right) * \overline{q}(x,t) = \frac{\partial^p \overline{q}(x,t)}{\partial x^p} * [A(t) \times \delta(x)] = \frac{\partial^p \overline{q}(x,t)}{\partial x^p} \underset{t}{*} A(t)$$

where the symbol $*_t$ is the convolution product with respect to variable $t \in \mathbb{R}$.

With this we can write

$$\begin{aligned}&\left(H(t)t^{2m+1}\exp\left(-\frac{b}{2\rho}t\right) \times \delta^{(4\alpha_2)}(x)\right) * \overline{q}(x,t)\\ &= \frac{\partial^{4\alpha_2}\overline{q}(x,t)}{\partial x^{4\alpha_2}} \underset{t}{*} H(t)t^{2m+1}\exp\left(-\frac{b}{2\rho}t\right) .\end{aligned}$$

Therefore, we have

$$\begin{aligned}&E(x,t) * \overline{q}(x,t)\\ &= \sum_{m=0}^{\infty} \sum_{\alpha_1+\alpha_2=m} \frac{(-1)^m}{(2m+1)!}\frac{m!}{\alpha_1!\alpha_2!}\left(\frac{c}{\rho} - \frac{b^2}{4\rho^2}\right)^{\alpha_1}\left(\frac{a}{\rho}\right)^{\alpha_2}\frac{\partial^{4\alpha_2}\overline{q}(x,t)}{\partial x^{4\alpha_2}} \underset{t}{*} G(t)\end{aligned} \tag{10.108}$$

where $G(t)$ has the expression (10.106).

The formulas (10.107) and (10.108) give us an explicit expression for the generalized Cauchy problem solution of (10.90) or (10.87).

10.6.2.1 **Particular Cases**

1. The initial conditions are zero, $v_0(x) = v_1(x) = 0$ and along the bar act external loads $\overline{q}(x, t) \in \mathcal{D}'(\mathbb{R}^2)$ representing distributed or concentrated loads

(forces and moments), then for the displacement $\overline{v}(x, t) \in \mathcal{D}'(\mathbb{R}^2)$ we obtain the expression

$$\overline{v}(x,t) = \sum_{m=0}^{\infty} \sum_{\alpha_1+\alpha_2=m} \frac{(-1)^m}{(2m+1)!} \frac{m!}{\alpha_1!\alpha_2!} \left(\frac{c}{\rho} - \frac{b^2}{4\rho^2}\right)^{\alpha_1} \left(\frac{a}{\rho}\right)^{\alpha_2} \frac{\partial^{4\alpha_2}\overline{q}(x,t)}{\partial x^{4\alpha_2}} \underset{t}{*} G(t) ,$$

where $G(t)$ has the expression (10.106).

2. If the external loads $\overline{q}(x, t)$, $(x, t) \in \mathbb{R}^2$ are function type distributions, then the convolution product $*_t$ represents the usual convolution with respect to the variable $t \in \mathbb{R}$ and thus for the displacement $\overline{v}(x, t)$, $(x, t) \in \mathbb{R}^2$ we obtain

$$\overline{v}(x,t) = \begin{cases} 0 , & x \in \mathbb{R} , \quad t < 0 , \\ G(x,t) , & x \in \mathbb{R} , \quad t \geq 0 , \end{cases}$$

where

$$G(x,t) = \sum_{m=0}^{\infty} \sum_{\alpha_1+\alpha_2=m} \frac{(-1)^m}{(2m+1)!} \frac{m!}{\alpha_1!\alpha_2!} \left(\frac{c}{\rho} - \frac{b^2}{4\rho^2}\right)^{\alpha_1} \left(\frac{a}{\rho}\right)^{\alpha_2} \frac{\partial^{4\alpha_2}\overline{q}(x,t)}{\partial x^{4\alpha_2}} g(t) ,$$

where

$$g(t) = \int_0^t \xi^{2m+1} \exp\left(-\frac{b}{2\rho}\xi\right) q(x, t-\xi) \mathrm{d}\xi .$$

11
Applications of the Distribution Theory in Physics

11.1
Applications of the Distribution Theory in Acoustics

11.1.1
Doppler Effect for a One-Dimensional Sound Source

Sound waves are longitudinal mechanical waves. They can propagate in solid, liquid and gaseous media. These waves have their origin in moving a portion of elastic medium from the normal balance position, leading to oscillations around the equilibrium position along the propagation direction of the wave.

Due to the elastic properties of the medium, the disturbance is passing from one layer to the next. This disturbance is the wave that propagates through the medium.

The factors that determine the wave speed in that medium are the inertia and its elasticity.

If the sound wave propagates in one direction (e. g., the Ox-axis) in a homogeneous, isotropic and linear elastic medium, then the differential equation of the wave propagation is

$$c^2 \frac{\partial^2 u(x,t)}{\partial x^2} - \frac{\partial^2 u(x,t)}{\partial t^2} = 0 , \tag{11.1}$$

where c is the propagation speed of the wave in the considered medium, and $u(x,t) \in C^{2,2}(\mathbb{R} \times \mathbb{R}_+)$ is the elongation.

In the case of the longitudinal vibrations one has

$$c = \sqrt{\frac{E}{\rho}} , \tag{11.2}$$

where ρ is the density and E is the longitudinal modulus of elasticity of the medium.

In the case of the transverse vibrations, (11.1) becomes the vibrating string equation, with

$$c = \sqrt{\frac{T}{\rho}} , \tag{11.3}$$

where T is the constant stress, tangent to the deformed shape of the string.

Distribution Theory, First Edition. Petre P. Teodorescu, Wilhelm W. Kecs, and Antonela Toma

In the two- or three-dimensional case, (11.1) will be written as

$$c^2 \Delta u(x,t) - \frac{\partial^2 u(x,t)}{\partial t^2} = 0 , \tag{11.4}$$

where Δ is the Laplace operator in two or three space variables.

Next, we consider that the sound wave propagates along the Ox-axis. We denote by S the sound source and by R the receiver (the observer).

Also, we denote by c, v_s and v_r, the elastic wave propagation speed in the medium, the speed of the sound source relatively to the medium S and the receiver speed relatively to the medium R, respectively (Figure 11.1).

The Doppler effect, [16] consists in the wave frequency variation recorded by the receiver R, when the source S (or the receiver) is moving with respect to the medium.

Thus, between the frequencies f' and f of the receiver and of the source, respectively, there exists the relation

$$f' = \frac{c \pm v_r}{c \mp v_s} f , \tag{11.5}$$

where the signs $+$ in the numerator and $-$ in the denominator correspond to the source and to the receiver which are approaching each other along the line that unites them; the lower signs correspond to the case when they move away from each other.

This relation is the mathematical expression of the Doppler effect.

We mention that in (11.5) the speed v_s is considered positive if the source approaches the receiver and negative if it moves away; also, the speed v_r is considered positive in the case of approaching the source and negative otherwise.

In establishing the formula (11.5), it was acknowledged that the source emits oscillations with period T, hence of the frequency $f = 1/T = \omega/2\pi$, and the receiver records the oscillations with the period T', that is, of the frequency $f' = 1/T' = \omega'/2\pi$, where ω and ω' are the pulsations (the circular frequencies).

Thus, if a person moves towards the sound source which is at rest, the sound heard is higher (has a higher frequency) than if it were at rest. If the person moves away from the sound source which is at rest, then they hear a sound less high than if they were at rest.

In the case in which v_r or v_s become comparable with c, the given formula (11.5) for the Doppler effect is no longer valid. If v_r and v_s exceed c, the formula (11.5) does not make sense.

Figure 11.1

11.1.2
Doppler Effect in the Presence of Wind

Let c_1 and c_2 be the longitudinal propagation speeds of the one-dimensional sound wave to the right or to the left, respectively; supposing that during the wave propagation energy losses do not occur, then the differential equation for acoustic waves in an elastic medium is

$$\left(\frac{\tilde{\partial}}{\partial t} + c_1\frac{\tilde{\partial}}{\partial x}\right)\left(\frac{\tilde{\partial}}{\partial t} - c_2\frac{\tilde{\partial}}{\partial x}\right) u(x,t) = 0 \,. \tag{11.6}$$

The fact that the acoustic phenomenon varies with the direction of propagation can be equated with a certain anisotropy of the medium. For this reason, we say that the acoustic (sound) wave propagates in an anisotropic medium.

If we consider the influence of external forces, (11.6) becomes

$$\left(\frac{\tilde{\partial}}{\partial t} + c_1\frac{\tilde{\partial}}{\partial x}\right)\left(\frac{\tilde{\partial}}{\partial t} - c_2\frac{\tilde{\partial}}{\partial x}\right) u(x,t) = X(x,t) \,, \tag{11.7}$$

where $u(x,t) \in C^{2,2}(\mathbb{R} \times \mathbb{R}_+)$, and $X(x,t) \in C^{2,2}(\mathbb{R} \times \mathbb{R}_+)$ is a quantity in direct proportion to the disturbing forces.

Such a case occurs when the one-dimensional sound wave is propagated in the atmosphere, in the presence of wind.

Consequently, we acknowledge that the medium in which the sound wave propagates is elastic and anisotropic.

Let c be the one-dimensional sound wave propagation speed and w the wind speed.

If the wave is propagated in the direction of the wind, then its speed of propagation is changed, being equal to

$$c_1 = c + w \,. \tag{11.8}$$

Similarly, the wave propagation speed in the opposite direction of the wind will be

$$c_2 = c - w \,. \tag{11.9}$$

Taking account of (11.6), the differential equation of the one-dimensional acoustic waves in the atmosphere, taking into account the wind speed w, will be

$$\left(\frac{\tilde{\partial}}{\partial t} + (c+w)\frac{\tilde{\partial}}{\partial x}\right)\left(\frac{\tilde{\partial}}{\partial t} - (c-w)\frac{\tilde{\partial}}{\partial x}\right) u(x,t) = X(x,t) \,. \tag{11.10}$$

In order to generalize (11.10), we consider it in the distributions space $\mathcal{D}'_+$ with respect to $x \in \mathbb{R}$ and where $t \in \mathbb{R}$ is a parameter.

Therefore, (11.10) in $\mathcal{D}'_+$ takes the form

$$\left(\frac{\partial}{\partial t} + (c+w)\frac{\partial}{\partial x}\right)\left(\frac{\partial}{\partial t} - (c-w)\frac{\partial}{\partial x}\right) u_t(x) = X_t(x) \,, \tag{11.11}$$

where the derivatives $\partial/\partial t$, $\partial/\partial x$ are considered in the sense of distributions and $u(x,t) = u_t(x) \in \mathcal{D}'_+$, $X(x,t) = X_t(x) \in \mathcal{D}'_+$ are distributions with respect to the spatial variable $x \in \mathbb{R}$, $t \in \mathbb{R}$ being a parameter.

Using (11.11), we can study the Doppler effect for one-dimensional sound waves in the presence of wind.

We begin with a point sound source S, which emits oscillations with the frequency $f = \omega/2\pi$ and is moving uniformly with the velocity v; the sound source motion starts at the initial moment $t = 0$, in the positive direction of the Ox-axis, in the sense in which the wave propagates.

In this case, the intensity of the sound source leads to an external force that is in direct proportion to $c^2 \exp(\mathrm{i}\omega t)$; we may write

$$X_t(x) = c^2 \exp(\mathrm{i}\omega t) H(t) \delta(x - vt) , \tag{11.12}$$

where δ is the Dirac delta distribution and H is the Heaviside distribution.

Hence, the external force $X_t(x) \in \mathcal{D}'_+$ is a distribution with respect to the variable $x \in \mathbb{R}$, t being a parameter.

We note that, because the sound source moves uniformly with the speed $v > 0$ for $t \geq 0$, the Dirac delta distribution $\delta(x - vt)$ is concentrated at the point $x = vt \geq 0$ on the Ox-axis, hence $\operatorname{supp}(\delta(x - vt)) \subset [0, \infty)$.

To study (11.11) with the condition (11.12), we apply successively the Laplace transforms, in distribution, with respect to the variables x and t.

We shall denote by $\mathrm{L}_x, \mathrm{L}_t$ the Laplace transforms with respect to x and t; the corresponding complex variables are denoted by p and q.

Applying successively these transforms to (11.11) and (11.12), we obtain

$$[(c + w)p + q][(c - w)p - q]\tilde{u}(p,q) = -\frac{c^2}{q + vp - \mathrm{i}\omega} , \tag{11.13}$$

where we have denoted

$$\tilde{u}(p,q) = \mathrm{L}_t \left[\mathrm{L}_x \left[u_t(x)\right]\right](p,q) \tag{11.14}$$

and have used the formulas

$$\begin{aligned}
\mathrm{L}_t \left[H(t) \exp(\lambda t)\right](q) &= \frac{1}{q - \lambda} , \\
\mathrm{L}_x \left[\delta(x - vt)\right](p) &= \exp(-pvt) , \\
\mathrm{L}_t \left[\mathrm{L}_x \left[X_t(x)\right]\right](p,q) &= \mathrm{L}_t \left[c^2 H(t) \exp(\mathrm{i}\omega t) \exp(-pvt)\right](q) \\
&= \frac{c^2}{q + pv - \mathrm{i}\omega} .
\end{aligned} \tag{11.15}$$

From (11.13) it follows

$$\begin{aligned}
\tilde{u}(p,q) &= -\frac{c^2}{[(c + w)p + q][(c - w)p - q][vp - \mathrm{i}\omega + q]} \\
&= \frac{A(p)}{q - q_1} + \frac{B(p)}{q - q_2} + \frac{C(p)}{q - q_3} ,
\end{aligned} \tag{11.16}$$

where

$$q_1 = -p(c+w)\,, \quad q_2 = p(c-w)\,, \quad q_3 = \mathrm{i}\omega - vp \tag{11.17}$$

and

$$\begin{aligned} A(p) &= \frac{c}{2p\left[p(c+w-v)+\mathrm{i}\omega\right]}\,, \\ B(p) &= \frac{c}{2p\left[p(c-w+v)-\mathrm{i}\omega\right]}\,, \\ C(p) &= -\frac{c^2}{\left[p(c+w-v)+\mathrm{i}\omega\right]\left[p(c-w+v)-\mathrm{i}\omega\right]}\,. \end{aligned} \tag{11.18}$$

By applying the inverse Laplace transform L_t^{-1} with respect to variable q to (11.16), we obtain

$$\begin{aligned} \tilde{u}(p) = \mathrm{L}_t^{-1}[\tilde{u}(p,q)] &= A(p)H(t)\exp(q_1 t) \\ &\quad + B(p)H(t)\exp(q_2 t) + C(p)H(t)\exp(q_3 t) \\ &= A(p)H(t)\exp(-p(c+w)t) + B(p)H(t)\exp(p(c-w)t) \\ &\quad + C(p)H(t)\exp((\mathrm{i}\omega - vp)t)\,. \end{aligned}$$

Using the formula

$$\mathrm{L}_x[f(x-\lambda)](p) = \exp(-\lambda p)\mathrm{L}_x[f(x)](p)$$

and applying the inverse Laplace transform L_x^{-1} with respect to the variable p, we get

$$\begin{aligned} u_t(x) = u(x,t) = \mathrm{L}_x^{-1}[\tilde{u}(p)] &= H(t)A_1[x-(c+w)t] \\ &\quad + H(t)B_1[x+(c-w)t] + H(t)\exp(\mathrm{i}\omega t)C_1(x-vt) \end{aligned} \tag{11.19}$$

where we have noted

$$A_1(x) = \mathrm{L}_x^{-1}[A(p)]\,, \quad B_1(x) = \mathrm{L}_x^{-1}[B(p)]\,, \quad C_1(x) = \mathrm{L}_x^{-1}[C(p)]\,. \tag{11.20}$$

Taking account of (11.18), we can write the decomposition in simple fractions

$$A(p) = \frac{A'}{p} + \frac{A''}{p-p_1}\,, \quad B(p) = \frac{B'}{p} + \frac{B''}{p-p_2}\,, \quad C(p) = \frac{C'}{p-p_3} + \frac{C''}{p-p_4}\,, \tag{11.21}$$

where

$$p_1 = p_3 = -\frac{\mathrm{i}\omega}{c+w-v}\,, \quad p_2 = p_4 = \frac{\mathrm{i}\omega}{c-w+v} \tag{11.22}$$

and

$$A' = B'' = C' = \frac{C}{2\mathrm{i}\omega}\,, \quad A'' = B' = C'' = -\frac{C}{2\mathrm{i}\omega}\,. \tag{11.23}$$

Taking into account the relations (11.20) and (11.21), we obtain

$$\begin{aligned}
A_1(x) &= A'H(x) + A''H(x)\exp(p_1 x) = \frac{cH(x)}{2\mathrm{i}\omega}\left[1-\exp\left(\frac{-\mathrm{i}\omega x}{c+w-v}\right)\right],\\
B_1(x) &= B'H(x) + B''H(x)\exp(p_2 x) = \frac{-cH(x)}{2\mathrm{i}\omega}\left[1-\exp\left(\frac{\mathrm{i}\omega x}{c-w+v}\right)\right],\\
C_1(x) &= C'H(x)\exp(p_3 x) + C''H(x)\exp(p_4 x)\\
&= \frac{c}{2\mathrm{i}\omega}\left[\exp\left(\frac{-\mathrm{i}\omega x}{c+w-v}\right) - \exp\left(\frac{\mathrm{i}\omega x}{c-w+v}\right)\right].
\end{aligned} \tag{11.24}$$

With this, the elongation (11.19) becomes

$$\begin{aligned}
&u_t(x) = u(x,t)\\
&= \frac{c}{2\mathrm{i}\omega}H(t)H(x-(c+w)t)\left[1-\exp\left(\frac{-\mathrm{i}\omega}{c+w-v}(x-(c+w)t)\right)\right]\\
&\quad - \frac{c}{2\mathrm{i}\omega}H(t)H(x+(c-w)t)\left[1-\exp\left(\frac{\mathrm{i}\omega}{c-w+v}(x+(c-w)t)\right)\right]\\
&\quad + \frac{c}{2\mathrm{i}\omega}H(t)\exp(\mathrm{i}\omega t)\,H(x-vt)\left[\exp\left(\frac{-\mathrm{i}\omega(x-vt)}{c+w-v}\right) - \exp\left(\frac{\mathrm{i}\omega(x-vt)}{c-w+v}\right)\right].
\end{aligned} \tag{11.25}$$

Starting from this result, we can deduce different particular solutions.

Let us assume that $x > 0$, $t > 0$, $w < c$ as well as $v < c + w = c_1$.

This last condition shows that the speed of the sound source is less than the waves speed in the direction of the source motion (the same with the wind direction).

Using the Euler formula, $\exp(\mathrm{i}\alpha) = \cos\alpha + \mathrm{i}\sin\alpha$, and considering the real part of the elongation (11.25), we obtain

$$\operatorname{Re} u(x,t) = \begin{cases} 0\,, & t < \dfrac{x}{c+w}\,,\\[2mm] \dfrac{-c}{2\omega}\sin\dfrac{\omega(x-(c+w)t)}{c+w-v}\,, & \dfrac{x}{c+w} < t < \dfrac{x}{v}\,,\\[2mm] \dfrac{c}{2\omega}\sin\dfrac{\omega(x+(c-w)t)}{c-w+v}\,, & t > \dfrac{x}{v}\,. \end{cases} \tag{11.26}$$

Hence, one observes that at $t = x/v$ occurs a discontinuity for the elongation (11.26) and the frequency undergoes a jump, from the value

$$f' = \frac{c+w}{c+w-v}\frac{\omega}{2\pi} = \frac{c+w}{c-(v-w)}f$$

to the value

$$f'' = \frac{c-w}{c-w+v}\frac{\omega}{2\pi} = \frac{c-w}{c+(v-w)}f\,, \tag{11.27}$$

according to the Doppler principle. In this way, a generalization of the relation (11.5) is obtained.

11.2 Applications of the Distribution Theory in Optics

Among the important phenomena accompanying the propagation of acoustic waves, optical waves, and so on, are their reflection, refraction and diffraction.

The knowledge of these phenomena is related to the study of the waves propagation equation in nonhomogeneous media.

For the study of the interference and diffraction of light waves it is sufficient to take into account the wave character of light propagation, neglecting the electromagnetic character of this process.

For this reason, a light wave will be characterized only by amplitude, phase, wavelength and propagation speed.

All the results presented are based on Huygens' principle, where all points on a wavefront can be regarded as point sources for the production of spherical secondary waves. After a time t, the new position of the wavefront will represent the envelope (tangent surface) to these secondary waves.

11.2.1 The Phenomenon of Diffraction at Infinity

A characteristic property of wave propagation is to curve when passing through a slit in a screen (diffraction).

A slit of plane diffraction is characterized by a distribution function $f(x, y)$, $(x, y) \in \Omega \subset \mathbb{R}^2$, of the light variable.

Starting from this, one can determine the diffraction figure at infinity.

To calculate the diffraction figure of a plane diffraction slit at infinity, Huygens's principle is used, [16, 50].

An element of area $\mathrm{d}S = \mathrm{d}x\mathrm{d}y$, in the $z = 0$-plane, emits an elementary wave of amplitude $f(x, y)\mathrm{d}x\mathrm{d}y$ in the direction $\boldsymbol{u}$ of cosine directors α, β, γ, of phase difference $2\pi(\alpha x + \beta y)/\lambda$ with respect to the elementary wave emitted at the origin O; here λ is the wavelength.

By superposing the effects, we obtain

$$\iint_{\Omega} f(x, y) \exp\left(\frac{2\pi\mathrm{i}(\alpha x + \beta y)}{\lambda}\right) \mathrm{d}x\mathrm{d}y\,, \tag{11.28}$$

where Ω is the domain occupied by the plane diffraction slit.

If we form the diffraction figure at infinity in the focus of a lens, with the focal length ℓ, then we may obtain a correspondence of a certain point of the focal plane to a direction (α, β, γ), by the relations

$$x' = \frac{\alpha}{\gamma}\ell\,, \quad y' = \frac{\beta}{\gamma}\ell\,. \tag{11.29}$$

Usually, the diffraction figure at infinity is contained in the neighborhood of the Oz-axis; we can consider $\alpha, \beta \ll 1$, so $\gamma \approx 1$, and

$$x' = \alpha\ell\,, \quad y' = \beta\ell\,. \tag{11.30}$$

If the wavelength λ is taken as unit length and if the reduced coordinates are used in the focal plane,

$$u = \frac{x'}{\ell}\,, \quad v = \frac{y'}{\ell}\,, \tag{11.31}$$

which is small enough compared to unity, we obtain for the figure of diffraction at infinity a repartition of the form

$$F(u, v) = \iint_{\mathbb{R}^2} f(x, y)\exp(2\pi\mathrm{i}(ux + vy))\mathrm{d}x\mathrm{d}y\,. \tag{11.32}$$

Because the Fourier transform of the function $f(x, y), (x, y) \in \mathbb{R}^2$, is

$$\hat{f}(\xi_1, \xi_2) = \mathrm{F}[f(x, y)](\xi_1, \xi_2) = \iint_{\mathbb{R}^2} f(x, y)\exp(\mathrm{i}(x\xi_1 + y\xi_2))\mathrm{d}x\mathrm{d}y\,,$$

then (11.32) can be written in the form

$$F(u, v) = \hat{f}(2\pi u, 2\pi v)\,. \tag{11.33}$$

Thus, for the diffraction figure at infinity, the repartition function $F(u, v)$ is the Fourier transform of the repartition $f(x, y)$ of the light variable at the point $(2\pi u, 2\pi v)$.

In some cases, the light variable distribution can be expressed by an ordinary function. Thus, in the case of a point source, the repartition function of the light variable should be zero everywhere, except at only one point, where its value is not determined.

In this case, the repartition function will be expressed by a distribution with point support, namely with the help of the Dirac delta distribution.

The representation used for point sources is the same form used for the point masses.

Thus, for the point light source at the point (x_0, y_0), the variable light repartition will be given by the distribution

$$f(x, y) = k\delta(x - x_0, y - y_0), k = \text{const}\,. \tag{11.34}$$

The diffraction figure at infinity will have the distribution

$$F(u,v) = \hat{f}(2\pi u, 2\pi v) = k\exp(2\pi \mathrm{i}(x_0 u + y_0 v)) ,$$

because

$$\hat{f}(\xi_1, \xi_2) = kF[\delta(x - x_0, y - y_0)](\xi_1, \xi_2) = k\exp(\mathrm{i}(x_0\xi_1 + y_0\xi_2)) .$$

Example 11.1 We consider a slit represented by the interval $[-a, a]$, $a > 0$ (Figure 11.2) on the Ox-axis, of negligible width.

We acknowledge that the distribution function of the light variable is equal to unity throughout the slit length. Then, this function will be expressed in the form

$$f(x) = \begin{cases} 1 , & x \in [-a, a] , \\ 0 , & \text{otherwise} . \end{cases}$$

Using the formula (11.33), we obtain for the distribution function of the diffraction figure at infinity the expression

$$F(u) = \hat{f}(2\pi u) = \int_{-a}^{a} \exp(2\pi \mathrm{i}ux)\mathrm{d}x = \frac{1}{2\pi \mathrm{i}u}\left[\exp(2\pi \mathrm{i}au) - \exp(-2\pi \mathrm{i}au)\right] .$$

Based on the Euler formula $\exp(\mathrm{i}z) = \cos z + \mathrm{i}\sin z$, we find

$$F(u) = \frac{\sin(2\pi au)}{\pi u} . \tag{11.35}$$

Example 11.2 We consider now a rectangular slit $\Omega = [-a, a] \times [-b, b]$, $a, b > 0$ (Figure 11.3) in the Oxy-plane. The repartition function of the light variable will be equal to unity on the domain $\Omega \subset \mathbb{R}^2$ of the slit.

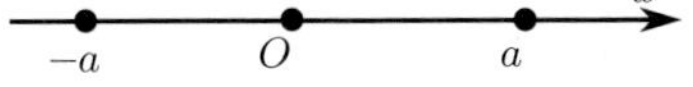

Figure 11.2

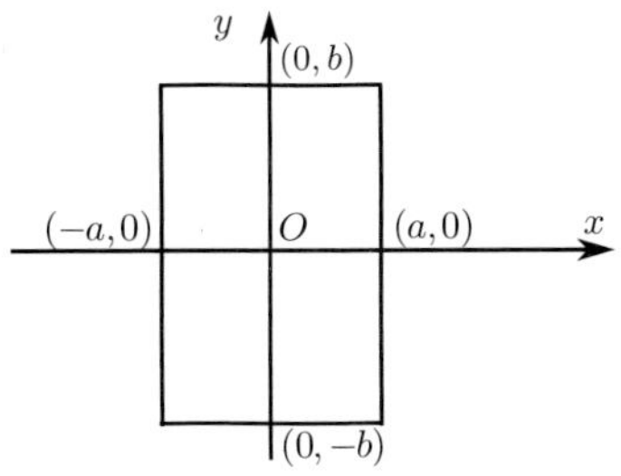

Figure 11.3

In this case we can write

$$f(x, y) = \begin{cases} 1\,, & (x, y) \in [-a, a] \times [-b, b]\,, \\ 0\,, & \text{otherwise}\,. \end{cases}$$

The repartition function for the diffraction figure at infinity will have the expression

$$F(u, v) = \hat{f}(2\pi u, 2\pi v) = \int_{-a}^{a}\int_{-b}^{b} \exp(2\pi \mathrm{i}(ux + vy))\mathrm{d}x\mathrm{d}y$$

$$= \int_{-a}^{a} \exp(2\pi \mathrm{i} ux)\mathrm{d}x \int_{-b}^{b} \exp(2\pi \mathrm{i} vy)\mathrm{d}y\,.$$

Taking into account (11.35), we obtain

$$F(u, v) = \frac{\sin(2\pi a u)\sin(2\pi b v)}{\pi^2 uv}\,.$$

11.2.2
Diffraction of Fresnel Type

In general, the phenomenon of diffraction occurs in the case of a partial obstruction of the beam of light, using opaque screens. We mention thus Fraunhofer and the Fresnel type of diffraction, namely the directly observed diffraction without the use of optical instruments (convergent lenses). The Fraunhofer diffraction is a limited case of Fresnel diffraction.

Let there exist a plane diffraction slit in a diffraction plane and $f(x, y)$ the repartition function of the corresponding light variable.

Due to the diffraction phenomenon, in an observation plane parallel to the first one and situated at a finite distance, will correspond to an image characterized by the repartition function $h(x, y)$ of the light variable (Figure 11.4).

Consequently, the distribution function $h(x, y)$ from the plane of observation (2) corresponds to the repartition function $f(x, y)$ from the plane of diffraction (1).

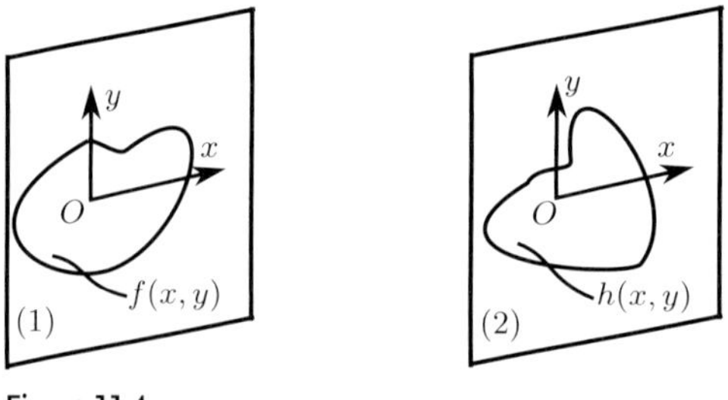

Figure 11.4

By changing the slit, hence the light variable $f(x, y)$, the diffracted figure from the plane of observation (2) will change as well.

Due to the linearity of Maxwell's equations, which describes the phenomenon of diffraction, the modification by translation of the slit leads to the same change by translation of the diffraction figure from the plane of observation (2).

Based on previous statements, there exists a function $g(x, y)$, independent of the slit and dependent on the distance between the two planes, such that between the two repartitions f and h gives the relation

$$h(x, y) = f(x, y) * g(x, y) = \iint_{\mathbb{R}^2} f(u, v) g(x - u, y - v) \mathrm{d}u\mathrm{d}v \,, \tag{11.36}$$

where the symbol $*$ corresponds to the convolution product.

We note that the formula (11.36) exists if $f, g \in L^1(\mathbb{R}^2)$, hence if they are absolutely integrable functions and also if f and g are distributions from the space $\mathcal{E}'(\mathbb{R}^2)$, hence with compact support.

Thus, if the point light source is located at the origin of the coordinate axes in the plane of diffraction, and has the intensity equal to unity, then the repartition of the light variable has the expression

$$f(x, y) = \delta(x, y) \,, \tag{11.37}$$

where δ is the Dirac delta distribution.

From (11.36) it follows

$$h(x, y) = \delta(x, y) * g(x, y) = g(x, y) \,. \tag{11.38}$$

Therefore, the function (or distribution) $g(x, y)$ is the response to a point slit, with the light variable of intensity equal to unity.

With the help of the function $g(x, y)$ once specified, the repartition $h(x, y)$ for a slit of any shape can be determined.

If in the diffraction plane there are n point slots, located at the points $A_i(x_i, y_i)$, with the light variable repartition $f_i(x, y) = k_i \delta(x - x_i, y - y_i)$, $i = \overline{1, n}$, then the image light variable will have the repartition

$$h(x, y) = \sum_{i=1}^{n} f_i(x, y) * g(x, y) = \sum_{i=1}^{n} k_i \delta(x - x_i, y - y_i) * g(x, y) \,.$$

Consequently, we get

$$h(x, y) = \sum_{i=1}^{n} k_i g(x - x_i, y - y_i) \,.$$

This case corresponds to what is called a diffraction grating.

References

1 Dirac, P. (1926–27) The physical interpretation of quantum dynamics. *Proc. R. Soc.*, **A113**, 621–641.

2 Schwartz, L. (1961) *Théorie des distributions*, Herman Éditeurs des Sciences et des arts.

3 Mikusiński, J., and Sikorski, R. (1961) *The Elementary Theory of Distributions*, Państwowe Wydawnictwo Naukowe, Warszawa.

4 Guelfand, I.M., and Chilov, G.E. (1962) *Les distributions*, vol. 1, Dunod, Paris.

5 Guelfand, I.M., and Chilov, G.E. (1964) *Les distributions*, vol. 2, Dunod, Paris.

6 Hörmander, L. (1964) *Linear Partial Differential Operators*, Springer Verlag, Berlin.

7 Hörmander, L. (1990) *The Analysis of Linear Partial Differential Operators I. Distribution Theory and Fourier Analysis*, Springer, Berlin.

8 Zemanian, A. (1968) *Generalized Integral Transformations*, John Wiley & Sons, Inc., New York.

9 Sato, M. (1959) Theory of hyperfunctions. *J. Fac. Sci. Univ. Tokyo*, **8**, 139–194.

10 Sato, M. (1960) Theory of hyperfunctions. *J. Fac. Sci. Univ. Tokyo*, **8**, 387–437.

11 Tréves, F. (1967) *Topological Vector Spaces, Distributions and Kernels*, Academic Press, New York, London.

12 Friedlander, F. (1982) *Introduction to the Theory of Distributions*, Cambridge University Press.

13 Kecs, W.W. (2003) *Teoria distributiilor cu aplicatii (Theory of Distributions with Applications)*, Academy Publishing House, Bucharest.

14 Toma, A. (2004) *Metode Matematice in elasticitate si vascoelasticitate (Mathematical Methods in Elasticity and Viscoelasticity)*, Technical Publishing House, Bucharest.

15 Vladimirov, V. (1979) *Distributions en Physique Mathématique*, Mir, Mouscou.

16 Kecs, W., and Teodorescu, P. (1975) *Introducere in Teoria Distributiilor cu Aplicatii in Tehnica (Introduction to the Theory of Distributions with Applications in Technique)*, Editura Tehnica, Bucuresti.

17 Kecs, W., and Teodorescu, P. (1974) *Applications of the Theory of Distributions in Mechanics*, Abacus Press, Tubridge Wells, Kent; Editura Academiei, Bucuresti.

18 Schwartz, L. (1961) *Méthodes mathématiques pour les sciences physiques*, Hermann, Paris.

19 Kecs, W. (1962) Sur les problems concernant la demispace élastique. *Bull. Math. de la Soc. Sci. Math. Phys. R.S. Roumanie*, **54** (3–4), 157–173.

20 Kecs, W. (1982) *The Convolution Product and some Applications*, D. Reidel Publishing Company.

21 Kecs, W.W., and Toma, A. (1995) Cauchy's problem for the generalized equation of the longitudinal vibrations of elastic rods. *Eur. J. Mech. A*, **14** (5–6), 827–835.

22 Kecs, W.W. (1988) The solution in displacement of the steady-state transverse vibration of inhomogeneous elastic rods. *Rev. Roum. Sci. Techn. Mec. Appl.*, **33** (1), 107–112.

Distribution Theory, First Edition. Petre P. Teodorescu, Wilhelm W. Kecs, and Antonela Toma

23 Toma, A. (1994) On the micropolar elasticity equations. *Rev. Roum. Sci. Tech., Ser. Mec. Appl.*, **39** (3–4), 353–59.

24 Toma, A. (1995) Properties of the partial convolution product. *Stud. Cerc. Mat.*, **47** (1), 61–69.

25 Mikusiński, J. (1983) *Operational Calculus*, vol. I., Pergamon Press.

26 Mikusiński, J., and Boehme, T. (1987) *Operational Calculus*, vol. II, Pergamon Press.

27 Doetsch, G. (1974) *Introduction to the Theory and Application of the Laplace Transformation*, Springer-Verlag, Berlin.

28 Kecs, W., and Toma, A. (2010) The generalized solution of the boundary-value problems regarding the bending of elastic rods on elastic foundationthe generalized solution of the boundary-value problems regarding the bending of elastic rods on elastic foundation. *Proc. Roum. Acad.*, **11** (1), 55–62.

29 Shilov, G.E. (1968) *Generalized Functions and Partial Differential Equations*, Gordon and Breach Science Publishers.

30 Kecs, W.W. (1990) Generalized solution of the cauchy problem for longitudinal vibrations in viscoelastic rods of maxwell type. *Rev. Roum. Sci. Tech.-Ser. Mec. Appl.*, **35** (3), 209–223.

31 Bateman, H., and Erdély, A. (1954) *Tables of Integral Transforms*, vol. I, McGraw-Hill Co.

32 Teodorescu, P. (2007–2009) *Mechanical Systems. Classical Models*, vol. 1–3, Springer Verlag, Dordrecht, Heidelberg, London, New York.

33 Kecs, W. (1986) *Elasticitate si Vascoelasticitate (Elasticity and Viscoelasticity)*, Bucuresti.

34 Nowacki, W. (1963) *Dynamics of Elastic Systems*, Chapman Hall LTd, London.

35 Rabotnov, I., and Iljuschin, A. (1970) *Methoden der Viscoelastizitätstheorie*, VEB Fachbuchverlag, Leipzig.

36 Gurtin, M., and Sternberg, E. (1961) A note on uniqueness in classical elastodynamics. *Quart. Appl. Math.*, (19), 169–171.

37 Gurtin, M., and Toupin, R. (1965) A unigueness theorem for the displacement boundary-value problem of linear elastodynamics. *Quart. Appl. Math.*, **23**, 79–81.

38 Kupradze, V., Gegelea, T., Bacseleivili, M., and Burciuladze, T. (1976) *Trehmerniie zadacii matematisceskoi teorii uprugostii i termouprugostii (Three-dimensional Problems of the Mathematical Theory of Elasticity and Thermoelasticity*, Nauka, Moskva.

39 Fichera, G. (1972) *Existence Theorems in Elasticity*, Springer, Berlin, Heidelberg, New York.

40 Gurtin, M.E., and Sternberg, E. (1962) On the linear theory of viscoelasticity. *Arch. Rational Mech. Anal.*, **11** (4), 291–356.

41 Rabotnov, Y.N. (1980) *Elements of Hereditary Solid Mechanics*, Mir Publisher, Moscow.

42 Alfrey, T. (1944) Non homogeneous stresses in viscoelastic media. *Quart. Appl. Math.*, **2**, 113–119.

43 Lee, E. (1955) Stress analysis in viscoelastic bodies. *Quart. Appl. Math.*, **13** (2), 183–173.

44 Read, W. (1950) Stress analysis for compressible viscoelastic materials. *J. Appl. Phys.*, **21**, 671–674.

45 Tsien, H. (1950) A generalization of alfrey's theorem for viscoelastic media. *Quart. Appl. Math.*, **8**, 104–106.

46 Love, A. (1944) *A Treatise on the Mathematical Theory of Elasticity*, Dover Publications.

47 Kecs, W.W. (1987) The generalized equation of longitudinal vibrations for elastic homogeneous isotropic rods of constant cross-section. *Mech. Res. Comm.*, **14** (5–6), 395–402.

48 Sneddon, I.N. (1972) *The Use of Integral Transforms*, Mc, Toronto.

49 Kecs, W.W. (1996) *Vibratiile Barelor Elastice si Vascoelastice (Vibrations of Elastic and Viscoelastic Bars)*, Editura Tehnica, Bucuresti.

50 Arsac, J. (1961) *Transformation de Fourier et Théorie des Distributions*, Dunod.

Index

Distribution Theory, First Edition. Petre P. Teodorescu, Wilhelm W. Kecs, and Antonela Toma